Methods of the Theory of Generalized Functions

Analytical Methods and Special Functions

Founding Editor: *A.P. Prudnikov*
Series Editors: *C.F. Dunkl (USA), H.-J. Glaeske (Germany), M. Saigo (Japan)*

Volume 1
Series of Faber Polynomials
P.K. Suetin

Volume 2
Inverse Spectral Problems for Differential Operators and their Applications
V.A. Yurko

Volume 3
Orthogonal Polynomials in Two Variables
P.K. Suetin

Volume 4
Fourier Transforms and Approximations
A. Sedletskii

Volume 5
Hypersingular Integrals and Applications
S. Samko

Volume 6
Methods of the Theory of Generalized Functions
V.S. Vladimirov

Methods of the Theory of Generalized Functions

V.S. Vladimirov
Steklov Mathematical Institute
Moscow, Russia

CRC Press
Taylor & Francis Group
Boca Raton London New York

CRC Press is an imprint of the
Taylor & Francis Group, an **informa** business

CRC Press
Taylor & Francis Group
6000 Broken Sound Parkway NW, Suite 300
Boca Raton, FL 33487-2742

First issued in paperback 2019

ISBN-13: 978-0-415-27356-5 (hbk)
ISBN-13: 978-0-367-39594-0 (pbk)

Visit the Taylor & Francis Web site at
http://www.taylorandfrancis.com

and the CRC Press Web site at
http://www.crcpress.com

CONTENTS

PREFACE

As physics advances, its theoretical statements require ever "higher" mathematics. In this connection it is well worth quoting what the eminent English mathematician and theoretical physicist Paul Dirac said in 1930 [**18**] when he gave a theoretical prediction of the existence of antiparticles:

"It seems likely that this process of increasing abstraction will continue in the future and that advance in physics is to be associated with a continual modification and generalization of the axioms at the base of the mathematics rather than with a logical development of any one mathematical scheme on a fixed foundation."

Subsequent development of theoretical physics, particularly of quantum field theory, fully corroborated this view. Again in this connection we quote the apt words of N.N. Bogolubov. In 1963 he said: "The basic concepts and methods of quantum field theory are becoming more and more mathematical."

The construction and investigation of mathematical models of physical phenomena constitute the subject of mathematical physics.

Since the time of Newton the search for and the study of mathematical models of physical phenomena — the problems of mathematical physics — have made it necessary to resort to a wide range of mathematical tools and have thus stimulated the development of various areas of mathematics. Traditional (classical) mathematical physics had to do with the problems of classical physics: mechanics, hydrodynamics, acoustics, diffusion, heat conduction, potential theory, electrodynamics, optics and so forth. These problems are reduced to boundary-value problems for differential equations (the equations of mathematical physics). The basic mathematical tool for investigating such problems is the theory of differential equations and allied fields of mathematics: integral equations, the calculus of variations, approximate and numerical methods. With the advent of quantum physics, the range of mathematical tools expanded considerably through the use of the theory of operators, the theory of generalized functions, the theory of functions of complex variables, topological, number-theoretical and algebraic methods, computational mathematics and computers. All these theories were pressed into service in addition to the traditional tools of mathematics. This intensive interaction of theoretical physics and mathematics gradually brought to the fore a new domain, that of modern mathematical physics.

To summarize, then: modern mathematical physics makes extensive use of the latest attainments of mathematics, one of which is the theory of generalized functions. The present monograph is devoted to a brief exposition of the fundamentals of that theory and of certain of its applications in mathematical physics.

At the end of the 1920's Paul Dirac (see [20]) introduced for the first time in his quantum mechanical studies the so-called delta-function (δ-function), which has the following properties:

$$\delta(x) = 0, \quad x \neq 0; \qquad \int \delta(x)\varphi(x)\,dx = \varphi(0), \quad \varphi \in \mathbb{C}. \qquad (*)$$

It was soon pointed out by mathematicians that from the purely mathematical point of view the definition is meaningless. It was of course clear to Dirac himself that the δ-function is not a function in the classical meaning and, what is important, it operates as an operator (more precisely, as a functional) that relates, via formula ($*$), to each continuous function φ a number $\varphi(0)$, which is its value at the point 0. It required quite a few years and the efforts of many mathematicians[1] in order to find a mathematically proper definition of the delta-function, of its derivatives and, generally, of generalized functions.

The foundations of the mathematical theory of generalized functions were laid by the Russian mathematician S.L. Sobolev in 1936 (see [96]) when he successfully applied generalized functions to a study of the Cauchy problem for hyperbolic equations. After World War II, the French mathematician L. Schwartz attempted, on the basis of an earlier created theory of linear locally convex topological spaces[2], a systematic construction of a theory of generalized functions and explained it in his well-known monograph entitled *Théorie des distributions* [89] (1950–51). From then on the theory of generalized functions was developed intensively by many mathematicians. This precipitate development of the theory of generalized functions received its main stimulus from the requirements of mathematical and theoretical physics, in particular, the theory of differential equations and quantum physics. At the present time, the theory of generalized functions has advanced substantially and has found numerous applications in physics and mathematics, and more and more is becoming a workaday tool of the physicist, mathematician and engineer[3]. Generalized functions possess a number of remarkable properties that extend the capabilities of classical mathematical analysis; for example, any generalized function turns out to be infinitely differentiable (in the generalized meaning), convergent series of generalized functions may be differentiated termwise an infinite number of times, there exists the Fourier transform of a generalized function, and so on. For this reason, the use of generalized function techniques substantially expands the range of problems that can be tackled and leads to appreciable simplifications that make elementary operations automatic.

In recent years, more general objects than generalized functions — so called hyperfunctions in sense of Sato [87] — have more and more applications (see Zharinov [**129**], Komatsu [**59, 60**]). Other approaches, leading to an extension of the notion of generalized function and admitting an operation of multiplication, were suggested by Colombeau [**13**] and Egorov [29]. We shall not consider here these new approaches.

---

[1]See the pioneering works of Bochner [5], Hadamard [44] and Riesz [85].

[2]See Dieudonné and Schwartz [17].

[3]See Arsac [3], Beltrami and Wholers [4], Bogolyubov, Logunov, Oksak, and Todorov [7], Bogolyubov, Medvedev and Polivanov [8], Bogolyubov and Shirkov [9], Bremermann [12], Ehrenpreis [30], Garding [36], Garsoux [37], Gel'fand and Shilov [38], Gel'fand and Vilenkin [39], Hörmander [46], Jost [53], Malgrange [72], Palamodov [79], Reed and Simon [84], Schwartz [89, 90], Sobolev [96, 97], Streater and Wightman [100], Treves [104], Vladimirov [105, 106], Zemanian [130], and others.

The theory of generalized functions can be especially effectively applied to the concept of *generalized solutions* of linear differential, integral, integro-differential, convolutional (more generally, pseudo-differential) equations. Generalized solutions appear when studying integral relations of the type of local balance, and the taking account of these solutions leads to *generalized settings of boundary-value problems* of the mathematical physics. The exact definition of a generalized solution is based on the notion of a generalized derivative of a (generalized) function.

A notion of a generalized solution for one or another problem of mathematical physics has been introduced by many mathematicians and physicists, beginning with Maxwell's, Kirchhoff's and Heaviside's works of XIX century. The history of this question is presented in detail in the book by Lützen (see Lützen [**71**]).

In connection with the concept of a generalized solution it is relevant to cite here the following phrase from the fundamental work of L. Euler "Integralrechnung" (Euler [**32**, p. 199]). After obtaining the general solution of the wave equation

$$u(x,t) = f(x - at) + g(x + at),$$

he writes:

"In this way, the highly astute man (D'Alembert — *author's note*) has obtained the complete integral, however he did not pay attention to the fact that as the functions introduced (f and g — *author's note*) one can take not only any continuous functions (analytically expressed — *author's note*) but also functions completely deprived of the property of continuity."

It follows from this citation that Euler came close to the notion of the generalized solution of the wave equation.

In German the phrase cited is the following:

"Auf diese Weise hat der höchst scharfsinnige Mann das vollständige Integral erhalten, hat aber nicht bemerkt, daß statt dieser eingeführten Funktionen, nich allein die continuirlichen Funktionen jeder Art, sondern aush jene Funktionen, by welchen durchaus das Gesetz der Continuität mangelt, genommen werden können."

The present monograph is an expanded version of a course of lectures that the author has been delivering to students, postgraduates, and associates of the Moscow Physics and Technology Institute and the Steklov Mathematical Institute.

This book appeared as a result of revision and complementation of the second edition of my monograph "Generalized Functions in Mathematical Physics", Nauka, Moscow, 1979 (see Vladimirov [**120**]). Section 14 is added which is devoted to Tauberian and Abelian theorems for generalized functions from the algebra $\mathcal{S}'(\Gamma)$. The sections considering convolution (Sections 4.1 and 6.5) and multiplication (Section 1.10) of generalized functions as well as divisors of identity in the algebra $H(C)$ (Section 12.5) and to the estimates of the growth of holomorphic functions with non-negative imaginary part in tubular regions over acute regular cones (Section 17.3) are essentially revised and extended. New sections are devoted to the study of the generalized functions $\mathcal{P}(\pi_V|x|^{\alpha-1})$ (Section 2.7) and homogeneous generalized functions (Section 5.7) and their Fourier transform (Section 6.6) as well as the Mellin transform of generalized functions (Section 6.7) and the sweeping principle (Section 15.8). The numerous changes and additions are made, inaccuracies and misprints are corrected.

In the large, the international terminology used in the theory of generalized functions is accepted in the book. The question only arose, which can be answered ambiguously, how to translate into English the main object of our investigation —

an element of the space $\mathcal{D}'$: "distribution" or "generalized function"? In French, beginning with L. Schwartz, the term "distribution" took root; at the same time in Russian the term "generalized function" is well customized (see I.M. Gel'fand *et al*), however the latter is often used for more general objects. In English literature the term "distribution" is used more frequently, although "generalized function" is used sometimes as well (see, for instance, Lighthill [67]).

As a result of this analysis, we chose the term "generalized function" the more so as this term was used when translating the first version of this book into English (Vladimirov [120]).

I take this opportunity to thank all my associates for their constructive criticism that has helped me to improve the presentation. In particular I wish to thank my pupils and collaborators Yu.N. Drozhzhinov, B.I. Zavialov, V.V. Zharinov and R.Kh. Galeev.

I express a special gratitude to Prof. A.P. Prudnikov for his active interest and support of the publication of this book and to Prof. Jacob Korevaar for his support and a series of valuable remarks.

V.S. Vladimirov

SYMBOLS AND DEFINITIONS

0.1. We denote the *points* of an n-dimensional real space $\mathbb{R}^n$ by x, y, ξ, ... ; $x = (x_1, x_2, \dots, x_n)$. The points of an n-dimensional complex space $\mathbb{C}^n$ are given as z, ζ, ... ; $z = (z_1, z_2, \dots, z_n) = x + iy$; $x = \Re z$ is the real part of z and $y = \Im z$ is the imaginary part of z, $\bar{z} = x - iy$ is the complex conjugate of z. In the usual manner we introduce in $\mathbb{R}^n$ and $\mathbb{C}^n$ the *scalar products*

$$(x, \xi) = x_1\xi_1 + \cdots + x_n\xi_n,$$
$$\langle z, \zeta \rangle = z_1\bar{\zeta}_1 + \cdots + z_n\zeta_n$$

and the *norms* (*lengths*)

$$|x| = \sqrt{(x, x)} = (x_1^2 + \cdots + +x_n^2)^{1/2},$$
$$|z| = \sqrt{\langle z, z \rangle} = (|z_1|^2 + \cdots + |z_n|^2)^{1/2}.$$

0.2. *Open sets* in $\mathbb{R}^n$ are denoted by $\mathcal{O}$, $\mathcal{O}'$, ... ; $\partial\mathcal{O}$ is the boundary of $\mathcal{O}$, or $\partial\mathcal{O} = \overline{\mathcal{O}} \setminus \mathcal{O}$. We will say that the set A *is compact in the open set* $\mathcal{O}$ (or *is strictly contained in* $\mathcal{O}$) if A is bounded and its closure $\overline{A}$ lies in $\mathcal{O}$; we then write $A \Subset \mathcal{O}$.

The following designations are used: $U(x_0; R)$ is an *open ball* of radius R with centre at the point x_0; $S(x_0; R) = \partial U(x_0; R)$ is a *sphere* of radius R with centre at the point x_0; $U_R = U(0; R)$, $S_R = S(0; R)$.

Below, all point sets A, B, $\mathcal{O}$, ... are assumed measurable.

We use $\Delta(A, B)$ to denote the *distance* between the sets A and B in $\mathbb{R}^n$, that is

$$\Delta(A, B) = \inf_{x \in A,\ y \in B} |x - y|.$$

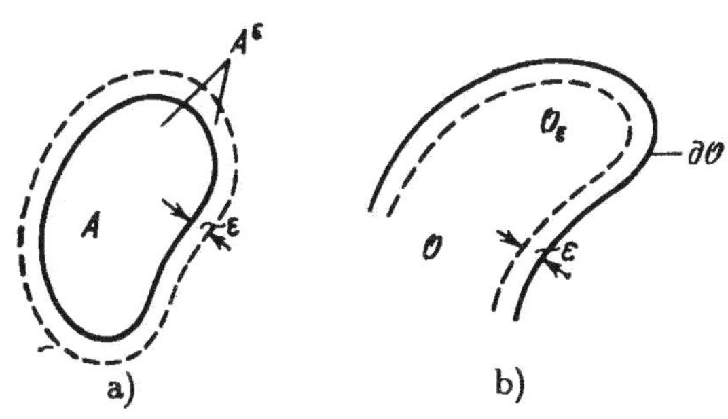

Figure 1

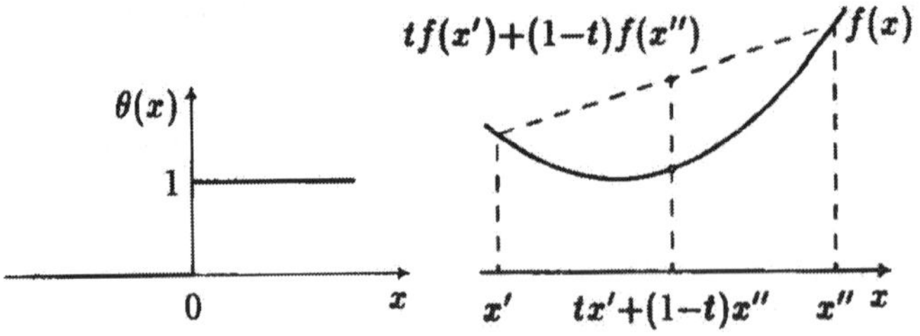

a) b)

Figure 2

We use A^ε to denote the *ε-neighbourhood* of a set A, $A^\varepsilon = A + U_\varepsilon$ (Fig. 1a). If $\mathcal{O}$ is an open set, then $\mathcal{O}_\varepsilon$ designates the set of those points of $\mathcal{O}$ which are separated from $\partial\mathcal{O}$ by a distance greater than ε (Fig. 1b):

$$\mathcal{O}_\varepsilon = [x : x \in \mathcal{O},\ \Delta(x, \partial\mathcal{O}) > \varepsilon].$$

We use $\operatorname{int} A$ to denote the set of interior points of the set A.

The *characteristic function of a set* A is the function $\theta_A(x)$ which is equal to 1 when $x \in A$ and is equal to 0 when $x \notin A$. The characteristic function $\theta_{[0,\infty)}(x)$ of the semiaxis $x \geq 0$ is called the *Heaviside unit function* and is denoted $\theta(x)$ (Fig 2a):

$$\theta(x) = 0, \quad x < 0,$$
$$\theta(x) = 1, \quad x \geq 0.$$

We write $\theta_n(x) = \theta(x_1) \dots \theta(x_n)$.

The set A is said to be *convex* if for any points x' and x'' in A the line segment joining them, $tx' + (1-t)x''$, $0 \leq t \leq 1$, lies entirely in A.

We will use $\operatorname{ch} A$ to denote the *convex hull* of a set A.

A real function $f(x) < +\infty$ is said to be *convex* on the set A if for any points x' and x'' in A such that the line segment $tx' + (1-t)x''$ joining them lies entirely in A the following inequality holds (Fig. 2b):

$$f(tx' + (1-t)x'') \leq tf(x') + (1-t)f(x'').$$

The function $f(x)$ is said to be *concave* if the function $-f(x)$ is convex.

0.3. The *Lebesgue integral* of a function f over an open set $\mathcal{O}$ is given as

$$\int_{\mathcal{O}} f(x)\,dx, \qquad \int_{\mathbb{R}^n} f(x)\,dx = \int f(x)\,dx.$$

The collection of all (complex-valued, measurable) functions f specified on $\mathcal{O}$ for which the norm

$$\|f\|_{\mathcal{L}^p(\mathcal{O})} = \begin{cases} \left[\int_{\mathcal{O}} |f(x)|^p\,dx\right]^{1/p}, & 1 \leq p < \infty, \\ \operatorname*{ess\,sup}_{x \in \mathcal{O}} |f(x)|, & p = \infty, \end{cases}$$

is finite will be denoted as $\mathcal{L}^p(\mathcal{O})$, $1 \le p \le \infty$; we write $\| \ \| = \| \ \|_{\mathcal{L}^2(\mathbb{R}^n)}$, $\mathcal{L}^p(\mathbb{R}^n) = \mathcal{L}^p$.

If $f \in \mathcal{L}^p(\mathcal{O}')$ for any $\mathcal{O}' \Subset \mathcal{O}$, then f is said to be *p-locally integrable* in $\mathcal{O}$ (for $p = 1$, we say it is *locally integrable in* $\mathcal{O}$). The collection of p-locally integrable functions in $\mathcal{O}$ is denoted $\mathcal{L}^p_{\text{loc}}(\mathcal{O})$, $\mathcal{L}^p_{\text{loc}}(\mathbb{R}^n) = \mathcal{L}^p_{\text{loc}}$.

A measurable function is said to have *compact support* in $\mathcal{O}$ if it vanishes almost everywhere outside a certain $\mathcal{O}' \Subset \mathcal{O}$. The set of all functions in $\mathcal{L}^p(\mathcal{O})$ that have compact support in $\mathcal{O}$ is denoted $\mathcal{L}^p_0(\mathcal{O})$.

0.4. Let $\alpha = (\alpha_1, \alpha_2, \ldots, \alpha_n)$ be a multi-index, that is to say its components α_j are nonnegative integers. We have the following symbolism:

$$\alpha! = \alpha_1!\alpha_2!\ldots\alpha_n!,$$
$$x^\alpha = x_1^{\alpha_1}x_2^{\alpha_2}\ldots x_n^{\alpha_n},$$
$$\binom{\alpha}{\beta} = \binom{\alpha_1}{\beta_1}\binom{\alpha_2}{\beta_2}\cdots\binom{\alpha_n}{\beta_n} = \frac{\alpha!}{\beta!(\alpha-\beta)!},$$
$$|\alpha| = \alpha_1 + \alpha_2 + \cdots + \alpha_n.$$

Let $\partial = (\partial_1, \partial_2, \ldots, \partial_n)$, $\partial_j = \frac{\partial}{\partial x_j}$, $j = 1, 2, \ldots, n$. Then

$$\partial^\alpha f(x) = \frac{\partial^{|\alpha|} f(x)}{\partial x_1^{\alpha_1} \partial x_2^{\alpha_2} \ldots \partial x_n^{\alpha_n}}.$$

It may sometimes happen that $\alpha = (\alpha_1, \alpha_2, \ldots, \alpha_n)$ will be used to denote a multi-index with components of any sign: $\alpha_j \gtreqless 0$.

0.5. We use $C^k(\mathcal{O})$ to denote the set of all functions $f(x)$that are continuous in $\mathcal{O}$ together with all derivatives $\partial^\alpha f(x)$, $|\alpha| \le k$; $C^\infty(\mathcal{O})$ is the collection of all functions infinitely differentiable in $\mathcal{O}$. The set of all functions $f(x)$ in $C^k(\mathcal{O})$ for which all derivatives $\partial^\alpha f(x)$, $|\alpha| \le k$, admit continuous extension onto $\overline{\mathcal{O}}$ will be denoted by $C^k(\overline{\mathcal{O}})$. We introduce the norm in $C^k(\overline{\mathcal{O}})$ for $k < \infty$ via the formula

$$\|f\|_{C^k(\overline{\mathcal{O}})} = \sup_{\substack{x \in \overline{\mathcal{O}} \\ |\alpha| \le k}} |\partial^\alpha f(x)|.$$

We also write $C(\mathcal{O}) = C^0(\mathcal{O})$, $C(\overline{\mathcal{O}}) = C^0(\overline{\mathcal{O}})$.

The *support* of a function $f(x)$ continuous in $\mathcal{O}$ is the closure in $\mathcal{O}$, of those points where $f(x) \ne 0$; the support of f is denoted by $\operatorname{supp} f$. If $\operatorname{supp} f \Subset \mathcal{O}$, then f has compact support in $\mathcal{O}$ (compare with Sec. 0.3).

We denote the collection of functions, with compact support in $\mathcal{O}$, of the class $C^k(\mathcal{O})$ by $C^k_0(\mathcal{O})$; $C_0(\mathcal{O}) = C^0_0(\mathcal{O})$. Finally, the set of all functions of the class $C^k(\overline{\mathcal{O}})$ that vanish on $\partial\mathcal{O}$ together with all derivatives up to order k inclusive will be denoted by $C^k_0(\overline{\mathcal{O}})$; $C_0(\overline{\mathcal{O}}) = C^0_0(\overline{\mathcal{O}})$. We write $C^k(\mathbb{R}^n) = C^k$; $C^k_0(\mathbb{R}^n) = C^k_0$; $C^k_0(\mathbb{R}^n) = \bar{C}^k_0$, $\bar{C}_0 = \bar{C}^0_0$ ($\bar{C}^k_0$ is the set of functions in C^k that vanish at infinity together with all their derivatives up to order k inclusive).

0.6. Symbolism: (a, b) is a bilinear form (linear in a and b separately); $\langle a, b\rangle$ is a sesquilinear form (linear in a and antilinear in b):

$$\langle\alpha a_1 + \beta a_2, \lambda b_1 + \mu b_2\rangle = \alpha\bar{\lambda}\langle a_1, b_1\rangle + \alpha\bar{\mu}\langle a_1, b_2\rangle + \beta\bar{\lambda}\langle a_2, b_1\rangle + \beta\bar{\mu}\langle a_2, b_2\rangle;$$

$$\sigma_n = \int\limits_{|x|=1} ds = \frac{2\pi^{n/2}}{\Gamma(n/2)}$$

is the surface area of a unit sphere in $\mathbb{R}^n$; A^T is the transpose of the matrix A.

We shall also use the alternative notation:

$$\nabla = \partial = \text{grad}, \qquad \Delta = \partial_1^2 + \partial_2^2 + \cdots + \partial_n^2,$$

$$\Box = \frac{\partial^2}{\partial x_0^2} - \nabla^2 = \partial_0^2 - \Delta.$$

We denote the uniform convergence of a sequence of functions, $\{\varphi_n(x)\}$, to a function $\varphi(x)$ on a set A thus:

$$\varphi_n(x) \overset{x\in A}{\Longrightarrow} \varphi(x), \qquad n \to \infty;$$

if $A = \mathbb{R}^n$, then instead of $\overset{x\in A}{\Longrightarrow}$ we write $\overset{x}{\Longrightarrow}$.

*

* *

The sections are numbered in a single sequence. Each section is made up of subsections, the numbers of which are included in any reference to a section. Formulae are numbered separately in each subsection; they contain the number of the formula and of the subsection. When reference is made to a formula in a different section, the number of that section is also given.

Symbol □ denotes that the proof is complete.

CHAPTER 1

GENERALIZED FUNCTIONS AND THEIR PROPERTIES

The exposition of the theory of generalized functions given in this chapter is tailored to the needs of theoretical and mathematical physics.

1. Test and Generalized Functions

1.1. Introduction. A generalized function is a generalization of the classical notion of a function. On the one hand, this generalization permits expressing in a mathematically proper form such idealized concepts as the density of a material point, the density of a point charge or dipole, the spatial density of a simple or double layer, the intensity of an instantaneous point source, the magnitude of an instantaneous force applied to a point, and so forth. On the other hand, the notion of a generalized function can reflect the fact that in reality one cannot measure the value of a physical quantity at a point but can only measure the mean values within sufficiently small neighbourhoods of the point and then proclaim the limit of the sequence of those mean values as the value of the physical quantity at the given point.

This can be explained by attempting to determine the density set up by a material point of mass 1. Assume that the point is the origin of coordinates. In order to determine the density, we distribute (or, as one often says, smear) the unit mass uniformly inside a sphere of radius ε centred at 0. We then obtain the mean density $f_\varepsilon(x)$ that is equal (see Fig. 3) to

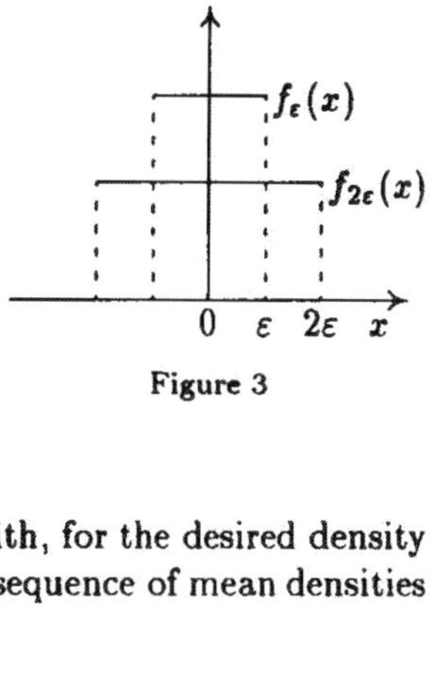

Figure 3

$$f_\varepsilon(x) = \begin{cases} \dfrac{1}{\frac{4}{3}\pi\varepsilon^3} & \text{if } |x| < \varepsilon, \\ 0 & \text{if } |x| > \varepsilon. \end{cases}$$

We are interested in the density at $\varepsilon = +\infty$. To begin with, for the desired density (which we denote by $\delta(x)$) we take the point limit of the sequence of mean densities $f_\varepsilon(x)$ as $\varepsilon \to +0$, that is, the function

$$\delta(x) = \begin{cases} +\infty & \text{if } x = 0, \\ 0 & \text{if } x \neq 0. \end{cases} \tag{1.1}$$

Of the density it is natural to require that the integral of the density over the entire space yield the total mass of substance, or

$$\int \delta(x)\, dx = 1. \tag{1.2}$$

But for the function $\delta(x)$ defined by (1.1), $\int \delta(x)\, dx = 0$. This means that the function does not restore the mass (it does not satisfy the requirement (1.2)) and therefore cannot be taken as the desired mass. Thus the point limit of a sequence of mean densities $f_\varepsilon(x)$ is unsuitable for our purposes. What is the way out?

Let us now find a somewhat different limit of the sequence of mean densities $f_\varepsilon(x)$, the so-called *weak limit*. It will readily be seen that for any continuous function $\varphi(x)$

$$\lim_{\varepsilon\to+0}\int f_\varepsilon(x)\varphi(x)\,dx = \varphi(0). \tag{1.3}$$

Formula (1.3) states that the weak limit of a sequence of functions $f_\varepsilon(x)$, $\varepsilon \to +0$, is a functional $\varphi(0)$ (and not a function!) that relates to every continuous function $\varphi(x)$ a number $\varphi(0)$, which is its value at the point $x = 0$. It is this functional that we take for our sought-for density $\delta(x)$. And this is the famous *delta-function* of Dirac. So now we can write

$$f_\varepsilon(x) \xrightarrow{\text{weak}} \delta(x), \qquad \varepsilon \to +0,$$

and we understand by this the limiting relation (1.3). The value of the functional δ on the function φ (the number $\varphi(0)$) will be denoted thus:

$$(\delta, \varphi) = \varphi(0). \tag{1.4}$$

This equality yields the exact meaning of the formula (*) (see Preface). The role of the "integral" $\int \delta(x)\varphi(x)\,dx$ is played here by the quantity (δ, φ), which is the value of the functional δ on the function φ.

Let us now check to see that the functional δ restores the total mass. Indeed, as we have just said, the role of the "integral" $\int \delta(x)\,dx$ is handled by the quantity $(\delta, 1)$, which, by virtue of (1.4), is equal to the value of the function identically equal to 1 at the point $x = 0$, that is, $(\delta, 1) = 1$.

Also, generally, if masses μ_k are concentrated at distinct points x_k, $k = 1, \ldots, N$, then the density that corresponds to such a mass distribution should be regarded as equal to

$$\sum_{1\le k\le N} \mu_k \delta(x - x_k). \tag{1.5}$$

The expression (1.5) is a linear functional that associates with each continuous function $\varphi(x)$ a number

$$\sum_{1\le k\le N} \mu_k \varphi(x_k).$$

Thus, the density corresponding to a point distribution of masses cannot be described within the framework of the classical concept of a function; to describe it requires resorting to entities of a more general mathematical nature, linear (continuous) functionals.

1.2. The space of test functions $\mathcal{D}(\mathcal{O})$. In the case of the delta-function we have already seen that it is determined by means of continuous functions as a linear (continuous) functional on those functions. Continuous functions are said to be *test functions* for the delta-function. It is this viewpoint that serves as the basis for defining an arbitrary generalized function as a continuous linear functional on a collection of sufficiently "good" so-called test functions. Clearly, the smaller the set of test functions, the more continuous linear functionals there are on it. On the other hand, the supply of test functions should be sufficiently large. In this subsection we introduce the important space of test functions $\mathcal{D}(\mathcal{O})$ for any open set $\mathcal{O} \subset \mathbb{R}^n$.

Included in the set of test functions $\mathcal{D}(\mathcal{O})$ are all functions with compact support infinitely differentiable in $\mathcal{O}$; $\mathcal{D}(\mathcal{O}) = C_0^\infty(\mathcal{O})$ (see Sec. 0.5). We define *convergence* in $\mathcal{D}(\mathcal{O})$ as follows. A sequence of functions $\varphi_1, \varphi_2, \ldots$ in $\mathcal{D}(\mathcal{O})$ converges to the function φ (in $\mathcal{D}(\mathcal{O})$) if there exists a set $\mathcal{O}' \Subset \mathcal{O}$ such that $\operatorname{supp}\varphi_k \subset \mathcal{O}'$ and for every α

$$\partial^\alpha \varphi_k(x) \overset{x\in\mathcal{O}}{\rightrightarrows} \partial^\alpha \varphi(x), \qquad k \to \infty.$$

We then write: $\varphi_k \to \varphi$, $k \to \infty$ in $\mathcal{D}(\mathcal{O})$.

A linear set $\mathcal{D}(\mathcal{O})$ equipped with this convergence is called the *space of test functions* $\mathcal{D}(\mathcal{O})$, and we have the following symbolism: $\mathcal{D} = \mathcal{D}(\mathbb{R}^n)$, $\mathcal{D}(a,b) = \mathcal{D}((a,b))$.

Clearly, if $\mathcal{O}_1 \subset \mathcal{O}_2$, then also $\mathcal{D}(\mathcal{O}_1) \subset \mathcal{D}(\mathcal{O}_2)$, and from the convergence in $\mathcal{D}(\mathcal{O}_1)$ there follows convergence in $\mathcal{D}(\mathcal{O}_2)$. An example of a nonzero test function is the "cap" in Fig. 4:

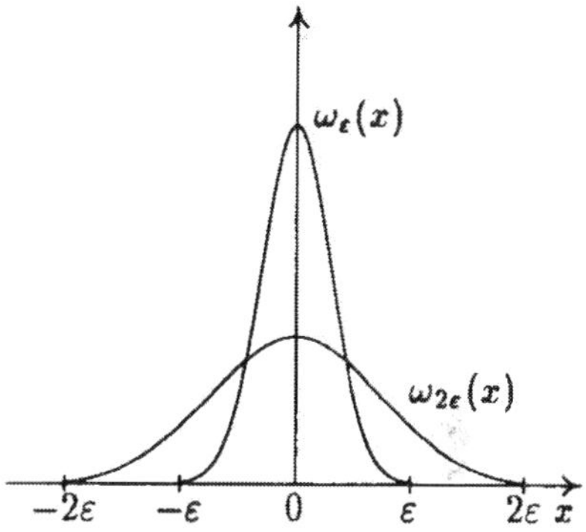

Figure 4

$$\omega_\varepsilon(x) = \begin{cases} C_\varepsilon e^{-\frac{\varepsilon^2}{\varepsilon^2-|x|^2}}, & |x| \le \varepsilon, \\ 0, & |x| > \varepsilon. \end{cases}$$

In what follows, the function ω_ε will play the part of an averaging function: and so we shall regard the constant C_ε as such that

$$\int \omega_\varepsilon(x)\,dx = 1, \qquad \text{that is,} \qquad C_\varepsilon \varepsilon^n \int\limits_{|\xi|<1} e^{-\frac{1}{1-|\xi|^2}}\,d\xi = 1.$$

The following lemma yields other examples of test functions.

LEMMA. *For any set A and any number $\varepsilon > 0$ there is a function $\eta_\varepsilon \in C^\infty$ such that*

$$\eta_\varepsilon(x) = 1, \quad x \in A^\varepsilon; \qquad \eta_\varepsilon(x) = 0, \quad x \notin A^{3\varepsilon};$$
$$0 \le \eta_\varepsilon(x) \le 1, \qquad |\partial^\alpha \eta_\varepsilon(x)| \le K_\alpha \varepsilon^{-|\alpha|}.$$

PROOF. Let $\theta_{A^{2\varepsilon}}$ be a characteristic function of the set $A^{2\varepsilon}$ (see Sec. 0.2). Then the function

$$\eta_\varepsilon(x) = \int \theta_{A^{2\varepsilon}}(y)\omega_\varepsilon(x-y)\,dy = \int\limits_{A^{2\varepsilon}} \omega_\varepsilon(x-y)\,dy,$$

where ω_ε is the "cap", has the required properties. □

COROLLARY. *Let $\mathcal{O}$ be an open set. Then for any $\mathcal{O}' \Subset \mathcal{O}$ there is a function $\eta \in \mathcal{D}(\mathcal{O})$ such that $\eta(x) = 1$, $x \in \mathcal{O}'$, $0 \le \eta(x) \le 1$.*

This follows from the lemma when $A = \mathcal{O}'$ and $\varepsilon = \frac{1}{3}\Delta(\mathcal{O}', \partial\mathcal{O}) > 0$.

Let $\mathcal{O}_k$, $k = 1, 2, \ldots$, be a countable system of open sets. We say that this system forms a *locally finite cover* of the open set $\mathcal{O}$ if $\mathcal{O} = \bigcup_{k\ge1} \mathcal{O}_k$, $\mathcal{O}_k \Subset \mathcal{O}$, and any compact $K \Subset \mathcal{O}$ intersects only a finite number of sets $\{\mathcal{O}_k\}$.

THEOREM I (partition of unity). *Let $\{\mathcal{O}_k\}$ be a locally finite cover of $\mathcal{O}$. Then there is a system of functions $\{e_k\}$ such that*

$$e_k \in \mathcal{D}(\mathcal{O}_k), \quad 0 \le e_k(x) \le 1, \quad \sum_{k\ge1} e_k(x) = 1, \quad x \in \mathcal{O}.$$

REMARK. For each $x \in \mathcal{O}$ there is only a finite number of nonzero summands $e_k(x)$; the set of functions $\{e_k\}$ is termed the *partition of unity corresponding to the given locally finite cover $\{\mathcal{O}_k\}$ of the open set $\mathcal{O}$.*

PROOF. We shall prove that there is another locally finite cover $\{\mathcal{O}'_k\}$ of the set $\mathcal{O}$ such that $\mathcal{O}'_k \Subset \mathcal{O}_k$. Construct $\mathcal{O}'_k$ and set

$$K_1 = \mathcal{O} \setminus \bigcup_{k\ge2} \mathcal{O}_k.$$

Then $K_1 \subset \mathcal{O}_1 \Subset \mathcal{O}$ and K_1 is closed in $\mathcal{O}$. Hence $K_1 \Subset \mathcal{O}_1$, for $\mathcal{O}'_1$ we take an open set such that $K_1 \Subset \mathcal{O}'_1 \Subset \mathcal{O}_1$. Then the sets $\mathcal{O}'_1, \mathcal{O}_2, \ldots$ form a locally finite cover of $\mathcal{O}$. In similar fashion we construct an open set $\mathcal{O}'_2 \Subset \mathcal{O}_2$, etc. We thus obtain the required cover $\{\mathcal{O}'_k\}$.

By the corollary to the lemma, there exist functions such that

$$\eta_k(x) = 1, \qquad x \in \mathcal{O}'_k, \qquad 0 \le \eta_k(x) \le 1.$$

Putting

$$e_k(x) = \frac{\eta_k(x)}{\sum_{k\ge1}\eta_k(x)}, \qquad \left(\sum_{k\ge1}\eta_k(x) \ge 1\right)$$

we obtain the required partition of unity. This completes the proof. □

We have thus seen that there are various functions in $\mathcal{D}(\mathcal{O})$. We will now see that there are a sufficiently large number of such functions.

Let f be a locally integrable function in $\mathcal{O}$, $f \in \mathcal{L}^p_{\text{loc}}(\mathcal{O})$. The convolution of f and the "cap", ω_ε,

$$f_\varepsilon(x) = \int f(y)\omega_\varepsilon(x-y)\,dy = \int \omega_\varepsilon(y) f(x-y)\,dy$$

(wherever it is defined) is called the *mean function of f*.

Let $f \in \mathcal{L}^p(\mathcal{O})$, $1 \le p \le \infty$ ($f(x)$ is regarded as zero outside $\mathcal{O}$). Then $f_\varepsilon \in C^\infty$ and the following inequality holds:

$$\|f_\varepsilon\|_{\mathcal{L}^p(\mathcal{O})} \le \|f\|_{\mathcal{L}^p(\mathcal{O})}. \tag{2.1}$$

Indeed, the fact that $f_\varepsilon \in C^\infty$ follows from the properties of the function f and from the definition of a mean function. When $1 \le p < \infty$ the inequality (2.1) follows from the Hölder inequality:

$$\|f_\varepsilon\|^p_{\mathcal{L}^p(\mathcal{O})} = \int_{\mathcal{O}} |f_\varepsilon(x)|^p\,dx = \int_{\mathcal{O}} \left|\int f(y)\omega_\varepsilon(x-y)\,dy\right|^p dx$$

$$\leq \int_{\mathcal{O}}\int_{\mathcal{O}} |f(y)|^p\omega_\varepsilon(x-y)\,dy\left[\int \omega_\varepsilon(x-y)\,dy\right]^{p-1} dx$$

$$= \int_{\mathcal{O}}\int_{\mathcal{O}} |f(y)|^p\omega_\varepsilon(x-y)\,dy\,dx$$

$$\leq \int_{\mathcal{O}} |f(y)|^p\,dy = \|f\|^p_{\mathcal{L}^p(\mathcal{O})}.$$

The case of $p = \infty$ is considered in a similar fashion. □

THEOREM II. *Let $f \in \mathcal{L}^1_0(\mathcal{O})$ and $f(x) = 0$ almost everywhere outside $K \Subset \mathcal{O}$. Then for all $\varepsilon < \Delta(K, \partial\mathcal{O})$ the mean function $f_\varepsilon \in \mathcal{D}(\mathcal{O})$ and (see Secs. 0.3 and 0.5 for the symbolism)*

$$f_\varepsilon \to f, \qquad \varepsilon \to +0 \begin{cases} \text{in } C(\overline{\mathcal{O}}) & \text{if } f \in C_0(\mathcal{O}), \\ \text{in } \mathcal{L}^p(\mathcal{O})\ (1 \leq p < \infty) & \text{if } f \in \mathcal{L}^p_0(\mathcal{O}), \\ \text{almost everywhere in } \mathcal{O} & \text{if } f \in \mathcal{L}^\infty_0(\mathcal{O}). \end{cases}$$

PROOF. If $\varepsilon < \Delta(K, \partial\mathcal{O})$, then $f_\varepsilon(x)$ is with compact support in $\mathcal{O}$, and since $f_\varepsilon \in C^\infty(\mathcal{O})$, it follows that $f_\varepsilon \in \mathcal{D}(\mathcal{O})$.

Let $f \in C_0(\mathcal{O})$. Then from the estimate

$$|f_\varepsilon(x) - f(x)| = \left|\int [f(y) - f(x)]\omega_\varepsilon(x-y)\,dy\right|$$

$$\leq \max_{|x-y|\leq\varepsilon} |f(x) - f(y)| \int \omega_\varepsilon(x-y)\,dy$$

$$= \max_{|x-y|\leq\varepsilon} |f(x) - f(y)|, \qquad x \in \mathcal{O},$$

and from the uniform continuity of the function f follows the uniform convergence, on $\mathcal{O}$, of $f_\varepsilon(x)$ to $f(x)$ as $\varepsilon \to +0$.

Let $f \in \mathcal{L}^p_0(\mathcal{O})$, $1 \leq p < \infty$. Take an arbitrary $\delta > 0$. There is a function $g \in C_0(\mathcal{O})$ such that

$$\|f - g\|_{\mathcal{L}^p(\mathcal{O})} < \frac{\delta}{3}.$$

From what has been proved, there will be an ε_0 such that

$$\|g - g_\varepsilon\|_{\mathcal{L}^p(\mathcal{O})} < \frac{\delta}{3} \quad \text{for all} \quad \varepsilon < \varepsilon_0.$$

From this, using the inequality (2.1), we obtain, for all $\varepsilon < \varepsilon_0$,

$$\|f - f_\varepsilon\|_{\mathcal{L}^p(\mathcal{O})} \leq \|f - g\|_{\mathcal{L}^p(\mathcal{O})} + \|g - g_\varepsilon\|_{\mathcal{L}^p(\mathcal{O})} + \|(g - f)_\varepsilon\|_{\mathcal{L}^p(\mathcal{O})}$$

$$\leq 2\|f - g\|_{\mathcal{L}^p(\mathcal{O})} + \|g - g_\varepsilon\|_{\mathcal{L}^p(\mathcal{O})} < \frac{2\delta}{3} + \frac{\delta}{3} = \delta.$$

And this means that $f_\varepsilon \to f$, $\varepsilon \to +0$ in $\mathcal{L}^p(\mathcal{O})$.

Now if $f \in \mathcal{L}^\infty_0(\mathcal{O})$, we can indicate a sequence of functions taken from $C_0(\mathcal{O})$ that converges to $f(x)$ almost everywhere in $\mathcal{O}$. From this fact and from what has been proved, it follows that $f_\varepsilon(x) \to f(x)$, $\varepsilon \to +0$, almost everywhere in $\mathcal{O}$. The theorem is proved. □

COROLLARY 1. *$\mathcal{D}(\mathcal{O})$ is dense in $\mathcal{L}^p(\mathcal{O})$, $1 \le p < \infty$.*

COROLLARY 2. *$\mathcal{D}(\mathcal{O})$ is dense in $C_0^k(\overline{\mathcal{O}})$ (in the norm $C^k(\overline{\mathcal{O}})$) if $\mathcal{O}$ is bounded or $\mathcal{O} = \mathbb{R}^n$.*

1.3. The space of generalized functions $\mathcal{D}'(\mathcal{O})$. A *generalized function* specified on an open set $\mathcal{O}$ is any continuous linear functional on the space of test functions $\mathcal{D}(\mathcal{O})$.

We will write the value of the functional (generalized function) f on the test function φ as (f, φ). By analogy with ordinary functions, we sometimes formally write $f(x)$ instead of f, and regard x as the argument of the test functions on which the functional f operates.

We now give an explanation of the definition of a generalized function.

(1) A generalized function f is a *functional* on $\mathcal{D}(\mathcal{O})$, that is, with each $\varphi \in \mathcal{D}(\mathcal{O})$ there is associated a (complex-valued) number (f, φ).

(2) A generalized function f is a *linear functional* on $\mathcal{D}(\mathcal{O})$, that is, if φ and ψ belong to $\mathcal{D}(\mathcal{O})$ and λ and μ are complex numbers, then

$$(f, \lambda\varphi + \mu\psi) = \lambda(f, \varphi) + \mu(f, \psi).$$

(3) A generalized function f is a *continuous functional* on $\mathcal{D}(\mathcal{O})$, that is, if $\varphi_k \to \varphi$, $k \to \infty$ in $\mathcal{D}(\mathcal{O})$, then

$$(f, \varphi_k) \to (f, \varphi), \qquad k \to \infty.$$

The generalized functions f and g specified in $\mathcal{O}$ are said to be *equal* in $\mathcal{O}$ if they are equal as functionals on $\mathcal{D}(\mathcal{O})$, that is, if for any φ in $\mathcal{D}(\mathcal{O})$, $(f, \varphi) = (g, \varphi)$. We will then write: $f = g$ in $\mathcal{O}$ or $f(x) = g(x)$, $x \in \mathcal{O}$.

Denote by $\mathcal{D}'(\mathcal{O})$ the set of all generalized functions specified in $\mathcal{O}$. The set $\mathcal{D}'(\mathcal{O})$ is a linear set if we define the linear combination $\lambda f + \mu g$ of the generalized functions f and g in $\mathcal{D}'(\mathcal{O})$ as a functional acting via the formula [thus, the form (f, φ) is a bilinear form (see Sec. 0.6)]

$$(\lambda f + \mu g, \varphi) = \lambda(f, \varphi) + \mu(g, \varphi), \qquad \varphi \in \mathcal{D}(\mathcal{O}).$$

Suppose $f \in \mathcal{D}'(\mathcal{O})$. We define a generalized function $\bar{f}$ in $\mathcal{D}'(\mathcal{O})$, which is the complex conjugate of f, as follows:

$$(\bar{f}, \varphi) = \overline{(f, \bar{\varphi})}, \qquad \varphi \in \mathcal{D}(\mathcal{O}).$$

The generalized functions

$$\Re f = \frac{f + \bar{f}}{2}, \qquad \Im f = \frac{f - \bar{f}}{2i}$$

are respectively the *real part* and the *imaginary part* of f so that

$$f = \Re f + i\Im f, \qquad \bar{f} = \Re f - i\Im f.$$

If $\Im f = 0$, then f is said to be a *real* generalized function.

EXAMPLE. The delta-function is real.

Let us define *convergence* in $\mathcal{D}'(\mathcal{O})$. A sequence of generalized functions $f_1, f_2, \ldots$ in $\mathcal{D}'(\mathcal{O})$ converges to a generalized function $f \in \mathcal{D}'(\mathcal{O})$ if for any test function $\varphi \in \mathcal{D}(\mathcal{O})$ $(f_k, \varphi) \to (f, \varphi)$, $k \to \infty$. We then write

$$f_k \to f, \qquad k \to \infty \quad \text{in} \quad \mathcal{D}'(\mathcal{O}).$$

This convergence is termed as a *weak convergence*. The linear set $\mathcal{D}'(\mathcal{O})$ together with the convergence with which it is equipped is called the *space* $\mathcal{D}'(\mathcal{O})$ *of generalized functions*. In symbols: $\mathcal{D}' = \mathcal{D}'(\mathbb{R}^n)$, $\mathcal{D}'(a,b) = \mathcal{D}'\big((a,b)\big)$.

Quite obviously, if $\mathcal{O}_1 \subset \mathcal{O}_2$, then $\mathcal{D}'(\mathcal{O}_2) \subset \mathcal{D}'(\mathcal{O}_1)$, and from convergence in $\mathcal{D}'(\mathcal{O}_2)$ follows convergence in $\mathcal{D}'(\mathcal{O}_1)$.

For this reason, for any generalized function f in $\mathcal{D}'(\mathcal{O})$ there is a (unique) restriction to any open set $\mathcal{O}' \subset \mathcal{O}$ such that $f \in \mathcal{D}'(\mathcal{O}')$.

REMARK. Linear functionals on $\mathcal{D}(\mathcal{O})$ need not necessarily be continuous on $\mathcal{D}(\mathcal{O})$. However, not a single discontinuous linear functional has been constructed explicitly on $\mathcal{D}(\mathcal{O})$; one can only prove their existence theoretically by using the axiom of choice.

THEOREM. *For a linear functional f on $\mathcal{D}(\mathcal{O})$ to belong to $\mathcal{D}'(\mathcal{O})$, that is, for it to be a generalized function in $\mathcal{O}$, it is necessary and sufficient that for any open set $\mathcal{O}' \Subset \mathcal{O}$ there exist numbers $K = K(\mathcal{O}')$ and $m = m(\mathcal{O}')$ such that*

$$|(f, \varphi)| \le K \|\varphi\|_{C^m(\overline{\mathcal{O}})}, \qquad \varphi \in \mathcal{D}(\mathcal{O}'). \tag{3.1}$$

PROOF. Sufficiency is obvious. We prove necessity. Suppose $f \in \mathcal{D}'(\mathcal{O})$ and $\mathcal{O}' \Subset \mathcal{O}$. If the inequality (3.1) does not hold, there will be a sequence φ_k, $k = 1, 2, \ldots$, of functions in $\mathcal{D}(\mathcal{O}')$ such that

$$|(f, \varphi_k)| \ge k \|\varphi_k\|_{C^k(\overline{\mathcal{O}}')}. \tag{3.2}$$

But the sequence

$$\psi_k = \frac{\varphi_k}{\sqrt{k}\|\varphi_k\|_{C^k(\overline{\mathcal{O}}')}} \to 0, \qquad k \to \infty \quad \text{in} \quad \mathcal{D}(\mathcal{O}),$$

since $\operatorname{supp} \psi_k \subset \mathcal{O}' \Subset \mathcal{O}$, and for $k \ge |\beta|$ we have

$$|\partial^\beta \psi_k(x)| = \left| \partial^\beta \frac{\varphi_k(x)}{\sqrt{k}\|\varphi_k\|_{C^k(\overline{\mathcal{O}}')}} \right| \le \frac{1}{\sqrt{k}} \to 0, \qquad k \to \infty.$$

Therefore, $(f, \psi_k) \to 0$, $k \to \infty$. On the other hand, by virtue of (3.2) we get

$$|(f, \psi_k)| = \frac{|(f, \varphi_k)|}{\sqrt{k}\|\varphi_k\|_{C^k(\overline{\mathcal{O}}')}} \ge \sqrt{k} \to \infty, \qquad k \to \infty.$$

This contradiction proves the theorem. □

Suppose $f \in \mathcal{D}'(\mathcal{O})$. If it is possible in the inequality (3.1) to choose the integer m as independent of $\mathcal{O}'$, then we say that the generalized function f is of *finite order*; the smallest such m is termed the *order* of f in $\mathcal{O}$. For example, the order of the delta-function is 0; the order of the generalized function

$$(f, \varphi) = \sum_{k \ge 1} \varphi^{(k)}(k)$$

in $(0, \infty)$ is infinite.

REMARK. The theorem just proved signifies that if we introduce in the space $\mathcal{D}(\mathcal{O})$ a topology of an inductive limit (union) of an increasing sequence of countable-normed spaces $C_0^\infty(\overline{\mathcal{O}}_k)$, where $\mathcal{O}_1 \Subset \mathcal{O}_2 \Subset \dots, \bigcup_{k\geq 1} \mathcal{O}_k = \mathcal{O}$, with norms

$$\|\varphi\|_{C^\nu(\overline{\mathcal{O}}_k)}, \qquad \nu = 0, 1, \dots, \quad \varphi \in C_0^\infty(\overline{\mathcal{O}}_k),$$

then $\mathcal{D}'(\mathcal{O})$ becomes the conjugate space of $\mathcal{D}(\mathcal{O})$ (see Bourbaki [**11**], and Dieudonné and Schwartz [**17**]). Here, the inequality (3.1) is preserved for all functions φ in $C_0^m(\overline{\mathcal{O}'})$ (see Corollary 2 to Theorem II of Sec. 1.2).

1.4. The completeness of the space of generalized functions $\mathcal{D}'(\mathcal{O})$. The property of the completeness of the space $\mathcal{D}'(\mathcal{O})$ is extremely important.

THEOREM. *Let there be a sequence of generalized functions $f_1, f_2, \dots$ in $\mathcal{D}'(\mathcal{O})$ such that for every function $\varphi \in \mathcal{D}(\mathcal{O})$, the numerical sequence (f_k, φ) converges as $k \to \infty$. Then the functional f on $\mathcal{D}(\mathcal{O})$ defined by $(f, \varphi) = \lim_{k\to\infty}(f_k, \varphi)$ is also linear and continuous on $\mathcal{D}(\mathcal{O})$, or $f \in \mathcal{D}'(\mathcal{O})$.*

PROOF. The linearity of the limiting functional f is obvious. Let us prove its continuity on $\mathcal{D}(\mathcal{O})$. Let $\varphi_\nu \to 0$, $\nu \to \infty$ in $\mathcal{D}(\mathcal{O})$; we have to prove that $(f, \varphi_\nu) \to 0$, $\nu \to \infty$. Assuming the contrary and passing, if necessary, to a subsequence, we may assume that for all $\nu = 1, 2, \dots$ the inequality $|(f, \varphi_k)| \geq 2a$ holds true for some $a > 0$. Since $(f, \varphi_\nu) = \lim_{k\to\infty}(f_k, \varphi_\nu)$, it follows that for every $\nu = 1, 2, \dots$ there is a number k_ν such that $|(f_{k_\nu}, \varphi_\nu)| \geq a$. But this is impossible due to the lemma which follows. This contradiction completes the proof of the continuity of f. The proof of the theorem is complete. □

LEMMA. *Given a sequence of functionals $f_1, f_2, \dots$ taken from a weakly bounded set $M' \subset \mathcal{D}'(\mathcal{O})$, that is, $|(f, \varphi)| < C_\varphi$, $f \in M'$ for all φ in $\mathcal{D}(\mathcal{O})$, and suppose the sequence of test functions $\varphi_1, \varphi_2, \dots$ in $\mathcal{D}(\mathcal{O})$ tends to 0 in $\mathcal{D}(\mathcal{O})$. Then $(f_k, \varphi_k) \to 0$ as $k \to \infty$.*

PROOF. Assume the lemma is not true. Then, passing, if necessary, to a subsequence, we can say that $|(f_k, \varphi_k)| \geq c > 0$. The convergence of φ_k to 0 in $\mathcal{D}(\mathcal{O})$ means that $\operatorname{supp} \varphi_k \subset \mathcal{O}' \Subset \mathcal{O}$ and for every α

$$\partial^\alpha \varphi_k(x) \overset{x\in\mathcal{O}}{\rightrightarrows} 0, \qquad k \to \infty.$$

Therefore, by again passing, if necessary, to a subsequence, we can assume that

$$|\partial^\alpha \varphi_k(x)| \leq \frac{1}{4^k}, \qquad |\alpha| \leq k = 0, 1, \dots.$$

Set $\psi_k = 2^k \varphi_k$; then $\operatorname{supp} \psi_k \subset \mathcal{O}' \Subset \mathcal{O}$ and

$$|\partial^\alpha \psi_k(x)| \leq \frac{1}{2^k}, \qquad |\alpha| \leq k = 0, 1, \dots, \tag{4.1}$$

so that

$$|(f_k, \psi_k)| = 2^k |(f_k, \varphi_k)| \geq 2^k c \to \infty, \qquad k \to \infty. \tag{4.2}$$

Now construct the subsequences $\{f_{k_\nu}\}$ and $\{\psi_{k_\nu}\}$ in the following manner. Choose f_{k_1} and ψ_{k_1} so that $|(f_{k_1}, \psi_{k_1})| \geq 2$. Suppose f_{k_j} and ψ_{k_j}, $j = 1, \dots, \nu - 1$, have already been constructed; construct f_{k_ν} and ψ_{k_ν}. Since $\psi_k \to 0$, $k \to \infty$ in

$\mathcal{D}(\mathcal{O})$, it follows that $(f_{k_j}, \psi_k) \to 0$, $k \to \infty$, $j = 1, \ldots, \nu - 1$, and therefore there is a number N such that for all $k \geq N$,

$$|(f_{k_j}, \psi_k)| \leq \frac{1}{2^{\nu-j}}, \qquad j = 1, \ldots, \nu - 1. \tag{4.3}$$

Now note that $|(f_k, \psi_{k_j})| \leq c_{k_j}$, $j = 1, \ldots, \nu - 1$. Finally, by virtue of (4.2) choose a number $k_\nu \geq N$ such that

$$|(f_{k_\nu}, \psi_{k_\nu})| \geq \sum_{1 \leq j \leq \nu-1} c_{k_j} + \nu + 1. \tag{4.4}$$

Thus, by (4.3)–(4.4), the (generalized) functions f_{k_ν} and ψ_{k_ν} thus constructed are such that

$$|(f_{k_j}, \psi_{k_\nu})| \leq \frac{1}{2^{\nu-j}}, \qquad j = 1, \ldots, \nu - 1, \tag{4.5}$$

$$|(f_{k_\nu}, \psi_{k_\nu})| \geq \sum_{1 \leq j \leq \nu-1} |(f_{k_j}, \psi_{k_j})| + \nu + 1. \tag{4.6}$$

Set $\psi = \sum_{j \geq 1} \psi_{k_j}$. By virtue of (4.1) this series converges in $\mathcal{D}(\mathcal{O})$ and, hence, $\psi \in \mathcal{D}(\mathcal{O})$ and

$$(f_{k_\nu}, \psi) = (f_{k\nu}, \psi_{k_\nu}) + \sum_{\substack{j=1 \\ j \neq \nu}} (f_{k_\nu}, \psi_{k_j}). \tag{4.7}$$

Whence, taking into account the inequalities (4.5) and (4.6), we get

$$\begin{aligned} |(f_{k_\nu}, \psi)| &\geq |(f_{k_\nu}, \psi_{k_\nu})| - \sum_{1 \leq j \leq \nu-1} |(f_{k_\nu}, \psi_{k_j})| - \sum_{j \geq \nu+1}^{\infty} |(f_{k_\nu}, \psi_{k_j})| \\ &\geq \nu + 1 - \sum_{j \geq \nu+1} \frac{1}{2^{j-\nu}} = \nu. \end{aligned}$$

That is, $(f_{k\nu}, \psi) \to \infty$, $\nu \to \infty$. But this contradicts the boundedness of the sequence (f_k, ψ), $k \to \infty$ $(f_k \in M')$. □

Corollary. *If a set $M' \subset \mathcal{D}'(\mathcal{O})$ is weakly bounded, then for any $\mathcal{O}' \Subset \mathcal{O}$ there exist numbers K and m such that the inequality (3.1) is preserved for all $\varphi \in \mathcal{D}(\mathcal{O}')$ and $f \in M'$.*

The proof is analogous to the proof of the theorem of Sec. 1.3 with the lemma invoked as well.

1.5. The support of a generalized function. Generally speaking, generalized functions do not have values at separate points. Nevertheless, one may speak of the vanishing of a generalized function in an open set. We say that a generalized function f in $\mathcal{D}'(\mathcal{O})$ *vanishes* in an open set $\mathcal{O}' \subset \mathcal{O}$ if its restriction to $\mathcal{O}'$ (see Sec. 1.3) is zero functional in $\mathcal{D}'(\mathcal{O}')$, that is, $(f, \varphi) = 0$ for all $\varphi \in \mathcal{D}(\mathcal{O}')$. We that write: $f(x) = 0$, $x \in \mathcal{O}'$.

Suppose a generalized function f in $\mathcal{D}'(\mathcal{O})$ vanishes in $\mathcal{O}$. It then obviously vanishes also in neighbourhoods of every point of the set $\mathcal{O}$.

Conversely, let f in $\mathcal{D}'(\mathcal{O})$ vanish in some neighbourhood $U(y) \subset \mathcal{O}$ of each point y of the open set $\mathcal{O}$.

Using the cover $\{U(y), y \in \mathcal{O}\}$ of the set $\mathcal{O}$, let us construct its locally finite cover $\{\mathcal{O}_k\}$ so that each $\mathcal{O}_k$ is contained in some $U(y)$. Let $\mathcal{O}_1' \Subset \mathcal{O}_2' \Subset \ldots$,

$\bigcup_{\nu\geq 1}\mathcal{O}'_\nu = \mathcal{O}$. By the Heine–Borel lemma, the compact $\overline{\mathcal{O}'_2}$ is covered by a finite number of neighbourhoods $U(y)$: $U(y_1),\ldots,U(y_{N_1})$; the compact $\overline{\mathcal{O}'_3}\setminus\mathcal{O}'_1$ is covered likewise by a finite number of such neighbourhoods: $U(y_{N_1+1}),\ldots,U(y_{N_1+N_2})$; and so forth. Setting $\mathcal{O}_k = U(y_k)\cap\mathcal{O}'_2$, $k=1,\ldots,N_1$, $\mathcal{O}_k = U(y_k)\cap(\mathcal{O}'_3\setminus\overline{\mathcal{O}'_1})$, $k = N_1+1,\ldots,N_1+N_2$, and so on, we obtain the required cover $\{\mathcal{O}_k\}$.

Let $\{e_k\}$ be the partition of unity that corresponds to the constructed cover $\{\mathcal{O}_k\}$ of the set $\mathcal{O}$ (see Sec. 1.2). Then for any φ in $\mathcal{D}(\mathcal{O})$, $\operatorname{supp}(\varphi e_k)\subset U(y)$ for some y and for that reason$(f,\varphi e_k)=0$; consequently

$$(f,\varphi) = \left(f, \sum_{k\geq 1} e_k\varphi\right) = \sum_{k\geq 1}(f,\varphi e_k) = 0.$$

Thus, we proved that the following lemma holds true.

LEMMA. *If a generalized function in $\mathcal{D}'(\mathcal{O})$ vanishes in some nighbourhood of every point of an open set $\mathcal{O}$, then it also vanishes in the whole set $\mathcal{O}$.*

Suppose $f\in\mathcal{D}'(\mathcal{O})$. The union of all neighbourhoods where $f=0$ forms an open set $\mathcal{O}_f$, which is called the *zero set* of the generalized function f. By the lemma, $f=0$ in $\mathcal{O}_f$; furthermore $\mathcal{O}_f$ is the largest open set in which f vanishes.

The ***support*** of a generalized function f is the complement of $\mathcal{O}_f$ to $\mathcal{O}$; the support of f is symbolized as $\operatorname{supp} f$, so that $\operatorname{supp} f = \mathcal{O}\setminus\mathcal{O}_f$; $\operatorname{supp} f$ is a closed set in $\mathcal{O}$. If $\operatorname{supp} f \Subset \mathcal{O}$, then f is said to be ***with compact support*** in $\mathcal{O}$.

From the foregoing follow the assertions:

(a) *If the supports of $f\in\mathcal{D}'(\mathcal{O})$ and $\varphi\in\mathcal{D}(\mathcal{O})$ do not have any points in common, then $(f,\varphi)=0$.*

(b) *To have $x\in\operatorname{supp} f$, it is necessary and sufficient that f should not vanish in any neighbourhood of the point x.*

Let A be a closed set in $\mathcal{O}$. We denote by $\mathcal{D}'(\mathcal{O},A)$ the collection of generalized functions in $\mathcal{D}'(\mathcal{O})$, whose supports are contained in A, and with convergence: $f_k\to 0$, $k\to\infty$ in $\mathcal{D}'(\mathcal{O},A)$ if $f_k\to 0$, $k\to\infty$ in $\mathcal{D}'(\mathcal{O})$ and $\operatorname{supp} f_k\subset A$. We write $\mathcal{D}'(A)=\mathcal{D}'(\mathbb{R}^n,A)$.

A similar meaning is also attributed to other spaces of generalized functions, for example, $\mathcal{S}'(A)$, $\mathcal{L}^2_s(A)$ and so on (see Secs. 5 and 7 below).

The lemma that was proved in this subsection admits of a generalization. In Sec. 1.3 we saw that any generalized function f in $\mathcal{D}'(\mathcal{O})$ induces, in each $\mathcal{O}'\subset\mathcal{O}$, its own local element $f\in\mathcal{D}'(\mathcal{O}')$. The converse is also true: it is possible to "join together" a single generalized function out of any collection of compatible local elements. To be more precise, the following theorem holds.

THEOREM ("of piecewise sewing"). *Suppose that for each point $y\in\mathcal{O}$ there exist a neighbourhood $U(y)\Subset\mathcal{O}$ and in it there is specified a generalized function f_y, wherein $f_{y_1}(x)=f_{y_2}(x)$ if $x\in U(y_1)\cap U(y_2)\neq\varnothing$. Then there is a unique generalized function f in $\mathcal{D}'(\mathcal{O})$ that coincides with f_y in $U(y)$ for all $y\in\mathcal{O}$.*

PROOF. Again, as in the proof of the lemma, we construct, with respect to the cover $\{U(y),\ y\in\mathcal{O}\}$, a locally finite cover $\{\mathcal{O}_k\}$, $\mathcal{O}_k\subset U(y_k)$, of the set $\mathcal{O}$ and the appropriate partition of unity $\{e_k\}$. Set

$$(f,\varphi) = \sum_{k\geq 1}(f_{y_k},\varphi e_k), \qquad \varphi\in\mathcal{D}(\mathcal{O}). \tag{5.1}$$

Since the number of summands in the right member of (5.1) is always finite and does not depend on $\varphi \in \mathcal{D}(\mathcal{O}')$ for any $\mathcal{O}' \Subset \mathcal{O}$, it follows that functional f defined by it is linear and continuous on $\mathcal{D}(\mathcal{O})$, or $f \in \mathcal{D}'(\mathcal{O})$. Furthermore, if $\varphi \in \mathcal{D}(U(y))$, then $\varphi e_k \in \mathcal{D}(U(y_k))$ and, hence, $(f_y, \varphi e_k) = (f_{y_k}, \varphi e_k)$ so that by virtue of (5.1)

$$(f, \varphi) = \sum_{k \geq 1} (f_{y_k}, \varphi e_k) = \left(f_y, \varphi \sum_{k \geq 1} e_k \right) = (f_y, \varphi).$$

That is, $f = f_y$ in $U(y)$. The uniqueness of the generalized function f thus constructed follows from the lemma. □

1.6. Regular generalized functions. The simplest example of a generalized function is a functional generated by a function $f(x)$ locally integrable in $\mathcal{O}$:

$$(f, \varphi) = \int f(x)\varphi(x)\,dx, \qquad \varphi \in \mathcal{D}(\mathcal{O}). \tag{6.1}$$

From the properties of the linearity of an integral and from the theorem on passage to the limit under the integral sign it follows that the functional in the right member of (6.1) is linear and continuous on $\mathcal{D}(\mathcal{O})$, that is, $f \in \mathcal{D}'(\mathcal{O})$.

Generalized functions determined, via (6.1), by functions locally integrable in $\mathcal{O}$ are termed *regular* generalized functions. All other generalized functions are said to be *singular*.

LEMMA (Du Bois Reymond). *For a function $f(x)$ that is locally integrable in $\mathcal{O}$ to vanish almost everywhere in $\mathcal{O}$, it is necessary and sufficient that the regular generalized function f generated by it vanish in $\mathcal{O}$.*

PROOF. Necessity is obvious. We prove sufficiency. Let

$$\int f(x)\varphi(x)\,dx = 0, \qquad \varphi \in \mathcal{D}(\mathcal{O}). \tag{6.2}$$

Take an arbitrary $\mathcal{O}' \Subset \mathcal{O}$; let $\theta_{\mathcal{O}'}$ be a characteristic function of $\mathcal{O}'$. By Theorem II of Sec. 1.2, there exists a sequence of functions $\varphi_k(x)$, $k = 1, 2, \ldots$, from $\mathcal{D}(\mathcal{O})$, which converges to the function $e^{-i \arg f(x)}\theta_{\mathcal{O}'}(x)$ almost everywhere in $\mathcal{O}$, and $|\varphi_k(x)| \leq 1$ almost everywhere in $\mathcal{O}$. From this, using the Lebesgue theorem on passage to the limit under the sign of the Lebesgue integral, we conclude, taking into account (6.2), that

$$\begin{aligned}\int_{\mathcal{O}'} |f(x)|\,dx &= \int f(x) e^{-i \arg f(x)} \theta_{\mathcal{O}'}(x)\,dx \\ &= \lim_{k\to\infty} \left\{ \int f(x)\varphi_k(x)\,dx + \int f(x) \left[e^{-i \arg f(x)}\theta_{\mathcal{O}'}(x) - \varphi_k(x) \right] dx \right\} \\ &= \int_{\mathcal{O}'} \lim_{k\to\infty} f(x) \left[e^{-i \arg f(x)} - \varphi_k(x) \right] dx = 0,\end{aligned}$$

so that $f(x) = 0$ almost everywhere in $\mathcal{O}'$. Due to the arbitrariness of the set $\mathcal{O}' \Subset \mathcal{O}$, we conclude that $f(x) = 0$ almost everywhere in $\mathcal{O}$. □

From the lemma just proved it follows that any regular generalized function in $\mathcal{O}$ is defined by a unique function (unique up to the values on a set of measure zero) that is locally integrable in $\mathcal{O}$. Consequently there is a one-to-one correspondence between functions locally integrable in $\mathcal{O}$ and regular generalized functions in $\mathcal{O}$.

For this reason we will henceforth identify a function $f(x)$ locally integrable in $\mathcal{O}$ with the generalized function from $\mathcal{D}'(\mathcal{O})$ that is generated by it via (6.1). In this sense, "ordinary" functions (that is, functions locally integrable in $\mathcal{O}$) are (regular) generalized functions taken from $\mathcal{D}'(\mathcal{O})$.

From the Du Bois Reymond lemma it follows likewise that both definitions of the support of a continuous function in $\mathcal{O}$ that were given in Sec. 0.5 and Sec. 1.5 coincide.

Also note that if the sequence $f_k(x)$, $k = 1, 2, \ldots$, of functions locally integrable in $\mathcal{O}$ converges uniformly to the function $f(x)$ on each compact $K \Subset \mathcal{O}$, then, also, $f_k \to f$, $k \to \infty$ in $\mathcal{D}'(\mathcal{O})$.

Suppose $f \in \mathcal{D}'(\mathcal{O})$ and $\mathcal{O}_1 \subset \mathcal{O}$. We will say that a generalized function f *belongs to a class* $C^k(\mathcal{O}_1)$ if, in $\mathcal{O}_1$, it coincides with the function f_1 of the class $C^k(\mathcal{O}_1)$, that is, for any $\varphi \in \mathcal{D}(\mathcal{O}_1)$,

$$(f, \varphi) = \int f_1(x)\varphi(x)\, dx.$$

If, besides, $f_1 \in C^k(\overline{\mathcal{O}}_1)$, then we will say that f belongs to the class $C^k(\overline{\mathcal{O}}_1)$.

1.7. Measures. A more general class of generalized functions that contains the regular generalized functions is generated by measures. A *measure* on a Borel set A is a completely additive (complex-valued) function of sets

$$\mu(E) = \int_E \mu(dx),$$

which function is specified and finite on all bounded Borel subsets E of the set A, $|\mu(E)| < \infty$.

For details of measure theory and integration see Kolmogorov and Fomin [58].

The measure $\mu(E)$ on A can be uniquely represented in terms of four nonnegative measures $\mu_j(E) \geq 0$ on A via the formula $\mu = \mu_1 - \mu_2 + i(\mu_3 - \mu_4)$; here,

$$\int_E \mu(dx) = \int_E \mu_1(dx) - \int_E \mu_2(dx) + i\int_E \mu_3(dx) - i\int_E \mu_4(dx). \tag{7.1}$$

The measure $\mu(E)$ on an open set $\mathcal{O}$ determines a generalized function μ in $\mathcal{O}$ by the formula

$$(\mu, \varphi) = \int \varphi(x)\mu(dx), \qquad \varphi \in \mathcal{D}(\mathcal{O}), \tag{7.2}$$

where the integral is the Lebesgue–Stieltjes integral. From the properties of this integral it follows that we do indeed have $\mu \in \mathcal{D}'(\mathcal{O})$.

REMARK. For measures μ on $\mathcal{O}$ that are absolutely continuous with respect to the Lebesgue measure, that is, $\mu(dx) = f(x)\, dx$, where $f \in \mathcal{L}^1_{\text{loc}}(\mathcal{O})$, formula (7.2) defines regular generalized functions f (see Sec. 1.6).

LEMMA. *For a measure $\mu(E)$ on $\mathcal{O}$ to be a zero measure, it is necessary and sufficient for the generalized function μ defined by it to vanish in $\mathcal{O}$.*

PROOF. The proof is based on the following assertion: for the measure $\mu(E)$ on $\mathcal{O}$ to be a zero measure, it is necessary and sufficient that

$$\int_{\mathcal{O}} \varphi(x)\mu(dx) = 0, \qquad \varphi \in C_0(\mathcal{O}) \tag{7.3}$$

whence immediately follows necessity. We now prove sufficiency. Assuming that (7.3) holds for all φ in $\mathcal{D}(\mathcal{O})$, we will prove that it holds also for an arbitrary φ in $C_0(\mathcal{O})$. Suppose $\operatorname{supp}\varphi \subset \mathcal{O}' \Subset \mathcal{O}$. By Theorem II, Sec. 1.2, there exists a sequence of functions φ_k, $k = 1, 2, \ldots$, in $\mathcal{D}(\mathcal{O})$ such that $\operatorname{supp}\varphi_k \subset \mathcal{O}' \Subset \mathcal{O}$ and $\varphi_k \to \varphi$, $k \to \infty$ in $C(\overline{\mathcal{O}'})$. Therefore

$$\int_{\mathcal{O}} \varphi(x)\mu(dx) = \int_{\mathcal{O}'} \lim_{k\to\infty} \varphi_k(x)\mu(dx) = \lim_{k\to\infty} \int_{\mathcal{O}} \varphi_k(x)\mu(dx) = 0,$$

which is what we set out to prove. The lemma is proved. □

From the lemma it follows that there exists a one-to-one correspondence between measures on $\mathcal{O}$ and the generalized functions generated by them via formula (7.2). For this reason, we will in future identify the measure $\mu(E)$ on $\mathcal{O}$ and the generalized function μ in $\mathcal{D}'(\mathcal{O})$ generated by that measure.

THEOREM I. *For a generalized function f in $\mathcal{D}'(\mathcal{O})$ to be a measure on $\mathcal{O}$, it is necessary and sufficient that its order in $\mathcal{O}$ be equal to 0.*

PROOF. *Necessity.* Let $f \in \mathcal{D}'(\mathcal{O})$ be a measure μ on $\mathcal{O}$. Then for any $\mathcal{O}' \Subset \mathcal{O}$ and any $\varphi \Subset \mathcal{D}(\mathcal{O}')$ we have

$$|(f, \varphi)| = \left| \int \varphi(x)\mu(dx) \right| \le \int_{\mathcal{O}'} |\mu(dx)| \max_{x \in \overline{\mathcal{O}'}} |\varphi(x)|,$$

whence we conclude that the order of f in $\mathcal{O}$ is 0 (see Sec. 1.3).

Sufficiency. Let the order of f in $\mathcal{O}$ be 0, that is, for all $\mathcal{O}' \Subset \mathcal{O}$

$$|(f, \varphi)| \le K(\mathcal{O}')\|\varphi\|_{C(\overline{\mathcal{O}'})}, \qquad \varphi \in \mathcal{D}(\mathcal{O}'). \tag{7.4}$$

Let $\mathcal{O}_k$, $k = 1, 2, \ldots$, be a strictly increasing sequence of open sets that exhausts $\mathcal{O}$: $\mathcal{O}_k \Subset \mathcal{O}_{k+1}$, $\bigcup_k \mathcal{O}_k = \mathcal{O}$. Since the set $\mathcal{D}(\mathcal{O}_k)$ is dense in $C_0(\overline{\mathcal{O}}_k)$ in the norm $C(\overline{\mathcal{O}}_k)$ (see Corollary 2 to Theorem II of Sec. 1.2), it follows from inequality (7.4) that the functional f admits of a (linear) continuous extension onto $C_0(\overline{\mathcal{O}}_k)$. By the Riesz–Radon theorem, there is a measure μ_k on $\overline{\mathcal{O}}_k$ such that

$$(f, \varphi) = \int \varphi(x)\mu_k(dx), \qquad \varphi \in C_0(\overline{\mathcal{O}}_k).$$

From this it follows that the measures μ_k and μ_{k+1} coincide on $\mathcal{O}_k$; therefore there exists a single measure μ on $\mathcal{O}$ that coincides with the measure μ_k in $\mathcal{O}_k$ and it coincides with the generalized function f in $\mathcal{O}$. The proof of the theorem is complete. □

A generalized function f in $\mathcal{D}'(\mathcal{O})$ is said to be *nonnegative* in $\mathcal{O}$ if $(f, \varphi) \ge 0$ for all $\varphi \in \mathcal{D}(\mathcal{O})$, $\varphi(x) \ge 0$, $x \in \mathcal{O}$.

THEOREM II. *For a generalized function f in $\mathcal{D}'(\mathcal{O})$ to be a nonnegative measure on $\mathcal{O}$, it is necessary and sufficient that it be nonnegative in $\mathcal{O}$.*

PROOF. Necessity is obvious. We prove sufficiency. Suppose $f \in \mathcal{D}'(\mathcal{O})$ is nonnegative in $\mathcal{O}$. Suppose $\varphi \in \mathcal{D}(\mathcal{O}')$, $\mathcal{O}' \Subset \mathcal{O}$. By the corollary to the lemma of Sec. 1.2 there is a function $\eta \in \mathcal{D}(\mathcal{O})$, $\eta(x) = 1$, $x \in \mathcal{O}'$. For this reason,

$$-\|\varphi\|_{C(\bar{\mathcal{O}}')}\eta(x) \le \varphi(x) \le \|\varphi\|_{C(\bar{\mathcal{O}}')}\eta(x), \qquad x \in \mathcal{O}.$$

Whence, using the nonnegative nature of the functional f in $\mathcal{O}$, we get

$$-(f,\eta)\|\varphi\|_{C(\bar{\mathcal{O}}')} \le (f,\varphi) \le (f,\eta)\|\varphi\|_{C(\bar{\mathcal{O}}')}$$

or

$$|(f,\varphi)| \le (f,\eta)\|\varphi\|_{C(\bar{\mathcal{O}}')}, \qquad \varphi \in \mathcal{D}(\mathcal{O}').$$

This inequality shows that the order of the generalized function f is 0. By Theorem I, f is a (nonnegative) measure on $\mathcal{O}$, and the proof is complete. □

The simplest example of a measure (and, what is more, a measure of a singular generalized function) is the delta-function of Dirac (see Sec. 1.1), which operates via the rule

$$(\delta,\varphi) = \varphi(0), \qquad \varphi \in \mathcal{D}.$$

Clearly, $\delta \in \mathcal{D}'$, $\delta(x) = 0$, $x \neq 0$ so that $\operatorname{supp}\delta = \{0\}$.

We will now prove that $\delta(x)$ is a *singular* generalized function. Suppose, on the contrary, that there is a function $f \in \mathcal{L}^1_{\text{loc}}(\mathbb{R}^n)$ such that for any $\varphi \in \mathcal{D}$

$$\int f(x)\varphi(x)\,dx = (\delta,\varphi) = \varphi(0). \tag{7.5}$$

Since $|x|^2\varphi \in \mathcal{D}$, it follows from (7.5) that

$$\int f(x)|x|^2\varphi(x)\,dx = |x|^2\varphi(x)\Big|_{x=0} = 0 = (|x|^2 f,\varphi)$$

for all $\varphi \in \mathcal{D}$. Thus the function $|x|^2 f(x)$ that is locally integrable in $\mathbb{R}^n$ is equal to zero in the sense of generalized functions. By the Du Bois Reymond lemma (see Sec. 1.6), $|x|^2 f(x) = 0$ almost everywhere and, hence, $f(x) = 0$ almost everywhere in $\mathbb{R}^n$. But this contradicts the equation (7.5). □

Suppose $\omega_\varepsilon(x)$ is a "cap" (see Sec. 1.2). We will prove that

$$\omega_\varepsilon(x) \to \delta(x), \qquad \varepsilon \to +0 \quad \text{in} \quad \mathcal{D}'. \tag{7.6}$$

The sequence $\omega_\varepsilon(x)$, $\varepsilon \to +0$ is depicted in Fig 4.

Indeed, by the definition of convergence in $\mathcal{D}'$ the relation (7.6) is equivalent to

$$\lim_{\varepsilon\to+0}\int \omega_\varepsilon(x)\varphi(x)\,dx = \varphi(0), \qquad \varphi \in \mathcal{D}.$$

This equation follows from the estimate

$$\begin{aligned}\left|\int \omega_\varepsilon(x)\varphi(x)\,dx - \varphi(0)\right| &\le \int \omega_\varepsilon(x)|\varphi(x)-\varphi(0)|\,dx \\ &\le \max_{|x|\le\varepsilon}|\varphi(x)-\varphi(0)|\int \omega_\varepsilon(x)\,dx \\ &= \max_{|x|\le\varepsilon}|\varphi(x)-\varphi(0)|\end{aligned}$$

and from the continuity of the function φ. □

The surface δ-function is a generalization of the point δ-function. Let S be a piecewise smooth surface in $\mathbb{R}^n$ and let μ be a continuous function on S. We introduce the generalized function $\mu\delta_S$ that operates via the rule

$$(\mu\delta_S, \varphi) = \int\limits_S \mu(x)\varphi(x)\, dS, \qquad \varphi \in \mathcal{D}.$$

Clearly $\mu\delta \in \mathcal{D}'$; $\mu\delta_S(x) = 0$, $x \notin S$ so that $\operatorname{supp} \mu\delta_S \subset S$; $\mu\delta_S$ is a singular measure if $\mu \not\equiv 0$.

The generalized function $\mu\delta_S(x)$ is termed a *simple layer* on the surface S. It describes the spatial density of masses or charges concentrated on the surface S with surface density μ. (Here, the density of the simple layer is defined as a weak limit of the densities that correspond to a discrete distribution on the surface S,

$$\sum_k \mu(x_k)\Delta S_k \delta(x - x_k), \qquad x_k \in S$$

when the surface S is unrestricted refinemented; compare Sec. 1.1.)

REMARK. Locally integrable functions and δ-functions describe the density distribution of masses, charges, forces and the like (see Sec. 1.1). For this reason, generalized functions are also termed *distributions* (see Schwartz [**89, 90**]). If, for example, a generalized function f is the density of masses or charges, then the expression $(f, 1)$ is the total mass or total charge (on the assumption that f is meaningful on a function identically equal to 1 because 1 is not with compact support!). In particular, $(\delta, 1) = 1$; $(f, 1) = \int f(x)\, dx$ if $f \in \mathcal{L}^1$.

1.8. Sochozki formulae. We now introduce another important singular generalized function $\mathcal{P}\frac{1}{x}$ that operates in accordance with the formula

$$\left(\mathcal{P}\frac{1}{x}, \varphi\right) = \mathbf{VP}\int \frac{\varphi(x)}{x}\, dx = \lim_{\varepsilon \to +0}\left(\int\limits_{-\infty}^{-\varepsilon} + \int\limits_{\varepsilon}^{\infty}\right)\frac{\varphi(x)}{x}\, dx.$$

The functional $\mathcal{P}\frac{1}{x}$ is a linear functional. Its continuity on $\mathcal{D} = \mathcal{D}(\mathbb{R}^n)$ follows from the equation

$$\begin{aligned} \left|\left(\mathcal{P}\frac{1}{x}, \varphi\right)\right| &= \left|\mathbf{VP}\int \frac{\varphi(x)}{x}\, dx\right| \\ &= \left|\mathbf{VP}\int\limits_{-R}^{R} \frac{\varphi(0) + x\varphi'(x')}{x}\, dx\right| \\ &\le \int\limits_{-R}^{R} |\varphi'(x')|\, dx \\ &\le 2R \max_x |\varphi'(x)|, \qquad \varphi \in \mathcal{D}(-R, R). \end{aligned} \tag{8.1}$$

Here, x' is some point in the interval $(-R, R)$. Thus $\mathcal{P}\frac{1}{x} \in \mathcal{D}'$.

The generalized function $\mathcal{P}\frac{1}{x}$ coincides with the function $\frac{1}{x}$ for $x \neq 0$ (in the meaning of Sec. 1.6). It is called the *finite part* or *principal value* of the integral of the function $\frac{1}{x}$. Let us now set up the equality

$$\lim_{\varepsilon\to+0}\int\frac{\varphi(x)}{x+i\varepsilon}\,dx=-i\pi\varphi(0)+\mathrm{VP}\int\frac{\varphi(x)}{x}\,dx. \tag{8.2}$$

Indeed, if $\varphi(x)=0$ for $|x|>R$, then

$$\begin{aligned}\lim_{\varepsilon\to+0}\int\frac{\varphi(x)}{x+i\varepsilon}\,dx&=\lim_{\varepsilon\to+0}\int_{-R}^{R}\frac{x-i\varepsilon}{x^2+\varepsilon^2}\varphi(x)\,dx\\&=\varphi(0)\lim_{\varepsilon\to+0}\int_{-R}^{R}\frac{x-i\varepsilon}{x^2+\varepsilon^2}\,dx+\lim_{\varepsilon\to+0}\int_{-R}^{R}\frac{x-i\varepsilon}{x^2+\varepsilon^2}[\varphi(x)-\varphi(0)]\,dx\\&=-2i\varphi(0)\lim_{\varepsilon\to+0}\arctan\frac{R}{\varepsilon}+\int_{-R}^{R}\frac{\varphi(x)-\varphi(0)}{x}\,dx\\&=-i\pi\varphi(0)+\mathrm{VP}\int\frac{\varphi(x)}{x}\,dx.\end{aligned}$$

The relation (8.2) means that there is a limit to the sequence $\frac{1}{x+i\varepsilon}$ as $\varepsilon\to+0$ in $\mathcal{D}'$, which limit we denote by $\frac{1}{x+i0}$; and this limit is equal to $-i\pi\delta(x)+\mathcal{P}\frac{1}{x}$. Thus

$$\frac{1}{x+i0}=-i\pi\delta(x)+\mathcal{P}\frac{1}{x}. \tag{8.3}$$

Similarly

$$\frac{1}{x-i0}=i\pi\delta(x)+\mathcal{P}\frac{1}{x}. \tag{8.3$'$}$$

The formulae (8.3) and (8.3') were actually first obtained in "integral" form of the type (8.2) in 1873 by the Russian mathematician Julian Sochozki (see Sochozki [98]). At the present time these formulae are widely used in quantum physics.

We will now prove that the order of $\mathcal{P}\frac{1}{x}$ in $\mathbb{R}^1$ is equal to 1. Indeed, from (8.1) it follows that its order in $\mathbb{R}^1$ does not exceed 1. If its order in $\mathbb{R}^1$ were equal to 0, then by Theorem I of Sec. 1.7, $\mathcal{P}\frac{1}{x}$ would be a measure on $\mathbb{R}^1$. But then the integral $\mathrm{VP}\int\frac{\varphi(x)}{x}\,dx$ would be defined on all continuous functions that are with compact support in $\mathbb{R}^1$, which, as we know, is not true (for example, it is not defined on functions equal to $\frac{\theta(x)}{\ln x}$ in the neighbourhood of 0; Fig. 5).

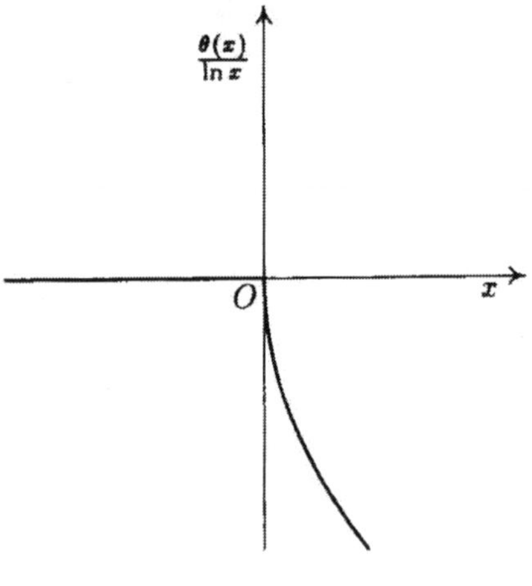

Figure 5

We note in passing that the order of $\mathcal{P}\frac{1}{x}$ in $\{x\neq0\}$ is equal to 0 because $\mathcal{P}\frac{1}{x}$ coincides with the locally integrable function $\frac{1}{x}$ when $x\neq0$.

The generalized function $\mathcal{P}\frac{1}{x}$ is a continuation of the regular generalized function $\frac{1}{x}$ from the set $\{x\neq0\}$ onto the whole axis $\mathbb{R}^1$.

The generalized function $\mathcal{P}\frac{1}{x}$ is called the *regularization* of the function $\frac{1}{x}$, $x \neq 0$. Similarly, the regularization of the functions $|x|^\alpha$ and $\operatorname{sgn} x|x|^{-\alpha}$. $x \neq 0$ for $\Re\alpha > 1$ are defined (see Sec. 2.7). For instance, the generalized function $\mathcal{P}f\frac{1}{|x|^N}$, $N = 1, 2, \ldots$, are defined by the formula

$$\left(\mathcal{P}f\frac{1}{|x|^N}, \varphi\right) = \int\limits_{|x|<1} \frac{\varphi(x) - S_{N-1}(x;\varphi)}{|x|^N}\,dx + \int\limits_{|x|>1} \frac{\varphi(x)}{|x|^N}\,dx, \qquad \varphi \in \mathcal{D}, \qquad (8.4)$$

where $S_N(x;\varphi)$ is the Taylor polynomial of the function φ at 0 of degree N,

$$S_N(x;\varphi) = \sum_{k=0}^{N} \frac{\varphi^{(k)}(0)}{k!} x^k.$$

The question now is: does any locally integrable function in $\mathcal{O} \neq \mathbb{R}^1$ admit a continuation onto the whole space $\mathbb{R}^n$ as a generalized function from $\mathcal{D}'(\mathbb{R}^1)$? The answer is negative, as will be seen from the following example:

$$e^{1/x} \in \mathcal{D}'(x \neq 0).$$

If there exists a function $f \in \mathcal{D}'(\mathbb{R}^1)$ that coincides with $e^{1/x}$ for $x \neq 0$, we would have

$$(f, \varphi) = \int e^{1/x}\varphi(x)\,dx, \qquad \varphi \in \mathcal{D}(x \neq 0). \qquad (8.5)$$

Let $\varphi_0 \in \mathcal{D}$, $\varphi_0(x) = 0$ for $x < 1$ and $x > 2$, $\varphi_0(x) \geq 0$ and

$$\int \varphi_0(x)\,dx = 1.$$

Then

$$\varphi_k(x) = e^{-k/2}k\varphi_0(kx) \to 0, \qquad k \to \infty \quad \text{in} \quad \mathcal{D},$$

$$\int e^{1/x}\varphi_k(x)\,dx = \int e^{\frac{1}{x}-\frac{k}{2}}k\varphi_0(kx)\,dx$$

$$= \int_1^2 e^{k\left(\frac{1}{y}-\frac{1}{2}\right)}\varphi_0(y)\,dy$$

$$\geq \int_1^2 \varphi_0(y)\,dy = 1,$$

$$\varphi_k \in \mathcal{D}(x \neq 0),$$

but this contradicts (8.5):

$$1 \leq \int e^{1/x}\varphi_k(x)\,dx = (f, \varphi_k) \to 0, \qquad k \to \infty.$$

1.9. Change of variables in generalized functions. Let $f \in \mathcal{L}^1_{\text{loc}}(\mathcal{O})$ and $x = Ay + b$ be a nonsingular linear transformation of $\mathcal{O}$ onto $\mathcal{O}_1$. Then for any $\varphi \in \mathcal{D}(\mathcal{O}_1)$ we have

$$\int\limits_{\mathcal{O}_1} f(Ay + b)\varphi(y)\,dy = \frac{1}{|\det A|}\int\limits_{\mathcal{O}} f(x)\varphi[A^{-1}(x - b)]\,dx.$$

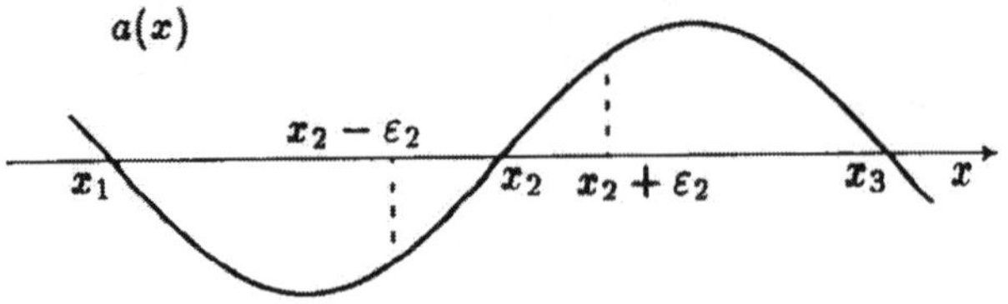

Figure 6

This equality is taken for the definition of the generalized function $f(Ay+b)$ for any $f(x)$ in $\mathcal{D}'(\mathcal{O})$:

$$\left(f(Ay+b), \varphi(y)\right) = \left(f(x), \frac{\varphi[A^{-1}(x-b)]}{|\det A|}\right), \qquad \varphi \in \mathcal{D}(\mathcal{O}_1). \tag{9.1}$$

Since the operation $\varphi(x) \to \varphi[A^{-1}(x-b)]$ is linear and continuous from $\mathcal{D}(\mathcal{O}_1)$ into $\mathcal{D}(\mathcal{O})$, the functional $f(Ay+b)$ defined by the right-hand side of (9.1) belongs to $\mathcal{D}'(\mathcal{O}_1)$.

In particular, if A is a rotation, so that $A^T = A^{-1}$ and $b=0$, then $(f(Ay), \varphi) = (f, \varphi(A^T x))$; if A is a similarity (with reflection), that is, $A = cI$, $c \neq 0$ and $b = 0$, then

$$\left(f(cy), \varphi\right) = \frac{1}{|c|^n}\left(f, \varphi\left(\frac{x}{c}\right)\right).$$

If $A = I$, then (a shift equal to b)

$$\left(f(y+b), \varphi\right) = \left(f, \varphi(x-b)\right).$$

The foregoing enables us to define **translation-invariant, spherically symmetrical, centrally symmetrical, homogeneous, periodic, Lorentz-invariant**, and other generalized functions.

EXAMPLES. (a) $\delta(-x) = \delta(x)$;
(b) $(\delta(x - x_0), \varphi) = \varphi(x_0)$.

Let $a \in C^1$. We define the generalized function $\delta(a(x))$ via the formula

$$\delta(a(x)) = \lim_{\varepsilon \to +0} \omega_\varepsilon(a(x)) \quad \text{in} \quad \mathcal{D}'(c,d) \tag{9.2}$$

where ω_ε is the "cap".

Suppose the function $a(x)$ has isolated and simple zeros which we denote by x_k, $k = 1, 2, \ldots$ (Fig. 6). Then $\delta(a(x))$ exists in $\mathcal{D}'(\mathbb{R}^1)$ and is given by the sum

$$\delta(a(x)) = \sum_k \frac{\delta(x - x_k)}{|a'(x_k)|}. \tag{9.3}$$

By virtue of the theorem of piecewise sewing (see Sec. 1.5), it suffices to prove formula (9.3) locally, in a sufficiently small neighbourhood of each point. Let $\varphi \in \mathcal{D}(x_k - \varepsilon_k, x_k + \varepsilon_k)$ and let the number ε_k be so small that in the interval $(x_k - \varepsilon_k, x_k + \varepsilon_k)$ the function $a(x)$ is monotonic. Making use of the limiting relation (7.6), we have the following chain of equalities:

$$\begin{aligned}\Big(\delta\big(a(x)\big),\varphi\Big) &= \lim_{\varepsilon\to+0}\int\limits_{x_k-\varepsilon_k}^{x_k+\varepsilon_k}\omega_\varepsilon[a(x)]\varphi(x)\,dx\\ &= \lim_{\varepsilon\to+0}\int\limits_{a(x_k-\varepsilon_k)}^{a(x_k+\varepsilon_k)}\omega_\varepsilon(y)\varphi[a^{-1}(y)]\frac{dy}{a'[a^{-1}(y)]}\\ &= \frac{\varphi(x_k)}{|a'(x_k)|}\\ &= \left(\frac{\delta(x-x_k)}{|a'(x_k)|},\varphi\right).\end{aligned}$$

Now if $\varphi\in\mathcal{D}(\alpha,\beta)$, where the interval (α,β) does not contain a single zero of x_k, then

$$\Big(\delta\big(a(x)\big),\varphi\Big) = \lim_{\varepsilon\to+0}\int\limits_{\alpha}^{\beta}\omega_\varepsilon[a(x)]\varphi(x)\,dx = 0.$$

The local elements $\frac{\delta(x-x_0)}{|a'(x_k)|}$ in $(x_k-\varepsilon_k, x_k+\varepsilon_k)$ and 0 in (α,β) are clearly in agreement. The proof of (9.3) is complete.

EXAMPLES. (a) $\delta(x^2-a^2)=\frac{1}{2a}[\delta(x-a)+\delta(x+a)]$;
(b) $\delta(\sin x)=\sum\limits_{k=-\infty}^{\infty}\delta(x-k\pi)$.

1.10. Multiplication of generalized functions. Suppose $f\in\mathcal{L}^1_{\text{loc}}(\mathcal{O})$ and $a\in C^\infty(\mathcal{O})$. Then for any φ in $\mathcal{D}(\mathcal{O})$ we have the equality

$$(af,\varphi)=\int a(x)f(x)\varphi(x)\,dx=(f,a\varphi).$$

This equality is then taken for the definition of the product of the generalized function f in $\mathcal{D}'(\mathcal{O})$ by the function a that is infinitely differentiable in $\mathcal{O}$:

$$(af,\varphi)=(f,a\varphi),\qquad \varphi\in\mathcal{D}(\mathcal{O}). \tag{10.1}$$

Since the operation $\varphi\to a\varphi$, $a\in C^\infty(\mathcal{O})$, is linear and continuous from $\mathcal{D}(\mathcal{O})$ into $\mathcal{D}(\mathcal{O})$, it follows that the functional af defined by the right-hand side of (10.1) is a generalized function in $\mathcal{D}'(\mathcal{O})$.

The following inclusion holds true:

$$\operatorname{supp}(af)\subset\operatorname{supp}a\cap\operatorname{supp}f,$$

because $\mathcal{O}_{af}\supset\mathcal{O}_a\cup\mathcal{O}_f$ (see Sec. 1.5) and

$$\begin{aligned}\operatorname{supp}(af)&=\mathcal{O}\setminus\mathcal{O}_{af}\subset\mathcal{O}\setminus(\mathcal{O}_a\cup\mathcal{O}_f)\\ &=(\mathcal{O}\setminus\mathcal{O}_a)\cap(\mathcal{O}\setminus\mathcal{O}_f)\\ &=\operatorname{supp}a\cap\operatorname{supp}f.\end{aligned}$$

If $f\in\mathcal{D}'(\mathcal{O})$, then we have the equality

$$f=\eta f \tag{10.2}$$

where η is any function of the class C^∞, that function being equal to 1 in the neighbourhood of the support of f.

Indeed, for any $\varphi \in \mathcal{D}(\mathcal{O})$, the supports of f and $(1-\eta)\varphi$ have no points in common, and for this reason (see Sec. 1.5)

$$(f,(1-\eta)\varphi) = 0 = (f(1-\eta),\varphi).$$

which is equivalent to (10.2). □

EXAMPLES. (a) $a(x)\delta(x) = a(0)\delta(x)$ since for all $\varphi \in \mathcal{D}$

$$(a\delta,\varphi) = (\delta, a\varphi) = a(0)\varphi(0) = \big(a(0)\delta,\varphi\big);$$

(b) $x\mathcal{P}\frac{1}{x} = 1$ since

$$\left(x\mathcal{P}\frac{1}{x},\varphi\right) = \left(\mathcal{P}\frac{1}{x}, x\varphi\right) = \int \varphi(x)\,dx = (1,\varphi).$$

The following question arises: Is it possible, in the case of generalized functions, to define the multiplication of any generalized functions and so that the multiplication is associative and commutative and agrees with the above-defined multiplication by an infinitely differentiable function? L. Schwartz demonstrated that no such multiplication can be defined. Indeed, if it existed, then, using examples (a) and (b), we would have the following contradictory chain of equalities:

$$0 = 0\mathcal{P}\frac{1}{x} = \big(x\delta(x)\big)\mathcal{P}\frac{1}{x} = \big(\delta(x)x\big)\mathcal{P}\frac{1}{x} = \delta(x)\left(x\mathcal{P}\frac{1}{x}\right) = \delta(x).$$

In order to define a product of two generalized functions f and g, it is necessary that they have (to put it crudely) the following properties: f must be just as "irregular" in the neighbourhood of an (arbitrary) point as g is "regular" in that neighbourhood, and conversely.

As for the general definition of the product of generalized functions, see Sec. 4.7 and Sec. 6.5. Here we give the definition of the product of the generalized function $f \in \mathcal{D}'(\mathbb{R}^n)$ with the characteristic function $\theta_\mathcal{O}(x)$ of an open set $\mathcal{O} \subset \mathbb{R}^n$. The problem consists in construction of the generalized function $\theta_\mathcal{O} f$ from $\mathcal{D}'$ which is equal to $f(x)$ for $x \in \mathcal{O}$ and equal to zero for $x \in \mathbb{R}^n \setminus \overline{\mathcal{O}}$.

The product $\theta_\mathcal{O} f$ exists if f is regular in a neighbourhood U of the boundary $\partial\mathcal{O}$ and it can be represented by the formula

$$(\theta_\mathcal{O} f,\varphi) = \big(f,\theta_\mathcal{O}(1-\eta)\varphi\big) + \int_{U\cap\mathcal{O}} f(x)\eta(x)\varphi(x)\,dx, \qquad \varphi \in \mathcal{D}, \tag{10.3}$$

where η is any C^∞-function, $\operatorname{supp}\eta \subset U$ *and $\eta = 1$ in a neighbourhood $U' \subset U$ of the boundary $\partial\mathcal{O}$.*

Indeed, noting that $\theta_\mathcal{O}(1-\eta) \in C^\infty$, we conclude that the right-hand side of equality (10.3) determines a linear continuous functional on $\mathcal{D}$, i.e., $\theta_\mathcal{O} f \in \mathcal{D}'$, and it does not depend on the auxiliary function η with the properties indicated. The other properties of the generalized function $\theta_\mathcal{O} f$ also follow from representation (10.3). □

EXAMPLE.

$$\theta(x-x_0)\delta(x) = \begin{cases} \delta(x), & \text{if } x_0 > 0, \\ 0, & \text{if } x_0 < 0. \end{cases}$$

2. Differentiation of Generalized Functions

2.1. Derivatives of generalized functions. Let $f \in C^k(\mathcal{O})$. Then for all α, $|\alpha| \leq k$, and $\varphi \in \mathcal{D}(\mathcal{O})$ we have the following integration-by-parts formula:

$$\begin{aligned}(\partial^\alpha f, \varphi) &= \int \partial^\alpha f(x)\varphi(x)\,dx \\ &= (-1)^{|\alpha|}\int f(x)\partial^\alpha\varphi(x)\,dx \\ &= (-1)^{|\alpha|}(f, \partial^\alpha\varphi).\end{aligned}$$

We take this equation for the definition of a (generalized) derivative $\partial^\alpha f$ of the generalized function f in $\mathcal{D}'(\mathcal{O})$:

$$(\partial^\alpha f, \varphi) = (-1)^{|\alpha|}(f, \partial^\alpha\varphi), \qquad \varphi \in \mathcal{D}(\mathcal{O}). \tag{1.1}$$

Since the operation $\varphi \to (-1)^{|\alpha|}\partial^\alpha\varphi$ is linear and continuous from $\mathcal{D}(\mathcal{O})$ into $\mathcal{D}(\mathcal{O})$, the functional $\partial^\alpha f$ defined by the right-hand side of (1.1) is a generalized function in $\mathcal{D}'(\mathcal{O})$.

In particular, when $f = \delta$, then (1.1) takes the form

$$(\partial^\alpha\delta, \varphi) = (-1)^{|\alpha|}\partial^\alpha\varphi(0), \qquad \varphi \in \mathcal{D}.$$

It follows from this definition that if a generalized function f in $\mathcal{D}'(\mathcal{O})$ belongs to the class $C^k(\mathcal{O}_1)$ in $\mathcal{O}_1 \subset \mathcal{O}$ (see Sec. 1.6), then its classical and generalized derivatives $\partial^\alpha f$, $|\alpha| \leq k$, coincide in $\mathcal{O}_1$.

The following properties of the operation of differentiation of generalized functions holds true:

2.1.1. *The operation of differentiation $f \to \partial^\alpha f$ is linear and continuous from $\mathcal{D}'(\mathcal{O})$ into $\mathcal{D}'(\mathcal{O})$.*

Linearity is obvious. We will prove continuity. Suppose $f_k \to 0$, $k \to \infty$ in $\mathcal{D}'(\mathcal{O})$. Then for all $\varphi \in \mathcal{D}(\mathcal{O})$ we have

$$(\partial^\alpha f_k, \varphi) = (-1)^{|\alpha|}(f_k, \partial^\alpha\varphi) \to 0, \qquad k \to \infty$$

and this signifies that $\partial^\alpha f_k \to 0$, $k \to \infty$ in $\mathcal{D}'(\mathcal{O})$. □

For example,

$$\partial^\alpha\omega_\varepsilon(x) \to \partial^\alpha\delta(x), \qquad \varepsilon \to +0 \quad \text{in} \quad \mathcal{D}'. \tag{1.2}$$

The relation (1.2) follows from the relation (7.6) of Sec. 1. The sequence $\omega'_\varepsilon(x)$, $\varepsilon \to +0$, is depicted in Fig. 7.

In particular, if the series

$$\sum_{k\geq 1} u_k(x) = S(x), \qquad u_k \in \mathcal{L}^1_{\text{loc}}(\mathcal{O})$$

converges uniformly on every compact $K \Subset \mathcal{O}$, then it may be differentiated term by term any number of times, and the resulting series will converge in $\mathcal{D}'(\mathcal{O})$,

$$\sum_{k\geq 1} \partial^\alpha u_k(x) = \partial^\alpha S(x).$$

True enough, the sequence of the partial sums of this series converges to $\partial^\alpha S(x)$ in $\mathcal{D}'(\mathcal{O})$ (see Sec. 1.6).

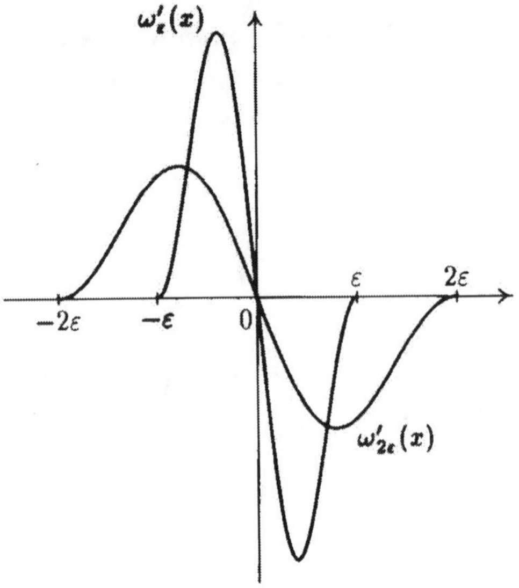

Figure 7

2.1.2. Any generalized function $f \in \mathcal{D}'(\mathcal{O})$ (in particular, any function locally integrable in $\mathcal{O}$) is infinitely differentiable (in the generalized sense).

Indeed, since $f \in \mathcal{D}'(\mathcal{O})$, it follows that $\frac{\partial f}{\partial x_j} \in \mathcal{D}'(\mathcal{O})$; in turn, $\frac{\partial}{\partial x_i}\left(\frac{\partial f}{\partial x_j}\right) \in \mathcal{D}'(\mathcal{O})$ and so forth. □

2.1.3. The result of differentiation does not depend on the order of differentiation:

$$\partial^{\alpha+\beta} f = \partial^\alpha(\partial^\beta f) = \partial^\beta(\partial^\alpha f). \tag{1.3}$$

Indeed,

$$\begin{aligned}(\partial^{\alpha+\beta} f, \varphi) &= (-1)^{|\alpha|+|\beta|}(f, \partial^{\alpha+\beta}\varphi)\\ &= (-1)^{|\alpha|}(\partial^\beta f, \partial^\alpha \varphi)\\ &= (\partial^\alpha(\partial^\beta f), \varphi)\\ &= (-1)^{|\beta|}(\partial^\alpha f, \partial^\beta \varphi)\\ &= (\partial^\beta(\partial^\alpha f), \varphi),\end{aligned}$$

whence follow the equalities (1.3). □

2.1.4. If $f \in \mathcal{D}'(\mathcal{O})$ and $a \in C^\infty(\mathcal{O})$, then the Leibniz formula holds true for the differentiation of a product af,

$$\partial^\alpha(af) = \sum_{\beta \leq \alpha} \binom{\alpha}{\beta} \partial^\beta a \partial^{\alpha-\beta} f. \tag{1.4}$$

Indeed, if $\varphi \in \mathcal{D}(\mathcal{O})$, then

$$\begin{aligned}\left(\frac{\partial(af)}{\partial x_1}, \varphi\right) &= -\left(af, \frac{\partial \varphi}{\partial x_1}\right) \\ &= -\left(f, a\frac{\partial \varphi}{\partial x_1}\right) \\ &= -\left(f, \frac{\partial(a\varphi)}{\partial x_1} - \frac{\partial a}{\partial x_1}\varphi\right) \\ &= -\left(f, \frac{\partial(a\varphi)}{\partial x_1}\right) + \left(f, \frac{\partial a}{\partial x_1}\varphi\right) \\ &= \left(\frac{\partial f}{\partial x_1}, a\varphi\right) + \left(\frac{\partial a}{\partial x_1}f, \varphi\right) \\ &= \left(a\frac{\partial f}{\partial x_1} + \frac{\partial a}{\partial x_1}f, \varphi\right),\end{aligned}$$

whence follows (1.4) for $\alpha = (1, 0, \ldots, 0)$. □

2.1.5.

$$\operatorname{supp} \partial^\alpha f \subset \operatorname{supp} f. \tag{1.5}$$

Indeed, if $f \in \mathcal{D}'(\mathcal{O})$, then for all $\varphi \in \mathcal{D}(\mathcal{O}_f)$ we have $\partial^\alpha \varphi \in \mathcal{D}(\mathcal{O}_f)$ and

$$(\partial^\alpha f, \varphi) = (-1)^{|\alpha|}(f, \partial^\alpha \varphi) = 0$$

so that $\mathcal{O}_{\partial^\alpha f} \supset \mathcal{O}_f$, whence follows the inclusion (1.5). □

2.2. The antiderivative (primitive) of a generalized function. Every function $f(x)$ continuous in an interval (a, b) has in (a, b) a unique (up to an additive constant) antiderivative $f^{(-1)}(x)$,

$$f^{(-1)}(x) = \int^x f(\xi)\, d\xi + C, \qquad f^{(-1)\prime}(x) = f(x).$$

The last equality is what we will start with to define the antiderivative or primitive of an arbitrary generalized function f (of one variable).

Suppose $f \in \mathcal{D}'(a, b)$. The generalized function $f^{(-1)}$ in $\mathcal{D}'(a, b)$ is termed the *antiderivative* (or *primitive*) of the generalized function f in (a, b) if $f^{(-1)\prime} = f$, that is,

$$(f^{(-1)}, \varphi') = -(f, \varphi), \qquad \varphi \in \mathcal{D}(a, b). \tag{2.1}$$

The equality (2.1) shows that the function $f^{(-1)}$ is not specified on all test functions taken from $\mathcal{D}(a, b)$, but only on their first derivatives. Our problem is to extend that functional onto the whole space $\mathcal{D}(a, b)$, and in a manner so that the extended functional $f^{(-1)}$ is linear and continuous on $\mathcal{D}(a, b)$, and to determine the degree of arbitrariness in such an extension.

First assume that $f^{(-1)}$ (the antiderivative of f) exists in $\mathcal{D}'(a, b)$. Construct it. Let $\varphi \in \mathcal{D}(a, b)$. (We assume the function φ to be continued by means of zero onto the entire axis $\mathbb{R}^1$.) We fix an arbitrary point $x_0 \in (a, b)$. Then

$$\varphi(x) = \psi'(x) + \omega_\varepsilon(x - x_0) \int \varphi(\xi)\, d\xi \tag{2.2}$$

Figure 8

where ω_ε is the "cap" when $\varepsilon < \min(x_0 - a, b - x_0)$ (see Sec. 1.2) and

$$\psi(x) = \int_{-\infty}^{\infty} \left[\varphi(x') - \omega_\varepsilon(x' - x_0) \int \varphi(\xi)\, d\xi\right] dx'. \tag{2.3}$$

We will prove that $\psi \in \mathcal{D}(a, b)$. Indeed, $\psi \in C^\infty$ and $\psi(x) = 0$ for $x < a'' = \min(a', x_0 - \varepsilon) > a$ if $\operatorname{supp}\varphi \subset [a', b'] \subset (a, b)$. Furthermore, for $x > b'' = \max(b', x_0 + \varepsilon) < b$,

$$\psi(x) = \int_{-\infty}^{\infty} \varphi(x')\, dx' - \int_{-\infty}^{\infty} \omega_\varepsilon(x' - x_0)\, dx' \int_{-\infty}^{\infty} \varphi(\xi)\, d\xi = 0.$$

Thus, $\operatorname{supp}\psi \subset [a'', b''] \subset (a, b)$ (Fig. 8). Hence $\psi \in \mathcal{D}(a, b)$.

Applying the functional $f^{(-1)}$ to (2.2), we obtain

$$(f^{(-1)}, \varphi) = (f^{(-1)}, \psi') + (f^{(-1)}, \omega_\varepsilon(x - x_0)) \int \varphi(\xi)\, d\xi.$$

That is to say, taking into account (2.1),

$$(f^{(-1)}, \varphi) = -(f, \psi) + C \int \varphi(\xi)\, d\xi \tag{2.4}$$

where $C = (f^{(-1)}, \omega_\varepsilon(x - x_0))$. Thus, if $f^{(-1)}$ exists, then it is expressed by (2.4), where ψ is defined by (2.3).

Let us now prove the converse: given an arbitrary constant C, the functional $f^{(-1)}$ defined by equalities (2.4) and (2.3) defines the antiderivative of f in (a, b).

Indeed, the functional $f^{(-1)}$ is clearly linear. Let us prove that it is continuous on $\mathcal{D}(a, b)$. Let $\varphi_k \to 0$, $k \to \infty$ in $\mathcal{D}(a, b)$, that is, $\operatorname{supp}\varphi_k \subset [a', b'] \subset (a, b)$ and $\varphi_k^{(\alpha)}(x) \overset{x}{\rightrightarrows} 0$, $k \to \infty$. Then, by what has already been proved,

$$\psi_k(x) = \int_{-\infty}^{x} \left[\varphi_k(x') - \omega_\varepsilon(x' - x_0) \int \varphi_k(\xi)\, d\xi\right] dx' = 0$$

$$\text{outside} \quad [a'', b''] \subset [a, b]$$

and, obviously, $\psi_k^{(\alpha)}(x) \overset{x}{\rightrightarrows} 0$, $k \to \infty$, that is, $\psi_k \to 0$, $k \to \infty$ in $\mathcal{D}(a, b)$. Therefore, by virtue of the continuity of f on $\mathcal{D}(a, b)$, we have

$$(f^{(-1)}, \varphi_k) = -(f, \psi_k) + C \int \varphi_k(\xi)\, d\xi \to 0, \qquad k \to \infty,$$

which is what was affirmed. Consequently, $f^{(-1)} \in \mathcal{D}'(a, b)$. It remains to verify that $f^{(-1)}$ is the antiderivative of f in (a, b). Indeed, substituting φ' for φ in (2.3) and noting that $\int \varphi'(\xi)\, d\xi = 0$, we get $\psi = \varphi$, and then from (2.4) there follows the equality (2.1), which is what we set out to prove. We have thus proved the following theorem.

THEOREM. *Every generalized function f in $\mathcal{D}'(a,b)$ has in (a,b) an antiderivative $f^{(-1)}$, and every antiderivative of it is expressed by the formula (2.4), where ψ is defined by (2.3) and C is an arbitrary constant.*

This theorem states that a solution of the differential equation

$$u' = f, \qquad f \in \mathcal{D}'(a,b) \tag{2.5}$$

exists in $\mathcal{D}'(a,b)$ and its general solution is of the form $u = f^{(-1)} + C$, where $f^{(-1)}$ is some antiderivative of f in (a,b) and C is an arbitrary constant. In particular, if $f \in C(a,b)$, then any solution in $\mathcal{D}'(a,b)$ of the equation (2.5) is a classical solution. For example, the general solution of the equation $u' = 0$ in $\mathcal{D}'(a,b)$ is the arbitrary constant.

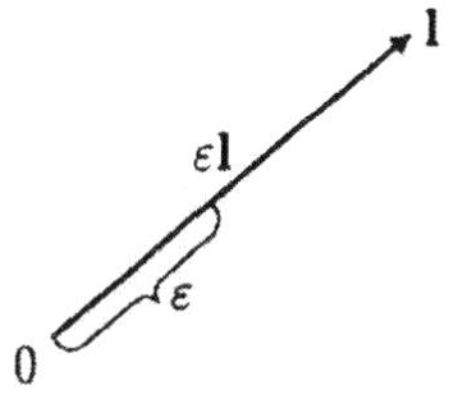

Figure 9

The definition of the antiderivative $f^{(-n)}$ of order n in (a,b) of the generalized function $f \in \mathcal{D}'(a,b)$, $f^{(-n)(n)} = f$, is similar. Applying this theorem to a recurrent chain for $f^{(-k)}$ (the antiderivatives of f of order k),

$$f^{(-1)'} = f, \quad f^{(-2)'} = f^{(-1)}, \quad \ldots, \quad f^{(-n)'} = f^{(-n+1)},$$

we conclude that $f^{(-n)}$ exists in $\mathcal{D}'(a,b)$ and is unique up to an arbitrary additive polynomial of degree $n-1$.

2.3. Examples.

2.3.1. Let us compute the density of charges corresponding to the dipole of the moment $+1$ located at the point $x = 0$ and oriented in a given direction $\mathbf{l} = (l_1, \ldots, l_n)$, $|\mathbf{l}| = 1$ (Fig. 9).

Approximately corresponding to this dipole is the charge density (see Secs. 1.1 and 1.7)

$$\frac{1}{\varepsilon}\delta(x - \varepsilon\mathbf{l}) - \frac{1}{\varepsilon}\delta(x), \qquad \varepsilon > 0.$$

Passing to the limit here as $\varepsilon \to +0$ in $\mathcal{D}'(\mathbb{R}^n)$,

$$\left(\frac{1}{\varepsilon}\delta(x - \varepsilon\mathbf{l}) - \frac{1}{\varepsilon}\delta(x), \varphi\right) = \frac{1}{\varepsilon}[\varphi(\varepsilon\mathbf{l}) - \varphi(0)]$$
$$\to \frac{\partial\varphi(0)}{\partial\mathbf{l}} = \left(\delta, \frac{\partial\varphi}{\partial\mathbf{l}}\right) = -\left(\frac{\partial\delta}{\partial\mathbf{l}}, \varphi\right),$$

we conclude that the desired density is equal to

$$-\frac{\partial\delta(x)}{\partial\mathbf{l}} = -(\mathbf{l}, \partial\delta(x)).$$

Let us now verify that the total charge of the dipole is 0:

$$\left(-\frac{\partial\delta}{\partial\mathbf{l}}, 1\right) = \left(\delta, \frac{\partial 1}{\partial\mathbf{l}}\right) = (\delta, 0) = 0,$$

and that its moment is equal to 1:

$$\left(-\frac{\partial\delta}{\partial\mathbf{l}}, (x, \mathbf{l})\right) = \left(\delta, \frac{\partial(x, \mathbf{l})}{\partial\mathbf{l}}\right) = (\delta, |\mathbf{l}|) = (\delta, 1) = 1.$$

2.3.2. A generalization of $-\frac{\partial\delta(x)}{\partial l}$ is a double layer on a surface. Let S be a piecewise smooth two-sided surface, $\mathbf{n}$ the normal to S (Fig. 10) and ν a continuous function on S. We introduce the generalized function $-\frac{\partial}{\partial\mathbf{n}}(\nu\delta_S)$, which operates via the rule

$$\left(-\frac{\partial}{\partial\mathbf{n}}(\nu\delta_S),\varphi\right) = \int_S \nu(x)\frac{\partial\varphi(x)}{\partial\mathbf{n}}\,dS, \qquad \varphi\in\mathcal{D}.$$

Clearly

$$-\frac{\partial}{\partial\mathbf{n}}(\nu\delta_S)\in\mathcal{D}', \qquad \operatorname{supp}\left[-\frac{\partial}{\partial\mathbf{n}}(\nu\delta_S)\right]\subset S.$$

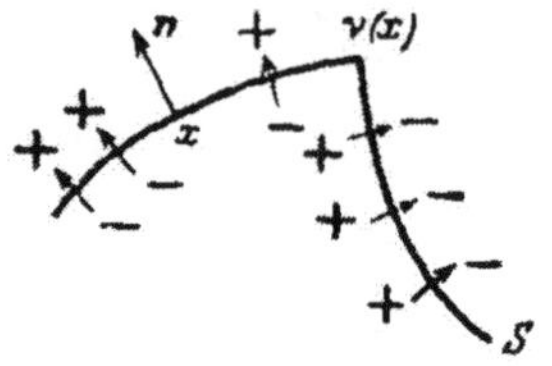

Figure 10

The generalized function $-\frac{\partial}{\partial\mathbf{n}}(\nu\delta_S)$ is called a *double layer* on the surface S. It describes the spatial density of charges corresponding to the distribution of dipoles on the surface S with surface moment density $\nu(x)$, the dipoles oriented in the given direction of the normal $\mathbf{n}$ on S. (Here, the density of the double layer is defined as the weak limit of the densities corresponding to the discrete arrangement of dipoles on the surface S,

$$-\sum_k \frac{\partial}{\partial\mathbf{n}_k}[\nu(x_k)\Delta S_k\delta(x-x_k)], \qquad x_k\in S,$$

by an unbounded refinement of the surface S; compare Sec. 1.7.)

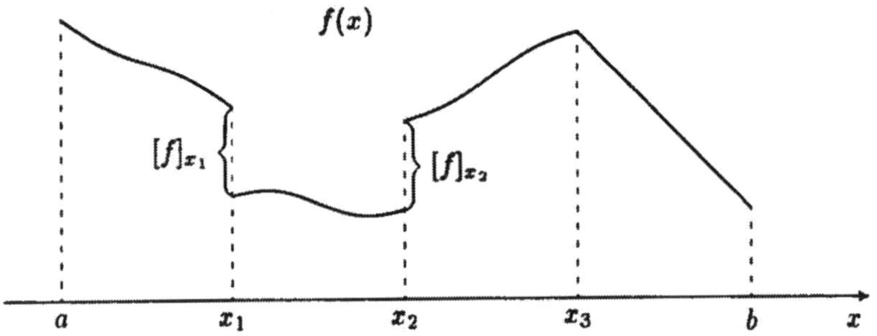

Figure 11

2.3.3. Let a function $f(x)$ be piecewise continuously differentiable in (a,b) and let $\{x_k\}$ be points in (a,b) at which it or its derivative has discontinuities of the first kind (Fig. 11). Then

$$f' = f'_{\mathrm{cl}}(x) + \sum_k [f]_{x_k}\delta(x-x_k) \tag{3.1}$$

where $f'_{\mathrm{cl}}(x)$ is the classical derivative of the function $f(x)$, equal to $f'(x)$ when $x\neq x_k$, and is not defined at the points $\{x_k\}$; $[f]_{x_k}$ is a jump of the function $f(x)$ at the point x_k,

$$[f]_{x_k} = f(x_k+0) - f(x_k-0).$$

Indeed, for any $\varphi \in \mathcal{D}(a,b)$ we have

$$\begin{aligned}
(f',\varphi) &= -(f,\varphi') \\
&= -\sum_k \int_{x_k}^{x_{k+1}} f(x)\varphi'(x)\,dx \\
&= \sum_k \int_{x_k}^{x_{k+1}} f'_{\mathrm{cl}}(x)\varphi(x)\,dx - \sum_k [f(x_{k+1}-0)\varphi(x_{k+1}) - f(x_k+0)\varphi(x_k)] \\
&= \int f'_{\mathrm{cl}}(x)\varphi(x)\,dx + \sum_k [f(x_k+0) - f(x_k-0)]\varphi(x_k) \\
&= (f'_{\mathrm{cl}},\varphi) + \sum_k [f]_{x_k}(\delta(x-x_k),\varphi),
\end{aligned}$$

which completes the proof of (3.1). □

In particular, if θ is the Heaviside unit function (see Sec. 0.2), then

$$\theta'(x) = \delta(x). \tag{3.2}$$

In the theory of electric circuits, the Heaviside unit function is called the ***unit-step function***, and the delta-function is called the ***unit-impulse function***. Formula (3.2) states that the unit-impulse function is a derivative of the unit-step function.

2.3.4. The following formulae hold true:

$$x^m\delta^{(k)}(x) = \begin{cases} 0, & k = 0,1,\ldots,m-1, \\ (-1)^m m! \binom{k}{m} \delta^{(k-m)}(x), & k \geq m. \end{cases} \tag{3.3}$$

Indeed,

$$\begin{aligned}
(x^m\delta^{(k)}(x),\varphi) &= (-1)^k (x^m\varphi)^{(k)}\Big|_{x=0} \\
&= (-1)^k \sum_{0\leq j\leq k} \binom{k}{j} (x^m)^{(j)}\varphi^{(k-j)}(x)\Big|_{x=0} \\
&= \begin{cases} 0, & k = 0,1,\ldots,m-1, \\ (-1)^k m! \binom{k}{m} \varphi^{(k-m)}(0) \end{cases} \\
&= \begin{cases} (0,\varphi), & k = 0,1,\ldots,m-1, \\ (-1)^m m! \binom{k}{m} (\delta^{(k-m)},\varphi), & k \geq m. \end{cases}
\end{aligned}$$

2.3.5. The trigonometric series

$$\sum_{k=-\infty}^{\infty} a_k e^{ikx}, \qquad |a_k| \leq A(1+|k|)^m \tag{3.4}$$

converges in $\mathcal{D}'$.

True enough, the series

$$\frac{a_0 x^{m+2}}{(m+2)!} + \sum_{k\neq 0} \frac{a_k}{(ik)^{m+2}} e^{ikx}$$

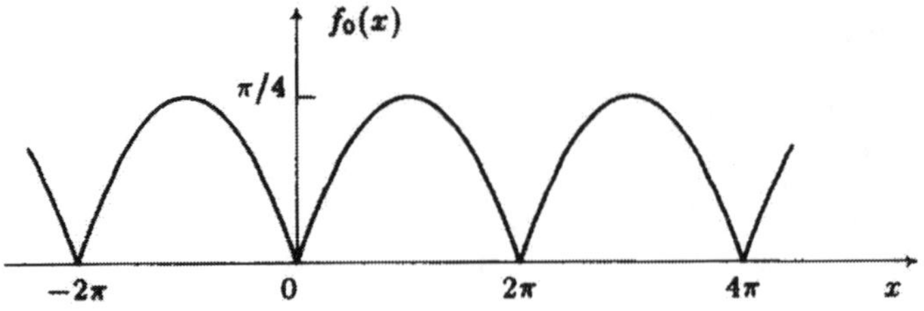

Figure 12

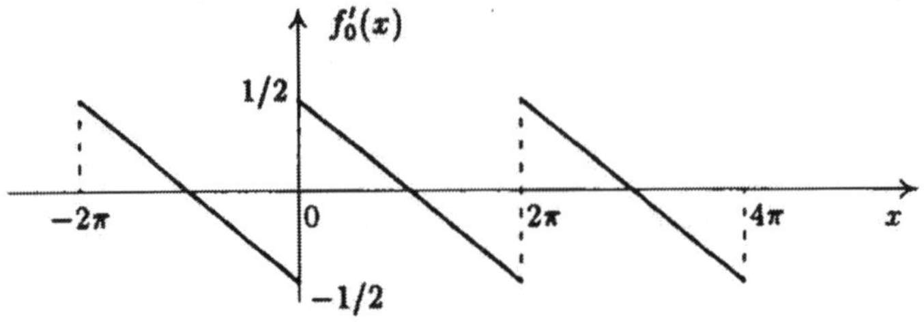

Figure 13

converges uniformly in $\mathbb{R}^1$; hence, the series which is a derivative of it of order $m+2$ converges in $\mathcal{D}'$ and its sum determines the sum of the series (3.4) (see Sec. 2.1.1).

2.3.6. Let us prove the formula

$$\frac{1}{2\pi}\sum_k e^{ikx} = \sum_k \delta(x - 2k\pi). \tag{3.5}$$

To do this we expand the 2π-periodic function (Fig. 12)

$$f_0(x) = \frac{x}{2} - \frac{x^2}{4\pi}, \qquad 0 \le x < 2\pi,$$

into a Fourier series that converges uniformly in $\mathbb{R}^1$:

$$f_0(x) = \frac{\pi}{6} - \frac{1}{2\pi}\sum_{k\neq 0}\frac{1}{k^2}e^{ikx}. \tag{3.6}$$

By virtue of 2.3.4, the series (3.5) can be differentiated termwise in $\mathcal{D}'$ any number of times. As a result, we get

$$f_0'(x) = \frac{1}{2} - \frac{x}{2\pi} = -\frac{i}{2\pi}\sum_{k\neq 0}\frac{1}{k}e^{ikx}, \qquad 0 \le x < 2\pi,$$

$$f_0''(x) = -\frac{1}{2\pi} + \sum_k \delta(x - 2k\pi) = \frac{1}{2\pi}\sum_{k\neq 0} e^{ikx},$$

whence follows formula (3.5). In differentiating the function $f_0'(x)$ (Fig. 13), we made use of the formula (3.1).

Note that the left-hand side of (3.5) is nothing other than the Fourier series of the 2π-periodic generalized function $\sum_k \delta(x - 2k\pi)$, the graph of which is symbolically depicted in Fig. 14 (see Sec. 7.2 for more details).

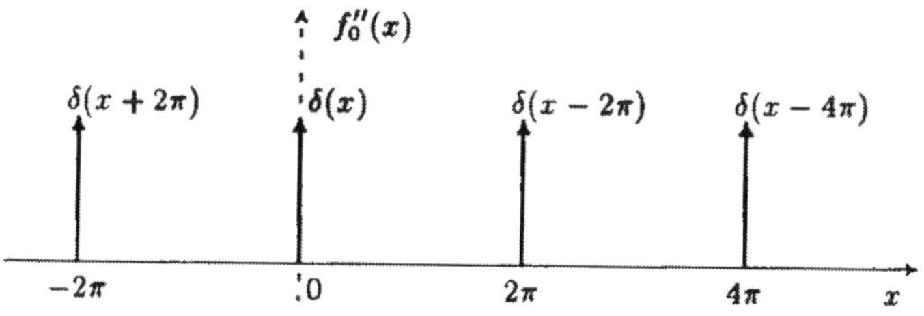

Figure 14

2.3.7. Let G be a domain in $\mathbb{R}^n$ with a piecewise smooth boundary S and let $\mathbf{n} = \mathbf{n}_x$ be an outer normal to S at the point $x \in S$ (Fig. 15). Suppose $f \in C^2(G) \cap C^1(\bar{G})$ and $f(x) = 0$ outside $\bar{G}$. Then for any $\varphi \in \mathcal{D}$ we have the following Green's formula:

$$\int_G (f\Delta\varphi - \varphi\Delta f)\,dx = \int_S \left(f\frac{\partial\varphi}{\partial\mathbf{n}} - \varphi\frac{\partial f}{\partial\mathbf{n}}\right) dS. \tag{3.7}$$

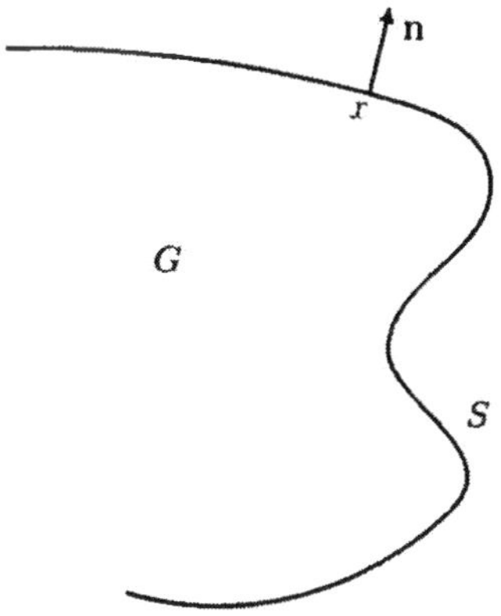

Figure 15

We can rewrite Green's formula (3.7) as follows in terms of the generalized functions (of a simple layer and a double layer) that were introduced in Sec. 1.7 and Sec. 2.3.2:

$$\Delta f = \Delta_{\text{cl}} f - \frac{\partial f}{\partial\mathbf{n}}\delta_S - \frac{\partial}{\partial\mathbf{n}}(f\delta_S) \tag{3.7'}$$

where $\Delta_{\text{cl}} f$ is the classical Laplacian of f:

$$\Delta_{\text{cl}} f(x) = \begin{cases} \Delta f(x), & x \in G, \\ 0, & x \notin \bar{G}, \\ \text{not defined}, & x \in S. \end{cases}$$

By f and $\frac{\partial f}{\partial\mathbf{n}}$ on S we mean the boundary values of f and $\frac{\partial f}{\partial\mathbf{n}}$ on S from within the domain G.

2.3.8. Let us verify that the function $\frac{1}{|x|}$ in $\mathbb{R}^3$ satisfies the Poisson equation

$$\Delta\frac{1}{|x|} = -4\pi\delta(x). \tag{3.8}$$

True enough, the function $1/|x|$ is locally integrable in $\mathbb{R}^3$ and

$$\Delta\frac{1}{|x|} = \frac{1}{r^2}\frac{\partial}{\partial r}\left(r^2\frac{\partial}{\partial r}\frac{1}{r}\right) = 0, \qquad x \neq 0. \tag{3.9}$$

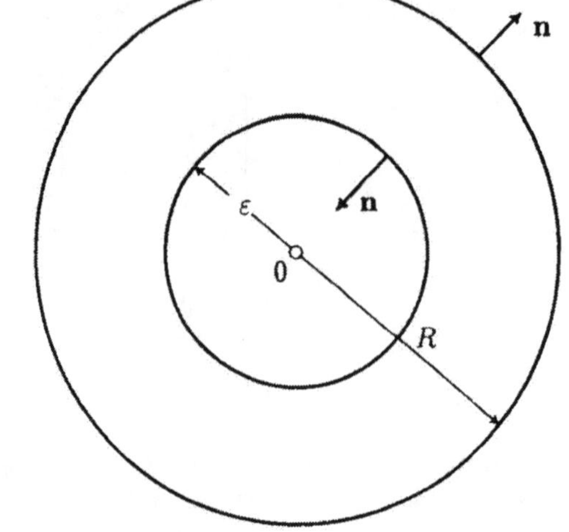

Figure 16

Let $\varphi \in \mathcal{D}$, $\operatorname{supp}\varphi \subset U_R$. Then

$$\left(\Delta\frac{1}{|x|}, \varphi\right) = \left(\frac{1}{|x|}, \Delta\varphi\right) = \int_{U_R} \frac{1}{|x|}\Delta\varphi(x)\,dx = \lim_{\varepsilon\to+0} \int_{\varepsilon<|x|<R} \frac{1}{|x|}\Delta\varphi(x)\,dx.$$

Applying Green's formula (3.7) for $f = 1/|x|$ and $G = [x : \varepsilon < |x| < R]$ (Fig. 16) and taking into account (3.9), we obtain the formula (3.8):

$$\begin{aligned}\left(\Delta\frac{1}{|x|}, \varphi\right) &= \lim_{\varepsilon\to+0}\left[\int_{\varepsilon<|x|<R} \Delta\frac{1}{|x|}\varphi(x)\,dx + \left(\int_{S_\varepsilon} + \int_{S_R}\right)\left(\frac{1}{|x|}\frac{\partial\varphi}{\partial\mathbf{n}} - \varphi\frac{\partial}{\partial\mathbf{n}}\frac{1}{|x|}\right) dS\right] \\ &= \lim_{\varepsilon\to+0}\int_{S_\varepsilon}\left(-\frac{1}{|x|}\frac{\partial\varphi}{\partial|x|} - \varphi\frac{1}{|x|^2}\right) dS \\ &= \lim_{\varepsilon\to+0}\frac{-1}{\varepsilon^2}\int_{S_\varepsilon}\varphi\,dS \\ &= \lim_{\varepsilon\to+0}\left\{\frac{1}{\varepsilon^2}\int_{S_\varepsilon}[\varphi(0) - \varphi(x)]\,dS - 4\pi\varphi(0)\right\} \\ &= -4\pi(\delta, \varphi). \quad \square\end{aligned}$$

The equation (3.8) may be interpreted as follows: the function $1/|x|$ is the Newtonian (Coulomb) potential generated by the charge $+1$ at the point $x = 0$.

Similarly,

$$\begin{aligned} \Delta \ln|x| &= 2\pi\delta(x), & n=2,\\ \Delta \frac{1}{|x|^{n-2}} &= -(n-2)\sigma_n\delta(x), & n\ge 3, \end{aligned} \tag{3.10}$$

where σ_n is the surface area of a unit sphere in $\mathbb{R}^n$ (see Sec. 0.6).

The function $\mathcal{E}_n(x)$, which is equal to

$$\mathcal{E}_n(x) = \begin{cases} -\dfrac{1}{(n-2)\sigma_n|x|^{n-2}}, & n\ge 3,\\ \dfrac{1}{2\pi}\ln|x|, & n=2,\\ \dfrac{1}{2}|x|, & n=1, \end{cases}$$

is termed the *fundamental solution* of the Laplace operator.

2.4. The local structure of generalized functions. We will now prove that the space $\mathcal{D}'(\mathcal{O})$ is locally a (smallest) extension of the space $\mathcal{L}^\infty_{\text{loc}}(\mathcal{O})$ such that, in it, differentiation is always possible.

THEOREM. *Let $f \in \mathcal{D}'(\mathcal{O})$ and let an open set $\mathcal{O}' \Subset \mathcal{O}$. Then there exist a function $g \in \mathcal{L}^\infty(\mathcal{O}')$ and an integer $m \ge 0$ such that*[1]

$$f(x) = \partial_1^m \ldots \partial_n^m g(x), \qquad x \in \mathcal{O}'. \tag{4.1}$$

PROOF. According to the theorem of Sec. 1.3, there exist numbers K and k such that the following inequality holds:

$$|(f,\varphi)| \le K\|\varphi\|_{C^k(\overline{\mathcal{O}'})}, \qquad \varphi \in \mathcal{D}(\mathcal{O}'). \tag{4.2}$$

Since, for $\psi \in \mathcal{D}(\mathcal{O}')$, $\psi(x) = \int_{-\infty}^{x_j} \partial_j\psi\, dx_j'$, it follows that

$$\max_{x\in\overline{\mathcal{O}'}} |\psi(x)| \le d \max_{x\in\overline{\mathcal{O}'}} |\partial_j\psi(x)|,$$

where d is the diameter of $\mathcal{O}'$. Therefore, applying this inequality a sufficient number of times, we obtain from (4.2) this inequality

$$|(f,\varphi)| \le C \max_{x\in\overline{\mathcal{O}'}} |\partial_1^k \ldots \partial_n^k \varphi(x)|, \qquad \varphi \in \mathcal{D}(\mathcal{O}'). \tag{4.3}$$

Furthermore, for $\psi \in \mathcal{D}(\mathcal{O}')$

$$\psi(x) = \int_{-\infty}^{x_1} \cdots \int_{-\infty}^{x_n} \frac{\partial^n \psi(y)}{\partial y_1 \ldots \partial y_n}\, dy_1 \ldots dy_n$$

and therefore

$$|\psi(x)| \le \int_{\mathcal{O}'} |\partial_1 \ldots \partial_n \psi(y)|\, dy.$$

From this fact and from (4.3) there follows the inequality (for $m = k+1$)

$$|(f,\varphi)| \le C \int_{\mathcal{O}'} |\partial_1^m \ldots \partial_n^m \varphi(x)|\, dx, \qquad \varphi \in \mathcal{D}(\mathcal{O}'). \tag{4.4}$$

[1]The derivative in (4.1) is to be understood in the sense of generalized functions.

From the Hahn–Banach theorem it follows that the continuous linear functional f^*:

$$\chi(x) = (-1)^{mn}\partial_1^m \ldots \partial_n^m \varphi(x) \to (f^*, \chi) = (f, \varphi), \qquad \varphi \in \mathcal{D}(\mathcal{O}') \tag{4.5}$$

admits of an extension to a continuous linear functional on $\mathcal{L}^1(\mathcal{O}')$ with norm $\leq C$ by virtue of the inequality (4.4):

$$|(f^*, \chi)| = |(f, \varphi)| \leq C\|\chi\|_{\mathcal{L}^1(\mathcal{O}')}.$$

By a theorem of F. Riesz, there exists a function $g \in \mathcal{L}^\infty(\mathcal{O}')$ with norm $\|g\|_{\mathcal{L}^\infty(\mathcal{O}')} \leq C$ such that

$$(f^*, \chi) = (-1)^{mn} \int_{\mathcal{O}'} g(x)\chi(x)\, dx.$$

From this and from (4.5) we derive, for all $\varphi \in \mathcal{D}(\mathcal{O}')$,

$$\begin{aligned}(f, \varphi) &= (-1)^{mn} \int_{\mathcal{O}'} g(x)\partial_1^m \ldots \partial_n^m \varphi(x)\, dx \\ &= (\partial_1^m \ldots \partial_n^m g, \varphi),\end{aligned}$$

which is equivalent to (4.1). This completes the proof of the theorem. □

COROLLARY. *Under the conditions of the theorem,*

$$f(x) = \partial_1^{m+1} \ldots \partial_n^{m+1} g_1(x) \quad \textit{in} \quad \mathcal{O}', \qquad g_1 \in C(\overline{\mathcal{O}'}). \tag{4.6}$$

The representation (4.6) follows from the representation (4.1) if the function $g(x)$ is continued via zero onto the whole of $\mathbb{R}^n$ and if we put

$$g_1(x) = \int_{-\infty}^{x_1} \cdots \int_{-\infty}^{x_n} g(y)\, dy_1 \ldots dy_n.$$

□

2.5. Generalized functions with compact support. We introduce convergence on a set of functions $C^\infty(\mathcal{O})$:

$$\varphi_k \to 0, \qquad k \to \infty \quad \text{in} \quad C^\infty(\mathcal{O})$$

$$\iff \partial^\alpha \varphi_k(x) \overset{x \in \mathcal{O}'}{\rightrightarrows} 0, \qquad k \to \infty \quad \text{for all } \alpha \text{ and } \quad \mathcal{O}' \Subset \mathcal{O}.$$

From this definition it follows that convergence in $\mathcal{D}(\mathcal{O})$ implies convergence in $C^\infty(\mathcal{O})$, but not vice versa.

Suppose a generalized function f in $\mathcal{D}'(\mathcal{O})$ has compact support in $\mathcal{O}$, $\operatorname{supp} f = K \Subset \mathcal{O}$. Suppose $\eta \in \mathcal{D}(\mathcal{O})$, $\eta(x) = 1$ in the neighbourhood of K (see Sec. 1.2). We will construct a functional $\tilde{f}$ on $C^\infty(\mathcal{O})$ via the rule

$$(\tilde{f}, \varphi) = (f, \eta\varphi), \qquad \varphi \in C^\infty(\mathcal{O}). \tag{5.1}$$

Clearly, $\tilde{f}$ is a linear functional on $C^\infty(\mathcal{O})$. Furthermore, since the operation $\varphi \to \eta\varphi$ is continuous from $C^\infty(\mathcal{O})$ into $\mathcal{D}(\mathcal{O})$, it follows that $\tilde{f}$ is a continuous functional on $C^\infty(\mathcal{O})$. The functional f is an extension of the functional from $\mathcal{D}(\mathcal{O})$ onto $C^\infty(\mathcal{O})$, since for $\varphi \in \mathcal{D}(\mathcal{O})$

$$(\tilde{f}, \varphi) = (f, \eta\varphi) = (\eta f, \varphi) = (f, \varphi)$$

by virtue of the equality (10.2) of Sec. 1.10.

We will show that there is a unique linear and continuous extension of f onto $C^\infty(\mathcal{O})$ (in particular, the extension (5.1) does not depend on the auxiliary function η). Let $\tilde{\tilde{f}}$ be another such extension of f. We introduce a sequence of functions $\{\eta_k\}$ in $\mathcal{D}(\mathcal{O})$ such that $\eta_k(x) = 1$, $x \in \mathcal{O}_k$ ($\mathcal{O}_1 \Subset \mathcal{O}_2 \Subset \ldots$, $\mathcal{O} = \bigcup_k \mathcal{O}_k$), so that $\eta_k \to 1$, $k \to \infty$ in $C^\infty(\mathcal{O})$. Therefore, for any $\varphi \in C^\infty(\mathcal{O})$ we will have $\eta_k\varphi \to \varphi$, $k \to \infty$ in $C^\infty(\mathcal{O})$. Hence,

$$(\tilde{f}, \varphi) = (\tilde{f}, \lim_{k\to\infty} \eta_k\varphi) = \lim_{k\to\infty} (\tilde{f}, \eta_k\varphi) = \lim_{k\to\infty} (f, \eta_k\varphi)$$
$$= \lim_{k\to\infty} (\tilde{\tilde{f}}, \eta_k\varphi) = (\tilde{\tilde{f}}, \lim_{k\to\infty} \eta_k\varphi) = (\tilde{\tilde{f}}, \varphi), \qquad \varphi \in C^\infty(\mathcal{O}),$$

so that $f = \tilde{\tilde{f}}$.

We have thus proved the necessity of the conditions in the following theorem.

THEOREM. *For a generalized function f in $\mathcal{D}'(\mathcal{O})$ to have compact support in $\mathcal{O}$, it is necessary and sufficient that it admit of a linear and continuous extension onto $C^\infty(\mathcal{O})$.*

PROOF OF SUFFICIENCY. Suppose $f \in \mathcal{D}'(\mathcal{O})$ admits of a linear and continuous extension $\tilde{f}$ onto $C^\infty(\mathcal{O})$. If f did not possess compact support in $\mathcal{D}$, then it would be possible to indicate a sequence of functions $\{\varphi_k\}$ in $\mathcal{D}(\mathcal{O})$ such that $\operatorname{supp}\varphi_k \subset \mathcal{O} \setminus \overline{\mathcal{O}}_k$ ($\mathcal{O}_1 \Subset \mathcal{O}_2 \Subset \ldots$, $\bigcup_k \mathcal{O}_k = \mathcal{O}$) and $(f, \varphi_k) = 1$. On the other hand $\varphi_k \to 0$, $k \to \infty$ in $C^\infty(\mathcal{O})$, and therefore $(\tilde{f}, \varphi_k) \to 0$, $k \to \infty$. But $(\tilde{f}, \varphi_k) = (f, \varphi_k) = 1$, which is contradictory. The proof is complete. □

Let f be a generalized function with compact support in $\mathcal{O}$. Then, by virtue of (5.1), we have

$$(f, \varphi) = (f, \eta\varphi), \qquad \varphi \in \mathcal{D}(\mathcal{O}).$$

Since $\eta \in \mathcal{D}(\mathcal{O})$ and $\operatorname{supp}\eta \Subset \mathcal{O}$, it follows that $\eta \in \mathcal{D}(\mathcal{O}')$ for some $\mathcal{O}' \Subset \mathcal{O}$. Therefore $\eta\varphi \in \mathcal{D}(\mathcal{O}')$ for all $\varphi \in \mathcal{D}(\mathcal{O})$. By the theorem of Sec. 1.3 there exist numbers $K = K(\mathcal{O}')$ and $m = m(\mathcal{O}')$ such that the following inequality holds:

$$|(f, \varphi)| = |(f, \eta\varphi)| \le K\|\eta\varphi\|_{C^m(\overline{\mathcal{O}'})}, \qquad \varphi \in \mathcal{D}(\mathcal{O}),$$

whence immediately follows the inequality

$$|(f, \varphi)| \le C\|\varphi\|_{C^m(\overline{\mathcal{O}'})}, \qquad \varphi \in \mathcal{D}(\mathcal{O}). \tag{5.2}$$

Inequality (5.2) implies the following assertion: any generalized function with compact support in $\mathcal{O}$ has a finite order in $\mathcal{O}$ (see Sec. 1.3).

We denote by $\mathcal{E}'$ the collection of generalized functions with compact support in $\mathbb{R}^n$. It has thus been proved that $\mathcal{E}' = C^\infty(\mathbb{R}^n)'$.

2.6. Generalized functions with point support. Generalized functions whose supports consist of isolated points admit of explicit description. This is given by the following theorem.

THEOREM. *If the support of a generalized function $f \in \mathcal{D}'$ consists of a single point $x = 0$, then it is uniquely representable in the form*

$$f(x) = \sum_{|\alpha| \le N} c_\alpha \partial^\alpha \delta(x) \tag{6.1}$$

where N is the order of f, and c_α are certain constants.

PROOF. Suppose $\eta \in \mathcal{D}(U_1)$, $\eta(x) = 1$, $|x| \le 1/2$. Then for any $\varepsilon > 0$ we have $f = \eta(\frac{x}{\varepsilon})f$ and, hence, for any $\varphi \in \mathcal{D}$

$$
\begin{aligned}
(f, \varphi) &= \left(\eta\left(\frac{x}{\varepsilon}\right)f, \varphi\right) \\
&= \left(f, \eta\left(\frac{x}{\varepsilon}\right)(\varphi - S_N)\right) + \left(f, \eta\left(\frac{x}{\varepsilon}\right)S_N\right)
\end{aligned}
\tag{6.2}
$$

where

$$S_N(x; \varphi) = \sum_{|\alpha| \le N} \frac{\partial^\alpha \varphi(0)}{\alpha!} x^\alpha$$

is the Taylor polynomial of φ at zero of degree N (see Sec. 1.8).

Since $\eta(\frac{x}{\varepsilon})(\varphi - S_N) \in \mathcal{D}(U_\varepsilon)$, by applying the inequality (5.2) we get

$$
\begin{aligned}
\left|\left(f, \eta\left(\frac{x}{\varepsilon}\right)(\varphi - S_N)\right)\right| &\le C \left\|\eta\left(\frac{x}{\varepsilon}\right)(\varphi - S_N)\right\|_{C^N(\bar{U}_\varepsilon)} \\
&= C \max_{\substack{|x| \le \varepsilon \\ |\alpha| \le N}} \left|\partial^\alpha \left\{\eta\left(\frac{x}{\varepsilon}\right)[\varphi(x) - S_N(x)]\right\}\right| \\
&= C \max_{\substack{|x| \le \varepsilon \\ |\alpha \le N}} \sum_{\beta \le \alpha} \binom{\alpha}{\beta} \left|\partial^\beta \eta\left(\frac{x}{\varepsilon}\right) \partial^{\alpha-\beta}[\varphi(x) - S_N(x)]\right| \\
&\le C' \max_{|\alpha| \le N} \sum_{\beta \le \alpha} \varepsilon^{-|\beta|} \varepsilon^{N - |\alpha - \beta|} \varepsilon \\
&\le C' \max_{|\alpha| \le N} \varepsilon^{N - |\alpha| + 1} = C'' \varepsilon.
\end{aligned}
$$

In the right-hand member of (6.2), let $\varepsilon \to +0$. By virtue of the resulting estimate, the first term will tend to zero. But the second term does not at all depend on ε and is equal to $(\tilde{f}, \varphi_N)$, where $\tilde{f}$ is the extension of f onto $C^\infty(\mathcal{O})$ (see Sec. 2.5). Therefore the equation (6.2) takes the form

$$(f, \varphi) = (\tilde{f}, S_N) = \sum_{|\alpha| \le N} \frac{\partial^\alpha \varphi(0)}{\alpha!} (\tilde{f}, x^\alpha).$$

Now set

$$c_\alpha = \frac{(-1)^{|\alpha|}}{\alpha!} (\tilde{f}, x^\alpha)$$

and we get the representation (6.1):

$$(f, \varphi) = \sum_{|\alpha| \le N} (-1)^{|\alpha|} c_\alpha \partial^\alpha \varphi(0) = \sum_{|\alpha| \le N} c_\alpha (\partial^\alpha \delta, \varphi), \qquad \varphi \in \mathcal{D}.$$

We now prove the uniqueness of the representation (6.1). If there is another such representation

$$f(x) = \sum_{|\alpha| \le N} c'_\alpha \partial^\alpha \delta(x),$$

then by subtracting we obtain

$$0 = \sum_{|\alpha| \le N} (c'_\alpha - c_\alpha) \partial^\alpha \delta(x),$$

whence

$$\begin{aligned}
0 &= \sum_{|\alpha|\le N} (c'_\alpha - c_\alpha)(\partial^\alpha \delta, x^k) \\
&= \sum_{|\alpha|\le N} (c'_\alpha - c_\alpha)(-1)^{|\alpha|}\partial^\alpha x^k\Big|_{x=0} \\
&= (-1)^{|k|}k!(c'_k - c_k),
\end{aligned}$$

that is, $c'_k = c_k$, and the proof of the theorem is complete. □

EXAMPLE. The general solution of the equation

$$x^m u(x) = 0 \tag{6.3}$$

in the class $\mathcal{D}'(\mathbb{R}^1)$ is given by the formula

$$u(x) = \sum_{0\le k\le m-1} c_k \delta^{(k)}(x) \tag{6.4}$$

where c_k are arbitrary constants.

Indeed, if $u \in \mathcal{D}'$ is a solution of the equation (6.3), then either $u = 0$ or $\operatorname{supp} u$ coincides with the point $x = 0$. By the theorem that has just been proved,

$$u(x) = \sum_{0\le k\le N} c_k \delta^{(k)}(x) \tag{6.5}$$

for certain numbers c_k and integer $N \ge 0$. Taking into account (3.3) and substituting (6.5) into (6.3), we have

$$0 = (-1)^m m! \sum_{m\le k\le N} \binom{k}{m} c_k \delta^{(k-m)}(x),$$

whence it follows that $c_k = 0$, $k \ge m$. Thus, in the representation (6.5) we can assume $N = m - 1$, and the formula (6.4) is proved. It remains to note that the right-hand side of (6.4) satisfies equation (6.3) for arbitrary constants c_k, $k = 0, 1, \ldots, m-1$. □

2.7. Generalized functions $\mathcal{P}(\pi_\nu|x|^{\alpha-1})$. Let $\pi_\nu(x) = \operatorname{sgn}^\nu x$, $\nu = 0, 1$, be a multiplicative characters of the field of real numbers $\mathbb{R}$. For $\Re\alpha > 0$, the function $\pi_\nu(x)|x|^{\alpha-1}$ is locally integrable in $\mathbb{R}^1$ and therefore it defines a regular generalized function $\pi_\nu|x|^{\alpha-1}$ by formula (6.1)

$$(\pi_\nu|x|^{\alpha-1}, \varphi) = \int \pi_\nu(x)|x|^{\alpha-1}\varphi(x)\,dx, \qquad \varphi \in \mathcal{D}. \tag{7.1}$$

The integral in the right-hand side of equality (7.1) is a holomorphic function of the complex variable α in the half-plane $\Re\alpha > 0$.

DEFINITION. We say that a generalized function $f_\alpha \in \mathcal{D}'$ dependent on the complex parameter α is *holomorphic* (*meromorphic*) with respect to α in a domain G if, for any $\varphi \in \mathcal{D}$, the function (f_α, φ) is holomorphic (meromorphic) with respect to α in G.

This definition implies that the generalized function $\pi_\nu|x|^{\alpha-1}$ is holomorphic with respect to α in $\Re\alpha > 0$.

Let us rewrite equality (7.1) for $\Re\alpha > 0$ and $N = 1, 2, \dots$ in the form

$$(\pi_\nu|x|^{\alpha-1}, \varphi) = \int\limits_{|x|<1} \pi_\nu(x)|x|^{\alpha-1}[\varphi(x) - S_{N-1}(x;\varphi)]\,dx + \sum_{k=0}^{N-1} \frac{\varphi^{(k)}(0)}{k!} \int\limits_{|x|<1} \pi_\nu(x)|x|^{\alpha-1}x^k\,dx + \int\limits_{|x|>1} \pi_\nu(x)|x|^{\alpha-1}\varphi(x)\,dx, \qquad \varphi \in \mathcal{D}, \tag{7.2}$$

where $S_N(x;\varphi)$ is the Taylor polynomial of the function φ at zero of degree N (see Sec. 1.8). Noting that, for $\Re\alpha > -N+1$,

$$\int\limits_{|x|<1} \pi_\nu(x)|x|^{\alpha-1}x^{N-1}\,dx = \begin{cases} \dfrac{2}{\alpha+N-1}, & \text{if } \nu+N \text{ is odd} \\ 0, & \text{if } \nu+N \text{ is even} \end{cases} \tag{7.3}$$

and for $\Re\alpha < -N+1$, $k = 0, 1, \dots, N-1$,

$$-\int\limits_{|x|>1} \pi_\nu(x)|x|^{\alpha-1}x^k\,dx = \begin{cases} \dfrac{2}{\alpha+k}, & \text{if } \nu+k \text{ is even} \\ 0, & \text{if } \nu+k \text{ is odd,} \end{cases}$$

we deduce from (7.2) and (7.3) the following theorem.

THEOREM. *The generalized function $\pi_\nu|x|^{\alpha-1}$, $\Re\alpha > 0$, admits the meromorphic continuation $\mathcal{P}(\pi_\nu|x|^\alpha)$ from $\mathcal{D}'$ onto the whole plane α with simple poles and residues*

$$\alpha = -2n, \qquad \frac{2}{(2n)!}\delta^{(2n)}(x), \qquad n = 0, 1, \dots, \qquad \textit{if } \pi_\nu(x) = 1;$$

$$\alpha = -2n-1, \qquad -\frac{2}{(2n+1)!}\delta^{(2n+1)}(x), \qquad n = 0, 1, \dots, \qquad \textit{if } \pi_\nu(x) = \operatorname{sgn} x.$$

In every half-plane $\Re\alpha > -N$, $N = 1, 2, \dots$, it admits the representation

$$(\mathcal{P}(\pi_\nu|x|^{\alpha-1}), \varphi) = \int\limits_{|x|<1} \pi_\nu(x)|x|^{\alpha-1}[\varphi(x) - S_{N-1}(x;\varphi)]\,dx + 2\sum_{\substack{\nu+k \text{ even}}}^{N-1} \frac{1}{(\alpha+k)k!}\varphi^{(k)}(0) + \int\limits_{|x|>1} \pi_\nu(x)|x|^{\alpha-1}\varphi(x)\,dx, \qquad \varphi \in \mathcal{D}; \tag{7.4}$$

and in any strip $-N+1 > \Re\alpha > -N$ it admits the representation

$$(\mathcal{P}(\pi_\nu|x|^{\alpha-1}), \varphi) = \int \pi_\nu(x)|x|^{\alpha-1}[\varphi(x) - S_{N-1}(x;\varphi)]\,dx, \qquad \varphi \in \mathcal{D}. \tag{7.5}$$

The generalized function $\mathcal{P}(\pi_\nu|x|^{\alpha-1})$ from $\mathcal{D}'$ considered outside its poles — the holomorphic continuation of the function $\pi_\nu(x)|x|^{\alpha-1}$ from the domain $\Re\alpha > 0$ — is called the *regularization* of this function (compare Sec. 1.8). It follows from this definition that the regularization is *unique*.

3. Direct Product of Generalized Functions

3.1. The definition of a direct product. Let $f(x)$ and $g(y)$ be locally integrable functions in open sets $\mathcal{O}_1 \subset \mathbb{R}^1$ and $\mathcal{O}_2 \subset \mathbb{R}^m$ respectively. The function $f(x)g(y)$ will also be locally integrable in $\mathcal{O}_1 \times \mathcal{O}_2$. It defines a (regular) generalized function $f(x)g(y) = g(y)f(x)$ in $\mathcal{D}'(\mathcal{O}_1 \times \mathcal{O}_2)$ operating on test functions $\varphi(x,y)$ in $\mathcal{D}(\mathcal{O}_1 \times \mathcal{O}_2)$ via the formula

$$\begin{aligned}(f(x)g(y),\varphi) &= \int_{\mathcal{O}_1\times\mathcal{O}_2} f(x)g(y)\varphi(x,y)\,dx\,dy\\ &= \int_{\mathcal{O}_1} f(x)\int_{\mathcal{O}_2} g(y)\varphi(x,y)\,dy\,dx\\ &= \int_{\mathcal{O}_1\times\mathcal{O}_2} g(y)f(x)\varphi(x,y)\,dx\,dy\\ &= \int_{\mathcal{O}_2} g(y)\int_{\mathcal{O}_2} f(x)\varphi(x,y)\,dx\,dy,\end{aligned}$$

that is

$$(f(x)g(y),\varphi) = \big(f(x),(g(y),\varphi(x,y))\big), \tag{1.1}$$

$$(g(y)f(x),\varphi) = \big(g(y),(f(x),\varphi(x,y))\big). \tag{1.1'}$$

These equations express the Fubini theorem on the coincidence of iterated integrals and a multiple integral.

We take (1.1) and (1.1′) as the starting equalities for defining the *direct products* $f(x)\times g(y)$ and $g(y)\times f(x)$ of the generalized functions $f \in \mathcal{D}'(\mathcal{O}_1)$ and $g \in \mathcal{D}'(\mathcal{O}_2)$:

$$(f(x)\times g(y),\varphi) = \big(f(x),(g(y),\varphi(x,y))\big), \tag{1.2}$$

$$(g(y)\times f(x),\varphi) = \big(g(y),(f(x),\varphi(x,y))\big), \tag{1.2'}$$

where $\varphi \in \mathcal{D}(\mathcal{O}_1 \times \mathcal{O}_2)$.

Now let us verify that this definition is proper, that is, that the right-hand side of (1.2) defined a continuous linear functional on $\mathcal{D}(\mathcal{O}_1 \times \mathcal{O}_2)$.

Since for every $x \in \mathcal{O}_1$ the functions $\varphi(x,y) \in \mathcal{D}(\mathcal{O}_2)$ while $g \in \mathcal{D}'(\mathcal{O}_2)$, it follows that the function

$$\psi(x) = (g(y),\varphi(x,y)), \qquad \varphi \in \mathcal{D}'(\mathcal{O}_1 \times \mathcal{O}_2), \tag{1.3}$$

is defined in $\mathcal{O}_1$. We now prove the following lemma.

LEMMA. *Let $\mathcal{O}' \Subset \mathcal{O}_1 \times \mathcal{O}_2$ and $g \in \mathcal{D}'(\mathcal{O}_2)$. Then there exist an open set $\tilde{\mathcal{O}}_1 = \tilde{\mathcal{O}}_1(\mathcal{O}') \Subset \mathcal{O}_1$ and numbers $C = C(\mathcal{O}',g) \geq 0$ and integer $m = m(\mathcal{O}',g) \geq 0$,* *such that*

$$\psi \in \mathcal{D}(\tilde{\mathcal{O}}_1) \quad \textit{if} \quad \varphi \in \mathcal{D}(\mathcal{O}'); \tag{1.4}$$

$$\partial^\alpha \psi(x) = \left(g(y), \partial_x^\alpha \varphi(x, y)\right), \qquad \varphi \in \mathcal{D}(\mathcal{O}_1 \times \mathcal{O}_2); \tag{1.5}$$

$$|\partial^\alpha \psi(x)| \le C \max_{\substack{(x,y)\in\overline{\mathcal{O}'} \\ |\beta|\le m}} |\partial_x^\alpha \partial_y^\beta \varphi(x, y)|, \qquad \varphi \in \mathcal{D}(\mathcal{O}'), \quad x \in \mathcal{O}_1. \tag{1.6}$$

PROOF. We will prove that the function $\psi(x)$ defined by (1.3) has compact support in $\mathcal{O}_1$. Since $\operatorname{supp}\varphi \subset \mathcal{O}' \Subset \mathcal{O}_1 \times \mathcal{O}_2$, it follows that there are open sets $\mathcal{O}_1' \Subset \mathcal{O}_1$ and $\mathcal{O}_2' \Subset \mathcal{O}_2$ such that $\mathcal{O}' \Subset \mathcal{O}_1' \times \mathcal{O}_2'$ (Fig. 17). Therefore, if $x \in \mathcal{O}_1 \setminus \mathcal{O}_1'$, then $\varphi(x, y) = 0$ for all $y \in \mathcal{O}_2$ and for this reason $\varphi(x) = (g, 0) = 0$, so that $\psi(x) = 0$ outside $\mathcal{O}_1'$. Choosing an open set $\tilde{\mathcal{O}}_1$ such that $\mathcal{O}_1' \Subset \tilde{\mathcal{O}}_1 \Subset \mathcal{O}_1$, we conclude that $\operatorname{supp}\psi \subset \tilde{\mathcal{O}}_1$.

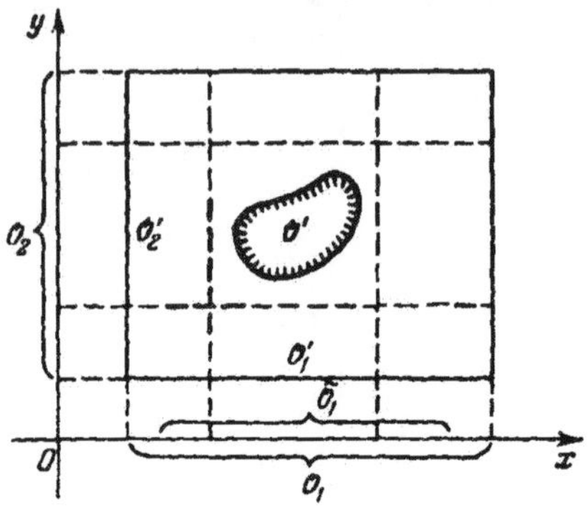

Figure 17

Now let us prove that ψ is continuous in $\mathcal{O}_1$. We fix an arbitrary point $x \in \mathcal{O}_1$ and let $x_k \to x$, $x_k \in \mathcal{O}_1$. Then

$$\varphi(x_k, y) \to \varphi(x, y), \qquad x_k \to x \quad \text{in} \quad \mathcal{D}(\mathcal{O}_2). \tag{1.7}$$

Indeed, $\operatorname{supp}\varphi(x_k, y) \subset \mathcal{O}_2' \Subset \mathcal{O}_2$ and

$$\partial_y^\alpha(x_k, y) \overset{y\in\mathcal{O}_2}{\Longrightarrow} \partial_y^\alpha \varphi(x, y), \qquad x_k \to x.$$

Taking advantage of the continuity of the functional g, we obtain from (1.3) and (1.7), as $x_k \to x$,

$$\psi(x_k) = \left(g(y), \varphi(x_k, y)\right) \to \left(g(y), \varphi(x, y)\right) = \psi(x),$$

which is to say that the function ψ is continuous at an arbitrary point x. Thus $\psi \in C(\mathcal{O}_1)$.

Now we will prove that $\psi \in C^\infty(\mathcal{O}_1)$ and that the differentiation formula (1.5) holds true. Let $e_1 = (1, 0, \ldots, 0)$. Then for every $x \in \mathcal{O}_1$

$$\chi_h(y) = \frac{1}{h}\left[\varphi(x + he_1, y) - \varphi(x, y)\right] \to \frac{\partial\varphi(x, y)}{\partial x_1}, \qquad h \to 0 \quad \text{in} \quad \mathcal{D}(\mathcal{O}_2). \tag{1.8}$$

Indeed, $\operatorname{supp}\chi_h \subset \mathcal{O}_2' \Subset \mathcal{O}_2$ for sufficiently small h and

$$\partial^\alpha \chi_h(y) \overset{y\in\mathcal{O}_2}{\Longrightarrow} \partial_y^\alpha \frac{\partial\varphi(x, y)}{\partial x_1}, \qquad h \to 0.$$

Since $g \in \mathcal{D}'(\mathcal{O}_2)$, then using (1.3) and (1.8), we get

$$\begin{aligned}
\frac{\psi(x + he_1) - \varphi(x)}{h} &= \frac{1}{h}\left[\left(g(y), \varphi(x + he_1, y)\right) - \left(g(y), \varphi(x, y)\right)\right] \\
&= \left(g(y), \frac{\varphi(x + he_1, y) - \varphi(x, y)}{h}\right) \\
&= (g, \chi_h) \to \left(g(y), \frac{\partial\varphi(x, y)}{\partial x_1}\right),
\end{aligned}$$

whence follows the truth of formula (1.5) for $\alpha = (1, 0, \ldots, 0)$ and, hence, for all first derivatives

$$\frac{\partial\psi(x)}{\partial x_j} = \left(g(y), \frac{\partial\varphi(x,y)}{\partial x_j}\right), \qquad j = 1, 2, \ldots, n.$$

Again applying the same reasoning to this formula, we see that (1.5) holds true for all second derivatives, and so forth; hence for all derivatives. And since the function $\partial^\alpha \varphi(x, y)$ also belongs to $\mathcal{D}(\mathcal{O}_1 \times \mathcal{O}_2)$, we conclude from (1.5) (by what has been proved) that $\partial^\alpha\psi(x)$ is a continuous function in $\mathcal{O}_1$ for all α so that $\psi \in C^\infty(\mathcal{O}_1)$.

From this and also from the fact that $\operatorname{supp}\psi \subset \tilde{\mathcal{O}}_1$ we conclude that $\psi \in \mathcal{D}(\mathcal{O}_1)$ and (1.4) is proved.

Let us prove the inequality (1.6). Let $x \in \mathcal{O}_1$. Then, by what has already been proved, $\partial_x^\alpha\varphi(x,y) \in \mathcal{D}(\mathcal{O}_2')$, $\mathcal{O}_2' \Subset \mathcal{O}_2$. By the theorem of Sec. 1.3 there exist a number $C \geq 0$ and an integer $m \geq 0$ that depend solely on g and $\mathcal{O}_2'$, such that

$$\left|\partial^\alpha\psi(x)\right| = \left|(g(y), \partial_x^\alpha\varphi(x,y))\right| \leq C \max_{\substack{y\in\overline{\mathcal{O}}_2' \\ |\beta|\leq m}} \left|\partial_y^\beta\partial_x^\alpha\varphi(x,y)\right|, \qquad x \in \mathcal{O}_1,$$

whence follows inequality (1.6). The proof of the lemma is complete. □

COROLLARY. *The operation*

$$\varphi(x,y) \to \psi(x) = (g(y), \varphi(x,y))$$

is linear and continuous from $\mathcal{D}(\mathcal{O}_1 \times \mathcal{O}_2)$ *into* $\mathcal{D}(\mathcal{O}_1)$.

Indeed, the linearity of the operation is obvious. Furthermore, if $\varphi \in \mathcal{D}(\mathcal{O}_1 \times \mathcal{O}_2)$, then, by the lemma, $\psi \in \mathcal{D}(\mathcal{O}_1)$ so that this operation carries $\mathcal{D}(\mathcal{O}_1 \times \mathcal{O}_2)$ into $\mathcal{D}(\mathcal{O}_1)$. Let us prove that it is continuous. Let $\varphi_k \to 0$, $k \to \infty$ in $\mathcal{D}(\mathcal{O}_1 \times \mathcal{O}_2)$. Then

$$\operatorname{supp}\varphi_k \subset \mathcal{O}' \Subset \mathcal{O}_1 \times \mathcal{O}_2, \qquad \partial_x^\alpha\partial_y^\beta\varphi_k(x,y) \overset{(x,y)}{\rightrightarrows} 0, \qquad k \to \infty.$$

From this and from (1.4) and (1.6) we derive the following for the sequence $\psi_k(x) = (g(y), \varphi_k(x,y))$, $k = 1, 2, \ldots$,:

$$\operatorname{supp}\psi_k \subset \tilde{\mathcal{O}}_1 \Subset \mathcal{O}_1, \qquad \partial^\alpha\psi_k(x) \overset{x}{\rightrightarrows} 0, \qquad k \to \infty,$$

so that $\psi_k \to 0$, $k \to \infty$ in $\mathcal{D}(\mathcal{O}_1)$. □

Let us return to formula (1.2), the definition of the direct product $f(x) \times g(y)$. By the corollary to the lemma that was just proved, the operation $\varphi \to \psi$ is linear and continuous from $\mathcal{D}(\mathcal{O}_1 \times \mathcal{O}_2)$ to $\mathcal{D}(\mathcal{O}_1)$ and, hence, the right-hand side of (1.2), which is equal to (f, ψ), defines a linear and continuous functional on $\mathcal{D}(\mathcal{O}_1 \times \mathcal{O}_2)$ so that $f(x) \times g(y) \in \mathcal{D}'(\mathcal{O}_1 \times \mathcal{O}_2)$.

Similarly, using (1.2′), we can prove that $g(y) \times f(x) \in \mathcal{D}'(\mathcal{O}_1 \times \mathcal{O}_2)$.

3.2. The properties of a direct product.

3.2.1. *Commutativity of a direct product.*

$$f(x) \times g(y) = g(y) \times f(x), \qquad f \in \mathcal{D}'(\mathcal{O}_1), \quad g \in \mathcal{D}'(\mathcal{O}_2). \tag{2.1}$$

Indeed, on the test functions $\varphi \in \mathcal{D}(\mathcal{O}_1 \times \mathcal{O}_2)$ of the form

$$\varphi(x,y) = \sum_{1\leq i\leq N} u_i(x)v_i(y), \qquad u_i \in \mathcal{D}(\mathcal{O}_1), \quad v_i \in \mathcal{D}(\mathcal{O}_2), \tag{2.2}$$

the equality (2.1) follows from the definitions (1.2) and (1.2′):

$$(f(x) \times g(y), \varphi) = \sum_{1\le i\le N} (f, u_i)(g, v_i) = (g(y) \times f(x), \varphi).$$

In order to extend (2.1) to any test functions in $\mathcal{D}(\mathcal{O}_1 \times \mathcal{O}_2)$ let us prove a lemma that states that the set of test functions of the form (2.2) is dense in $\mathcal{D}(\mathcal{O}_1 \times \mathcal{O}_2)$.

LEMMA. *For any $\varphi \in \mathcal{D}(\mathcal{O}_1 \times \mathcal{O}_2)$ there exists a sequence of test functions in $\mathcal{D}(\mathcal{O}_1 \times \mathcal{O}_2)$,*

$$\varphi_k(x, y) = \sum_{1\le i\le N_k} u_{ik}(x) v_{ik}(y), \qquad u_{ik} \in \mathcal{D}(\mathcal{O}_1), \quad v_{ik} \in \mathcal{D}(\mathcal{O}_2), \quad k = 1, 2, \ldots,$$

that converges to φ in $\mathcal{D}(\mathcal{O}_1 \times \mathcal{O}_2)$.

PROOF. Suppose $\operatorname{supp}\varphi \Subset \tilde{\mathcal{O}}_1 \times \tilde{\mathcal{O}}_2 \Subset \mathcal{O}_1' \times \mathcal{O}_2' \Subset \mathcal{O}_1 \times \mathcal{O}_2$. By the Weierstrass theorem, there exists a sequence of polynomials $P_k(x, y)$, $k = 1, 2, \ldots$, such that

$$|\partial^\alpha \varphi(x, y) - \partial^\alpha P_k(x, y)| < \frac{1}{k}, \qquad |\alpha| \le k, \qquad (x, y) \in \overline{\mathcal{O}}_1' \times \overline{\mathcal{O}}_2'. \tag{2.3}$$

Suppose $\xi(x) \in \mathcal{D})\mathcal{O}_1')$, $\xi(x) = 1$, $x \in \tilde{\mathcal{O}}_1$; $\eta(y) \in \mathcal{D}(\mathcal{O}_2')$, $\eta(y) = 1$, $y \in \tilde{\mathcal{O}}_2$. Then the sequence of functions

$$\varphi_k(x, y) = \xi(x)\eta(y)P_k(x, y), \qquad k = 1, 2, \ldots,$$

is the required sequence. Indeed, $\operatorname{supp}\varphi_k \subset \mathcal{O}_1' \times \mathcal{O}_2' \Subset \mathcal{O}_1 \times \mathcal{O}_2$ and of all $k \ge |\alpha|$, by virtue of (2.3), we have

$$|\partial^\alpha \varphi(x, y) - \partial^\alpha \varphi_k(x, y)| \le \begin{cases} \dfrac{1}{k} & \text{if } (x, y) \in \tilde{\mathcal{O}}_1 \times \tilde{\mathcal{O}}_2, \\ \dfrac{c_\alpha}{k} & \text{if } (x, y) \in \mathcal{O}_1' \times \mathcal{O}_2' \setminus (\tilde{\mathcal{O}}_1 \times \tilde{\mathcal{O}}_2), \end{cases}$$

for certain c_α estimated in terms of $\max|\partial^\beta \xi|$ and $\max|\partial^\beta \eta|$, $\beta \le \alpha$. And that means that $\varphi_k \to \varphi$, $k \to \infty$ in $\mathcal{D}(\mathcal{O}_1 \times \mathcal{O}_2)$. The proof of the lemma is complete. □

Let φ be an arbitrary test function in $\mathcal{D}(\mathcal{O}_1 \times \mathcal{O}_2)$. By the lemma, there exists a sequence $\{\varphi_k\}$ of test functions of the form (2.2) that converges to φ in $\mathcal{D}(\mathcal{O}_1 \times \mathcal{O}_2)$. From this, taking advantage of the continuity, on $\mathcal{D}(\mathcal{O}_1 \times \mathcal{O}_2)$, of the functionals $f(x) \times g(y)$ and $g(y) \times f(x)$ (see Sec. 3.1) and also taking advantage of the above-proved equation (2.1) on test functions of the form (2.2), we obtain (2.1) in the general case:

$$\begin{aligned} (f(x) \times g(y), \varphi) &= \lim_{k\to\infty} (f(x) \times g(y), \varphi_k) \\ &= \lim_{k\to\infty} (g(y) \times f(x), \varphi_k) \\ &= (g(y) \times f(x), \varphi) \end{aligned}$$

3.2.2. *Associativity of a direct product.* If $f \in \mathcal{D}'(\mathcal{O}_1)$, $g \in \mathcal{D}'(\mathcal{O}_2)$ and $h \in \mathcal{D}'(\mathcal{O}_2)$, then

$$[f(x) \times g(y)] \times h(z) = f(x) \times [g(y) \times h(z)]. \tag{2.4}$$

Indeed, if $\varphi \in \mathcal{D}(\mathcal{O}_1 \times \mathcal{O}_2 \times \mathcal{O}_3)$, then

$$\begin{aligned}\Big([f(x) \times g(y)] \times h(z), \varphi\Big) &= \Big(f(x) \times g(y), (h(z), \varphi(x,y,z))\Big)\\ &= \Big(f(x), (g(y), (h(z), \varphi(x,y,z)))\Big)\\ &= \Big(f(x), (g(y) \times h(z), \varphi(x,y,z))\Big)\\ &= \Big(f(x) \times [g(y) \times h(z)], \varphi\Big).\end{aligned}$$

Henceforth, taking into account the commutativity and associativity of the operation of a direct product, we will write

$$(f \times g) \times h = f \times g \times h.$$

EXAMPLE. $\delta(x) = \delta(x_1) \times \delta(x_2) \times \cdots \times \delta(x_n)$.

3.2.3. If $g \in \mathcal{D}'(\mathcal{O}_2)$, the operation $f \to f \times g$ is linear and continuous from $\mathcal{D}'(\mathcal{O}_1)$ into $\mathcal{D}'(\mathcal{O}_1 \times \mathcal{O}_2)$.

The linearity of the operation is obvious. Let us prove continuity. Suppose $f_k \to 0$, $k \to \infty$ in $\mathcal{D}(\mathcal{O}_1)$. Then for all $\varphi \in \mathcal{D}(\mathcal{O}_1 \times \mathcal{O}_2)$ we have

$$(f_k(x) \times g(y), \varphi) = \Big(f_k(x), (g(y), \varphi(x,y))\Big) = (f_k, \psi) \to 0, \qquad k \to \infty.$$

That is, $f_k(x) \times g(y) \to 0$, $k \to \infty$ in $\mathcal{D}'(\mathcal{O}_1 \times \mathcal{O}_2)$. Here we made use of the fact that $\psi \in \mathcal{D}(\mathcal{O}_1)$ by virtue of the lemma of Sec. 3.1. □

3.2.4. The following formula holds:

$$\operatorname{supp}(f \times g) = \operatorname{supp} f \times \operatorname{supp} g, \qquad f \in \mathcal{D}'(\mathcal{O}), \quad g \in \mathcal{D}'(\mathcal{O}_2). \tag{2.5}$$

COROLLARY. *If $f(x) \times 1(y) = 0$ in $\mathcal{O}_1 \times \mathcal{O}_2$, then $f(x) = 0$ in $\mathcal{O}_1$.*

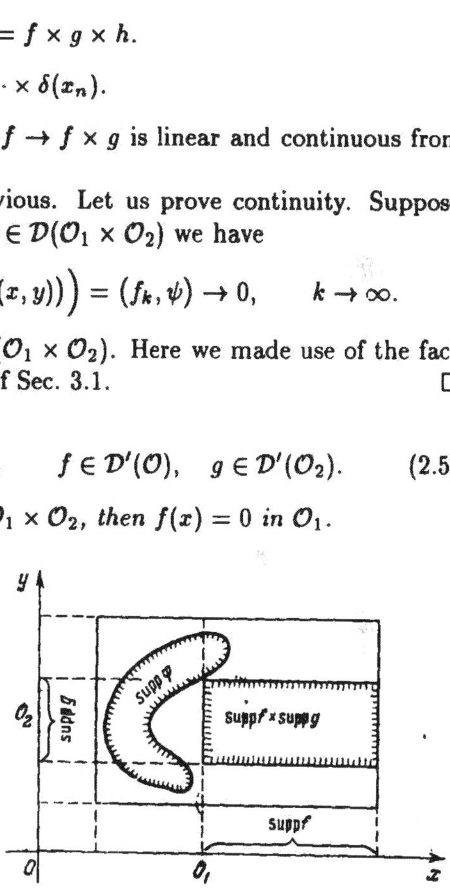

Figure 18

Indeed, suppose $(x_0, y_0) \in \operatorname{supp} f \times \operatorname{supp} g$ and $U(x_0, y_0)$ is the neighbourhood of the point (x_0, y_0) lying in $\mathcal{O}_1 \times \mathcal{O}_2$. There exist neighbourhoods U_1 and U_2 of the points x_0 and y_0 respectively such that $U_1 \times U_2 \subset U(x_0, y_0)$. From the definition of the support of a generalized function (see Sec. 1.5(b)) it follows that there are functions $\varphi_1 \in \mathcal{D}(U_1)$ and $\varphi_2 \in \mathcal{D}(U_2)$ such that $(f, \varphi_1) \neq 0$ and $(g, \varphi_2) \neq 0$. And so $(f \times g, \varphi_1\varphi_2) = (f, \varphi_1)(g, \varphi_2) \neq 0$. From this fact, due to the arbitrariness of the neighbourhood $U(x_0, y_0)$, it follows that $(x_0, y_0) \in \operatorname{supp}(f \times g)$ so that

$$\operatorname{supp} f \times \operatorname{supp} g \subset \operatorname{supp}(f \times g). \tag{2.6}$$

Let us now prove the converse inclusion. Take a test function φ in $\mathcal{D}(\mathcal{O}_1 \times \mathcal{O}_2)$ such that

$$\operatorname{supp} \varphi \subset (\mathcal{O}_1 \times \mathcal{O}_2) \setminus (\operatorname{supp} f \times \operatorname{supp} g)$$

(Fig. 18). Then there is a neighbourhood U of the set $\operatorname{supp} f$ such that for every $x \in U$, $\operatorname{supp} \varphi(x, y) \subset \mathcal{O}_2 \setminus \operatorname{supp} g$. Therefore (see Sec. 1.5(a))

$$\psi(x) = (g(y), \varphi(x, y)) = 0, \qquad x \in U,$$

and, hence, $\operatorname{supp} \psi \cap \operatorname{supp} f = \varnothing$, and so

$$(f \times g, \varphi) = (f, \psi) = 0.$$

Thus the zero set $\mathcal{O}_{f\times g}$ contains $(\mathcal{O}_1 \times \mathcal{O}_2) \setminus (\operatorname{supp} f \times \operatorname{supp} g)$ and, hence, the following inclusion holds true:

$$\operatorname{supp}(f \times g) \subset f \times \operatorname{supp} g.$$

This, together with the converse inclusion (2.6), proves the equality (2.5). □

3.2.5. The following formulae, which are readily verifiable, hold true: if $f \in \mathcal{D}'(\mathcal{O}_1)$ and $g \in \mathcal{D}'(\mathcal{O}_2)$, then

$$\partial_x^\alpha \partial_y^\beta [f(x) \times g(y)] = \partial^\alpha f(x) \times \partial^\beta g(y), \tag{2.7}$$

$$a(x)b(y)[f(x) \times g(y)] = [a(x)f(x)] \times [b(y)g(y)], \tag{2.8}$$

$$(f \times g)(x + x_0, y + y_0) = f(x + x_0) \times g(y + y_0). \tag{2.9}$$

3.3. Some applications. We will say that a generalized function $F(x, y)$ in $\mathcal{D}'(\mathcal{O}_1 \times \mathcal{O}_2)$ does not depend on the variables y if it can be represented in the form

$$F(x, y) = f(x) \times 1(y), \qquad f \in \mathcal{D}'(\mathcal{O}_1) \tag{3.1}$$

(and then $F \in \mathcal{D}'(\mathcal{O}_1 \times \mathbb{R}^m)$). The generalized function $f(x) \times 1(y) = 1(y) \times f(x)$ acts on the test functions φ in $\mathcal{D}(\mathcal{O}_1 \times \mathbb{R}^m)$ via the rule

$$\begin{aligned}(f(x) \times 1(y), \varphi) &= \left(f(x), \int \varphi(x, y)\, dy\right) \\ &= (1(y) \times f(x), \varphi) \\ &= \int (f(x), \varphi(x, y))\, dy.\end{aligned}$$

We have thus obtained the equality

$$\left(f(x), \int \varphi(x, y)\, dy\right) = \int (f(x), \varphi(x, y))\, dy \tag{3.2}$$

which holds for all $f \in \mathcal{D}'(\mathcal{O}_1)$ and $\varphi \in \mathcal{D}(\mathcal{O}_1 \times \mathbb{R}^m)$.

The formula (3.2) may be regarded as a peculiar kind of generalization of the Fubini theorem.

Suppose $F \in \mathcal{D}'(\mathcal{O} \times (a, b))$. The following three statements are equivalent:

(1) $F(x, y)$ does not depend on the variable y;

(2) $F(x, y)$ is invariant in $\mathcal{O} \times (a, b)$ with respect to translations along y, that is,

$$F(x, y + h) = F(x, y); \qquad a < y, \quad y + h < b; \tag{3.3}$$

(3) $f(x, y)$ satisfies the following equation in $\mathcal{O} \times (a, b)$:

$$\frac{\partial F(x, y)}{\partial y} = 0. \tag{3.4}$$

COROLLARY. *If $f \in \mathcal{D}'(\mathcal{O})$ and $\frac{\partial f}{\partial x_j} = 0$, $j = 1, \ldots, n$, in $\mathcal{O}$, then $f = \text{const}$ in $\mathcal{O}$; if f is invariant with respect to translations along all arguments in $\mathcal{O}$, then $f = \text{const}$ in $\mathcal{O}$.*

PROOF. (1) → (2). It follows from (3.1) by virtue of (2.9).

(2) → (3). Passing to the limit in

$$\frac{F(x, y+h) - F(x,y)}{h} = 0$$

as $h \to 0$ in $\mathcal{D}'(\mathcal{O} \times (a,b))$, we conclude that F satisfies the equation (3.4).

(3) → (1). Let F satisfy the equation (3.4) in $\mathcal{O} \times (a,b)$. Then, proceeding as in Sec. 2.2, for any φ in $\mathcal{D}(\mathcal{O} \times (a,b))$ we obtain the representation

$$\varphi(x,y) = \frac{\partial \psi(x,y)}{\partial y} + \omega_\varepsilon(y - y_0) \int \varphi(x,\xi)\, d\xi, \tag{3.5}$$

where $y_0 \in (a,b)$, $\varepsilon < \min(y_0 - a, b - y_0)$ and

$$\psi(x,y) = \int_{-\infty}^{y} \left[\varphi(x,y') - \omega_\varepsilon(y' - y_0) \int \varphi(x,\xi)\, d\xi\right] dy' \in \mathcal{D}(\mathcal{O} \times (a,b)).$$

By introducing the generalized function $f(x)$ taken from $\mathcal{D}'(\mathcal{O})$, which function acts on the test functions χ taken from $\mathcal{D}(\mathcal{O})$ via the rule

$$(f,\chi) = \big(F(x,y), \omega_\varepsilon(y - y_0)\chi(x)\big),$$

and by taking into account (3.4), we obtain from (3.5)

$$\begin{aligned}(F,\varphi) &= \left(F(x,y), \frac{\partial \psi(x,y)}{\partial y} + \omega_\varepsilon(y - y_0) \int \varphi(x,\xi)\, d\xi\right) \\ &= \left(f(x), \int \varphi(x,\xi)\, d\xi\right).\end{aligned}$$

That is, $F(x,y) = f(x) \times 1(y)$, which is what we set out to prove. □

The proof is similar for the following assertion (compare Sec. 2.6).

Suppose $f \in \mathcal{D}'(\mathcal{O} \times \mathbb{R}^1)$. The equation

$$yu(x,y) = F(x,y) \tag{3.6}$$

is always solvable and its general solution is of the form

$$(u,\varphi) = (F,\psi) + \big(f(x) \times \delta(y), \varphi\big), \qquad \varphi \in \mathcal{D}(\mathcal{O} \times \mathbb{R}^1) \tag{3.7}$$

where f is an arbitrary generalized function in $\mathcal{D}'(\mathcal{O})$,

$$\psi(x,y) = \frac{1}{y}\big[\varphi(x,y) - \eta(y)\varphi(x,0)\big], \tag{3.8}$$

$\eta(y)$ is an arbitrary function in $\mathcal{D}(\mathbb{R}^1)$ equal to 1 in a neighbourhood of $y = 0$.

Indeed, since the operation $\varphi \to \psi$ given by (3.8) is linear and continuous from $\mathcal{D}(\mathcal{O} \times \mathbb{R}^1)$ into $\mathcal{D}(\mathcal{O} \times \mathbb{R}^1)$, the right-hand side of (3.7) is a generalized function in $\mathcal{D}'(\mathcal{O} \times \mathbb{R}^1)$ and the first term (F,ψ) satisfies equation (3.6) while the second term, the generalized function $f(x) \times \delta(y)$, satisfies the homogeneous equation

$$yu(x,y) = 0 \tag{3.9}$$

which corresponds to the equation (3.6).

It remains to prove that $f(x) \times \delta(y)$, $f \in \mathcal{D}'(\mathcal{O})$, is the general solution of the equation (3.9) in $\mathcal{D}'(\mathcal{O} \times \mathbb{R}^1)$. Suppose $u(x, y)$ is a solution of the equation (3.9) in $\mathcal{D}'(\mathcal{O} \times \mathbb{R}^1)$. Then, by virtue of (3.8)

$$\varphi(x, y) = y\psi(x, y) + \eta(y)\varphi(x, 0), \qquad \psi \in \mathcal{D}(\mathcal{O} \times \mathbb{R}^1)$$

and therefore

$$\begin{aligned}(u, \varphi) &= (u, y\psi) + \big(u, \eta(y)\varphi(x, 0)\big)\\ &= \big(u, \eta(y)\varphi(x, 0)\big).\end{aligned} \tag{3.10}$$

By introducing the generalized function $f(x)$, taken from $\mathcal{D}'(\mathcal{O})$, that acts on the test functions χ in $\mathcal{D}(\mathcal{O})$ via the rule

$$(f, \chi) = \big(u(x, y), \eta(y)\chi(x)\big),$$

we obtain from (3.10)

$$(u, \varphi) = \big(f, \varphi(x, 0)\big) = \big(f(x) \times \delta(y), \varphi\big), \qquad \varphi \in \mathcal{D}(\mathcal{O} \times \mathbb{R}^1),$$

that is, $u(x, y) = f(x) \times \delta(y)$, which is what we set out to prove. □

3.4. Generalized functions that are smooth with respect to some of the variables. Suppose $f(x, y)$ is a generalized function in $\mathcal{D}'(\mathcal{O}_1 \times \mathcal{O}_2)$ and $\varphi(x)$ is a test function in $\mathcal{D}(\mathcal{O}_1)$. We introduce the generalized function $f_\varphi(y)$ in $\mathcal{D}'(\mathcal{O}_2)$ via the formula

$$(f_\varphi, \psi) = \big(f, \varphi(x)\psi(y)\big) \qquad \psi \in \mathcal{D}(\mathcal{O}_2).$$

From this definition there follows the differentiation formula

$$\partial^\alpha f_\varphi(y) = \left(\partial_y^\alpha f\right)_\varphi(y). \tag{4.1}$$

Indeed

$$\begin{aligned}((\partial_y^\alpha f)_\varphi, \psi) &= (\partial_y^\alpha f.\psi\varphi) = (-1)^{|\alpha|}(f, \varphi\partial^\alpha\psi)\\ &= (-1)^{|\alpha|}(f_\varphi, \partial^\alpha\psi) = (\partial^\alpha f_\varphi, \psi), \qquad \psi \in \mathcal{D}(\mathcal{O}_2).\end{aligned}$$

□

We will say that the generalized function $f(x, y)$ taken from $\mathcal{D}'(\mathcal{O}_1 \times \mathcal{O}_2)$ *belongs to the class* $C^p(\mathcal{O}_2)$ *with respect to* y, $p = 0, 1, \ldots$, if for any $\varphi \in \mathcal{D}(\mathcal{O}_1)$ the generalized function $f_\varphi \in C^p(\mathcal{O}_2)$; but if $f_\varphi \in C^p(\overline{\mathcal{O}}_2)$, we then say that $f \in C^p(\overline{\mathcal{O}}_2)$ with respect to y (compare Sec. 1.6).

Suppose $f \in C(\mathcal{O}_2)$ with respect to y. It then follows that *for every* $y \in \mathcal{O}_2$ *there exists a restriction* $f_y(x)$ *in* $\mathcal{D}'(\mathcal{O}_1)$ *of the generalized function* $f(x, y)$, *and*

$$f_\varphi(y) = (f_y, \varphi), \qquad \varphi \in \mathcal{D}(\mathcal{O}_1), \quad y \in \mathcal{O}_2. \tag{4.2}$$

Indeed, for a fixed $y_0 \in \mathcal{O}_2$, $\varphi \to f_\varphi(y_0)$ is a linear functional on $\mathcal{D}(\mathcal{O}_1)$. Let us now prove that it is continuous. To do this, note that for all sufficiently large $k \geq N(y_0)$, the functional

$$\varphi \to \big(f_\varphi(y), \omega_{1/k}(y - y_0)\big),$$

where $\omega_{1/k}$ is the "cap" (see Sec. 1.2), belongs to $\mathcal{D}'(\mathcal{O}_2)$. But by virtue of (7.6), Sec. 1.7,

$$\big(f_\varphi(y), \omega_{1/k}(y - y_0)\big) \to \big(f_\varphi(y), \delta(y - y_0)\big) = f_\varphi(y_0), \qquad k \to \infty.$$

By the theorem on the completeness of the space $\mathcal{D}'(\mathcal{O}_1)$ (see Sec. 1.4) we conclude therefrom that the functional $f_\varphi(y_0)$ belongs to $\mathcal{D}'(\mathcal{O}_1)$. Denoting it by f_{y_0}, we obtain (4.2). □

Taking into account formula (4.1), we obtain, from (4.2),

$$\partial^\alpha(f_y, \varphi) = ((\partial_y^\alpha f)_y, \varphi), \qquad \varphi \in \mathcal{D}(\mathcal{O}_1), \quad y \in \mathcal{O}_2, \quad |\alpha| \le p. \tag{4.3}$$

EXAMPLE. Suppose a generalized function F does not depend on the variable y, $F(x,y) = f(x) \times 1(y)$, $f \in \mathcal{D}'(\mathcal{O}_1)$ (see Sec. 3.3). Then $f \in C^\infty(\mathbb{R}^m)$ with respect to y and $F_y(x) = f(x)$.

Recall that the class $C_0(\mathcal{O})$, which is defined in Sec. 0.5, consists of continuous functions with compact support in $\mathcal{O}$. We introduce convergence thus: $f_k \to 0$, $k \to \infty$ in $C_0(\mathcal{O})$ if

$$\operatorname{supp} f_k \subset \mathcal{O}' \Subset \mathcal{O} \quad \text{and} \quad f_k(x) \overset{x \in \mathcal{O}}{\rightrightarrows} 0, \quad k \to \infty.$$

LEMMA. *If $f(x,y) \in C(\mathcal{O}_2)$ with respect to y, then the operation*

$$\chi \to (f_y, \chi(x,y))$$

is continuous from $\mathcal{D}(\mathcal{O}_1 \times \mathcal{O}_2)$ into $C_0(\mathcal{O}_2)$.

PROOF. Let $\chi \in \mathcal{D}(\mathcal{O}_1 \times \mathcal{O}_2)$. Then $\operatorname{supp}\chi \subset \mathcal{O}_1' \times \mathcal{O}_2'$, where $\mathcal{O}_1' \Subset \mathcal{O}_1$ and $\mathcal{O}_2' \Subset \mathcal{O}_2$. Put $\psi(y) = (f_y, \chi(x,y))$. We have $\operatorname{supp}\psi \subset \mathcal{O}_2' \Subset \mathcal{O}_2$. We now prove that $\psi \in C(\mathcal{O}_2)$. Let y_0 be an arbitrary point in $\mathcal{O}_2$ and let $y_k \to y_0$, $k \to \infty$. Then

$$\begin{aligned}|\psi(y_k) - \psi(y_0)| \le& \Big|(f_{y_k}, \chi(x,y_0)) - (f_{y_0}, \chi(x,y_0))\Big| \\ &+ \Big|(f_{y_k}, \chi(x,y_k) - \chi(x,y_0))\Big|.\end{aligned} \tag{4.4}$$

The first summand on the right of (4.4) tends to 0 as $k \to \infty$ by virtue of the continuity of the function $(f_y, \chi(x,y_0))$, and the second summand, by the weak boundedness of the set $\{f_{y_k}\} \subset \mathcal{D}'(\mathcal{O}_1)$ and by virtue of the fact that

$$\chi(x,y_k) - \chi(x,y_0) \to 0, \quad k \to \infty \quad \text{in} \quad \mathcal{D}(\mathcal{O}_1)$$

via the lemma of Sec. 1.4. Thus $\psi \in C(\mathcal{O}_2)$ and for this reason $\psi \in C_0(\mathcal{O}_2)$.

Let $\chi_k \to 0$, $k \to \infty$ in $\mathcal{D}(\mathcal{O}_1 \times \mathcal{O}_2)$. Then $\operatorname{supp}\chi_k \subset \mathcal{O}_1' \times \mathcal{O}_2'$ where $\mathcal{O}_1' \Subset \mathcal{O}_1$ and $\mathcal{O}_2' \Subset \mathcal{O}_2$. Putting $\psi_k(y) = (f_y, \chi_k(x,y))$, we have $\operatorname{supp}\psi_k \subset \mathcal{O}_2' \Subset \mathcal{O}_2$. Furthermore, the set of generalized functions $\{f_y, y \in \mathcal{O}_2'\}$ in $\mathcal{D}'(\mathcal{O}_1)$ is weakly bounded. For this reason, by applying the inequality (3.1) of Sec. 1.3 (see also the corollary to the lemma of Sec. 1.4), for certain $K > 0$, $m \ge 0$, and for all $y \in \mathcal{O}_2$ we obtain

$$\begin{aligned}|\psi_k(y)| = \Big|(f_y, \chi_k(x,y))\Big| &\le K\|\chi_k(x,y)\|_{C^m(\bar{\mathcal{O}}_1')} \\ &\le K\|\chi_k\|_{C^m(\bar{\mathcal{O}}_1' \times \bar{\mathcal{O}}_2')} \to 0, \qquad k \to \infty.\end{aligned}$$

This completes the proof of the lemma. □

Let us now prove the following formula: *if $f(x,y) \in C(\mathcal{O}_2)$ with respect to y, then*

$$(f, \chi) = \int (f_y, \chi(x,y))\, dy, \qquad \chi \in \mathcal{D}(\mathcal{O}_1 \times \mathcal{O}_2) \tag{4.5}$$

i.e. $f(x,y) = f_y(x)$.

Indeed, by virtue of (4.2) the equality (4.5) holds true on the test functions χ of the form $\sum \varphi(x)\psi(y)$, where $\varphi \in \mathcal{D}(\mathcal{O}_1)$ and $\psi \in \mathcal{D}(\mathcal{O}_2)$:

$$\begin{aligned}\Big(f, \sum \varphi(x)\psi(y)\Big) &= \sum \int f_\varphi(y)\psi(y)\,dy \\ &= \sum \int (f_y, \varphi)\psi(y)\,dy \\ &= \int \Big(f_y, \sum \varphi(x)\psi(y)\Big)\,dy. \qquad (4.6)\end{aligned}$$

But the set of such functions is dense in $\mathcal{D}(\mathcal{O}_1 \times \mathcal{O}_2)$ (see Sec. 3.2.1) and, besides, by the lemma,

$$\Big(f_y, \sum \varphi(x)\psi(y)\Big) \to (f_y, \chi(x,y)) \quad \text{in} \quad C_0(\mathcal{O}_2)$$

if $\sum \varphi(x)\psi(y) \to \chi(x,y)$ in $\mathcal{D}(\mathcal{O}_1 \times \mathcal{O}_2)$. It is from this and from (4.6) that the formula (4.5) follows. □

4. The Convolution of Generalized Functions

4.1. The definition of convolution. Let f and g be locally integrable functions in $\mathbb{R}^n$. If the integral $\int f(y)g(x-y)\,dy$ exists for almost all $x \in \mathbb{R}^n$ and defines a locally integrable function in $\mathbb{R}^n$, then it is called the *convolution* of the functions f and g and is symbolized as $f * g$ so that

$$\begin{aligned}(f*g)(x) &= \int f(y)g(x-y)\,dy \\ &= \int g(y)f(x-y)\,dy = (g*f)(x). \qquad (1.1)\end{aligned}$$

We note two cases where the convolution $f * g$ definitely exists.

4.1.1. Let $f \in \mathcal{L}^1_{\text{loc}}$, $g \in \mathcal{L}^1_{\text{loc}}$, $\operatorname{supp} f \subset A$, $\operatorname{supp} g \subset B$, and the sets A and B are such that for any $R > 0$ the set

$$T_R = \big[(x,y) : x \in A,\ y \in B,\ |x+y| \le R\big]$$

is bounded in $\mathbb{R}^{2n}$ (Fig. 19). Then $f*g \in \mathcal{L}^1_{\text{loc}}$.

Indeed, using the Fubini theorem, we have, for all $R > 0$,

$$\begin{aligned}\int_{|x|<R} |f*g|\,dx &\le \iint_{|x|<R} |f(y)|\,|g(x-y)|\,dy\,dx \\ &\le \int_{T_R} |f(y)|\,|g(\xi)|\,dy\,d\xi < \infty.\end{aligned}$$

In particular, if f of g is with compact support, then T_R is bounded.

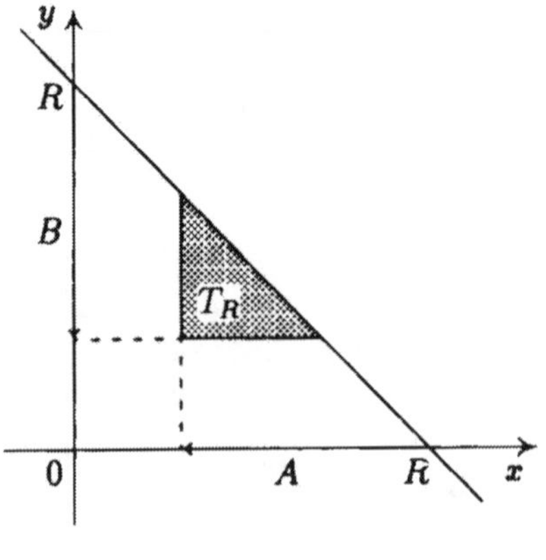

Figure 19

4.1.2. Let $f \in \mathcal{L}^p$ and $g \in \mathcal{L}^q$ if $\frac{1}{p}+\frac{1}{q} \geq 1$. Then $f*g \in \mathcal{L}^r$, where $\frac{1}{r} = \frac{1}{p}+\frac{1}{q}-1$. Indeed, choosing the numbers $\alpha \geq 0$, $\beta \geq 0$, $s \geq 1$ and $t \geq 1$ such that

$$\frac{1}{r} + \frac{1}{s} + \frac{1}{t} = 1, \qquad \alpha r = p = (1-\alpha)s, \qquad \beta r = q = (1-\beta)t,$$

and then

$$p + \frac{pr}{s} = r = q + \frac{qr}{t},$$

and making use of the Hölder inequality and the Fubini theorem, we obtain the required estimate

$$\begin{aligned}
\|f * g\|^r_{\mathcal{L}^r} &= \int \left| \int f(y) g(x-y)\, dy \right|^r dx \\
&\leq \int \left[\int |f(y)|^\alpha |g(x-y)|^\beta |f(y)|^{1-\alpha} |g(x-y)|^{1-\beta}\, dy \right]^r dx \\
&\leq \int \int |f(y)|^{\alpha r} |g(x-y)|^{\beta r}\, dy \left[\int |f(y)|^{(1-\alpha)s}\, dy \right]^{r/s} \\
&\quad \times \left[\int |g(x-y)|^{(1-\beta)t}\, dy \right]^{r/t} dx \\
&\leq \|f\|^r_{\mathcal{L}^p} \|g\|^r_{\mathcal{L}^q}.
\end{aligned}$$

i.e.

$$\|f * g\|_{\mathcal{L}^r} \leq \|f\|_{\mathcal{L}^p} \|g\|_{\mathcal{L}^q}.$$

This estimate is called the *Young inequality*.

The convolution $f * g$ defines a regular functional on $\mathcal{D}(\mathbb{R}^n)$ via the rule

$$\begin{aligned}
(f * g, \varphi) &= \int (f * g) \varphi(x)\, dx \\
&= \int \varphi(x) \int f(y) g(x-y)\, dy\, dx \\
&= \int f(y) \int g(x-y) \varphi(x)\, dx\, dy \\
&= \int f(y) \int g(\xi) \varphi(y+\xi)\, d\xi\, dy.
\end{aligned}$$

That is

$$(f * g, \varphi) = \int f(x) g(y) \varphi(x+y)\, dx\, dy, \qquad \varphi \in \mathcal{D}. \tag{1.2}$$

(In deriving (1.2) we made repeatedly use of the Fubini theorem.)

We will say that the sequence $\{\eta_k\}$ of functions taken from $\mathcal{D}(\mathbb{R}^n)$ *converges to 1 in* $\mathbb{R}^n$ if

(a) for any compact K there is a number $N = N(K)$ such that $\eta_k(x) = 1$, $x \in K$, $k \geq N$, and
(b) the functions $\{\eta_k\}$ are uniformly bounded together with all their derivatives, $|\partial^\alpha \eta_k(x)| < c_\alpha$, $x \in \mathbb{R}^n$, $k = 1, 2, \ldots$.

Note that there always exist such sequences, for example:

$$\eta_k(x) = \eta\left(\frac{x}{k}\right), \quad \text{where} \quad \eta \in \mathcal{D}, \quad \eta(x) \equiv 1, \quad |x| < 1.$$

Let us now prove that the equality (1.2) can be rewritten as

$$(f * g, \varphi) = \lim_{k\to\infty} \big(f(x) \times g(y), \eta_k(x;y)\varphi(x+y)\big), \qquad \varphi \in \mathcal{D}, \tag{1.3}$$

where $\{\eta_k\}$ is any sequence of functions taken from $\mathcal{D}(\mathbb{R}^{2n})$ that converges to 1 in $\mathbb{R}^{2n}$.

Indeed, the function $c_0|f(x)g(y)\varphi(x+y)|$ is integrable on $\mathbb{R}^{2n}$ and dominates the sequence of functions $f(x)g(y)\eta_k(x;y)\varphi(x+y)$, $k = 1, 2, \ldots$, that converges almost everywhere in $\mathbb{R}^{2n}$ to the function $f(x)g(y)\varphi(x+y)$. From this, making use of the Lebesgue theorem, we obtain

$$\int f(x)g(y)\varphi(x+y)\,dx\,dy = \lim_{k\to\infty} \int f(x)g(y)\eta_k(x;y)\varphi(x+y)\,dx\,dy$$

which is equivalent to (1.3) by virtue of (1.2). □

Proceeding from the equalities (1.3) and (1.2), we define a convolution of generalized functions as follows. Suppose f and g taken from $\mathcal{D}'(\mathbb{R}^n)$ are such that their direct product $f(x) \times g(y)$ admits of an extension $\big(f(x) \times g(y), \varphi(x+y)\big)$ to functions of the form $\varphi(x+y)$, where φ is any function in $\mathcal{D}(\mathbb{R}^n)$, in the following sense: no matter what sequence $\{\eta_k\}$ there is of functions from $\mathcal{D}(\mathbb{R}^{2n})$, which sequence converges to 1 in $\mathbb{R}^{2n}$, there exists a limit to the numerical sequence,

$$\lim_{k\to\infty} \big(f(x) \times g(y), \eta_k(x;y)\varphi(x+y)\big) = \big(f(x) \times g(y), \varphi(x+y)\big);$$

in fact that limit does not depend on the sequence $\{\eta_k\}$. Note that for every k the function $\eta_k(x;y)\varphi(x+y) \in \mathcal{D}(\mathbb{R}^{2n})$ and so our numerical sequence is defined.

The *convolution* $f * g$ is the functional

$$\begin{aligned}(f * g, \varphi) &= \big(f(x) \times g(y), \varphi(x+y)\big)\\ &= \lim_{k\to\infty} \big(f(x) \times g(y), \eta_k(x;y)\varphi(x+y)\big), \qquad \varphi \in \mathcal{D}(\mathbb{R}^n).\end{aligned} \tag{1.4}$$

Let us prove that the functional $f * g$ belongs to $\mathcal{D}'(\mathbb{R}^n)$, that is, it is a generalized function. For this purpose, it suffices, by virtue of the theorem on the completeness of the space $\mathcal{D}'$ (see Sec. 1.4), to establish the continuity of the linear functionals

$$\big(f(x) \times g(y), \eta_k(x;y)\varphi(x+y)\big), \qquad k = 1, 2, \ldots, \tag{1.5}$$

on $\mathcal{D}(\mathbb{R}^n)$. Let $\varphi_\nu \to 0$, $\nu \to \infty$ in $\mathcal{D}(\mathbb{R}^n)$. Then

$$\eta_k(x;y)\varphi_\nu(x+y) \to 0, \qquad \nu \to \infty \quad \text{in} \quad \mathcal{D}(\mathbb{R}^{2n})$$

since $\eta_k \in \mathcal{D}(\mathbb{R}^{2n})$. From this, since the functional $f(x) \times g(y)$ on $\mathcal{D}(\mathbb{R}^{2n})$ (see Sec. 3.1) is continuous, we obtain

$$\big(f(x) \times g(y), \eta_k(x;y)\varphi_\nu(x+y)\big) \to 0, \qquad \nu \to \infty$$

and this completes the proof of the continuity of the functionals (1.5) on $\mathcal{D}(\mathbb{R}^n)$. □

Note that since $\varphi(x+y)$ does not belong to $\mathcal{D}(\mathbb{R}^{2n})$ (it is not with compact support in $\mathbb{R}^{2n}$), the right-hand side of (1.4) does not exist for any pairs of generalized functions f and g and, thus, the convolution does not always exist.

The convolution of any number of generalized functions is defined in similar fashion. For example, let f, g and h be generalized functions taken from $\mathcal{D}'(\mathbb{R}^n)$

and let $\{\eta_k\}$ be the sequence of functionals from $\mathcal{D}(\mathbb{R}^{3n})$ that converges to 1 in $\mathbb{R}^{3n}$. The convolution $f * g * h$ is the functional

$$\begin{aligned}(f * g * h, \varphi) &= \big(f(x) \times g(y) \times h(z), \varphi(x+y+z)\big)\\ &= \lim_{k\to\infty} \big(f(x) \times g(y) \times h(z), \eta_k(x;y;z)\varphi(x+y+z)\big), \qquad \varphi \in \mathcal{D}(\mathbb{R}^n).\end{aligned} \tag{1.6}$$

if that functional exists

In applications, another definition of the convolution of generalized functions, which is equivalent to the definition above, is useful (see Kaminski [54]). Let $f, g \in \mathcal{D}'$. By their convolution we call the limit

$$f * g = \lim_{k\to\infty} (f\eta_k) * g \quad \text{in} \quad \mathcal{D}', \tag{1.7}$$

if this limit exists for any sequence $\{\eta_k\}$ converging to 1 in $\mathbb{R}^n$. (In this case this limit does not depend on the sequence $\{\eta_k\}$.)

Other definitions of the convolution can be found in Schwartz [89], Hirata, Ogata [45], Shiraishi [93], Mikusinski [76], Dierolf, Voigt [16], Kaminski [54].

4.2. The properties of a convolution.

4.2.1. *Commutativity of convolution.* If the convolution $f * g$ exists, then so also does the convolution $g * f$, and they are equal:

$$f * g = g * f.$$

This statement follows from the definition of a convolution (see Sec. 4.1) and from the commutativity of a direct product (see Sec. 3.2.1):

$$\begin{aligned}(f * g, \varphi) &= \lim_{k\to\infty} \big(f(x) \times g(y), \eta_k(x;y)\varphi(x+y)\big)\\ &= \lim_{k\to\infty} \big(g(y) \times f(x), \eta_k(x;y)\varphi(x+y)\big)\\ &= (g * f, \varphi), \qquad \varphi \in \mathcal{D}.\end{aligned}$$

Similarly, from the definition (1.6) we obtain

$$f * g * h = f * h * g = h * f * g = \dots \qquad \text{and so forth.}$$

4.2.2. *Convolution with the delta-function.* The convolution of any generalized function f in $\mathcal{D}'$ with the δ-function exists and is equal to f:

$$f * \delta = \delta * f = f. \tag{2.1}$$

True enough, let $\varphi \in \mathcal{D}(\mathbb{R}^n)$ and let $\{\eta_k\}$ be a sequence of functions taken from $\mathcal{D}(\mathbb{R}^{2n})$ that converges to 1 in $\mathbb{R}^{2n}$. Then

$$\eta_k(x;0)\varphi(x) \to \varphi, \qquad k \to \infty \quad \text{in} \quad \mathcal{D}(\mathbb{R}^n)$$

and so

$$\begin{aligned}(f * \delta, \varphi) &= \lim_{k\to\infty} \big(f(x) \times \delta(y), \eta_k(x;y)\varphi(x+y)\big)\\ &= \lim_{k\to\infty} \big(f(x), \eta_k(x;0)\varphi(x)\big)\\ &= (f, \varphi),\end{aligned}$$

which is what we set out to prove. □

REMARK. The meaning of the formula $f = f*\delta$ is that any generalized function f may be expressed in terms of δ-functions, which, formally, is often written thus:

$$f(x) = \int f(\xi)\delta(x-\xi)\,d\xi.$$

It is precisely this formula which one has in mind when we say that every material body consists of mass points, every source consists of source points, and so on (compare Sec. 1.1).

4.2.3. *The shift of convolution.* If the convolution $f*g$ exists, then so also does the convolution $f(x+h)*g(x)$ for all $h \in \mathbb{R}^n$, and

$$f(x+h)*g(x) = (f*g)(x+h). \tag{2.2}$$

That is, the operations of shift and convolution commute; in other words, the convolution operator

$$f \to f*g$$

is a translation invariant operator.

Indeed, let $\{\eta_k\}$ be a sequence of functions in $\mathcal{D}(\mathbb{R}^{2n})$ that converges to 1 in $\mathbb{R}^{2n}$. Then for any $h \in \mathbb{R}^n$,

$$\eta_k(x-h;y) \to 1, \qquad k\to\infty \quad \text{in} \quad \mathbb{R}^{2n}.$$

Now, using the definition of a shift (see Sec. 1.9) and of a convolution (see Sec. 4.1), we obtain, for all $\varphi \in \mathcal{D}(\mathbb{R}^n)$,

$$\begin{aligned}((f*g)(x+h),\varphi) &= (f*g,\varphi(x-h)) \\ &= \lim_{k\to\infty}\left(f(x)\times g(y),\eta_k(x-h;y)\varphi(x-h+y)\right) \\ &= \lim_{k\to\infty}\left(f(x+h)\times g(y),\eta_k(x;y)\varphi(x+y)\right) \\ &= \left(f(x+h)*g,\varphi\right),\end{aligned}$$

which is what we set out to prove. □

Here we made use of formula (2.9) of Sec. 3.2 for the shift of a direct product.

4.2.4. *The reflection of convolution.* If the convolution $f*g$ exists, then so also does the convolution $f(-x)*g(-x)$, and

$$f(-x)*g(-x) = (f*g)(-x). \tag{2.3}$$

The proof is similar to that of 4.2.3.

4.2.5. *Differentiating of convolution.* If the convolution $f*g$ exists, then there exist the convolutions $\partial^\alpha f*g$ and $f*\partial^\alpha g$, and we have

$$\partial^\alpha f*g = \partial^\alpha(f*g) = f*\partial^\alpha g. \tag{2.4}$$

It will suffice to prove this assertion for the first derivatives ∂_j, $j=1,\dots,n$. Let $\varphi \in \mathcal{D}(\mathbb{R}^n)$ and let $\{\eta_k\}$ be a sequence of functions taken from $\mathcal{D}(\mathbb{R}^{2n})$ that converges to 1 in $\mathbb{R}^{2n}$. Then the sequence $\{\eta_k + \partial_j\eta_k\}$ also converges to 1 in $\mathbb{R}^{2n}$. From this fact, taking advantage of the existence of the convolution $f*g$ (see

Sec. 4.1), we obtain the following chain of equalities (with $\eta_k \equiv \eta_k(x,y)$):

$$\begin{aligned}
(\partial_j(f * g), \varphi) &= -(f * g, \partial_j \varphi) \\
&= -\lim_{k\to\infty}\left(f(x) \times g(y), \eta_k \frac{\partial \varphi(x+y)}{\partial x_j}\right) \\
&= -\lim_{k\to\infty}\left(f(x) \times g(y), \frac{\partial}{\partial x_j}[\eta_k \varphi(x+y)] - \varphi(x+y)\frac{\partial \eta_k}{\partial x_j}\right) \\
&= \lim_{k\to\infty}\left(\frac{\partial}{\partial x_j}[f(x) \times g(y)], \eta_k \varphi(x+y)\right) \\
&\quad + \lim_{k\to\infty}\left(f(x) \times g(y), \left(\eta_k + \frac{\partial \eta_k}{\partial x_j}\right)\varphi(x+y)\right) \\
&\quad - \lim_{k\to\infty}(f(x) \times g(y), \eta_k \varphi(x+y)) \\
&= \lim_{k\to\infty}(\partial_j f(x) \times g(y), \eta_k \varphi(x+y)) + f * g, \varphi) - (f * g, \varphi) \\
&= (\partial_j f * g, \varphi),
\end{aligned}$$

whence follows the first equality (2.4) for ∂_j. The second one of (2.4) follows from the first one and from the commutativity of a convolution:

$$\partial_j(f * g) = \partial_j(g * f) = \partial_j g * f = f * \partial_j g.$$

From (2.1) and (2.4) there follow the equalities

$$\partial^\alpha f = \partial^\alpha \delta * f = \delta * \partial^\alpha f, \qquad f \in \mathcal{D}'.$$

Note that the existence of the convolutions $\partial^\alpha f * g$ and $f * \partial^\alpha g$ for $|\alpha| \geq 1$ is not yet enough for the existence of a convolution $f * g$ and for the truth of (2.4). For example,

$$\theta' * 1 = \delta * 1 = 1, \quad \text{but} \quad \theta * 1' = \theta * 0 = 0.$$

4.2.6. *The operation $f \to f*g$ is linear on the set of those generalized functions f for which the convolution with g exists.*

This property of a convolution follows directly from the definition of a convolution (1.4) and from the linearity of the operation $f \to f \times g$ (see Sec. 3.2.3).

In passing we may note that the operation $f \to f * g$ is not, generally speaking, continuous from $\mathcal{D}'$ into $\mathcal{D}'$, as the following example shows:

$$\delta(x-k) \to 0, \qquad k \to \infty \quad \text{in} \quad \mathcal{D}'; \quad \text{however,} \quad 1 * \delta(x-k) = 1.$$

4.2.7. *If the convolution $f * g$ exists, then*

$$\operatorname{supp}(f * g) \subset \overline{\operatorname{supp} f + \operatorname{supp} g}. \tag{2.5}$$

Indeed, suppose $\{\eta_k\}$ is a sequence of functions taken from $\mathcal{D}(\mathbb{R}^{2n})$ that converges to 1 in $\mathbb{R}^{2n}$ and $\varphi \in \mathcal{D}(\mathbb{R}^n)$ is such that

$$\operatorname{supp} \varphi \cap \overline{\operatorname{supp} f + \operatorname{supp} g} = \varnothing. \tag{2.6}$$

Since $\operatorname{supp}(f \times g) = \operatorname{supp} f \times \operatorname{supp} g$ (see Sec. 3.2.4), we conclude that

$$\begin{aligned}
&\operatorname{supp}[f(x) \times g(y)] \cap \operatorname{supp}[\eta_k(x;y)\varphi(x+y)] \\
&\qquad \subset [\operatorname{supp} f \times \operatorname{supp} g] \cap [(x,y) : x + y \in \operatorname{supp} \varphi] = \varnothing.
\end{aligned}$$

And so, due to Sec. 1.5(a), we have

$$(f * g, \varphi) = \lim_{k\to\infty} \big(f(x) \times g(y), \eta_k(x; y)\varphi(x + y)\big) = 0$$

for all test functions φ in $\mathcal{D}(\mathbb{R}^n)$ that satisfy the condition (2.6). And this means that the inclusion (2.5) holds. □

REMARK. The set $\operatorname{supp} f + \operatorname{supp} g$ may also be not closed. Generally speaking, there is no equality in the inclusion (2.5). For example, for the convolution $\delta' * \theta = \delta$ it takes the form $\{0\} \subset \{x \geq 0\}$.

4.2.8. *Associativity of convolution.* Generally, the operation of convolution is not associative; for example,

$$(1 * \delta') * \theta = 1' * \theta = 0 * \theta = 0, \quad \text{but} \quad 1 * (\delta' * \theta) = 1 * \delta = 1.$$

However, this unpleasantness does not arise if the convolution $f * g * h$ exists. To be more precise, the following assertion holds true.

If the convolutions $f * g$ and $f * g * h$ exist, then so also does the convolution $(f * g) * h$, and we have

$$(f * g) * h = f * g * h. \tag{2.7}$$

Indeed, suppose $\{\eta_k\}$ and $\{\xi_k\}$ are sequences of functions from $\mathcal{D}(\mathbb{R}^{2n})$ that converge to 1 in $\mathbb{R}^{2n}$. Then the sequence

$$\eta_i(x; y)\xi_k(x + y; z), \qquad i \to \infty, \quad k \to \infty,$$

of functions taken from $\mathcal{D}(\mathbb{R}^{3n})$ converges to 1 in $\mathbb{R}^{3n}$. From this fact and from the existence of the convolution $f * g * h$ (see Sec. 4.1) there follows that existence of a double limit:

$$\lim_{\substack{i\to\infty\\ k\to\infty}} \big(f(x) \times g(y) \times h(z), \eta_i(x; y)\xi_k(x + y; z)\varphi(x + y + z)\big) = (f * g * h, \varphi),$$
$$\varphi \in \mathcal{D}(\mathbb{R}^n),$$

and, consequently, of the repeated limit

$$\begin{aligned}
(f * g * h, \varphi) &= \lim_{k\to\infty}\lim_{i\to\infty} \big(f(x) \times g(y) \times h(z), \eta_i(x; y)\xi_k(x + y; z)\varphi(x + y + z)\big) \\
&= \lim_{k\to\infty}\lim_{i\to\infty} \big(f(x) \times g(y), \eta_i(x; y)\big(h(z), \xi_k(x + y; z)\varphi(x + y + z)\big)\big) \\
&= \lim_{k\to\infty} \big((f * g)(t), \big(h(z), \xi_k(t; z)\varphi(t + z)\big)\big) \\
&= \lim_{k\to\infty} \big((f * g)(t) \times h(z), \xi_k(t; z)\varphi(t + z)\big) \\
&= \big((f * g) * h, \varphi\big),
\end{aligned}$$

which proves (2.7) and the existence of the convolution $(f * g) * h$ (see Sec. 4.1). □

COROLLARY. *If there exist convolutions $f * g * h$, $f * g$, $g * h$ and $f * h$, then there exist convolutions $(f * g) * h$, $f * (g * h)$ and $(f * h) * g$ and we have*

$$f * g * h = (f * g) * h = f * (g * h) = (f * h) * g,$$

which in this case means the convolution is associative.

4.3. The existence of a convolution. Let us establish certain sufficient conditions (besides those given in Sec. 4.1), under which a convolution definitely exists in $\mathcal{D}'$. Recall (see Fig. 19) that

$$T_R = \left[(x,y) : x \in A,\ y \in B,\ |x+y| \le R\right].$$

For the definition of the space $\mathcal{D}'(A)$ see Sec. 1.5.

THEOREM. *Let $f \in \mathcal{D}'(A)$, $g \in \mathcal{D}'(B)$ and suppose that for any $R > 0$ the set T_R is bounded in $\mathbb{R}^{2n}$. Then the convolution $f * g$ exists in $\mathcal{D}'(\overline{A+B})$ and may be represented as*

$$(f * g, \varphi) = \big(f(x) \times g(y), \xi(x)\eta(y)\varphi(x+y)\big), \qquad \varphi \in \mathcal{D} \tag{3.1}$$

where ξ and η are any functions in C^∞ that are equal to 1 in A^ε and B^ε and are equal to 0 outside $A^{2\varepsilon}$ and $B^{2\varepsilon}$ respectively (ε is any positive number). Here the operation $f \to f * g$ is continuous from $\mathcal{D}'(A)$ into $\mathcal{D}'(\overline{A+B})$.

PROOF. Let $\varphi \in \mathcal{D}(U_R)$ and let $\{\eta_k\}$ be a sequence of functions in $\mathcal{D}(\mathbb{R}^{2n})$ that converges to 1 in $\mathbb{R}^{2n}$. Since

$$\operatorname{supp}(f \times g) = \operatorname{supp} f \times \operatorname{supp} g \subset A \times B$$

(see Sec. 3.2.4), it follows that

$$\operatorname{supp}\Big\{\big[f(x) \times g(y)\big]\varphi(x+y)\Big\} \subset \big[(x,y) : x \in A,\ y \in B\ |x+y| \le R\big] = T_R.$$

And since T_R is a bounded set, there is a number $N = N(R)$ such that $\eta_k(x;y) = 1$ in the neighbourhood of T_R for all $k \ge N$. For this reason,

$$\begin{aligned}(f * g, \varphi) &= \lim_{k\to\infty} \big(f(x) \times g(y), \eta_k(x;y)\varphi(x+y)\big)\\ &= \lim_{k\to\infty} \big([f(x) \times g(y)]\varphi(x+y), \eta_k(x;y)\big)\\ &= \big(f(x) \times g(y), \eta_N(x;y)\varphi(x+y)\big)\end{aligned}$$

and the representation

$$(f * g, \varphi) = \big(f(x) \times g(y), \eta_N(x;y)\varphi(x+y)\big), \qquad \varphi \in \mathcal{D}(U_R), \tag{3.2}$$

is proved. Clearly, the representation (3.2) does not depend on the auxiliary function $\eta_N(x;y)$. It can be replaced by the function $\xi(x)\eta(y)$. Indeed, the function $\xi(x)\eta(y)\varphi(x+y) \in \mathcal{D}(\mathbb{R}^{2n})$, since the set

$$T_{R,\varepsilon} = \left[(x,y) : x \in A^{2\varepsilon},\ y \in B^{2\varepsilon}, |x+y| \le R\right] \subset T_R^{4\varepsilon}$$

for any $R > 0$ and $\varepsilon > 0$ is bounded in $\mathbb{R}^{2n}$; furthermore, the function

$$\big[\eta_N(x;y) - \xi(x)\eta(y)\big]\varphi(x+y), \qquad \varphi \in \mathcal{D}(U_R),$$

vanishes in a neighbourhood of T_R. This completes the proof of the representation (3.1).

From (2.5) it follows that

$$\operatorname{supp}(f * g) \subset \overline{A+B},$$

so that the operation $f \to f * g$ carries $\mathcal{D}'(A)$ into $\mathcal{D}'(\overline{A+B})$. Its continuity follows from the continuity of the direct product $f \times g$ with respect to f (see Sec. 3.2.3) and from the representation (3.1): if $f_k \to 0$, $k \to \infty$ in $\mathcal{D}'(A)$, then

$$(f_k * g, \varphi) = \big(f_k(x) \times g(y), \xi(x)\eta(y)\varphi(x+y)\big) \to 0$$

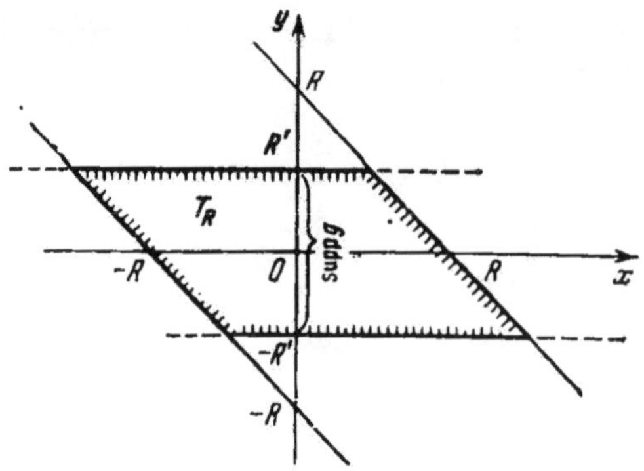

Figure 20

for all $\varphi \in \mathcal{D}$, that is, $f_k * g \to 0$, $k \to \infty$ in $\mathcal{D}'(\overline{A+B})$. The proof of the theorem is complete. □

Note that the continuity of the convolution $f * g$ relative to the collection of f and g may not occur, as the following simple example illustrates:

$$\delta(x+k) \to 0, \quad k \to +\infty, \qquad \delta(x-k) \to 0, \quad k \to +\infty.$$

However,

$$\delta(x+k) * \delta(x-k) = \delta * \delta = \delta.$$

We note here an important special case of the theorem just proved.

If $f \in \mathcal{D}'$ and $g \in \mathcal{E}'$, then the convolution $f * g$ exists and can be represented in the form

$$(f * g, \varphi) = \big(f(x) \times g(y), \eta(y)\varphi(x+y)\big), \qquad \varphi \in \mathcal{D}, \tag{3.3}$$

where η is any test function taken from $\mathcal{D}$ that is equal to 1 in a neighbourhood of the support of g.

Indeed, in this case the boundedness condition of the set T_R is fulfilled for all $R > 0$ (Fig. 20): if $\operatorname{supp} g \subset \overline{U}_{R'}$, then

$$T_R = \big[(x,y) : x \in R^n,\ y \in \operatorname{supp} g,\ |x+y| \le R\big] \subset \overline{U}_{R+R'} \times \overline{U}_{R'}.$$

Similarly, *if $f \in \mathcal{D}'$ and $g_1, \ldots, g_m \in \mathcal{E}'$, then there exists a convolution $f * g_1 * \cdots * g_m$ (see Sec. 4.1) that is associative and commutative (see Sec. 4.2.1, 4.2.8) and the formula* (3.3) *is generalized thus:*

$$(f * g_1 * \cdots * g_m, \varphi) = \big(f(x) \times g_1(y) \times \cdots \times g_m(z),$$
$$\eta_1(y) \ldots \eta_m(z)\varphi(x+y+\cdots+z)\big), \qquad \varphi \in \mathcal{D}. \tag{3.4}$$

But if $f \in C^\infty$ and $g \in \mathcal{E}'$, then the convolution $f * g \in C^\infty$, and the formula (3.3) *takes on the form*

$$(f * g)(x) = \big(\tilde{g}(y), f(x-y)\big), \tag{3.5}$$

where $\tilde{g}$ is the extension of g onto $C^\infty = C^\infty(\mathbb{R}^n)$ (see Sec. 2.5).

True enough, as in the proof of the lemma of Sec. 3.1, it is established that the function

$$(\tilde{g}(y), f(x-y)) = (g(y), \eta(y) f(x-y)) \in C^\infty.$$

Then, from the representation (3.3) we have, for all $\varphi \in \mathcal{D}$,

$$(f * g, \varphi) = \left(g(y), \eta(y) \int f(\xi)\varphi(\xi + y)\, d\xi\right)$$
$$= \left(g(y), \int \eta(y) f(x-y)\varphi(x)\, dx\right).$$

Noticing now that $\eta(y)f(x-y)\varphi(x) \in \mathcal{D}(\mathbb{R}^{2n})$ and using the formula (3.2) of Sec. 3, we obtain

$$(f * g, \varphi) = \int (g(y), \eta(y) f(x-y))\varphi(x)\, dx,$$

whence follows formula (3.5).

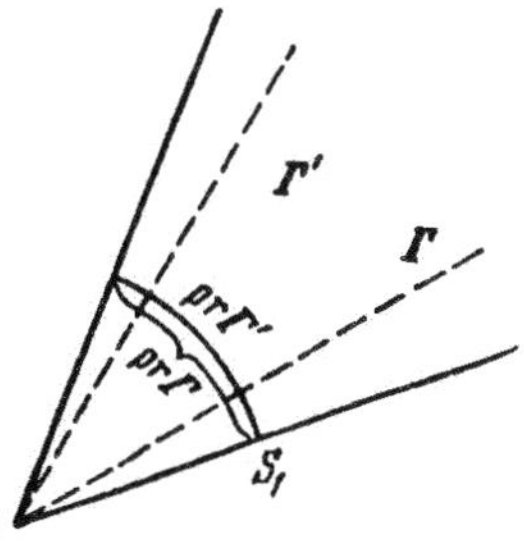

Figure 21

In similar fashion, *if* $f \in C^\infty(\mathbb{R}^n \setminus \{0\})$ *and* $g \in \mathcal{E}'$, *then the convolution* $f * g$ *in* $\mathbb{R}^n \setminus \operatorname{supp} g$ *is expressed by the formula* (3.5); *in particular,* $f * g \in C^\infty(\mathbb{R}^n \setminus \operatorname{supp} g)$.

4.4. Cones in $\mathbb{R}^n$. A *cone* in $\mathbb{R}^n$ (with vertex at 0) is a set Γ with the property that if $x \in \Gamma$, then λx too belongs to Γ for all $\lambda > 0$. Denote by $\operatorname{pr}\Gamma$ the intersection of Γ and the unit sphere with centre at 0 (Fig. 21). The cone Γ' is said to be *compact* in the cone Γ if $\operatorname{pr}\bar{\Gamma}' \subset \operatorname{pr}\Gamma$ (Fig. 21); we then write $\Gamma' \Subset \Gamma$.

The cone

$$\Gamma^* = [\xi : (\xi, x) \ge 0,\ \forall x \in \Gamma]$$

is said to be *conjugate* to the cone Γ. Clearly, Γ^* is a closed convex cone with vertex at 0 (Fig. 22) and $(\Gamma^*)^* = \overline{\operatorname{ch}\Gamma}$; here $\operatorname{ch}\Gamma$ is the convex hull of Γ (see Sec. 0.2).

A cone Γ is said to be *acute* if there exists a plane of support for $\overline{\operatorname{ch}\Gamma}$ that has a unique common point with $\overline{\operatorname{ch}\Gamma}$ (Fig. 22).

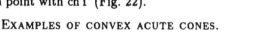

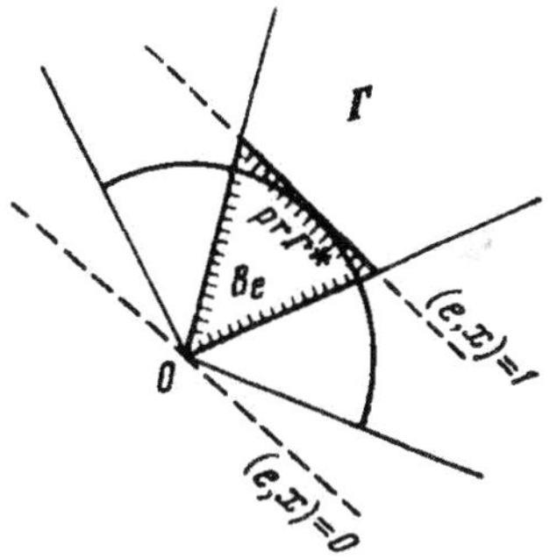

Figure 22

EXAMPLES OF CONVEX ACUTE CONES.

(a) an n-hedral cone in $\mathbb{R}^n$:

$$C = [x : (e_1, x) > 0, \ldots, (e_n, x) > 0]$$

is acute (convex and open) if and only if the vectors $e_1, \ldots, e_n$ form a basis in $\mathbb{R}^n$. Then

$$C^* = \left[\xi : \xi = \sum_{1 \le k \le n} \lambda_k e_k,\ \lambda_k \ge 0\right].$$

In particular, the positive quadrant

$$\mathbb{R}^n_+ = [x : x_1 > 0, \ldots, x_n > 0], \qquad (\mathbb{R}^n_+)^* = \overline{\mathbb{R}^n_+}.$$

(b) The future light cone in $\mathbb{R}^{n+1}$.

$$V^+ = [x : (x_0, x) : x_0 > |\mathbf{x}|], \qquad (V^+)^* = \bar{V}^+.$$

(c) The origin of coordinates $\{0\}$, $\{0\}^* = \mathbb{R}^n$.

Note, however, that the cone $\mathbb{R}^1_+ \times \mathbb{R}^{n-1} = [x : x_1 > 0]$ is not acute.

(d) The cone $P_n \subset \mathbb{R}^{n^2}$ of positive (Hermitian) $n \times n$ matrices $X = (x_{pq})$, $P_n^* = P_n^+$ is the cone of nonnegative matrices. This follows from the assertion that, for $X \in P_n$, it is necessary and sufficient that for all $\Xi \in P_n^+$, $\Xi \neq 0$,

$$(X, \Xi) = \operatorname{Tr}(X\Xi) = \sum_{p,q} x_{pq}\xi_{qp} > 0.$$

Lemma 1. *The following statements are equivalent:*

(1) *the cone* Γ *is acute;*

(2) *the cone* $\overline{\operatorname{ch}\Gamma}$ *does not contain an integral straight line;*

(3) $\operatorname{int}\Gamma^* \neq \varnothing$;

(4) *for any* $C' \Subset \operatorname{int}\Gamma^*$ *there exists a number* $\sigma = \sigma(C') > 0$ *such that*

$$(\xi, x) \geq \sigma|\xi|\,|x|, \qquad \xi \in C', \qquad x \in \overline{\operatorname{ch}\Gamma}; \tag{4.1}$$

(5) *for any* $e \in \operatorname{pr}\operatorname{int}\Gamma^*$ *the set*

$$B_e[x : 0 \leq (e, x) \leq 1, \ x \in \overline{\operatorname{ch}\Gamma}]$$

is bounded in $\mathbb{R}^n$ *(Fig. 22).*

Proof. (1) $\to$ (2). If the cone $\overline{\operatorname{ch}\Gamma}$ contains an integral straight line $x = x^0 + te$, $-\infty < t < \infty$ $(|e| = 1)$, then it also contains the straight line $x = te$, $-\infty < t < \infty$. Consequently, any plane of support for $\overline{\operatorname{ch}\Gamma}$ must contain that straight line, but this contradicts (1).

(2) $\to$ (3). If $\operatorname{int}\Gamma^* = \varnothing$, then, since Γ^* is a convex cone with vertex at 0, it lies in some $(n-1)$-dimensional plane $(e, x) = 0$ $(|e| = 1)$. For this reason, $\pm e \in \Gamma^{**} = \overline{\operatorname{ch}\Gamma}$. But then the integral straight line $y = te$, $-\infty < t < \infty$ too lies in $\overline{\operatorname{ch}\Gamma}$, but this contradicts (2).

(3) $\to$ (4). Since all points of the cone C' different from 0 are interior points relative to Γ^*, it follows that $(\xi, x) > 0$ for all $\xi \in C'$ and $x \in \overline{\operatorname{ch}\Gamma}$. From this fact and also from the continuity and the homogeneity of the form (ξ, x) follows the existence of a number $\sigma > 0$ for which the inequality (4.1) holds true.

(4) $\to$ (5). Let us take an arbitrary $e \in \operatorname{pr}\operatorname{int}\Gamma^*$. Then, by applying the inequality (4.1), $(e, x) \geq \sigma|x|$, $x \in \overline{\operatorname{ch}\Gamma}$, we conclude that the set B_e is bounded: $|x| \leq \frac{(e,x)}{\sigma} \leq \frac{1}{\sigma}$.

(5) $\to$ (1). If for some $e \in \operatorname{pr}\operatorname{int}\Gamma^*$ the set B_e is bounded, then the plane $(e, x) = 0$ cannot have any other points in common with $\overline{\operatorname{ch}\Gamma}$, with the exception of 0. □

Lemma 2. *Let* Γ *be a convex cone. Then* $\Gamma = \Gamma + \Gamma$.

Proof. The inclusion $\Gamma \subset \Gamma + \Gamma$ is obvious. Let $x \in \Gamma + \Gamma$ so that $x = y + z$, where $y \in \Gamma$ and $z \in \Gamma$. Then for all $\lambda \in (0, 1)$ we have $x = \lambda\frac{y}{\lambda} + (1-\lambda)\frac{z}{1-\lambda} \in \Gamma$ and for this reason $\Gamma + \Gamma \subset \Gamma$, thus completing the proof of the lemma. □

The *indicator* of the cone Γ is the function

$$\mu_\Gamma(\xi) = -\inf_{x \in \operatorname{pr}\Gamma}(\xi, x).$$

From the definition of the indicator it follows that $\mu_\Gamma(\xi)$ is a convex (see Sec. 0.2) and, hence, continuous (see, for example, Vladimirov [**105**, Chapter II]) and homogeneous first-degree function defined on the whole of $\mathbb{R}^n$. Besides,

$$\mu_\Gamma(\xi) \le \mu_{\operatorname{ch}\Gamma}(\xi),$$
$$\mu_\Gamma(\xi) = -\Delta(\xi, \partial\Gamma^*), \qquad \xi \in \Gamma^*$$

and $\mu_\Gamma(\xi) > 0$ for $\xi \notin \Gamma^*$. Thus

$$\Gamma^* = [\xi : \mu_\Gamma(\xi) \le 0]$$

so that the indicator of a cone fully defines only the closure of its convex hull, by virtue of $\overline{\operatorname{ch}\Gamma} = \Gamma^{**}$.

EXAMPLE.

$$\mu_{V^+}(\xi) = \begin{cases} \dfrac{1}{\sqrt{2}}(|\xi| - \xi_0), & \xi \notin -V^+, \\ |\xi|, & \xi \in -V^+. \end{cases}$$

LEMMA 3. *If Γ is a convex cone, then for any $a \ge 0$*

$$[\xi : \mu_\Gamma(\xi) \le a] = \Gamma^* + \overline{U}_a. \tag{4.2}$$

PROOF. The inclusion

$$\Gamma^* + \overline{U}_a \subset [\xi : \mu_\Gamma(\xi) \le a] \tag{4.3}$$

is trivial: if $\xi = \xi_1 + \xi_2$, $\xi_1 \in \Gamma^*$, $|\xi_2| \le a$, then

$$\mu_\Gamma(\xi) = -\inf_{x \in \operatorname{pr}\Gamma}(\xi, x) = -\inf_{x \in \operatorname{pr}\Gamma}[(\xi_1, x) + (\xi_2, x)] \le -\inf_{x \in \operatorname{pr}\Gamma}(\xi_2, x) \le a,$$

since $(\xi_1, x) \ge 0$, $x \in \Gamma$.

Now let us prove the inverse inclusion of (4.3). Let the point ξ_0 be such that $\mu_\Gamma(\xi_0) \le a$. If $\xi_0 \in \Gamma^*$ or $|\xi_0| \le a$, then $\xi_0 \in \Gamma^* + \overline{U}_a$. Now let $\xi_0 \notin \Gamma^*$ and $|\xi_0| > a$. Let the point $\xi_1 \in \Gamma^*$ realize the distance from ξ_0 to Γ^*, $\Delta(\xi_0, \Gamma^*) = |\xi_0 - \xi_1|$. Then, since Γ^* is a convex cone (Fig. 23), it follows that

(a) $(\xi_1 - \xi_0, \xi) \ge 0, \qquad \xi \in \Gamma^*;$

(b) $(\xi_1 - \xi_0, \xi_1) = 0.$

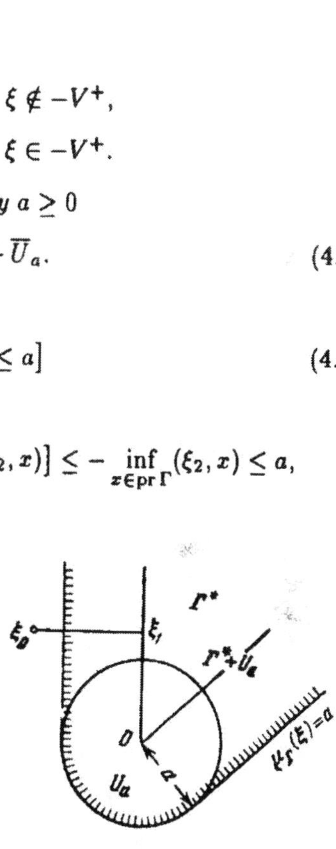

Figure 23

From the inequality (a) it follows that $\xi_1 - \xi_0 \in \Gamma^{**} = \bar{\Gamma}$ and therefore

$$a \ge \mu_\Gamma(\xi_0) = -\inf_{x \in \operatorname{pr}\Gamma}(\xi_0, x) \ge -\left(\xi_0, \frac{\xi_1 - \xi_0}{|\xi_1 - \xi_0|}\right).$$

Now the latter is equivalent, by virtue of (b), to the inequality $|\xi_1 - \xi_0| \le a$. Thus, the point $\xi_0 = \xi_1 + (\xi_0 - \xi_1)$ is represented in the form of a sum of two terms $\xi_1 \in \Gamma^*$ and $\xi_0 - \xi_1 \in \overline{U}_a$ that is, $\xi_0 \in \Gamma^* + \overline{U}_a$. This completes the proof of the inverse inclusion of (4.3) and also the equality (4.2). The proof of Lemma 3 is complete. □

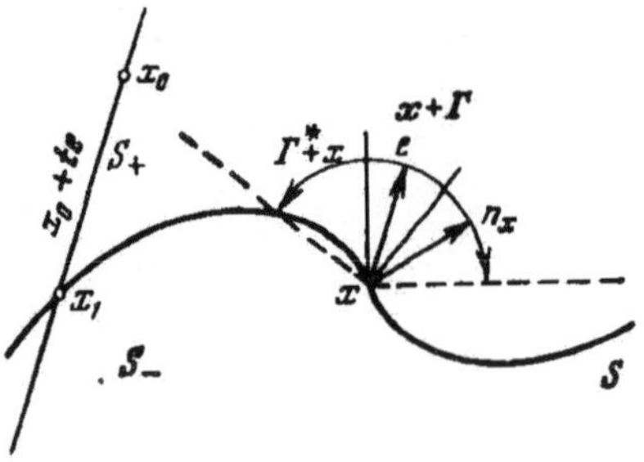

Figure 24

Suppose Γ is a closed convex acute cone. Set $C = \operatorname{int}\Gamma^*$ (via Lemma 1, $C \neq \varnothing$). The smooth $(n-1)$-dimensional surface S without an edge is said to be *C-like* if each straight line $x = x_0 + te$, $-\infty < t < \infty$, $e \in \operatorname{pr}\Gamma$, intersects it in a unique point; in other words: for any $x \in S$ the cone $\Gamma + x$ intersects S in a unique point x (Fig. 24). Thus, the C-like surface S cuts $\mathbb{R}^n$ into two infinite regions S_+ and S_-: S_+ lies above S and S_- lies below S; $S_+ \cup S_- \cup S = \mathbb{R}^n$. At every point x of the surface S, the normal $\mathbf{n}_x$ is contained in the cone $\Gamma^* + x$ (Fig. 24).

EXAMPLE. The surface S in $\mathbb{R}^{n+1}$, which surface is specified by the equation

$$x_0 = f(\mathbf{x}), \qquad |\nabla f(\mathbf{x})| \leq \sigma < 1, \qquad \mathbf{x} \in \mathbb{R}^n, \qquad f \in C^1,$$

is V^+-like (*space-like*).

LEMMA 4. *If S is a C-like surface, than*

$$\overline{S}_+ = S + \Gamma. \tag{4.4}$$

PROOF. Suppose $x_0 \in \overline{S}_+$. The straight line $x = x_0 + te$, $|t| < \infty$, $e \in \operatorname{pr}\Gamma$, intersects S at some point $x_1 = x_0 - t_1 e$, $t_1 \geq 0$ (Fig. 24) so that $x_0 = x_1 + t_1 e$, $x_1 \in S$, $t_1 e \in \Gamma$, and the inclusion $\overline{S}_+ \subset S + \Gamma$ is proved. Clearly, the inclusion $S + \Gamma \subset \overline{S}_+$ is true; together with the inverse inclusion it leads to the equality (4.4), thus completing the proof of the lemma. □

LEMMA 5. *Let S be a C-like surface. Then for any $R > 0$ there is a number $R'(R) > 0$ such that the set*

$$T_R = \big[(x, y) : x \in S,\ y \in \Gamma,\ |x + y| \leq R\big]$$

is contained in the ball $U_{R'} \subset \mathbb{R}^{2n}$.

PROOF. Since S is a C-like surface, it follows that any point $x \in S$ that can be represented as $\xi - y$, $y \in \Gamma$, $|\xi| \leq R$, is of the form $x = \xi - eT$, $e \in \operatorname{pr}\Gamma$, where the number $T = T(e, \xi)$ is uniquely determined by e and ξ and constitutes a continuous function of the argument (e, ξ) on the compact $e \in \operatorname{pr}\Gamma$, $|\xi| \leq R$. Hence the set $\big[(y, \xi) : y = eT(e, \xi),\ e \in \operatorname{pr}\Gamma,\ |\xi| \leq R\big]$ is bounded and so also is the set T_R. The lemma is proved. □

We will say that a C-like surface S is a *strictly C-like* surface if, under the conditions of Lemma 5,

$$R'(R) \leq a(1 + R)^{\nu}, \qquad \nu \geq 1, \quad a > 0. \tag{4.5}$$

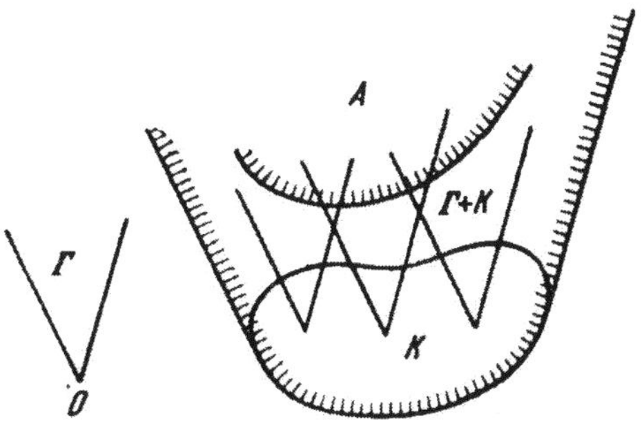

Figure 25

EXAMPLE. The plane $(e, x) = 0$, $e \in \operatorname{pr} C$, is strictly C-like with $\nu = 1$ (by virtue of Lemma 1).

4.5. Convolution algebras $\mathcal{D}'(\Gamma+)$ and $\mathcal{D}'(\Gamma)$. We will say that a set A is *bounded on the side of the cone* Γ if $A \subset \Gamma + K$, where K is a certain compact (Fig. 25). It is clear that the sets bounded on the side of the cone $\{0\}$ are compacts in $\mathbb{R}^n$.

Suppose Γ is a closed cone in $\mathbb{R}^n$. The collection of generalized functions in $\mathcal{D}'$ whose supports are bounded on the side of the cone Γ will be denoted by $\mathcal{D}'(\Gamma+)$. We define *convergence* in $\mathcal{D}'(\Gamma+)$ in the following manner: $f_k \to 0$, $k \to \infty$ in $\mathcal{D}'(\Gamma+)$, if $f_k \to 0$, $k \to \infty$ in $\mathcal{D}'$, and $\operatorname{supp} f_k \subset \Gamma + K$, where the compact K does not depend on k.[2] Set $\mathcal{D}'(\{0\}+) = \mathcal{E}'$; $\mathcal{E}'$ is the space of generalized functions with compact supports (compare Sec. 2.5).

Let Γ be a closed convex acute cone, $C = \operatorname{int} \Gamma^*$, S a C-like surface, and S_+ the region lying above S (see Sec. 4.4).

If $f \in \mathcal{D}'(\Gamma+)$ and $g \in \mathcal{D}'(\overline{S}_+)$, then the convolution $f * g$ exists in $\mathcal{D}'$ and can be represented as

$$(f * g, \varphi) = \big(f(x) \times g(y), \xi(x)\eta(y)\varphi(x+y)\big), \qquad \varphi \in \mathcal{D}, \tag{5.1}$$

where ξ and η are any functions in C^∞ that are equal to 1 in $(\operatorname{supp} f)^\epsilon$ and $(\operatorname{supp} g)^\epsilon$ and are equal to 0 outside $(\operatorname{supp} f)^{2\epsilon}$ and $(\operatorname{supp} g)^{2\epsilon}$, respectively ($\epsilon$ is any positive number). Here, if $\operatorname{supp} f \subset \Gamma + K$, where K is a compact, then $\operatorname{supp}(f * g) \subset \overline{S}_+ + K$ and the corresponding operations $f \to f * g$ and $g \to f * g$ are continuous.

This assertion follows from the theorem of Sec. 4.3 for $A = \Gamma + K$ and $B = \overline{S}_+$ if we note that by Lemmas 2, 4 and 5 of Sec. 4.4 the set

$$\begin{aligned} T_R &= \big[(x,y) : x \in \Gamma + K,\ y \in \overline{S}_+,\ |x+y| \le R\big] \\ &= \big[(x,y) : x \in \Gamma + K,\ y \in S + \Gamma,\ |x+y| \le R\big] \end{aligned}$$

is bounded for all $R > 0$ and

$$\overline{\Gamma + K + \overline{S}_+} = \overline{\Gamma + K + \Gamma + S} = \overline{\Gamma + S + K} = \overline{S}_+ + K.$$

□

[2] A similar meaning will be attached to other spaces of generalized functions as well; for example, $\mathcal{S}'(\Gamma+)$, $\mathcal{L}_s^2(\Gamma+)$ and so forth (see Secs. 5 and 7 below).

We now note an important special case of the last criterion for the existence of a convolution.

THEOREM. *Let* Γ *be a closed convex acute cone. If* $f \in \mathcal{D}'(\Gamma+)$ *and* $g \in \mathcal{D}'(\Gamma+)$, *then the convolution* $f*g$ *exists in* $\mathcal{D}'(\Gamma+)$ *and can be represented as* (5.1); *here, the operation* $f \to f*g$ *is continuous from* $\mathcal{D}'(\Gamma+)$ *into* $\mathcal{D}'(\Gamma+)$.

PROOF. Since $\Gamma + K$, where K is a compact, is contained in $\overline{S}_+$ for some C-like surface (which depends on K), it follows, by the preceding criterion, that the convolution $f*g$ exists in $\mathcal{D}'$ and can be represented by the formula (5.1). Let us prove that $f*g \in \mathcal{D}'(\Gamma+)$. Suppose $\operatorname{supp} f \subset \Gamma + K_1$ and $\operatorname{supp} g \subset \Gamma + K_2$, where K_1 and K_2 are certain compacts in $\mathbb{R}^n$. Then, using the inclusion (2.5) and Lemma 2 of Sec. 4.4, we obtain

$$\operatorname{supp}(f*g) \subset \overline{\Gamma + K_1 + \Gamma + K_2} = \Gamma + K_1 + K_2$$

so that $f*g \in \mathcal{D}'(\Gamma+)$. The continuity of the operation $f \to f*g$ from $\mathcal{D}'(\Gamma+)$ into $\mathcal{D}'(\Gamma+)$ also follows from this inclusion. The proof of the theorem is complete. □

In similar fashion we can prove that the convolution of any number of generalized functions taken from $\mathcal{D}'(\Gamma+)$ (see Sec. 4.1) exists in $\mathcal{D}'(\Gamma+)$ and can be expressed by a formula similar to (5.1).

From this and from the results of Sec. 4.2.8 it follows that the convolution of generalized functions taken from $\mathcal{D}'(\Gamma+)$ is *associative*.

A linear set is termed an *algebra* if the operation of multiplication is defined on it, and the operation is linear with respect to every factor separately. An algebra is said to be *associative* if $x(yz) = (xy)z$ and *commutative* if $xy = yx$.

The results established in this subsection enables us to assert that the set of generalized functions $\mathcal{D}'(\Gamma+)$ forms an algebra that is associative and commutative if for the operation of multiplication we take the convolution operation $*$. Such algebras are called *convolution algebras*; the unit element here is the δ-function (see Sec. 4.2.2).

Finally, note that the set of generalized functions $\mathcal{D}'(\Gamma)$ also forms a convolution algebra, a subalgebra of the algebra $\mathcal{D}'(\Gamma+)$.

Indeed, if $f \in \mathcal{D}'(\Gamma)$ and $g \in \mathcal{D}'(\Gamma)$, then

$$\operatorname{supp}(f*g) \subset \overline{\operatorname{supp} f + \operatorname{supp} g} \subset \Gamma + \Gamma = \Gamma$$

so that $f*g \in \mathcal{D}'(\Gamma)$. (Here, we again took advantage of the inclusion (2.5) and Lemma 2 of Sec. 4.4.)

4.6. Mean functions of generalized functions. Let us extend the concept of a convolution $f*g$ when f and g are generalized functions taken from $\mathcal{D}'(\mathcal{O})$ and g is with a compact and sufficiently small support in $\mathcal{O}$: $\operatorname{supp} g \subset U_\varepsilon$ and $\mathcal{O}_\varepsilon \neq \varnothing$ (see Sec. 0.2 and Fig. 26). In accordance with formula (3.3) we set, by definition

$$(f*g,\varphi) = \big(f(x)\times g(y), \eta(y)\varphi(x+y)\big), \qquad \varphi \in \mathcal{D}(\mathcal{O}_\varepsilon), \tag{6.1}$$

where $\eta \in \mathcal{D}(\mathcal{O}_\varepsilon)$, $\eta(y) = 1$ in a neighbourhood of $\operatorname{supp} g$.

By construction, the operation $\varphi \to \eta(y)\varphi(x+y)$ is linear and continuous from $\mathcal{D}(\mathcal{O}_\varepsilon)$ into $\mathcal{D}(\mathcal{O}\times U_\varepsilon)$. From this it follows that the right-hand side of (6.1) defines a continuous linear functional on $\mathcal{D}(\mathcal{O}_\varepsilon)$ so that $f*g \in \mathcal{D}'(\mathcal{O}_\varepsilon)$. Furthermore, it is easy to see (compare Sec. 4.3) that the right-hand side of (6.1) is not dependent on the auxiliary function η. Finally, as in Sec. 4.2, it can be established that the convolution $f * g$ is commutative and continuous with respect to f and g separately, and $f * \delta = f$.

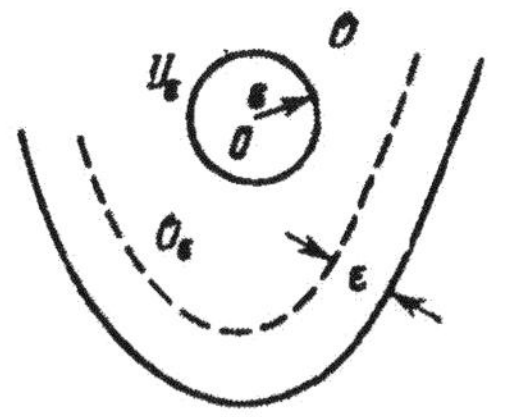

Figure 26

In particular, if $\alpha \in \mathcal{D}(U_\varepsilon)$, then, using the representation (6.1) and acting in a manner similar to that of Sec. 4.3 when deriving (3.5) we obtain a representation for the convolution $f * \alpha$;

$$(f * \alpha)(x) = (f(y), \alpha(x-y)), \qquad x \in \mathcal{O}_\varepsilon, \tag{6.2}$$

whence follows $f * \alpha \in C^\infty(\mathcal{O}_\varepsilon)$ and

$$(f * \alpha)(0) = (f(y), \alpha(-y)) = (\delta, f * \alpha). \tag{6.3}$$

By virtue of (6.2), the formula (6.1) takes the form

$$(f * g, \varphi) = (f, g(-y) * \varphi), \qquad \varphi \in \mathcal{D}(\mathcal{O}_\varepsilon). \tag{6.4}$$

Note that when $\mathcal{O} = \mathbb{R}^n$, formula (6.4) also follows from (3.3) and (6.2).

Let $\omega_\varepsilon(x)$ be the "cap" (see Sec. 1.2) and let f be a generalized function in $\mathcal{D}'(\mathcal{O})$. The convolution

$$f_\varepsilon(x) = (f * \omega_\varepsilon)(x) = (f(y), \omega_\varepsilon(x-y))$$

is termed a *mean function of* f (compare this with the definition of a mean function for the case of $f \in \mathcal{L}^1_{\text{loc}}(\mathcal{O})$, see Sec. 1.2). By what has been proved, the mean function $f_\varepsilon \in C^\infty(\mathcal{O}_\varepsilon)$.

Now let us prove that

$$f_\varepsilon \to f, \qquad \varepsilon \to +0 \quad \text{in} \quad \mathcal{D}'(\mathcal{O}). \tag{6.5}$$

True enough, the limiting relation (6.5) follows from the relation $\omega_\varepsilon(x) \to \delta(x)$, $\varepsilon \to +0$ in $\mathcal{D}'$, (see Sec. 1.7) and from the continuity of a convolution, by virtue of

$$f_\varepsilon = f * \omega_\varepsilon \to f * \delta = f, \qquad \varepsilon \to +0 \quad \text{in} \quad \mathcal{D}'(\mathcal{O}).$$

To summarize: *every generalized function taken from $\mathcal{D}'(\mathcal{O})$ is a weak limit of its mean function.*

Let us use this statement and establish a stronger result.

THEOREM. *Every generalized function f in $\mathcal{D}'(\mathcal{O})$ is a weak limit of the test functions in $\mathcal{D}(\mathcal{O})$, that is, $\mathcal{D}(\mathcal{O})$ is dense in $\mathcal{D}'(\mathcal{O})$.*

PROOF. Let $f_\varepsilon(x)$ be a mean function of f. Furthermore, let $\mathcal{O}_1 \Subset \mathcal{O}_2 \Subset \dots$, $\bigcup_k \mathcal{O}_k = \mathcal{O}$, $\varepsilon_k = \Delta(\mathcal{O}_k, \partial\mathcal{O}) > 0$ and $\eta_k \in \mathcal{D}(\mathcal{O}_k)$, $\eta_k(x) = 1$, $x \in \mathcal{O}_{k-1}$. We will prove that the sequence $\eta_k(x) f_{\varepsilon_k}(x)$, $k = 1, 2, \dots$, of test functions taken from $\mathcal{D}(\mathcal{O})$ converges to f in $\mathcal{D}'(\mathcal{O})$. Indeed, $\varepsilon_k \to 0$, as $k \to \infty$ and by (6.5) for all $\varphi \in \mathcal{D}(\mathcal{O})$ we have

$$\lim_{k\to\infty} (\eta_k f_{\varepsilon_k}, \varphi) = \lim_{k\to\infty} (f_{\varepsilon_k}, \eta_k \varphi) = \lim_{k\to\infty} (f_{\varepsilon_k}, \varphi) = (f, \varphi),$$

which completes the proof. □

REMARK. From the completeness of the space $\mathcal{D}'(\mathcal{O})$ (see Sec. 1.4) there follows a converse statement to the theorem that has just been proved: any weak limit of locally integrable functions in $\mathcal{O}$ is a generalized function in $\mathcal{D}'(\mathcal{O})$. Therefore, it is possible to construct a theory of generalized functions by proceeding from weakly convergent sequences of ordinary, locally integrable functions. With regard to this approach, see Antosik, Mikusinski and Sikorski [2].

It is appropriate at this point to mention the following analogy. The relation of generalized functions to test functions is reminiscent, in a certain sense, of the relation of irrational numbers to rational numbers: by completing the set of rational numbers by means of all possible limits of sequences of rational numbers, we obtain real numbers; by completing the set of test functions by all weak limits of sequences of test functions, we obtain generalized functions.

4.7. Multiplication of generalized functions. In order to give a formal definition of the product of generalized functions, we introduce the following definition. Let $\eta \in \mathcal{D}$, $\int \eta(t)\,dt = 1$ and $\lambda_k \to \infty$, $k \to \infty$, $\lambda_k > 0$. The sequence $\delta_k(x) = \lambda_k^n \eta(x\lambda_k)$, $k \to \infty$, of functions from $\mathcal{D}$ is called a *special δ-sequence.*

Obviously, $\delta_k(x) \to \delta(x)$, $k \to \infty$ in $\mathcal{D}'$ (compare (7.6) in Sec. 1.7).

Let $f \in \mathcal{D}'(\mathcal{O})$. The convolution (see (6.2) in Sec. 4.6)

$$f_k(x) = f * \delta_k = \big(f(y), \delta_x(x - y)\big) \in C(\mathcal{O}_k), \qquad \mathcal{O}_k \subset \mathcal{O}$$

is called the *mean function* for f and (see (6.5) in Sec. 4.6)

$$f_k(x) \to f(x), \qquad k \to \infty \quad \text{in} \quad \mathcal{D}'(\mathcal{O}). \tag{7.1}$$

By the *product* $f \cdot g$ and $g \cdot f$ of generalized functions f and g from $\mathcal{D}'(\mathcal{O})$ we call the limits, respectively,

$$(f \cdot g, \varphi) = \lim_{k\to\infty} \big((f * \delta_k)g, \varphi\big), \qquad (g \cdot f, \varphi) = \lim_{k\to\infty} \big((g * \delta_k)f, \varphi\big), \tag{7.2}$$

if these limits exist for any $\varphi \in \mathcal{D}(\mathcal{O})$ and for any special δ-sequences $\{\delta_k\}$ (and do not depend on $\{\delta_k\}$). By virtue of the completeness of the space $\mathcal{D}'(\mathcal{O})$ (see Sec. 1.4), $f \cdot g$ and $g \cdot f \in \mathcal{D}'(\mathcal{O})$.

If $f \cdot g$ exists, then $g \cdot f$ also exists and they are equal,

$$f \cdot g = g \cdot f, \tag{7.3}$$

i.e., the product is *commutative* (see Itano [52]).

Note, in particular, that if f, g and fg are locally integrable functions in $\mathcal{O}$, then $f \cdot g = fg$. This fact follows from Theorem I of Sec. 1.2. If $f \in \mathcal{D}'$ and $a \in C^\infty$, then $af = a \cdot f$.

There exist other, more general, definitions of products of generalized functions (see Schwartz [89], Shiraishi, Itano [94], Vladimirov [105], Itano [52], Mikusinski [76], Kaminski [54]).

4.8. Convolution as a continuous linear translation-invariant operator. An operator L acting from $\mathcal{D}'$ to $\mathcal{D}'$ is said to be *translation-invariant* if $Lf(x+h) = (Lf)(x+h)$ for all $f \in \mathcal{D}'$ and for all translations $h \in \mathbb{R}^n$.

Recall that the definition of convergence in the space $C^\infty = C^\infty(\mathbb{R}^n)$ is given in Sec. 2.5 and in the space $\mathcal{E}'$ in Sec. 4.5; $\mathcal{E}'$ is a collection of continuous linear functionals on C^∞ (see Sec. 2.5).

THEOREM. *For an operator L to be linear, continuous and translation-invariant from $\mathcal{E}'$ to $\mathcal{D}'$, it is necessary and sufficient that it be a convolution operator, that is to say, that it be representable in the form $L = f_0 *$, where f_0 is some generalized function taken from $\mathcal{D}'$; then f_0, the kernel of the operator L, is unique and is expressed by the formula $f_0 = L\delta$.*

PROOF. Sufficiency follows from the results of Sec. 4.3 and Sec. 4.2, according to which the convolution operator $f \to f_0 * f$, $f_0 \in \mathcal{D}'$, is linear, continuous and translation-invariant from $\mathcal{E}'$ to $\mathcal{D}'$, and $f_0 * \delta = f_0$. To prove necessity let us first establish the following lemma.

LEMMA. *For an operator L to be linear, continuous and translation-invariant from $\mathcal{D}$ to C^∞, it is necessary and sufficient that it be a convolution operator $L = f_0 *$, $f_0 \in \mathcal{D}'$; here, the kernel f_0 is unique.*

PROOF. To prove sufficiency, it remains to establish the continuity of the operation

$$\varphi \to f_0 * \varphi = \big(f_0(y), \varphi(x-y)\big)$$

(see (6.2)) from $\mathcal{D}$ to C^∞. But this follows from the inequality (see the theorem of Sec. 1.3)

$$\big|\partial^\alpha(f_0 * \varphi)(x)\big| = \big|\big(f_0(y), \partial^\alpha \varphi(x-y)\big)\big| \le K\|\varphi\|_{C^{m+|\alpha|}}, \tag{8.1}$$

which holds true for all $\varphi \in \mathcal{D}(U_R)$ and $|x| \le R_1$ (the numbers K and m in (8.1) depend on R and R_1).

Necessity. From the assumed conditions it follows that the functional $(L\varphi)(0)$ is linear and continuous on $\mathcal{D}$. For this reason there exists an (obviously) unique generalized function $f_0 \in \mathcal{D}'$ such that $L\varphi)(0) = \big(f_0(-x), \varphi\big)$. From this, by the property of translational invariance of the operator L, for all $x_0 \in \mathbb{R}^n$ we derive

$$\begin{aligned}\big(L\varphi(x+x_0)\big)(0) &= (L\varphi)(x_0) = \big(f_0(-x), \varphi(x+x_0)\big) \\ &= \big(f_0(x), \varphi(x_0 - x)\big) = (f_0 * \varphi)(x_0),\end{aligned}$$

thus completing the proof of the lemma. □

PROOF OF NECESSITY OF THE HYPOTHESIS OF THE THEOREM. The operator $L_1 = L - L\delta *$ is linear, continuous and translation-invariant from $\mathcal{E}'$ to $\mathcal{D}'$ (see proof of sufficiency). Besides, for all $x_0 \in \mathbb{R}^n$ we have

$$\begin{aligned}L_1\delta(x+x_0) &= (L\delta - L\delta * \delta)(x+x_0) \\ &= (L\delta - L\delta)(x+x_0) = 0\end{aligned}$$

so that L_1 vanishes on all translations of the δ-function. Now let φ be an arbitrary test function in $\mathcal{D}$. Then

$$\frac{1}{N^n} \sum_{0 \le |k| \le N} \varphi\left(\frac{k}{N}\right)\delta\left(x - \frac{k}{N}\right) \to \varphi(x), \qquad N \to \infty \quad \text{in} \quad \mathcal{E}',$$

because for any $\psi \in C^\infty$

$$\frac{1}{N^n} \sum_{0 \le |k| \le N} \varphi\left(\frac{k}{N}\right)\psi\left(\frac{k}{N}\right) \to \int \varphi(x)\psi(x)\,dx, \qquad N \to \infty.$$

Therefore, by virtue of the linearity and the continuity from $\mathcal{E}'$ to $\mathcal{D}'$ of the operator L_1,

$$L_1\varphi = \lim_{N\to\infty}\left[\frac{1}{N^n} \sum_{0 \le |k| \le N} \varphi\left(\frac{k}{N}\right)\delta\left(x - \frac{k}{N}\right)\right] = 0, \qquad \varphi \in \mathcal{D}.$$

Now let f be any generalized function in $\mathcal{E}'$. There exists a sequence $\{f_k\}$ of functions in $\mathcal{D}$ that converges to f in $\mathcal{E}'$ (see Sec. 4.6). From this fact and from the continuity from $\mathcal{E}'$ to $\mathcal{D}'$ of the operator L_1 we conclude that $L_1 f = \lim_{k\to\infty} L_1 f_k = 0$ for all $f \in \mathcal{E}'$ so that $L_1 = 0$ and, hence, $L = L\delta * = f_0 *$.

The uniqueness of the kernel f_0 of the operator L stems from the following reasoning: if $f_1 \in \mathcal{D}'$ is such that $f_1 * f = 0$ for all $f \in \mathcal{E}'$ and, hence, for all $f \in \mathcal{D}$, then, by the above-proved lemma, $f_1 = 0$. This completes the proof of the theorem. □

4.9. Some applications.

4.9.1. *Newtonian potential.* Let $f \in \mathcal{D}'$. The convolutions

$$V_n = \frac{1}{|x|^{n-1}} * f, \quad n \geq 3; \qquad V_2 = \ln\frac{1}{|x|} * f, \quad n = 2$$

(if they exist) are called the *Newtonian* (for $n = 2$, the *logarithmic*) *potential* with density f.

The potential V_n satisfies the Poisson equation

$$\Delta V_n = -(n-2)\sigma_n f, \qquad n \geq 3; \qquad \Delta V_2 = -2\pi f.$$

Indeed, using the formula (4.2) of Sec. 2.4 and (2.4), we obtain, for $n \geq 3$,

$$\Delta V_n = \Delta\left(\frac{1}{|x|^{n-1}} * f\right) = \Delta\frac{1}{|x|^{n-2}} * f$$
$$= -(n-2)\sigma_n \delta * f = -(n-2)\sigma_n f.$$

We proceed in similar fashion in the case of $n = 2$ as well.

If $f = \rho(x)$ is a function with compact support integrable on $\mathbb{R}^n$, $n \geq 3$, then the corresponding Newtonian potential V_n is called the *volume potential*. In this case, V_n is a locally integrable function in $\mathbb{R}^n$ and is given by the integral

$$V_n(x) = \int \frac{\rho(y)\,dy}{|x-y|^{n-2}} \tag{9.1}$$

in accordance with formula (1.1) for the convolution of a function $\rho(x)$ with compact support integrable in $\mathbb{R}^n$ with the function $|x|^{-n+2}$ locally integrable in $\mathbb{R}^n$.

Let $f = \mu\delta_n$ and $f = -\frac{\partial}{\partial \mathbf{n}}(\nu\delta_S)$ be a simple layer and a double layer on a piecewise-smooth surface $S \subset \mathbb{R}^n$, $n \geq 3$, with surface densities μ and ν (see Secs. 1.7 and 2.3). The corresponding Newtonian potentials

$$V_n^{(0)} = \frac{1}{|x|^{n-2}} * \mu\delta_S,$$
$$V_n^{(1)} = -\frac{1}{|x|^{n-2}} * \frac{\partial}{\partial \mathbf{n}}(\nu\delta_S)$$

are, respectively, the *surface potentials of a simple layer* and *a double layer* with densities μ and ν.

If S is a bounded surface, then the surface potentials $V_n^{(0)}$ and $V_n^{(1)}$ are locally integrable functions in $\mathbb{R}^n$ and can be represented by the integrals

$$V_n^{(0)}(x) = \int_S \frac{\mu(y)}{|x-y|^{n-2}}\,dS_y,$$
$$V_n^{(1)}(x) = \int_S \nu(y)\frac{\partial}{\partial \mathbf{n}_y}\frac{1}{|x-y|^{n-2}}\,dS_y. \tag{9.2}$$

For the sake of definiteness, let us prove the representation (9.2) for the potential $V_n^{(1)}$. Using the representation (3.3) and the definition of a double layer (see

Sec. 2.3), for all $\varphi \in \mathcal{D}$ we obtain a chain of equalities (the function $\eta \in \mathcal{D}$ and $\eta(x) \equiv 1$ in a neighbourhood of S):

$$\begin{aligned}
\left(V_n^{(1)}, \varphi\right) &= -\left(\frac{1}{|x|^{n-2}} * \frac{\partial}{\partial \mathbf{n}}(\nu\delta_S), \varphi\right) \\
&= -\left(\frac{1}{|\xi|^{n-2}} \times \frac{\partial}{\partial \mathbf{n}}(\nu\delta_S)(y), \eta(y)\varphi(y+\xi)\right) \\
&= -\left(\frac{\partial}{\partial \mathbf{n}}(\nu\delta_S), \eta(y)\int \frac{\varphi(y+\xi)}{|\xi|^{n-2}}\, d\xi\right) \\
&= \int_S \nu(y)\frac{\partial}{\partial \mathbf{n}}\left[\eta(y)\int \frac{\varphi(y+\xi)}{|\xi|^{n-2}}\, d\xi\right] dS_y \\
&= \int_S \nu(y)\frac{\partial}{\partial \mathbf{n}}\int \frac{\varphi(x)}{|x-y|^{n-2}}\, dx\, dS_y \\
&= \int_S \nu(y)\int \varphi(x)\frac{\partial}{\partial \mathbf{n}_y}\frac{1}{|x-y|^{n-2}}\, dx\, dS_y \\
&= \int \varphi(x)\int_S \nu(y)\frac{\partial}{\partial \mathbf{n}_y}\frac{1}{|x-y|^{n-2}}\, dS_y\, dx,
\end{aligned}$$

whence follows the required formula (9.2) for $V_n^{(1)}$. The change in the order of integration is ensured by the Fubini theorem, by virtue of the existence of the iterated integral

$$\int_S |\nu(y)| \int |\varphi(x)| \left|\frac{\partial}{\partial \mathbf{n}_y}\frac{1}{|x-y|^{n-2}}\right| dx\, dS_y.$$

4.9.2. *Green's formula.* Let the domain $G \subset \mathbb{R}^n$, $n \geq 3$, be bounded by a piecewise-smooth boundary S and let the function $\mu \in C^2(G) \cap C^1(\bar{G})$. Then it can be represented in the form of a sum of three Newtonian potentials via *Green's formula* ($\mathbf{n}$ is an outer normal to S):

$$-\frac{1}{(n-2)\sigma_n}\left\{\int_G \frac{\Delta u(y)}{|x-y|^{n-2}}\, dy - \int_S \left[\frac{1}{|x-y|^{n-2}}\frac{\partial u(y)}{\partial \mathbf{n}} - u(y)\frac{\partial}{\partial \mathbf{n}_y}\frac{1}{|x-y|^{n-2}}\right] dS_y\right\}$$
$$= \begin{cases} u(x), & x \in G, \\ 0, & x \notin \bar{G}. \end{cases} \qquad (9.3)$$

Indeed, assuming that the function $u(x)$ has been continued by zero for $x \notin \bar{G}$ and taking advantage of the formula (3.7′) and (3.10) of Sec. 2.3, we conclude that

$$\begin{aligned}
u = \delta * u &= -\frac{1}{(n-2)\sigma_n}\Delta\frac{1}{|x|^{n-2}} * u \\
&= -\frac{1}{(n-2)\sigma_n|x|^{n-2}} * \Delta u \\
&= -\frac{1}{(n-2)\sigma_n|x|^{n-2}} * \left[\Delta_{\mathrm{cl}} u - \frac{\partial u}{\partial n}\delta_S - \frac{\partial}{\partial n}(u\delta_S)\right].
\end{aligned}$$

Whence, using (9.1) and (9.3), we convince ourselves that the representation (9.3) holds true. □

In particular, if the function $u(x)$ is harmonic in the region G, then the representation (9.3) transforms into the *Green's formula for harmonic functions:*

$$\frac{1}{(n-2)\sigma_n}\int\limits_S\left[\frac{1}{|x-y|^{n-2}}\frac{\partial u(y)}{\partial \mathbf{n}} - u(y)\frac{\partial}{\partial \mathbf{n}_y}\frac{1}{|x-y|^{n-2}}\right] dS_y = \begin{cases} u(x), & x \in G, \\ 0, & x \notin \bar{G}. \end{cases} \quad (9.4)$$

Formulas similar to (9.3) and (9.4) occur in the case of $n = 2$ as well. In this case the fundamental solution $-\frac{1}{(n-2)\sigma_n|x|^{n-2}}$ must be replaced by $\frac{1}{2\pi}\ln|x|$.

REMARK. Green's formula (9.4) expresses the values of the harmonic function in the domain in terms of its values and the values of its normal derivative on the boundary of that domain. In that sense, it is similar to Cauchy's formula for analytic functions.

4.9.3. A *convolution equation* has the form

$$a * u = f \quad (9.5)$$

where a and f are specified generalized functions in $\mathcal{D}'$ and u is an unknown generalized function in $\mathcal{D}'$. Convolution equations involve all linear partial differential equations with constant coefficients:

$$a(x) = \sum_{|\alpha|\le m} a_\alpha \partial^\alpha \delta(x),$$

$$a * u = \sum_{|\alpha|\le m} a_\alpha \partial^\alpha u(x);$$

linear difference equations:

$$a(x) = \sum_\alpha a_\alpha \delta(x - x_\alpha),$$

$$a * u = \sum_\alpha a_\alpha u(x - x_\alpha);$$

linear integral equations of the first kind:

$$a \in \mathcal{L}^1_{\text{loc}}, \qquad a * u = \int u(y)a(x-y)\,dy;$$

linear integral equations of the second kind:

$$a = \delta + \mathcal{K}, \qquad \mathcal{K} \in \mathcal{L}^1_{\text{loc}},$$

$$a * u = u(x) + \int u(y)\mathcal{K}(x-y)\,dy;$$

linear integro-differential equations; and so forth.

The *fundamental solution* of the convolution operator $a*$ is a generalized function $\mathcal{E} \in \mathcal{D}'$ that satisfies the equation (9.5) for $f = \delta$,

$$a * \mathcal{E} = \delta. \quad (9.6)$$

Generally speaking, the fundamental solution $\mathcal{E}$ is not unique; it is determined up to the summand $\mathcal{E}_0$, which is an arbitrary solution in $\mathcal{D}'$ of the homogeneous equation $a * \mathcal{E}_0 = 0$. Indeed,

$$a * (\mathcal{E} + \mathcal{E}_0) = a * \mathcal{E} + a * \mathcal{E}_0 = \delta.$$

EXAMPLES. (1) The function $\mathcal{E}_n(x)$ defined in Sec. 2.3.8 is a fundamental solution of the Laplace operator: $\Delta\mathcal{E}_n = \delta$.

(2) The formula $\theta(x) + C$ yields the general form of the fundamental solution in $\mathcal{D}'$ of the operator $\frac{d}{dx} = \delta' *$ (see Sec. 2.2 and Sec. 2.3.3).

Let the fundamental solution $\mathcal{E}$ of the operator $a*$ in $\mathcal{D}'$ exist. We denote by $A(a, \mathcal{E})$ the collection of those generalized functions f taken from $\mathcal{D}'$ for which the convolutions $\mathcal{E} * f$ and $a * \mathcal{E} * f$ exist in $\mathcal{D}'$.

The following theorem holds.

THEOREM. *Suppose $f \in A(a, \mathcal{E})$. Then the solution u of the equation* (9.5) *exists and can be expressed by the formula*

$$u = \mathcal{E} * f. \tag{9.7}$$

The solution of (9.5) *is unique in the class $A(a, \mathcal{E})$.*

PROOF. The generalized function $u = \mathcal{E} * f$ satisfies (9.5) since, by virtue of the commutativity and the associativity of a convolution (see Sec. 4.2.8) (the convolutions $\mathcal{E} * f$, $a * \mathcal{E} * f$ and $a * \mathcal{E} = \delta$ exist):

$$a * u = a * (\mathcal{E} * f) = a * \mathcal{E} * f = (a * \mathcal{E}) * f = \delta * f = f.$$

Uniqueness: if $a * u = 0$ and $u \in A(a, \mathcal{E})$, then

$$u = u * \delta = u * (a * \mathcal{E}) = u * a * \mathcal{E} = (u * a) * \mathcal{E} = 0 * \mathcal{E} = 0,$$

which is what we set out to prove. The proof is complete. □

REMARK. We can give the solution $u = \mathcal{E} * f$, (9.7), the following physical interpretation. Let us represent the source $f(x)$ in the form of a "sum" of point sources $f(\xi)\delta(x - \xi)$ (see Sec. 4.2.2),

$$f(x) = f * \delta = \int f(\xi)\delta(x - \xi)\, d\xi.$$

The fundamental solution $\mathcal{E}(x)$ is the disturbance due to the point source $\delta(x)$. Whence, by virtue of the linearity and translational invariance of the convolution operator $a*$ (see Sec. 4.8) it follows that each point source $f(\xi)\delta(x - \xi)$ generates a disturbance $f(\xi)\mathcal{E}(x - \xi)$. It is therefore natural to expect that the "sum" (superposition) of these disturbances

$$\int f(\xi)\mathcal{E}(x - \xi)\, d\xi = \mathcal{E} * f$$

will yield a total disturbance due to the source f, that is, the solution u of the equation (9.5). This nonrigorous reasoning is brought into shape by the theorem proved above.

4.9.4. *Equations in convolution algebras.* Let A be a convolution algebra, for example $\mathcal{D}'(\Gamma+)$, $\mathcal{D}'(\Gamma)$ (see Sec. 4.5). Let us consider the equation (9.5) in the algebra A, that is, we will assume that $a \in A$ and $f \in A$; the solution u will also be sought in A. In the algebra A, the above theorem takes the following form: *if the fundamental solution $\mathcal{E}$ of the operator $a*$ exists in A, then the solution u of equation* (9.5) *is unique in A, exists for any f taken from A, and can be expressed by the formula $u = \mathcal{E} * f$.*

The fundamental solution $\mathcal{E}$ of the operator $a*$ in the algebra A is conveniently denoted as a^{-1} so that, by (9.6),

$$a * a^{-1} = \delta. \tag{9.8}$$

In other words, a^{-1} is the *inverse element* of a in the algebra A.

The following proposition is very useful when constructing fundamental solutions in the A algebra:
if a_1^{-1} and a_2^{-1} exist in A, then

$$(a_1 * a_2)^{-1} = a_1^{-1} * a_2^{-1}. \tag{9.9}$$

Indeed,

$$\begin{aligned}(a_1 * a_2) * (a_1^{-1} * a_2^{-1}) &= (a_2 * a_1) * (a_1^{-1} * a_2^{-1}) \\ &= a_2 * \left((a_1 * a_1^{-1}) * a_2^{-1}\right) \\ &= a_2 * (\delta * a_2^{-1}) = a_2 * a_2^{-1} = \delta.\end{aligned}$$

Formula (9.9) forms the basis of operational calculus.

4.9.5. *Fractional differentiation and integration.* Denote by $\mathcal{D}'_+$ the algebra $\mathcal{D}'(\overline{\mathbb{R}^1_+})$.

We introduce the generalized function f_α, taken from $\mathcal{D}'_+$, that depends on a real parameter α, $-\infty < \alpha < \infty$, via the formula

$$f_\alpha(x) = \begin{cases} \dfrac{\theta(x)x^{\alpha-1}}{\Gamma(\alpha)}, & \alpha > 0, \\ f'_{\alpha+1}, & \alpha \le 0. \end{cases}$$

Let us verify that

$$f_\alpha * f_\beta = f_{\alpha+\beta}. \tag{9.10}$$

Indeed, if $\alpha > 0$ and $\beta > 0$, then (see Sec. 4.1)

$$\begin{aligned} f_\alpha * f_\beta &= \frac{\theta(x)}{\Gamma(\alpha)\Gamma(\beta)} \int_0^x y^{\alpha-1}(x-y)^{\beta-1}\,dy \\ &= \frac{\theta(x)x^{\alpha+\beta-1}}{\Gamma(\alpha)\Gamma(\beta)} \int_0^1 t^{\alpha-1}(1-t)^{\beta-1}\,dt \\ &= \frac{\theta(x)x^{\alpha+\beta-1}}{\Gamma(\alpha)\Gamma(\beta)} \mathrm{B}(\alpha,\beta) \\ &= \frac{\theta(x)x^{\alpha+\beta-1}}{\Gamma(\alpha+\beta)} = f_{\alpha+\beta}. \end{aligned}$$

Now if $\alpha \le 0$ or $\beta \le 0$, then, by choosing integers $m > -\alpha$ and $n > -\beta$, we obtain

$$\begin{aligned} f_\alpha * f_\beta &= f_{\alpha+m}^{(m)} * f_{\beta+n}^{(n)} = (f_{\alpha+m} * f_{\beta+n})^{(m+n)} \\ &= f_{\alpha+\beta+m+n}^{(m+n)} = f_{\alpha+\beta}, \end{aligned}$$

which is what we set out to prove. □

Let us consider the convolution operator $f_\alpha *$ in the algebra $\mathcal{D}'_+$. Since $f_0 = \theta' = \delta$, it follows from (9.10) that the fundamental solution f_α^{-1} of the operator $f_\alpha *$ exists and is equal to $f_{-\alpha}$: $f_\alpha^{-1} = f_{-\alpha}$. Furthermore, for integer $n < 0$, $f_n = \delta^{(-n)}$, and for this reason $f_n * u = \delta^{(n)} * u = u^{(n)}$, which means the operator $f_n *$ is the operator of n-fold differentiation. Finally, for integer $n > 0$,

$$(f_n * u)^{(n)} = f_{-n} * (f_n * u) = (f_{-n} * f_n) * u = \delta * u = u,$$

which is to say that $f_n * u$ is an antiderivative of order n of the generalized function u (see Sec. 2.2).

By virtue of what has been said, the operator $f_\alpha *$ is termed the *operator of fractional differentiation of order* $-\alpha$ for $\alpha < 0$ and the *operator of fractional integration of order* α for $\alpha > 0$ (it is also called the *Riemann–Liouville* operator).

EXAMPLE. Let $f \in \mathcal{D}'_+$. Then

$$\partial^{1/2} f = \partial(f_{1/2} * f) = \frac{1}{\sqrt{\pi}} \frac{d}{dx} \int_0^x \frac{f(y)\,dy}{\sqrt{x-y}}.$$

4.9.6. *Heaviside's operational calculus* is nothing but analysis in the convolution algebra $\mathcal{D}'_+$. To illustrate, let us calculate in the algebra $\mathcal{D}'_+$ the fundamental solution $\mathcal{E}(t)$ of the differential operator

$$P\left(\frac{d}{dt}\right) = \frac{d^m}{dt^m} + a_1 \frac{d^{m-1}}{dt^{m-1}} + \cdots + a_m,$$

where a_j are constants. In the $\mathcal{D}'_+$ algebra the corresponding equation takes the form

$$\begin{aligned} &P(\delta) * \mathcal{E} = \delta, \\ &P(\delta)(t) = \delta^{(m)}(t) + a_1 \delta^{(m-1)}(t) + \cdots + a_m \delta(t). \end{aligned}$$

If, in the $\mathcal{D}'_+$ algebra[3], we factor $P(\delta)$,

$$P(\delta) = * \prod_j (\delta' - \lambda_j \delta)^{k_j},$$

and take advantage of (9.9), we obtain

$$P^{-1}(\delta)(t) = \mathcal{E}(t) = \left[* \prod_j (\delta' - \lambda_j \delta)^{k_j} \right]^{-1} = * \prod_j (\delta' - \lambda_j \delta)^{-k_j}. \tag{9.11}$$

But it is easy to verify that

$$*(\delta' - \lambda\delta)^{-k} = *\left[(\delta' - \lambda\delta)^{-1}\right]^k = * \frac{\theta(t) t^{k-1}}{(k-1)!} e^{\lambda t}, \tag{9.12}$$

[3]The symbol $* \prod_{1 \le j \le l} a_j$ stands for $a_1 * a_2 * \cdots * a_l$.

whence, by continuing the equalities (9.11), we derive

$$\mathcal{E}(t) = *\prod_j \frac{\theta(t)t^{k_j-1}}{(k_j-1)!}e^{\lambda_j t}. \tag{9.13}$$

The convolution (9.13) admits of explicit calculation. By decomposing the right-hand side of (9.11) into partial fractions in the $\mathcal{D}'_+$ algebra, we obtain

$$\begin{aligned}\mathcal{E}(t) &= *\prod_j(\delta' - \lambda_j\delta)^{-k_j}\\ &= \sum_j\left[c_{j,k_j} * (\delta' - \lambda_j\delta)^{-k_j} + \cdots + c_{j,1} * (\delta' - \lambda_j\delta)^{-1}\right],\end{aligned}$$

whence, using (9.12), we finally derive

$$\mathcal{E}(t) = \theta(t)\sum_j\left[c_{j,k_j}\frac{t^{k_j-1}}{(k_j-1)!} + \cdots + c_{j,1}\right]e^{\lambda_j t}. \tag{9.14}$$

We thus have the following rule for finding the fundamental solution of the operator $P(\frac{d}{dt})$: substitute p for $\frac{d}{dt}$, set up the polynomial $P(p)$, decompose the expression $\frac{1}{P(p)}$ into partial fractions:

$$\frac{1}{P(p)} = \prod_j(p-\lambda_j)^{-k_j} = \sum_j\left[c_{j,k_j}(p-\lambda_j)^{-k_j} + \cdots + c_{j,1}(p-\lambda_j)^{-1}\right],$$

and with each partial fraction $(p-\lambda)^{-k}$ associate the right-hand side of formula (9.12). As a result, we obtain the formula (9.14).

EXAMPLE. Find $\mathcal{E}$ if $\mathcal{E}'' + \omega^2\mathcal{E} = \delta$.

We have

$$\frac{1}{p^2+\omega^2} = \frac{1}{2\omega i}\left(\frac{1}{p-i\omega} - \frac{1}{p+i\omega}\right) \leftrightarrow \frac{\theta(t)}{2\omega i}\left(e^{i\omega t} - e^{-i\omega t}\right) = \theta(t)\frac{\sin\omega t}{\omega} = \mathcal{E}(t).$$

5. Tempered Generalized Functions

5.1. The space $\mathcal{S}$ of test (rapidly decreasing) functions. We refer to the space of test functions $\mathcal{S} = \mathcal{S}(\mathbb{R}^n)$ all functions infinitely differentiable in $\mathbb{R}^n$ that decrease together with all their derivatives, as $|x| \to \infty$, faster than any power of $|x|^{-1}$. We introduce in $\mathcal{S}$ a countable number of norms via the formula

$$\|\varphi\|_p = \sup_{x,\ |\alpha|\le p}(1+|x|^2)^{p/2}|\partial^\alpha\varphi(x)|, \qquad \varphi\in\mathcal{S}, \quad p = 0,1,\dots.$$

Clearly,

$$\|\varphi\|_0 \le \|\varphi\|_1 \le \|\varphi\|_2 \le \cdots, \qquad \varphi\in\mathcal{S}. \tag{1.1}$$

We define convergence in $\mathcal{S}$ as follows: the sequence of functions $\varphi_1, \varphi_2, \dots$ in $\mathcal{S}$ converges to 0, $\varphi_k \to 0$, $k\to\infty$ in $\mathcal{S}$, if for all $p = 0,1,\dots$ $\|\varphi_k\|_p \to 0$, $k\to\infty$. In other words, $\varphi_k \to 0$, $k\to\infty$ in $\mathcal{S}$, if for all α and β

$$x^\alpha\partial^\beta\varphi_k(x) \overset{x\in\mathbb{R}^n}{\rightrightarrows} 0, \qquad k\to\infty.$$

It is clear that $\mathcal{D}\subset\mathcal{S}$, and if $\varphi_k\to 0$, $k\to\infty$ in $\mathcal{D}$, then $\varphi_k\to 0$, $k\to\infty$ in $\mathcal{S}$.

However, $\mathcal{S}$ does not coincide with $\mathcal{D}$; for example, the function $e^{-|x|^2}$ belongs to $\mathcal{S}$ but not to $\mathcal{D}$ (it is not with compact support).

Yet $\mathcal{D}$ is dense in $\mathcal{S}$, that is, for any $\varphi \in \mathcal{S}$ there is a sequence $\{\varphi_k\}$ of functions in $\mathcal{D}$ such that $\varphi_k \to \varphi$, $k \to \infty$ in $\mathcal{S}$.

Indeed, the sequence of functions, in $\mathcal{D}$,

$$\varphi_k(x) = \varphi(x)\eta\left(\frac{x}{k}\right), \qquad k = 1, 2, \ldots,$$

where $\eta \in \mathcal{D}$, $\eta(x) = 1$, $|x| < 1$, converges to φ in $\mathcal{S}$. □

Let us denote by $\mathcal{S}_p$ the completion of $\mathcal{S}$ in the pth norm; $\mathcal{S}_p$ is a Banach space. The following imbeddings hold:

$$\mathcal{S}_0 \supset \mathcal{S}_1 \supset \ldots . \tag{1.2}$$

Each imbedding

$$\mathcal{S}_{p+1} \subset \mathcal{S}_p, \qquad p = 0, 1, \ldots,$$

is *continuous*, by (1.1). We will now prove that this imbedding is *totally continuous* (*compact*), that is, it is possible, from each infinite bounded set in $\mathcal{S}_{p+1}$, to choose a sequence that converges in $\mathcal{S}_p$.

Indeed, let M be an infinite set bounded in $\mathcal{S}_{p+1}$, $\|\varphi\|_{p+1} < C$, $\varphi \in M$. From this, for all $\varphi \in M$ and $|\alpha| \leq p$, we obtain

$$\left|\frac{\partial}{\partial x_j}\partial^\alpha \varphi(x)\right| < C, \qquad j = 1, \ldots, n;$$

$$(1 + |x|^2)^{p/2}\partial^\alpha \varphi(x) \to 0, \qquad |x| \to \infty.$$

Suppose R_k, $k = 1, 2, \ldots$, is an increasing sequence of positive numbers such that

$$(1 + |x|^2)^{p/2}\,|\partial^\alpha \varphi(x)| < \frac{1}{k}, \qquad |x| > R_k, \quad |\alpha| \leq p. \tag{1.3}$$

By the Ascoli–Arzelá lemma there is a sequence $\{\varphi_j^{(1)}\}$ of functions in M that converges in $C^p(\overline{U}_{R_1})$; furthermore, by the same lemma there is a subsequence $\{\varphi_j^{(2)}\}$ of the functions $\{\varphi_j^{(1)}\}$ that converges in $C^p(\overline{U}_{R_2})$, and so on. It remains to remark that by virtue of (1.3) the diagonal sequence $\{\varphi_k^{(k)}\}$ converges in $\mathcal{S}_p$. □

The following lemma gives an exact characterisation of functions taken from the space $\mathcal{S}_p$.

LEMMA. *So that $\varphi \in \mathcal{S}_p$, it is necessary and sufficient that $\varphi \in C^p$ and $|x|^p\partial^\alpha\varphi(x) \to 0$ for $|x| \to \infty$ and $|\alpha| \leq p$, so that $\varphi \in \overline{C}_0^p$ (see Sec. 0.5).*

PROOF. Necessity is obvious. Let us prove sufficiency. Suppose $\varphi \in \overline{C}_0^p$ and $\varphi_\varepsilon = \varphi * \omega_\varepsilon$ is a mean function of φ (see Sec. 4.6). Furthermore, let $\{\eta_k\}$ be a sequence of functions taken from $\mathcal{D}$ that converges to 1 in $\mathbb{R}^n$ (see Sec. 4.1). Then the sequence $\{\varphi_{1/k}\eta_k\}$ of functions taken from $\mathcal{D} \subset \mathcal{S}$ converges to φ in $\mathcal{S}_p$. Indeed, let $\varepsilon > 0$; there exists a number $R = R(\varepsilon)$ such that

$$(1 + |x|^2)^{p/2}\,|\partial^\alpha \varphi(x)| < \varepsilon, \qquad |x| > R, \qquad |\alpha| \leq p. \tag{1.4}$$

Let N_1 be a number such that $\eta_k(x) = 1$, $|x| \leq R + 1$, $k \geq N_1$. Finally, from Theorem II of Sec. 1.2 it follows the existence of a number $N \geq N_1$ such that for all $k \geq N$, $|x| \leq R + 1$, and $|\alpha| \leq p$, the following inequality holds true:

$$(1 + |x|^2)^{p/2}\,|\partial^\alpha \varphi(x) - \partial^\alpha \varphi_{1/k}(x)| < \varepsilon. \tag{1.5}$$

Now, using (1.4) and (1.5) for $k \geq N$, we obtain

$$\begin{aligned}
\|\varphi - \varphi_{1/k}\eta_k\|_p &= \sup_{x, |\alpha| \leq p} (1+|x|^2)^{p/2} \left|\partial^\alpha[\varphi(x) - \varphi_{1/k}\eta_k(x)]\right| \\
&\leq \varepsilon + \sup_{\substack{|x|>R+1 \\ |\alpha|\leq p}} (1+|x|^2)^{p/2} \left[|\partial^\alpha \varphi(x)| + \sum_{\beta\leq\alpha} \binom{\alpha}{\beta} |\partial^\beta \varphi_{1/k}\partial^{\alpha-\beta}\eta_k(x)|\right] \\
&\leq 2\varepsilon + C_p' \sup_{\substack{|x|>R+1 \\ |\beta|\leq p}} (1+|x|^2)^{p/2} \left|\partial^\beta \int \omega_{1/k}(y)\varphi(x-y)\,dy\right| \\
&\leq 2\varepsilon + C_p' \sup_{\substack{|x|>R+1 \\ |\beta|\leq p}} \int \omega_{1/k}(y)(1+|x|^2)^{p/2} |\partial^\beta \varphi(x-y)|\,dy \\
&\leq 2\varepsilon + C_p' \sup_{\substack{|x|>R+1 \\ |\beta|\leq p}} \int \omega_{1/k}(y) \left[(1+|x-y|^2)^{p/2} + |y|^p\right] |\partial^\beta \varphi(x-y)|\,dy \\
&\leq 2\varepsilon + C_p\varepsilon + C_p\varepsilon \int \omega_{1/k}(y)(1+|y|^2)\,dy \\
&\leq (2+3C_p)\varepsilon,
\end{aligned}$$

which is what we set out to prove. The proof of the lemma is complete. □

It follows from the lemma that $\mathcal{S}$ is a complete space and

$$\mathcal{S} = \cap_{p\geq 0}\mathcal{S}_p. \tag{1.6}$$

The operations of differentiation $\varphi \to \partial^\alpha\varphi$ and of the nonsingular linear change of variables $\varphi(x) \to \varphi(Ax+b)$ are linear and continuous from $\mathcal{S}$ to $\mathcal{S}$.

This follows directly from the definition of convergence in the space $\mathcal{S}$.

On the other hand, multiplication by an infinitely differentiable function may take one outside the domain of $\mathcal{S}$, for example, $e^{-|x|^2}e^{|x|^2} = 1 \notin \mathcal{S}$.

Suppose the function $a \in C^\infty$ grows at infinity together with all its derivatives not faster than the polynomial

$$|\partial^\alpha a(x)| \leq C_\alpha(1+|x|)^{m_\alpha}. \tag{1.7}$$

We denote by θ_M the set of all such functions. This is called the set of *multipliers in $\mathcal{S}$.*

The operation $\varphi \to a\varphi$, where $a \in \theta_M$, is continuous (and, obviously, linear) from $\mathcal{S}$ to $\mathcal{S}$.

Indeed, if $\varphi \to \mathcal{S}$, then $a\varphi \in C^\infty$ and, by virtue of (1.7),

$$\begin{aligned}
\|a\varphi\|_p &= \sup_{x, |\alpha|\leq p} (1+|x|^2)^{p/2} |\partial^\alpha(a\varphi)| \\
&\leq \sup_{x, |\alpha|\leq p} (1+|x|^2)^{p/2} \sum_{\beta\leq\alpha} \binom{\alpha}{\beta} |\partial^\beta\varphi(x)\partial^{\alpha-\beta}a(x)| \\
&\leq K_p \sup_{x, |\alpha|\leq p} (1+|x|^2)^{p/2+N_p/2} |\partial^\alpha\varphi(x)| \\
&= K_p\|\varphi\|_{p+N_p}, \qquad p = 0, 1, \ldots,
\end{aligned}$$

where N_p is the smallest integer not less than $\max_{|\alpha|\leq p} m_\alpha$. These inequalities signify that $a\varphi \in \mathcal{S}$ and the operation $\varphi \to a\varphi$ is continuous from $\mathcal{S}$ to $\mathcal{S}$. □

5.2. The space $\mathcal{S}'$ of tempered generalized functions. A *tempered generalized function* is any continuous linear functional on the space $\mathcal{S}$ of test functions. We denote by $\mathcal{S}' = \mathcal{S}'(\mathbb{R}^n)$ the set of all tempered generalized functions. Clearly, $\mathcal{S}'$ is a linear set and $\mathcal{S}' \subset \mathcal{D}'$.

We define convergence in $\mathcal{S}'$ as *weak* convergence of a sequence of functionals: a sequence of generalized functions $f_1, f_2, \dots$ taken from $\mathcal{S}'$ converges to the generalized function $f \in \mathcal{S}'$, $f_k \to f$, $k \to \infty$ in $\mathcal{S}'$, if for any $\varphi \in \mathcal{S}$, $(f_k, \varphi) \to (f, \varphi)$ $k \to \infty$. The linear set $\mathcal{S}'$ equipped with convergence is termed the *space $\mathcal{S}'$ of tempered generalized functions.*

From the definition it follows that convergence in $\mathcal{S}'$ implies convergence in $\mathcal{D}'$.

THEOREM (L. Schwartz). *Let M' be a weakly bounded set of functionals from $\mathcal{S}'$, that is, $|(f, \varphi)| < C_\varphi$ for all $f \in M'$ and $\varphi \in \mathcal{S}$. Then there are numbers $K \geq 0$ and integer $m \geq 0$ such that*

$$|(f, \varphi)| \leq K\|\varphi\|_m, \qquad f \in M', \quad \varphi \in \mathcal{S}. \tag{2.1}$$

PROOF. If the inequality (2.1) does not hold, then there will be sequence $\{f_k\}$ of functionals from M' and sequences $\{\varphi_k\}$ of functions taken from $\mathcal{S}$ such that

$$|(f_k, \varphi_k)| \geq k\|\varphi_k\|_k, \qquad k = 1, 2, \dots . \tag{2.2}$$

The sequence of functions

$$\psi_k(x) = \frac{\varphi_k(x)}{\sqrt{k}\|\varphi_k\|_k}, \qquad k = 1, 2, \dots,$$

tends to 0 in $\mathcal{S}$ because for $k \geq p$

$$\|\psi_k\|_p = \frac{\|\varphi_k\|_p}{\sqrt{k}\|\varphi_k\|_k} \leq \frac{1}{\sqrt{k}}.$$

The sequence of functionals $\{f_k\}$ is bounded on every test function φ taken from $\mathcal{S}$. For this reason, we have an analogue of the lemma of Sec. 1.4 according to which $(f_k, \psi_k) \to 0$, $k \to \infty$. On the other hand, the inequality (2.2) yields

$$|(f_k, \psi_k)| = \frac{1}{\sqrt{k}\|\varphi\|_k} |(f_k, \varphi_k)| \geq \sqrt{k}.$$

The resulting contradiction proves the theorem. □

From the Schwartz theorem we have just proved there follow a number of corollaries.

COROLLARY 1. *Every tempered generalized function has a finite order (compare Sec. 1.3), that is to say, it admits of an extension as a continuous linear functional from some (least) conjugate space $\mathcal{S}'_m$; then, for f, the inequality* (2.1) *takes the form*

$$|(f, \varphi)| \leq \|f\|_{-m}\|\varphi\|_m, \qquad \varphi \in \mathcal{S}_m, \tag{2.3}$$

where $\|f\|_{-m}$ is the norm of the functional f in $\mathcal{S}'_m$ and m is the order of f.

Thus, the following relations hold true:

$$\mathcal{S}'_0 \subset \mathcal{S}'_1 \subset \mathcal{S}'_2 \subset \dots, \qquad \mathcal{S}' = \bigcup_{p \geq 0} \mathcal{S}'_p. \tag{2.4}$$

They are duals of (1.2) and (1.6).

Also note that every imbedding

$$S'_p \subset S'_{p+1}, \qquad p = 0, 1, \dots,$$

is totally continuous (see Sec. 5.1); in particular, every (weakly) convergent sequence of functionals taken from S'_p converges in norm in S'_{p+1}.

COROLLARY 2. *Every (weakly) convergent sequence of tempered generalized functions converges weakly in some space S'_p and, hence, converges in norm in S'_{p+1}.*

This follows from the Schwartz theorem since every (weakly) convergent sequence of functionals taken from S' is a weakly bounded set in S'; it also follows from the remark referring to Corollary 1. □

COROLLARY 3. *The space of tempered generalized functions is complete.*

This follows from the completeness of the conjugate spaces S'_p and from Corollary 2. □

5.3. Examples of tempered generalized functions and elementary operations in S'. A function $f(x)$ is called a *tempered* function in $\mathbb{R}^n$, if, for some $m \geq 0$

$$\int |f(x)|(1 + |x|)^{-m}\, dx < \infty.$$

A tempered function f defines a regular functional f in S' via the formula (6.1) of Sec. 1.6,

$$(f, \varphi) = \int f(x)\varphi(x)\, dx, \qquad \varphi \in S.$$

Not every locally integrable function defines a tempered generalized function, for example, $e^x \notin S'$. On the other hand, not every locally integrable function taken from S' is tempered. For example, the function $(\cos e^x)' = -e^x \sin e^x$ is not a tempered function, yet it defines a generalized function from S' via the formula

$$((\cos e^x)', \varphi) = -\int \cos e^x \varphi'(x)\, dx, \qquad \varphi \in S.$$

However, there can be no such unpleasantness as regards nonnegative functions (and even measures), as we shall now see.

A measure μ specified on $\mathbb{R}^n$ (see Sec. 1.7) is said to be a *tempered measure* if for some $m \geq 0$

$$\int (1 + |x|)^{-m} \mu(dx) < \infty.$$

It defines a generalized function in S' via formula (7.2) of Sec. 1.7,

$$(\mu, \varphi) = \int \varphi(x)\mu(dx), \qquad \varphi \in S.$$

If a nonnegative measure μ defines a generalized function in S' then μ is tempered.

Indeed, since $\mu \in S'$, it follows from the Schwartz theorem that it is of finite order m so that

$$\left|\int \varphi(x)\, \mu(dx)\right| \leq K\|\varphi\|_m, \qquad \varphi \in S. \tag{3.1}$$

Let $\{\eta_k\}$ be a sequence of nonnegative functions in $\mathcal{D}$ that tend to 1 in $\mathbb{R}^n$ (see Sec. 4.1). Substituting into (3.1)

$$\varphi(x) = \eta_k(x)\left(1+|x|^2\right)^{-m/2}$$

and making use of the nonnegativity of the measure μ, we obtain

$$\int \eta_k(x)\left(1+|x|^2\right)^{-m/2}\mu(dx) \le C$$

where C does not depend on k. From this, by virtue of the Fatou lemma, it follows that the measure μ is tempered. □

If $f \in \mathcal{E}'$, then $f \in \mathcal{S}'$, and

$$(f,\varphi) = (f,\eta\varphi), \qquad \varphi \in \mathcal{S}, \tag{3.2}$$

where $\eta \in \mathcal{D}$ and $\eta = 1$ in the neighbourhood of the support of f (compare (10.2) of Sec. 1.10).

Indeed, since the operation $\varphi \to \eta\varphi$ is linear and continuous from $\mathcal{S}$ to $\mathcal{D}$, the functional $(f,\eta\varphi)$ on the right-hand side of (3.2) is linear and continuous on $\mathcal{S}$ so that $f \in \mathcal{S}'$. The uniqueness of the extension follows from the density of $\mathcal{D}$ in $\mathcal{S}$ (see Sec. 5.1); in particular, it is independent of the auxiliary function η. □

If $f \in \mathcal{S}'$, then every derivative $\partial^\alpha f \in \mathcal{S}'$ as well; here, the operation $f \to \partial^\alpha f$ is continuous (and linear) from $\mathcal{S}'$ to $\mathcal{S}'$.

Indeed, since the operation $\varphi \to \partial^\alpha\varphi$ is linear and continuous from $\mathcal{S}$ to $\mathcal{S}$ (see Sec. 5.1), it follows that the right-hand side of

$$(\partial^\alpha f,\varphi) = (-1)^{|\alpha|}(f,\partial^\alpha\varphi), \qquad \varphi \in \mathcal{S},$$

is a continuous linear functional an $\mathcal{S}$ (compare Sec. 2.1). □

If $f \in \mathcal{S}'$ and $\det A \neq 0$, then $f(Ax+b) \in \mathcal{S}'$, and the operation $f(x) \to f(Ax+b)$ is continuous (and linear) from $\mathcal{S}'$ to $\mathcal{S}'$.

True enough, since the operation $\varphi(x) \to \varphi\left[A^{-1}(x-b)\right]$ is linear and continuous from $\mathcal{S}$ to $\mathcal{S}$ (see Sec. 5.1), the right-hand side of

$$(f(Ay+b),\varphi) = \left(f, \frac{\varphi[A^{-1}(x-b)]}{|\det A|}\right), \qquad \varphi \in \mathcal{S},$$

is a continuous linear functional on $\mathcal{S}$ (compare Sec. 1.9). □

If $f \in \mathcal{S}'$ and $a \in \theta_M$, then $af \in \mathcal{S}'$, and the operation $f \to af$ is continuous (and linear) from $\mathcal{S}'$ to $\mathcal{S}'$.

Indeed, since the operation $\varphi \to a\varphi$ is linear and continuous from $\mathcal{S}$ to $\mathcal{S}$ (see Sec. 1.5), it follows that the right-hand side of the equality

$$(af,\varphi) = f(,a\varphi), \qquad \varphi \in \mathcal{S},$$

is a continuous linear functional on $\mathcal{S}$ (compare Sec. 1.10). □

Thus, the set θ_M contains all multipliers in $\mathcal{S}'$ (actually, it consists of them; prove it).

EXAMPLE. If $|a_k| \le C\left(1+|k|\right)^N$, then

$$\sum_k a_k\delta(x-k) \in \mathcal{S}'.$$

5.4. The structure of tempered generalized functions. We will now prove that the space $\mathcal{S}'$ is a (smallest) extension of the collection of tempered functions in $\mathbb{R}^n$ such that in it differentiation is always possible (compare Sec. 2.4). Hence this explains the name of $\mathcal{S}'$ as the space of tempered generalized functions (tempered distributions, according to L. Schwartz [89]).

THEOREM. *If $f \in \mathcal{S}'$, then there exist a tempered continuous function g in $\mathbb{R}^n$ and an integer $m \geq 0$ such that*

$$f(x) = \partial_1^m \dots \partial_n^m g(x). \tag{4.1}$$

PROOF. Let $f \in \mathcal{S}'$. By the theorem of L. Schwartz (see Sec. 5.2) there exist numbers K and p such that for all $\varphi \in \mathcal{S}$

$$\begin{aligned}|(f,\varphi)| &\leq K\|\varphi\|_p \\ &\leq K \max_{|\alpha|\leq p} \int \left|\partial_1 \dots \partial_n \left[(1+|x|^2)^{p/2}\partial^\alpha\varphi(x)\right]\right| dx,\end{aligned}$$

that is

$$|(f,\varphi)| \leq K \max_{|\alpha|\leq p} \left\|\partial_1 \dots \partial_n \left[(1+|x|^2)^{p/2}\partial^\alpha\varphi(x)\right]\right\|_{\mathcal{L}^1}. \tag{4.2}$$

With every function φ of $\mathcal{S}$ we associate a vector function $\{\psi_\alpha\}$ with components

$$\psi_\alpha(x) = \partial_1 \dots \partial_n \left[(1+|x|^2)^{p/2}\partial^\alpha\varphi(x)\right], \qquad |\alpha| \leq p. \tag{4.3}$$

In this way we define a one-to-one mapping $\varphi \to \{\psi_\alpha\}$ of the space $\mathcal{S}$ into the direct sum $\bigoplus_{|\alpha|\leq p} \mathcal{L}^1$ with norm

$$\|\{f_\alpha\}\| = \max_{|\alpha|\leq p} \|f_\alpha\|_{\mathcal{L}^1}.$$

On the linear subset $[\{\psi_\alpha\}, \varphi \in \mathcal{S}]$ of the space $\bigoplus_{|\alpha|\leq p} \mathcal{L}^1$, in which the components ψ_α are defined by (4.3), we introduce a linear functional f^*:

$$(f^*, \{\psi_\alpha\}) = (f, \varphi). \tag{4.4}$$

By virtue of the estimate (4.2),

$$|(f^*, \{\psi_\alpha\})| = |(f,\varphi)| \leq K \max_{|\alpha|\leq p} \|\psi\|_{\mathcal{L}^1} = K\|\{\psi_\alpha\}\|,$$

the functional f^* is continuous. By the Hahn–Banach and F. Riesz theorems there exists a vector function $\{\chi_\alpha\} \in \bigoplus_{|\alpha|\leq p} \mathcal{L}^\infty$ such that

$$(f^*, \{\psi_\alpha\}) = \sum_{|\alpha|\leq p} \int \chi_\alpha(x)\psi_\alpha(x)\, dx.$$

That is to say, by virtue of (4.3) and (4.4), we have

$$(f,\varphi) = \sum_{|\alpha|\leq p} \int \chi_\alpha(x)\partial_1 \dots \partial_n \left[(1+|x|^2)^{p/2}\partial^\alpha\varphi(x)\right] dx, \qquad \varphi \in \mathcal{S}.$$

Integrating the right-hand side of this equation by parts, we are convinced of the existence of continuous tempered functions g_α, $|\alpha| \leq p+2$, such that

$$(f,g) = (-1)^{pn} \int \sum_{|\alpha|\leq(p+2)n} g_\alpha(x)\partial_1^{p+2} \dots \partial_n^{p+2}\varphi(x)\, dx,$$

whence follows the representation (4.1) for $m = p+2$. The proof of the theorem is complete. □

COROLLARY. *If $f \in \mathcal{S}'$, then there exists an integer $p \geq 0$ such that for any $\varepsilon > 0$ there are functions $g_{\alpha,\varepsilon}$, $|\alpha| \leq p$, which are continuous tempered in $\mathbb{R}^n$ and vanish outside the ε-neighbourhood of the support of f, so that*

$$f(x) = \sum_{|\alpha|\leq p} \partial^\alpha g_{\alpha,\varepsilon}(x). \tag{4.5}$$

Indeed, suppose $\varepsilon > 0$ and $\eta \in \theta_M$, $\eta(x) = 1$, $x \in (\operatorname{supp} f)^{\varepsilon/3}$, and $\eta(x) = 0$, $x \notin (\operatorname{supp} f)^\varepsilon$. (By the lemma of Sec. 1.2, such functions exist.) Then, taking into account the representation (4.1) and using the Leibniz formula (see Sec. 2.1), we have

$$\begin{aligned} f(x) &= \eta(x) f(x) \\ &= \eta(x)\partial_1^m \ldots \partial_n^m g(x) \\ &= \partial_1^m \ldots \partial_n^m [\eta(x) g(x)] + \sum_{|\alpha|\leq mn-1} \eta_\alpha(x)\partial^\alpha g(x), \end{aligned}$$

where $\eta_\alpha \in \theta_M$ and $\eta_\alpha(x) = 0$, $x \notin (\operatorname{supp} f)^\varepsilon$. Each term in the last sum is again transformed in that fashion, and so on. Then, in a finite number of steps, we arrive at the representation (4.5) with $p = mn$ and $g_{\alpha,\varepsilon} = \chi_\alpha g$, where χ_α are certain functions taken from θ_M with support in $\overline{(\operatorname{supp} f)^\varepsilon}$. □

5.5. The direct product of tempered generalized functions. Let $f(x) \in \mathcal{S}'(\mathbb{R}^n)$ and $g(y) \in \mathcal{S}'(\mathbb{R}^m)$. Since $\mathcal{S}' \subset \mathcal{D}'$, the direct product $f(x) \times g(y)$ is a generalized function in $\mathcal{D}'(\mathbb{R}^{n+m})$ (see Sec. 3.1). We will prove that $f(x) \times g(y) \in \mathcal{S}'(\mathbb{R}^{n+m})$.

By the definition of the functional $f(x) \times g(y)$ (see Sec. 3.1),

$$(f(x) \times g(y), \varphi) = \big(f(x), (g(y), \varphi(x,y))\big). \tag{5.1}$$

We will now prove that the right-hand side of (5.1) is a continuous linear functional on $\mathcal{S}(\mathbb{R}^{n+m})$.

To do this, we set up the following lemma that is similar to the lemma of Sec. 3.1.

LEMMA. *If $\in \mathcal{S}'$, then for all α*

$$\partial^\alpha \psi(x) = \big(g(y), \partial_x^\alpha \varphi(x,y)\big), \qquad \varphi \in \mathcal{S}(\mathbb{R}^{n+m}), \tag{5.2}$$

and there is an integer $q \geq 0$ such that

$$\|\psi\|_p \leq \|g\|_{-q} \|\varphi\|_{p+q}, \qquad p = 0, 1, \ldots, \tag{5.3}$$

so that the operation $\varphi \to \psi = \big(g(y), \varphi(x,y)\big)$ is continuous (and linear) from $\mathcal{S}(\mathbb{R}^{n+m})$ to $\mathcal{S}(\mathbb{R}^n)$.

PROOF. As in the proof of the lemma of Sec. 3.1, we establish the truth of the equality (5.2) for all α and the continuity of the right-hand side. Consequently, $\psi \in C^\infty$. Let q be the order of g. Applying (2.3) to the right-hand side of (5.2), we obtain for all $x \in \mathbb{R}^n$ the estimate

$$\left|\partial^\alpha \psi(x)\right| \leq \|g\|_{-q} \sup_{y, |\beta|\leq q} (1+|y|^2)^{q/2} \left|\partial_x^\alpha \partial_y^\beta \varphi(x,y)\right|,$$

whence follows (5.3):

$$\begin{aligned}\|\psi\|_p &= \sup_{x,\ |\alpha|\le p}(1+|x|^2)^{p/2}|\partial^\alpha\psi(x)|\\ &\le \|g\|_{-q}\sup_{(x,y),\ |\alpha|\le p,\ |\beta|\le q}(1+|x|^2)^{p/2}(1+|y|^2)^{q/2}|\partial_x^\alpha\partial_y^\beta\varphi(x,y)|\\ &\le \|g\|_{-q}\|\varphi\|_{p+q}, \qquad p=0,1,\dots .\end{aligned}$$

The proof of the lemma is complete. □

From this lemma it follows that the right-hand side of (5.1), which is equal to (f,ψ) is a continuous and linear functional on $\mathcal{S}(\mathbb{R}^{n+m})$ so that $f(x)\times g(y)\in\mathcal{S}'(\mathbb{R}^{n+m})$ (compare Sec. 3.1).

All the properties of a direct product that are listed in Sec. 3.2 for the space $\mathcal{D}'$ hold true also for the space $\mathcal{S}'$. This assertion follows from the density of $\mathcal{D}$ in $\mathcal{S}$ (see Sec. 5.1). In particular, the operation $f(x)\to f(x)\times g(y)$ is continuous from $\mathcal{S}'(\mathbb{R}^n)$ to $\mathcal{S}'(\mathbb{R}^{n+m})$.

Finally, the formula (3.2) of Sec. 3.3 holds true for $f\in\mathcal{S}'(\mathbb{R}^n)$ and $\varphi\in\mathcal{S}(\mathbb{R}^{n+m})$:

$$\left(f(x),\int\varphi(x,y)\,dy\right)=\int(f(x),\varphi(x,y))\,dy. \tag{5.4}$$

5.6. The convolution of tempered generalized functions. Let $f\in\mathcal{S}'$, $g\in\mathcal{S}'$ and let their convolution $f*g$ exist in $\mathcal{D}'$ (see Sec. 4.1). Now: When does $f*g\in\mathcal{S}'$ and when is the operation $f\to f*g$ continuous from $\mathcal{S}'$ to $\mathcal{S}'$?

In accordance with (1.7) of Sec. 4.1, we can assume the following definition of the convolution of tempered generalized functions, which is equivalent to (1.4) of Sec. 4.1 and is convenient for computations. Let $f,g\in\mathcal{S}'$. By their convolution $f*g\in\mathcal{S}'$ we call the limit

$$f*g=\lim_{k\to\infty}(f\eta_k)*g \quad \text{in} \quad \mathcal{S}'$$

if this limit exists for any sequence $\{\eta_k\}$ converging to 1 in $\mathbb{R}^n$ (in this case it does not depend on $\{\eta_k\}$). Then $f*g\in\mathcal{S}'$ and there exists the convolution $g*f$ and they are equal each other,

$$f*g=g*f.$$

We state three sufficient criteria for the existence of a convolution in $\mathcal{S}'$.

5.6.1. *Let $f\in\mathcal{S}'$ and $g\in\mathcal{E}'$. Then the convolution $f*g$ belongs to $\mathcal{S}'$ and can be represented as*

$$(f*g,\varphi)=(f(x)\times g(y),\eta(y)\varphi(x+y)), \qquad \varphi\in\mathcal{S}, \tag{6.1}$$

where η is any function from $\mathcal{D}$ equal to 1 in a neighbourhood of the support of g; here, the operation $f\to f*g$ is continuous from $\mathcal{S}'$ to $\mathcal{S}'$, and the operation $g\to f*g$ is continuous from $\mathcal{E}'$ to $\mathcal{S}'$.

Indeed, the convolution $f*g\in\mathcal{D}'$ and the representation (3.3) of Sec. 4.3 holds true on the test functions in $\mathcal{D}$. Since $f(x)\times g(y)\in\mathcal{S}'(\mathbb{R}^{2n})$ (see Sec. 5.5), and the operation $\varphi \to \eta(y)\varphi(x+y)$ is linear and continuous from $\mathcal{S}(\mathbb{R}^n)$ to $\mathcal{S}(\mathbb{R}^{2n})$:

$$\|\eta(y)\varphi(x+y)\|_p \leq \sup_{\substack{(x,y)\\|\alpha|\leq p}} (1+|x|^2+|y|^2)^{p/2} |\partial^\alpha[\eta(y)\varphi(x+y)]|$$
$$\leq C_p \sup_{\substack{(x,y)\\|\alpha|\leq p}} (1+|x+y|^2)^{p/2} |\partial^\alpha\varphi(x+y)| = C_p\|\varphi\|_p,$$

it follows that the right-hand side of (6.1) defines a continuous linear functional on $\mathcal{S}$ so that $f*g \in \mathcal{S}'$. □

5.6.2. Let Γ be a closed convex acute cone in $\mathbb{R}^n$ with vertex at 0, $C = \operatorname{int}\Gamma^*$, S a strictly C-like surface, and S_+ the domain lying above S (see Sec. 4.4).

If $f \in \mathcal{S}'(\Gamma+)$ and $g \in \mathcal{S}'(\overline{S}_+)$, then the convolution $f*g$ exists in $\mathcal{S}'$ and can be represented as

$$(f*g,\varphi) = (f(x)\times g(y), \xi(x)\eta(y)\varphi(x+y)), \qquad \varphi \in \mathcal{S} \tag{6.2}$$

where ξ and η are any C^∞-functions, $|\partial^\alpha\xi(x)| \leq c_\alpha$, $|\partial^\alpha\eta(y)| \leq c_\alpha$, equal to 1 in $(\operatorname{supp} f)^\epsilon$ and $(\operatorname{supp} g)^\epsilon$ and equal to 0 outside $(\operatorname{supp} f)^{2\epsilon}$ and $(\operatorname{supp} g)^{2\epsilon}$ respectively (ϵ is any positive number[4]. Here, if $\operatorname{supp} f \subset \Gamma + K$, where K is a compact, then the operation $f \to f*g$ is continuous from $\mathcal{S}'(\Gamma+K)$ to $\mathcal{S}'(\overline{S}_+ + K)$.

To prove this assertion, it remains — by using the representation (5.1) of Sec. 4.5 and by reasoning as in Sec. 5.6.1 — to establish the continuity of the operation $\varphi \to \chi = \xi(x)\eta(y)\varphi(x+y)$ from $\mathcal{S}(\mathbb{R}^n)$ to $\mathcal{S}(\mathbb{R}^{2n})$. For all $\varphi \in \mathcal{S}$ we have

$$\|\chi\|_p = \sup_{\substack{(x,y)\\|\alpha|\leq p}} (1+|x|^2+|y|^2)^{p/2} \left|\partial^\alpha_{(x,y)}[\xi(x)\eta(y)\varphi(x+y)]\right|$$
$$\leq C'_p \sup_{\substack{x\in\Gamma+K+\overline{U}_{2\epsilon}\\ y\in\overline{S}_+,\ |\alpha|\leq p}} (1+|x|^2+|y|^2)^{p/2} \left|\partial^\alpha_{(x,y)}\varphi(x+y)\right|$$
$$\leq 2^p C'_p \sup_{\substack{z\\ x\in T(z)\\ |\alpha|\leq p}} (1+|x|^2+|y|^2)^{p/2} |\partial^\alpha\varphi(\xi)|,$$

where

$$T(z) = [x : x \in \Gamma + K + \overline{U}_{2\epsilon},\ x = z-y,\ y \in \overline{S}_+].$$

Since S is assumed to be a strictly C-like surface, it follows that the set $T(z)$ is contained in a ball of radius $a(1+|\xi|)^\nu$, $\nu \geq 1$ (see Sec. 4.4). Therefore, continuing our estimates, we obtain

$$\|\chi\|_p \leq C''_p \sup_{\substack{z\\|\alpha|\leq p}} [1+|z|^2+a^2(1+|z|)^{2\nu}]^{p/2} |\partial^\alpha\varphi(z)|$$
$$\leq C_p\|\varphi\|_{p(\nu+1)}, \qquad p = 0,1,\ldots,$$

which is what we set out to do. □

From the criterion obtained it follows, in particular, that the set of generalized functions $\mathcal{S}'(\Gamma+)$ forms a convolution algebra, a subalgebra of the algebra $\mathcal{D}'(\Gamma+)$; in the same way, $\mathcal{S}'(\Gamma)$ also forms a convolution algebra, a subalgebra of the algebras $\mathcal{S}'(\Gamma+)$ and $\mathcal{D}'(\Gamma)$.

[4]According to the lemma of Sec. 1.2, such functions exist.

5.6.3. *Let $f \in \mathcal{S}'$ and $\eta \in \mathcal{S}$. Then the convolution $f * \eta$ exists in θ_M and can be represented in the form* [compare (6.2) of Sec. 4.6]

$$(f * \eta, \varphi) = \big(f, \eta * \varphi(-x)\big), \qquad \varphi \in \mathcal{S}, \tag{6.3}$$

that is

$$f * g = \big(f(y), \eta(x-y)\big). \tag{6.3'}$$

Here, there is an integer $m \geq 0$ (an order f) such that

$$\left|\partial^\alpha (f * \eta)(x)\right| = \|f\|_{-m} \left(1 + |x|^2\right)^{m/2} \|\eta\|_{m+|\alpha|}, \qquad x \in \mathbb{R}^n. \tag{6.4}$$

Indeed, suppose $\{\eta_k(x; y)\}$ is a sequence of functions taken from $\mathcal{D}(\mathbb{R}^{2n})$ that converges to 1 in $\mathbb{R}^{2n}$, and $\varphi \in \mathcal{S}(\mathbb{R}^n)$. Then

$$\int \eta(y)\eta_k(x;y)\varphi(x+y)\,dy \to \int \eta(y)\varphi(x+y)\,dy, \qquad k \to \infty \quad \text{in} \quad \mathcal{S}.$$

From this, if we make use of the definitions of a convolution (see Sec. 4.1) and of a direct product (see Sec. 3.1), we obtain, for all $\varphi \in \mathcal{S}$, the representation (6.3):

$$\begin{aligned}
(f * \eta, \varphi) &= \lim_{k\to\infty} \big(f(x) \times \eta(y), \eta_k(x;y)\varphi(x+y)\big) \\
&= \lim_{k\to\infty} \left(f(x), \int \eta(y)\eta_k(x;y)\varphi(x+y)\,dy\right) \\
&= \left(f(x), \int \eta(y)\varphi(x+y)\,dy\right) \\
&= \left(f(x), \int \varphi(\xi)\eta(\xi - x\,d\xi\right) \\
&= \big(f, \eta * \varphi(-x)\big).
\end{aligned}$$

Noting that $\varphi(\xi)\eta(\xi - x) \in \mathcal{S}(\mathbb{R}^{2n})$ and taking advantage of (5.4), we continue our chain of equalities

$$(f * \eta, \varphi) = \int \big(f(x), \eta(\xi - x)\big)\varphi(\xi)\,d\xi,$$

whence follows the representation (6.3′). □

As in the proof of the lemma of Sec. 3.1, we conclude from the representation (6.3) that $f * \eta \in C^\infty$ and the following formula holds true:

$$\partial^\alpha (f * \eta)(x) = \big(f(y), \partial_x^\alpha \eta(x-y)\big). \tag{6.5}$$

Let m be the order of f. Applying the inequality (2.3) to the right-hand side of (6.5), we obtain the inequality (6.4):

$$\begin{aligned}|\partial^\alpha(f*\eta)(x)| &\le \|f\|_{-m}\,\|\partial_x^\alpha\eta(x-y)\|_m\\ &= \|f\|_{-m}\sup_{y,\ |\beta|\le m}(1+|y|^2)^{m/2}\,|\partial_x^\alpha\partial_y^\beta\eta(x-y)|\\ &= \|f\|_{-m}\sup_{\xi,\ |\beta|\le m}(1+|x-\xi|^2)^{m/2}\,|\partial^{\alpha+\beta}\eta(\xi)|\\ &\le \|f\|_{-m}(1+|x|^2)^{m/2}\sup_{\xi,\ |\beta|\le m}(1+|\xi|^2)^{m/2}\,|\partial^{\alpha+\beta}\eta(\xi)|\\ &\le \|f\|_{-m}(1+|x|^2)^{m/2}\|\eta\|_{m+|\alpha|}.\end{aligned}$$

□

COROLLARY. *$\mathcal{S}$ is dense in $\mathcal{S}'$.*

From what has been proved, if $f\in\mathcal{S}'$, then its mean function $f_\varepsilon = f*\omega_\varepsilon\in\theta_M$ and $f_\varepsilon\to f$, $\varepsilon\to+0$ in $\mathcal{S}'$ (see Sec. 5.6.1). Therefore θ_M is dense in $\mathcal{S}'$. But $\mathcal{S}$ is dense in θ_m because if $a\in\theta_M$, then

$$\mathcal{S}\ni e^{-\varepsilon|x|^2}a\to a,\qquad \varepsilon\to+0\quad\text{in}\quad \mathcal{S}'.$$

5.7. Homogeneous generalized functions. The functions $\pi_\nu(x)|x|^{\alpha-1}$ are defined for any $x\ne0$ and complex $\alpha\in\mathbb{C}$ (see Sec. 2.7).

DEFINITION. A generalized function $f\in\mathcal{D}'(\mathbb{R}^+)$ is called *homogeneous* of degree of homogeneity $\alpha-1$ if

$$f(tx)=\pi_\nu(t)|t|^{\alpha-1}f(x),\qquad t\ne0,\tag{7.1}$$

i.e. (see Sec. 1.9)

$$\left(f(x),\varphi\left(\frac{x}{t}\right)\right)=\pi_\nu(t)|t|^\alpha(f,\varphi),\qquad t\ne0,\tag{7.1'}$$

EXAMPLE (see Sec. 1.8).

$$\mathcal{P}\frac{1}{x}=\mathcal{P}(\operatorname{sgn}x|x|^{-1}).$$

One can readily see that the homogeneous generalized functions of different degrees are linearly independent.

The definition implies that homogeneous generalized functions for $\nu=0$ are even and for $\nu=-1$ are odd. Furthermore, $\delta^{(2n)}(x)$, $n=0,1,\dots$, are even homogeneous functions of degree of homogeneity $-2n-1$; $\delta^{(2n+1)}(x)$, $n=0,1,\dots$, are odd homogeneous functions of degree of homogeneity $-2n$; finally, the generalized functions $\mathcal{P}(\pi_\nu|x|^{\alpha-1})$ (see Sec. 2.7) are even homogeneous of degree $\alpha-1$, $\alpha\ne-2n$, $n=0,1,\dots$, for $\nu=0$ and odd homogeneous of degree $\alpha-1$, $\alpha\ne-2n-1$, $n=0,1,\dots$, for $\nu=1$. We are going to show that these are all the homogeneous generalized functions (in one variable).

THEOREM. *Any even homogeneous generalized function of degree of homogeneity $\alpha-1$ has the form $C\mathcal{P}|x|^{\alpha-1}$ for $\alpha\ne-2n$ and $C\delta^{(2n)}(x)$ for $\alpha=-2n$, $n=0,1,\dots$; any odd homogeneous generalized function of degree of homogeneity $\alpha-1$ has the form $\mathcal{P}(\operatorname{sgn}x|x|^{\alpha-1})$ for $\alpha\ne-2n-1$ and $C\delta^{(2n+1)}(x)$ for $\alpha=-2n-1$, $n=0,1,\dots$, where C is an arbitrary constant.*

PROOF. Let us prove the theorem for even homogeneous generalized functions $f \not\equiv 0$ of degree $\alpha - 1$ (the proof for the odd generalized functions is analogous).

Suppose that $\operatorname{supp} f$ contains points distinct from 0. Then there exists a function $\omega \in \mathcal{D}(x \neq 0)$ such that $(f, \omega) = 1$ and $\int |x|^{\alpha-1}\omega(x)\,dx \neq 0$. By virtue of (7.1′), this implies

$$\left(f(x), \omega\left(\frac{x}{t}\right)\right) = |t|^{\alpha}, \qquad t \neq 0,$$

hence, the following equality holds

$$\int \left(f(x), \omega\left(\frac{x}{t}\right)\right) \frac{\varphi(t)}{|t|}\,dt = \left(|t|^{\alpha-1}, \varphi\right), \qquad \varphi \in \mathcal{D}(t \neq 0). \tag{7.2}$$

Since $\omega\left(\frac{x}{t}\right)\frac{\varphi(t)}{|t|} \in \mathcal{D}(\mathbb{R}^2)$, it follows from (7.2) by virtue of (3.2) of Sec. 3.3 that

$$\left(|t|^{\alpha-1}, \varphi\right) = \left(f(x), \int \omega\left(\frac{x}{t}\right)\frac{\varphi(t)}{|t|}\,dt\right). \tag{7.3}$$

We make the change of the variable of integration in the inner integral in (7.3) (for every $x \neq 0$)

$$t = \frac{x}{x'}, \qquad dt = \frac{|x|}{x'^2}\,dx'.$$

As a result, we obtain

$$\left(|t|^{\alpha-1}, \varphi\right) = \left(f(x), \int \omega(x')\varphi\left(\frac{x}{x'}\right)\frac{dx'}{|x'|}\right) = \int \frac{\omega(x')}{|x'|}\left(f(x), \varphi\left(\frac{x}{x'}\right)\right)dx'. \tag{7.4}$$

Here, we once again used formula (3.2) of Sec. 3.3, since $\frac{\omega(x')}{|x'|}\varphi\left(\frac{x}{x'}\right) \in \mathcal{D}(\mathbb{R}^2)$. Applying again property (7.1′) ($x' \neq 0$!) to the right-hand side of equality (7.4), we obtain the equality

$$\left(|x|^{\alpha-1}, \varphi\right) = \int \omega(x')|x'|^{\alpha-1}\,dx'(f, \varphi),$$

so, for $C = \left[\int \omega(x)|x|^{\alpha-1}\,dx\right]^{-1} \neq 0$, the following representation holds

$$(f, \varphi) = C\left(|x|^{\alpha-1}, \varphi\right), \qquad \varphi \in \mathcal{D}(x \neq 0). \tag{7.5}$$

The generalized function $|x|^{\alpha-1}$ from $\mathcal{D}'(x \neq 0)$ admits a regularization $\operatorname{reg}|x|^{\alpha-1}$, which is equal to $\mathcal{P}|x|^{\alpha-1}$ for $\alpha \neq -2n$ (see Sec. 2.7) and to $\mathcal{P}f|x|^{\alpha-1}$ for $\alpha = -2n$, $n = 0, 1, \dots$ (see Sec. 1.8). Therefore, equality (7.5) implies

$$\operatorname{supp}\left[f - C\operatorname{reg}|x|^{\alpha-1}\right] \subset \{0\}, \qquad \alpha \in C.$$

By the theorem of Sec. 2.6, we conclude

$$f(x) - C\operatorname{reg}|x|^{\alpha-1} = \sum_k c_k\delta^{(k)}(x), \qquad \alpha \in C. \tag{7.6}$$

For $\alpha \neq -2n$, $n = 0, 1, \dots$, we deduce from (7.6) that $c_k = 0$ and, hence, $f = C\mathcal{P}|x|^{\alpha-1}$ what is required to prove.

Now let $\alpha = -2n$, $n = 0, 1, \dots$. We choose a function $\omega \in \mathcal{D}$ with the properties: $\operatorname{supp}\omega \subset [-1, 1]$, $\omega \equiv 1$ in a neighbourhood of 0 and $(f, \omega) \neq 0$. From representation (7.5), by virtue of (7.1′) and (8.4) of Sec. 1.8, for all $t > 1$ we obtain the equalities

$$\left(f(x), \omega\left(\frac{x}{t}\right)\right) - \left(\mathcal{P}f|x|^{-2n-1}, \omega\left(\frac{x}{t}\right)\right) = C_0$$
$$= t^{-2n}(f, \omega) - t^{-2n} \int\limits_{|x|<1/t} \frac{\omega(x) - 1}{|x|^{2n+1}}\, dx. \quad (7.7)$$

For sufficiently large $t \geq T$, we have

$$\int\limits_{|x|<1/t} \frac{\omega(x) - 1}{|x|^{2n+1}}\, dx = 0.$$

This and (7.7) imply the equality

$$(f, \omega) = t^{2n} C_0, \qquad t \geq T,$$

hence $C_0 = 0$ and $(f, \omega) = 0$ that contradicts the assumption.

Consider now the case $\operatorname{supp} f = \{0\}$. By the theorem of Sec. 2.6,

$$f(x) = \sum_k c_k \delta^{(k)}(x).$$

From this, by virtue of evenness and homogeneity of f, we deduce that $f(x) = c_{2n}\delta^{(2n)}(x)$ for some $n = 0, 1, \ldots$ The theorem is proved. □

As a consequence of the theorem we note that the homogeneous generalized functions are tempered.

It follows from the proof of the theorem that the generalized functions $\mathcal{P}f|x|^{-N}$, $N = 1, 2, \ldots$, introduced in (8.4) of Sec. 1.8 are not homogeneous.

CHAPTER 2

INTEGRAL TRANSFORMATIONS OF GENERALIZED FUNCTIONS

One of the most powerful tools of investigation of problems in mathematical physics is the method of integral transformation. In this chapter we consider the theories of the Fourier transformation and the Laplace transformation that is closely linked with it; we also consider the transformations of Mellin, Cauchy–Bochner, Hilbert and Poisson for the class of tempered generalized functions.

6. The Fourier Transform of Tempered Generalized Functions

A remarkable property of the class of tempered generalized functions is that the operation of the Fourier transform does not take one outside that class.

6.1. The Fourier transform of test functions in $\mathcal{S}$. Since the test functions $\varphi(x)$ in $\mathcal{S}$ are integrable in $\mathbb{R}^n$, the (classical) operation of the Fourier transform $F[\varphi]$ is defined on them:

$$F[\varphi](\xi) = \int \varphi(x) e^{i(\xi,x)}\, dx, \qquad \varphi \in \mathcal{S}.$$

In this case, the function $F[\varphi](\xi)$, which is the *Fourier* transform of the function $\varphi(x)$, is bounded and continuous in $\mathbb{R}^n$. The test function $\varphi(x)$ decreases at infinity faster than any power of $|x|^{-1}$. Therefore, its Fourier transform may be differentiated under the integral sign any number of times:

$$\begin{aligned} \partial^\alpha F[\varphi](\xi) &= \int (ix)^\alpha \varphi(x) e^{i(\xi,x)}\, dx \\ &= F\big[(ix)^\alpha \varphi\big](\xi), \end{aligned} \tag{1.1}$$

whence it follows that $F[\varphi] \in C^\infty$. Furthermore, every derivative $\partial^\alpha \varphi(x)$ has the same properties and so

$$\begin{aligned} F[\partial^\alpha \varphi](\xi) &= \int \partial^\alpha \varphi(x) e^{i(\xi,x)}\, dx \\ &= (-i\xi)^\alpha F[\varphi](\xi), \end{aligned} \tag{1.2}$$

From (1.2) it follows, for one thing, that $F[\varphi](\xi)$ is an integrable function on $\mathbb{R}^n$.

From the general theory of the Fourier transformation it follows that the function $\varphi(x)$ is expressed in terms of its Fourier transform $F[\varphi](\xi)$ with the aid of the inverse Fourier transform, F^{-1}:

$$\varphi = F^{-1}\big[F[\varphi]\big] = F\big[F^{-1}[\varphi]\big] \tag{1.3}$$

where

$$
\begin{aligned}
F^{-1}[\psi](x) &= \frac{1}{(2\pi)^n}\int \psi(\xi)e^{-i(x,\xi)}\,d\xi \\
&= \frac{1}{(2\pi)^n}F[\psi](-x) \\
&= \frac{1}{(2\pi)^n}\int \psi(-\xi)e^{i(x,\xi)}\,d\xi \\
&= \frac{1}{(2\pi)^n}F[\psi(-\xi)].
\end{aligned} \tag{1.4}
$$

LEMMA. *The operation of the Fourier transform F carries $\mathcal{S}$ onto itself in reciprocal one-to-one fashion and reciprocal continuous fashion*[1].

PROOF. Let $\varphi \in \mathcal{S}$. Then, using (1.1) and (1.2), for all $p = 0, 1, \ldots$ and all α we obtain

$$
\begin{aligned}
(1+|\xi|^2)^{p/2}\,|\partial^\alpha F[\varphi](\xi)| &\le (1+|\xi|^2)^{\left[\frac{p+1}{2}\right]}\,|\partial^\alpha F[\varphi](\xi)| \\
&\le \left|\int (1-\Delta)^{\left[\frac{p+1}{2}\right]}\left[(ix)^\alpha\varphi(x)\right]e^{i(\xi,x)}\,dx\right| \\
&\le C\sup_x (1+|x|^2)^{(n+1)/2}\left|(1-\Delta)^{\left[\frac{p+1}{2}\right]}|x^\alpha\varphi(x)|\right|,
\end{aligned}
$$

whence we derive the estimates (see Sec. 5.1)

$$
\|F[\varphi]\|_p \le C_p\|\varphi\|_{p+n+1}, \qquad p = 0, 1, \ldots, \tag{1.5}
$$

for certain C_p that does not depend on φ. (Here, $[x]$ is the integer part of the number $x \ge 0$.) The estimate (1.5) shows that the operation $\varphi \to F[\varphi]$ transforms $\mathcal{S}$ to $\mathcal{S}$ and is continuous. Furthermore, from (1.3) and (1.4) it follows that any function of φ taken from $\mathcal{S}$ is a Fourier transform of the function $\psi = F^{-1}[\varphi]$ taken from $\mathcal{S}$, $\varphi = F[\psi]$, and if $F[\varphi] = 0$, then $\varphi = 0$ as well. This means that the mapping $\varphi \to F[\varphi]$ carries $\mathcal{S}$ onto $\mathcal{S}$ in a reciprocal one-to-one fashion. The properties are similar for the operation of the inverse Fourier transform, F^{-1}. This complete the proof of the lemma. □

6.2. The Fourier transform of tempered generalized functions. First let $f(x)$ be an integrable function on $\mathbb{R}^n$. Then its Fourier transform

$$
F[f](\xi) = \int f(x)e^{i(\xi,x)}\,dx, \qquad |F[f](\xi)| \le \int |f(x)|\,dx < \infty,
$$

is a (continuous) bounded function in $\mathbb{R}^n$ and, hence, determines a regular tempered generalized function via the formula (see Sec. 5.3)

$$
(F[f], \varphi) = \int F[f](\xi)\varphi(\xi)\,d\xi, \qquad \varphi \in \mathcal{S}.
$$

[1] We say that the mapping F is a (linear) *isomorphism of $\mathcal{S}$ onto $\mathcal{S}$*.

Using the Fubini theorem on changing the order of integration, we transform the last integral:

$$\begin{aligned}\int F[f](\xi)\varphi(\xi)\,d\xi &= \int\left[\int f(x)e^{i(\xi,x)}\,dx\right]\varphi(\xi)\,d\xi \\ &= \int f(x)\int \varphi(\xi)e^{i(x,\xi)}\,d\xi\,dx \\ &= \int f(x)F[\varphi](x)\,dx\end{aligned}$$

that is,

$$(F[f],\varphi) = (f, F[\varphi]), \qquad \varphi \in S.$$

It is this equation that we take for the definition of the *Fourier transform* $F[f]$ *of any tempered generalized function* f:

$$(F[f],\varphi) = (f, F[\varphi]), \qquad f \in S', \qquad \varphi \in S. \tag{2.1}$$

Since by the lemma of Sec. 6.1 the operation $\varphi \to F[\varphi]$ is linear and continuous from S to S, the functional $F[f]$ defined by the right-hand side of (2.1) is a generalized function taken from S' and, what is more, the operation $f \to F[f]$ is linear and continuous from S' ıto S'.

Let us now introduce in S' yet another Fourier transform operation, which we will denote by F^{-1}, via the formula [compare (1.4)]

$$F^{-1}[f] = \frac{1}{(2\pi)^n}F[f(-x)], \qquad f \in S', \tag{2.2}$$

where $f(-x)$ is a reflection of $f(x)$ (see Sec. 1.9). Clearly, F^{-1} is a linear and continuous operation from S' to S'.

Now we will prove that F^{-1} is the inverse of F, that is,

$$F^{-1}[F[f]] = f, \qquad F[F^{-1}[f]] = f, \qquad f \in S'. \tag{2.3}$$

Indeed, by virtue of (1.3) and (1.4), the formulas (2.3) hold on the set S, which is dense in S' (see Sec. 5.6); the operations F and F^{-1} are continuous from S' to S'. Hence, the formulas (2.3) hold true for all f in S' as well.

From (2.3) it follows that any f in S' is a Fourier transform of some $g = F^{-1}[f]$ in S', $f = F[g]$, and if $F[f] \to 0$, then $f = 0$. Thus, we have proved that

the operation $f \to F[f]$ transforms S' to S' in a reciprocal one-to-one fashion and a reciprocal continuous fashion, that is, we have a (linear) isomorphism of S' onto S'.

Suppose $f(x,y) \in S'(\mathbb{R}^{n+m})$, where $x \in \mathbb{R}^n$ and $y \in \mathbb{R}^m$. We introduce the Fourier transform $F_x[f]$ with respect to the variables $x = (x_1, \ldots, x_n)$ by putting, for any test function $\varphi(\xi, y)$ in $S(\mathbb{R}^{n+m})$,

$$(F_x[f],\varphi) = (f, F_\xi[\varphi]). \tag{2.4}$$

As in Sec. 6.1, we establish that the operation

$$\varphi(\xi,y) \to F_\xi[\varphi](x,y) = \int \varphi(\xi,y)e^{i(x,\xi)}\,d\xi$$

accomplishes a (linear) isomorphism of $S(\mathbb{R}^{n+m})$ onto $S(\mathbb{R}^{n+m})$ so that the formula (2.4) does indeed defines a generalized function $F_x[f](\xi,y)$ in $S'(\mathbb{R}^{n+m})$. The operation of Fourier inversion is defined in a manner similar to that of (2.2):

$$F_\xi^{-1}[g] = \frac{1}{(2\pi)^n} F_\xi\left[g(-\xi, y)\right](x, y), \qquad g \in \mathcal{S}'(\mathbb{R}^{n+m}). \tag{2.5}$$

The operation $f \to F_x[f]$ is a (linear) isomorphism of $\mathcal{S}'(\mathbb{R}^{n+m})$ onto $\mathcal{S}'(\mathbb{R}^{n+m})$.

EXAMPLE.

$$F\left[\delta(x - x_0)\right] = e^{i(\xi, x_0)}. \tag{2.6}$$

Indeed,

$$\begin{aligned}\left(F\left[\delta(x - x_0)\right], \varphi\right) &= \left(\delta(x - x_0), F[\varphi]\right) = F[\varphi](x_0)\\ &= \int \varphi(\xi) e^{i(x_0, \xi)}\, d\xi = \left(e^{i(x_0, \xi)}, \varphi\right), \qquad \varphi \in \mathcal{S}.\end{aligned}$$

Putting $x_0 = 0$ in (2.6), we get

$$F[\delta] = 1 \tag{2.7}$$

whence, by (2.2), we derive

$$\delta = F^{-1}[1] = \frac{1}{(2\pi)^n} F[1]$$

so that

$$F[1] = (2\pi)^n \delta(\xi). \tag{2.8}$$

6.3. Properties of the Fourier transform. The formulas for the Fourier transform given in this subsection hold true on the test functions in $\mathcal{S}$. But $\mathcal{S}$ is dense in $\mathcal{S}'$. Therefore, these formulas remains true also for all generalized functions in $\mathcal{S}'$.

6.3.1. *Differentiating a Fourier transform:*

$$\partial^\alpha F[f] = F\left[(ix)^\alpha f\right], \qquad f \in \mathcal{S}'. \tag{3.1}$$

In particular, putting $f = 1$ in (3.1) and using (2.8), we obtain

$$F[x^\alpha] = (-i)^{|\alpha|} \partial^\alpha F[1] = (2\pi)^n (-i)^{|\alpha|} \partial^\alpha \delta(\xi). \tag{3.2}$$

6.3.2. *The Fourier transform of a derivative:*

$$F[\partial^\alpha f] = (-i\xi)^\alpha F[f], \qquad f \in \mathcal{S}'. \tag{3.3}$$

Putting $f = \delta$ in (3.3) and using (2.7), we obtain

$$F[\partial^\alpha \delta] = (-i\xi)^\alpha F[\delta] = (-i\xi)^\alpha. \tag{3.4}$$

6.3.3. *The Fourier transform of a translation:*

$$F\left[f(x - x_0)\right] = e^{i(\xi, x_0)} F[f], \qquad f \in \mathcal{S}'. \tag{3.5}$$

6.3.4. *The translation of a Fourier transform:*

$$F[f](\xi + \xi_0) = F\left[e^{i(\xi_0, x)} f\right](\xi), \qquad f \in \mathcal{S}'. \tag{3.6}$$

6.3.5. *The Fourier transform under a linear transformation of the argument* (see Sec. 5.3):

$$F\left[f(Ax)\right](\xi) = \frac{1}{|\det A|} F[f]\left((A^{-1})^T \xi\right), \qquad \det A \neq 0. \tag{3.7}$$

Here, $A \to A^T$ denotes the transpose operation of the matrix A.

6.3.6. *The Fourier transform of a direct product:*

$$\begin{aligned} F[f(x) \times g(y)] &= F_x[f(x) \times F[g](\eta)] \\ &= F_y[F[f](\xi) \times g(y)] \\ &= F[f](\xi) \times F[g](\eta). \end{aligned} \tag{3.8}$$

6.3.7. Analogous formulas hold true also for the Fourier transform F_x (see Sec. 6.2), for example:

$$\begin{aligned} \partial_\xi^\alpha \partial_y^\beta F_x[f] &= F_x[(ix)^\alpha \partial_y^\beta f], \\ F_x[\partial_\xi^\alpha \partial_y^\beta f] &= (-i\xi)^\alpha \partial_y^\beta F_x[f]. \end{aligned} \tag{3.9}$$

6.4. The Fourier transform of generalized functions with compact support. If f is a generalized function with compact support, $f \in \mathcal{E}'$, then it is tempered, $f \in \mathcal{S}'$ (see Sec. 5.3), and therefore its Fourier transform exists. What is more, the following theorem holds true.

THEOREM. *If $f \in \mathcal{E}'$, then the Fourier transform $F[f]$ exists in θ_M and can be represented as*

$$F[f](\xi) = (f(x), \eta(x)e^{i(\xi,x)}), \tag{4.1}$$

where η is any function in $\mathcal{D}$ equal to 1 in a neighbourhood of the support of f. And there exist numbers $C_\alpha \geq 0$ and $m \geq 0$ such that

$$|\partial^\alpha F[f](\xi)| \leq \|f\|_{-m} C_\alpha (1 + |\xi|^2)^{m/2}, \qquad \xi \in \mathbb{R}^n. \tag{4.2}$$

PROOF. Taking into account the equalities (3.2) of Sec. 5.3 and (3.3), we obtain, for all $\varphi \in \mathcal{S}$,

$$\begin{aligned} (\partial^\alpha F[f], \varphi) &= (-1)^{|\alpha|}(F[f], \partial^\alpha \varphi) \\ &= (-1)^{|\alpha|}(f, F[\partial^\alpha \varphi]) \\ &= (-1)^{|\alpha|}(f, \eta(x)(-ix)^\alpha F[\varphi]) \\ &= \left(f(x), \int \eta(x)(ix)^\alpha \varphi(\xi) e^{i(x,\xi)}\, d\xi\right). \end{aligned}$$

Now, noting that

$$\eta(x)(ix)^\alpha \varphi(\xi) e^{i(x,\xi)} \in \mathcal{S}(\mathbb{R}^{2n})$$

and using (5.4) of Sec. 5.5,

$$\left(f(x), \int \eta(x)(ix)^\alpha \varphi(\xi) e^{i(x,\xi)}\, d\xi\right) = \int (f(x), \eta(x)(ix)^\alpha e^{i(x,\xi)}) \varphi(\xi)\, d\xi,$$

we derive the following equation from the preceding ones:

$$(\partial^\alpha F[f], \varphi) = \int (f(x), \eta(x)(ix)^\alpha e^{i(x,\xi)}) \varphi(\xi)\, d\xi.$$

It follows from this equation that

$$\partial^\alpha F[f](\xi) = (f(x), \eta(x)(ix)^\alpha e^{i(x,\xi)}). \tag{4.3}$$

And from (4.3), for $\alpha = 0$, follows the formula (4.1).

From the representation (4.3), as in the proof of the lemma of Sec. 5.5, we derive that $F[f] \in C^\infty$. Let m be the order of f. Applying to the right-hand side of (4.3) the inequality (2.3) of Sec. 5.2, we obtain, for all $\xi \in \mathbb{R}^n$, the estimate (4.2):

$$\begin{aligned}|\partial^\alpha F[f](\xi)| &= \left|\left(f(x), \eta(x)(ix)^\alpha e^{i(x,\xi)}\right)\right| \\ &\le \|f\|_{-m}\|\eta(x)(ix)^\alpha e^{i(x,\xi)}\|_m \\ &= \|f\|_{-m} \sup_{x, |\beta|\le m} \left(1+|x|^2\right)^{m/2} |\partial_x^\beta[\eta(x)x^\alpha e^{i(x,\xi)}]| \\ &\le \|f\|_{-m} C_\alpha\left(1+|\xi|^2\right)^{m/2}\end{aligned}$$

for certain $C_\alpha \ge 0$. Thus, $F[f] \in \theta_M$, and the proof of the theorem is complete. □

REMARK. As may be seen from the proof of the theorem, the numbers C_α that appear in the inequality (4.2) may be chosen as being independent of the family of generalized functions f if all supports of that family are uniformly bounded.

6.5. The Fourier transform of a convolution. Let $f \in \mathcal{S}'$ and $g \in \mathcal{E}'$. Then their convolution $f * g \in \mathcal{E}'$ (see Sec. 5.6.1) and its Fourier transform can be calculated from the formula

$$F[f * g] = F[f]F[g]. \tag{5.1}$$

True enough, by virtue of (6.1) of Sec. 5.6, the convolution $f * g \in \mathcal{S}'$ can be represented in the form

$$(f * g, \varphi) = (f(x), (g(y), \eta(y)\varphi(x+y))), \qquad \varphi \in \mathcal{S},$$

where $\eta \in \mathcal{D}$, $\eta(y) = 1$ in a neighbourhood of supp g. Taking this representation into account and making use of the definition of the Fourier transform (see Sec. 6.2), we obtain

$$\begin{aligned}(F[f * g], \varphi) &= (f * g, F[\varphi]) \\ &\left(f(x), \left(g(y), \eta(y)\int \varphi(\xi)e^{i(x+y,\xi)}\, d\xi\right)\right).\end{aligned}$$

Using the formulae (5.4) of Sec. 5.5 and (4.1) and taking into account that $F[g] \in \theta_M$, we transform the resulting equation:

$$\begin{aligned}(F[f * g], \varphi) &= \left(f(x), \int \left(g(y), \eta(y)e^{i(\xi,y)}\right)e^{i(x,\xi)}\varphi(\xi)\, d\xi\right) \\ &= \left(f(x), \int F[g](\xi)e^{i(x,\xi)}\varphi(\xi)\, d\xi\right) \\ &= (f, F[F[g]\varphi]) \\ &= (F[f], F[g]\varphi) \\ &= (F[g]F[f], \varphi),\end{aligned}$$

whence follows (5.1). □

The chain of equalities in the proof of formula (5.1) implies the formula of the *Fourier transform of the product*: if $f \in \mathcal{S}'$ and $F[g] \in \mathcal{E}'$, then

$$F[f \cdot g] = \frac{1}{(2\pi)^n} F[f] * F[g]. \tag{5.2}$$

Some other cases follow in which (5.1) holds true:

6.5.1. Let $f \in \mathcal{S}'$, $g \in \mathcal{S}$. Then $f * g \in \theta_M$. This follows from Sec. 5.6.3.

6.5.2. Let f and $g \in \mathcal{L}^2$. Then $f * g \in C$ and $(f * g)(x) = o(1)$, $|x| \to \infty$, that is, $f * g \in \overline{C}_0$ (see Sec. 0.5).

Indeed, in this case, $F[f]$ and $F[g] \in \mathcal{L}^2$ and, hence, $F[f]F[g] \in \mathcal{L}^1$. Besides, $f(y)g(x-y)\varphi(x) \in \mathcal{L}^1(\mathbb{R}^{2n})$ for all $\varphi \in \mathcal{S}$ by virtue of the Cauchy–Bunyakovsky inequality:

$$\left[\int |f(y)g(x-y)|\,|\varphi(x)|\,dx\,dy\right]^2$$
$$\leq \left[\int |f(y)|^2|\varphi(x)|\,dx\,dy\right]\left[\int |g(x-y)|^2|\varphi(x)|\,dx\,dy\right]$$
$$\leq \|f\|^2\|g\|^2\left[\int |\varphi(x)|\,dx\right]^2 < \infty.$$

Therefore, using (1.1) for the convolution $f * g$ (see Sec. 4.1.2), we obtain the following equalities with the aid of the Fubini theorem for all $\varphi \in \mathcal{S}$:

$$\begin{aligned}(F[f * g], \varphi) &= (f * g, F[\varphi]) \\ &= \int F[\varphi](x) \int f(y)g(x-y)\,dy\,dx \\ &= \int f(y) \int g(x-y)F[\varphi](x)\,dx\,dy \\ &= \int f(y) \int F[g(x-y)](\xi)\varphi(\xi)\,d\xi\,dy \\ &= \int f(y) \int F[g]\xi)\varphi(\xi)e^{i(y,\xi)}\,d\xi\,dy \\ &= F[g]F[f]\varphi\,d\xi.\end{aligned}$$

From these follows formula (5.1). Therefore

$$f * g = F^{-1}[F[f]F[g]] \in C$$

and by the Riemann–Lebesgue theorem $(f * g)(x) \to 0$, $|x| \to \infty$. □

REMARK. If it is known that the convolution $f * g$ exists in $\mathcal{S}'$ [for example, for $f \in \mathcal{S}'(\Gamma+)$ and $g \in \mathcal{S}'(\overline{S}_+)$ (see Sec. 5.6.2)], then (5.1) may serve as a definition of the product of the generalized functions $F[f]$ and $F[g]$ (compare Sec. 1.10).

Product of tempered generalized functions. In order to define such a product we use formula (1.7) of Sec. 4.1: if f and $g \in \mathcal{S}'$, then their product is defined by the formula

$$f \cdot g = \lim_{k\to\infty} (f * \delta_k)g \quad \text{in} \quad \mathcal{S}',$$

if the limit exists for any special δ-sequence $\{\delta_k\}$ and does not depend on this sequence.

If $f \cdot g$ exists, then $g \cdot f$ also exists and they are equal

$$f \cdot g = g \cdot f. \tag{5.3}$$

The question arises: what are the most general conditions under which formula (5.1) of the Fourier transform of a convolution and the inverse formula (5.2) of the Fourier transform of a product hold? In order to obtain these formulae, it is necessary to extend the notion of the convolution of tempered generalized functions.

Let $\eta \in \mathcal{S}$ ($\eta \in \mathcal{D}$), $\eta(0) = 1$ and $\lambda_k \to \infty$, $k \to \infty$. We call the sequence of the functions $e_k(x) = \eta(x/\lambda_k)$, $k \to \infty$, the *special 1-sequence* in $\mathcal{S}$ (in $\mathcal{D}$).

By the *convolution* $f \circledast g$ of generalized functions f and g taken from $\mathcal{S}'$ we call the limit

$$f \circledast g = \lim (e_k f) * g \quad \text{in} \quad \mathcal{S}',$$

if this limit exists for any special 1-sequence $\{\eta_k\}$ in $\mathcal{S}$ and does not depend on it. Then $f \circledast g \in \mathcal{S}'$.

If $f \circledast g$ exists, then $g \circledast f$ also exists and they are equal

$$f \circledast g = g \circledast f. \tag{5.4}$$

The convolution $f \circledast g$ is more general than the convolution $f * g$ as the following example shows:

$$f = \theta_{(-1,1)}, \qquad g = \delta, \qquad f \cdot g = \delta, \qquad F[f] = 2\frac{\sin\xi}{\xi}, \qquad F[g] = 1.$$

The convolution $1 * \frac{\sin\xi}{\xi}$ does not exist (see Kaminski [54]), however, the convolution $1 \circledast \frac{\sin\xi}{\xi}$ exists and is equal to π. (Prove this fact without using formula (5.6).)

Let $f, g \in \mathcal{S}'$ and the convolution $f \circledast g$ exist in $\mathcal{S}'$. Then there exists $F[f] \cdot F[g]$ in $\mathcal{S}'$ and the formula of the Fourier transform of the convolution

$$F[f \circledast g] = F[f] \cdot F[g] \tag{5.5}$$

holds.

Let $f, g \in \mathcal{S}'$ and the product $f \cdot g$ exist in $\mathcal{S}'$. Then there exists the convolution $F[f] \circledast F[g]$ in $\mathcal{S}'$ and the formula of the Fourier transform of the product

$$F[f \cdot g] = \frac{1}{(2\pi)^n} F[f] \circledast F[g] \tag{5.6}$$

holds.

See the details and the proofs in Kaminski [54], Hirata, Ogata [45], Dierolf, Voigt [16].

6.6. Examples.

6.6.1.

$$F\left[e^{-\alpha^2 x^2}\right] = \frac{\sqrt{\pi}}{|\alpha|} e^{-\frac{\xi^2}{4\alpha^2}}, \qquad \alpha \neq 0, \qquad n = 1. \tag{6.1}$$

True enough, the function $e^{-\alpha^2 x^2}$ is integrable on $\mathbb{R}^1$ and therefore ($\alpha > 0$)

$$\begin{aligned} F\left[e^{-\alpha^2 x^2}\right] &= \int e^{-\alpha^2 x^2 + i\xi x}\, dx = \frac{1}{\alpha} \int e^{-\sigma^2 + i\frac{\xi}{\alpha}\sigma}\, d\sigma \\ &= \frac{1}{\alpha} e^{-\frac{\xi^2}{4\alpha^2}} \int e^{-\left(\sigma + \frac{i\xi}{2\alpha}\right)^2} d\sigma \\ &= \frac{1}{\alpha} e^{-\frac{\xi^2}{4\alpha^2}} \int\limits_{\Im\zeta = \xi/(2\alpha)} e^{-\zeta^2}\, d\zeta. \end{aligned}$$

In the last integral, the line of integration may be shifted onto the real axis and therefore

$$F\left[e^{-\alpha^2 x^2}\right] = \frac{1}{\alpha} e^{-\frac{\xi^2}{4\alpha^2}} \int_{-\infty}^{\infty} e^{-\sigma^2}\, d\sigma = \frac{\sqrt{\pi}}{\alpha} e^{-\frac{\xi^2}{4\alpha^2}}.$$

□

6.6.2. A multi-dimensional analogue of formula (6.1) is

$$F\left[e^{-(Ax,x)}\right] = \frac{\pi^{n/2}}{\sqrt{\det A}} e^{-\frac{1}{4}(A^{-1}\xi,\xi)}, \tag{6.2}$$

where A is a real positive-definite matrix.

To obtain (6.2) with the aid of a nonsingular real linear transformation $x = By$, let us reduce the quadratic form (Ax, x) to a sum of squares

$$(Ax, x) = (ABy, By) = (B^T ABy, y) = |y|^2.$$

Note that

$$A^{-1} = BB^T, \qquad \det A|\det B|^2 = 1.$$

From this, using the formula (6.1), we obtain

$$\begin{aligned} F\left[e^{-(Ax,x)}\right] &= \int e^{-(Ax,x)+i(\xi,x)}\,dx \\ &= |\det B| \int e^{-(ABy,By)+i(\xi,By)}\,dy \\ &= \frac{1}{\sqrt{\det A}} \int e^{-|y|^2+i(B^T\xi,y)}\,dy \\ &= \frac{1}{\sqrt{\det A}} \prod_{1\le j\le n} \int e^{-y_j^2+i(B^T\xi)_j y_j}\,dy_j \\ &= \frac{\pi^{n/2}}{\sqrt{\det A}} e^{-\frac{1}{4}|B^T\xi|^2} \\ &= \frac{\pi^{n/2}}{\sqrt{\det A}} e^{-\frac{1}{4}(\xi,BB^T\xi)} \\ &= \frac{\pi^{n/2}}{\sqrt{\det A}} e^{-\frac{1}{4}(\xi,A^{-1}\xi)}. \end{aligned}$$

□

6.6.3. Let the function $f(x)$ be tempered in $\mathbb{R}^n$ (see Sec. 5.3). Then

$$F[f](\xi) = \lim_{R\to\infty} \int_{|x|<R} f(x)e^{i(\xi,x)}\,dx \quad \text{in} \quad \mathcal{S}'. \tag{6.3}$$

Indeed,

$$\Theta(R - |x|)f(x) \to f(x), \qquad R \to \infty \quad \text{in} \quad \mathcal{S}'.$$

whence by virtue of the continuity, in $\mathcal{S}'$, of the Fourier transform operation F, follows the equation (6.3). □

In particular, for $f \in \mathcal{L}^2$ the following theorem of Plancherel holds true: *The Fourier transform $F[f]$ is expressed by the equation*

$$F[f](\xi) = \lim_{R\to\infty} \int_{|x|<R} f(x)e^{i(\xi,x)}\,dx \quad \textit{in} \quad \mathcal{L}^2.$$

It maps $\mathcal{L}^2$ onto $\mathcal{L}^2$ reciprocally in one-to-one fashion and reciprocally continuously; the Parseval–Steklov equation

$$(2\pi)^n (f,\varphi) = (F[f], F[\varphi]), \qquad f, \varphi \in \mathcal{L}^2$$

holds true so that

$$(2\pi)^n \|f\|^2 = \|F[f]\|^2, \qquad f \in \mathcal{L}^2$$

(the scalar product $\langle \cdot, \cdot \rangle$ is defined in Sec. 0.5).

6.6.4. Let f be an arbitrary tempered generalized function. By the theorem of Sec. 5.4 there exist a function $g(x)$, which is continuous and tempered in $\mathbb{R}^n$, and an integer $m \geq 0$ such that

$$f(x) = \partial_1^m \ldots \partial_n^m g(x).$$

From this, using (3.3), we get

$$F[f] = (-i)^{mn} \xi_1^m \ldots \xi_n^m F[g], \tag{6.4}$$

and the Fourier transform $F[g]$ may be computed via (6.3).

6.6.5.

$$F\left[e^{ix^2}\right] = \sqrt{\pi} e^{-\frac{i}{4}(\xi^2 - \pi)}, \qquad n = 1. \tag{6.5}$$

True enough, from the convergence of the improper integral (Fresnel's integral)

$$\int_{-\infty}^{\infty} e^{iy^2}\, dy = \sqrt{\pi} e^{-\frac{i\pi}{4}}$$

it follows that the sequence of Fourier transforms

$$\int_{-R}^{R} e^{ix^2 + i\xi x}\, dx = e^{-\frac{i}{4}\xi^2} \int_{-R+\frac{\xi}{2}}^{R+\frac{\xi}{2}} e^{iy^2}\, dy, \qquad R \to \infty,$$

converges uniformly with respect to ξ on every interval to the function

$$e^{-\frac{i}{4}\xi^2} \int_{-\infty}^{\infty} e^{iy^2}\, dy = \sqrt{\pi} e^{-\frac{i}{4}(\xi^2 - \pi)}.$$

From this, by virtue of 6.6.3, we conclude that (6.5) holds true on all test functions in $\mathcal{D}$. But $\mathcal{D}$ is dense in $\mathcal{S}$ (see Sec. 6.1) and so (6.5) holds true in $\mathcal{S}'$. □

6.6.6. A multi-dimensional analogue of (6.5) is the equation [compare Sec. 6.6.2]

$$F\left[e^{i(Ax,x)}\right] = \frac{\pi^{n/2}}{\sqrt{\det A}} e^{i\frac{\pi n}{4} - \frac{i}{4}(A^{-1}\xi, \xi)} \tag{6.6}$$

where A is a real positive-definite matrix.

6.6.7.

$$F\left[\frac{1}{|x|^2}\right] = \frac{2\pi^2}{|\xi|}, \qquad n = 3. \tag{6.7}$$

We have

$$\int\limits_{|x|<R} \frac{e^{i(\xi,x)}}{|x|^2}\,dx = \int\limits_0^R \int\limits_0^\pi \int\limits_0^{2\pi} \frac{e^{i|\xi|\rho\cos\theta}}{\rho^2}\rho^2\,d\psi\,\sin\theta\,d\theta\,d\rho$$

$$= 2\pi \int\limits_0^R \int\limits_{-1}^1 e^{i|\xi|\rho\mu}\,d\mu\,d\rho$$

$$= \frac{4\pi}{|\xi|}\int\limits_0^R \frac{\sin(|\xi|\rho)}{\rho}\,d\rho.$$

Since

$$\left|\int\limits_R^\infty \frac{\sin(|\xi|\rho)}{\rho}\,d\rho\right| = \left|\frac{\cos(|\xi|R)}{|\xi|R} - \frac{1}{|\xi|}\int\limits_R^\infty \frac{\cos(|\xi|\rho)}{\rho^2}\,d\rho\right| \le \frac{2}{|\xi|R},$$

$$\int\limits_0^\infty \frac{\sin(|\xi|\rho)}{\rho}\,d\rho = \frac{\pi}{2}, \qquad |\xi| \ne 0,$$

it follows that

$$\frac{4\pi}{|\xi|}\int\limits_0^R \frac{\sin(|\xi|\rho)}{\rho}\,d\rho \to \frac{2\pi^2}{|\xi|}, \qquad R\to\infty \quad \text{in} \quad S'$$

and, by virtue of Sec. 6.6.3, the equation (6.7) holds true. □

6.6.8. Let $n = 2$. We introduce the generalized function $\mathrm{Pf}\,\frac{1}{|x|^2}$ taken from S', which function operates via the rule (cf. (8.4) Sec. 1.8)

$$\left(\mathrm{Pf}\,\frac{1}{|x|^2},\varphi\right) = \int\limits_{|x|<1} \frac{\varphi(x)-\varphi(0)}{|x|^2}\,dx + \int\limits_{|x|>1} \frac{\varphi(x)}{|x|^2}\,dx.$$

Obviously, $\mathrm{Pf}\,\frac{1}{|x|^2} = \frac{1}{|x|^2}$ for $x \ne 0$. Let us prove that formula

$$F\left[\mathrm{Pf}\,\frac{1}{|x|^2}\right] = -2\pi\ln|\xi| - 2\pi c_0, \tag{6.8}$$

where

$$c_0 = \int\limits_0^1 \frac{1-J_0(u)}{u}\,du - \int\limits_1^\infty \frac{J_0(u)}{u}\,du,$$

(J_0 is the Bessel function).

Indeed, for all $\varphi \in S$ the following chain of equalities holds true:

$$\left(F\left[\operatorname{Pf}\frac{1}{|x|^2}\right], \varphi\right) = \left(\operatorname{Pf}\frac{1}{|x|^2}, F[\varphi]\right)$$

$$= \int\limits_{|x|<1} \frac{F[\varphi](x) - F[\varphi](0)}{|x|^2}\, dx + \int\limits_{|x|>1} \frac{F[\varphi](x)}{|x|^2}\, dx$$

$$= \int\limits_{|x|<1} \frac{1}{|x|^2} \int \varphi(\xi)\left[e^{i(x,\xi)} - 1\right] d\xi\, dx$$

$$+ \int\limits_{|x|>1} \frac{1}{|x|^2} \int \varphi(\xi) e^{i(x,\xi)}\, d\xi\, dx$$

$$= \int\limits_0^1 \frac{1}{r} \int \varphi(\xi) \int\limits_0^{2\pi} \left(e^{ir|\xi|\cos\theta} - 1\right) d\theta\, d\xi\, dr$$

$$+ \int\limits_1^\infty \frac{1}{r} \int \varphi(\xi) \int\limits_0^{2\pi} e^{ir|\xi|\cos\theta}\, d\theta\, d\xi\, dr$$

$$= 2\pi \int\limits_0^1 \frac{1}{r} \int \varphi(\xi)\left[J_0(r|\xi|) - 1\right] d\xi\, dr$$

$$+ 2\pi \int\limits_1^\infty \frac{1}{r} \int \varphi(\xi) J_0(r|\xi|)\, d\xi\, dr$$

$$= 2\pi \int \varphi(\xi) \left[\int\limits_0^1 \frac{J_0(r|\xi|) - 1}{r}\, dr + \int\limits_1^\infty \frac{J_0(r|\xi|)}{r}\, dr\right] d\xi$$

$$= 2\pi \int \varphi(\xi) \left[\int\limits_0^{|\xi|} 1\frac{J_0(u) - 1}{u}\, du + \int\limits_{|\xi|}^\infty \frac{J_0(u)}{u}\, du\right] d\xi$$

$$= -2\pi \int \varphi(\xi)(\ln|\xi| + c_0)\, d\xi.$$

And formula (6.8) follows from this. □

6.6.9. Let Γ be a closed convex acute cone in $\mathbb{R}^n$ (with vertex at 0) and let $f \in S'(\Gamma+)$ (see Sec. 5.6.2). Then the following formula holds true in the sense of convergence in S':

$$F[f](\xi) = \lim_{\substack{\xi' \to 0 \\ \xi' \in \operatorname{int}\Gamma^*}} \left(f(x), \eta(x) e^{i(x,\xi) - i(x,\xi')}\right), \tag{6.9}$$

where η is any C^∞-function with the following properties:

$$\left|\partial^\alpha \eta(x)\right| \le C_\alpha; \qquad \eta(x) = 1, \quad x \in (\operatorname{supp} f)^\varepsilon;$$
$$\eta(x) = 0, \quad x \notin (\operatorname{supp} f)^{2\varepsilon}$$

(ε is any positive number).

To prove (6.9), we first note that

$$\eta(x)e^{-(x,\xi')} \in \mathcal{S} \quad \text{for all} \quad \xi' \in \operatorname{int}\Gamma^*, \tag{6.10}$$

$$\eta(x)f(x)e^{-(x,\xi')} \to f(x), \qquad \xi' \to 0, \quad \xi' \Subset \operatorname{int}\Gamma^* \quad \text{in} \quad \mathcal{S}'. \tag{6.11}$$

Indeed, if $x \notin (\operatorname{supp} f)^{2\epsilon}$, then $\eta(x) = 0$; but if $x \in (\operatorname{supp} f)^{2\epsilon}$, then $x = x' + x''$, where $x' \in \Gamma$, $|x''| \le R$ for some $R > 0$. Let $\xi' \in C' \Subset \operatorname{int}\Gamma^*$. Then by Lemma 1 of Sec. 4.4 there is a number $\sigma = \sigma(C') > 0$ such that $(x', \xi') \ge \sigma|x'|\,|\xi'|$ and therefore

$$\begin{aligned} -(x,\xi') &= -(x',\xi') - (x'',\xi') \\ &\le -\sigma|x'|\,|\xi'| + R|\xi'| \\ &\le (-\sigma|x| + \sigma R + R)|\xi'|. \end{aligned}$$

The relations (6.10) and (6.11) follow from the resulting estimate and from the properties of the function $\eta(x)$.

Now, for all $\varphi \in \mathcal{S}$, we have a chain of equalities:

$$\begin{aligned} (F[f],\varphi) &= (f, F[\varphi]) \\ &= \lim_{\substack{\xi' \to 0 \\ \xi' \Subset \operatorname{int}\Gamma^*}} \left(\eta(x)f(x)e^{-(x,\xi')}, \int \varphi(\xi)e^{i(x,\xi)}\,d\xi\right) \\ &= \lim_{\substack{\xi' \to 0 \\ \xi' \Subset \operatorname{int}\Gamma^*}} \int \left(f(x), \eta(x)e^{i(x,\xi)-i(x,\xi')}\right)\varphi(\xi)\,d\xi, \\ &= \lim_{\substack{\xi' \to 0 \\ \xi' \Subset \operatorname{int}\Gamma^*}} \left(\eta(x)f(x)e^{-i(x,\xi')}, \int \varphi(\xi)e^{i(x,\xi)}\,d\xi\right), \end{aligned}$$

whence follows formula (6.9). Here we made use of (5.4) of Sec. 5.5, since

$$\eta(x)\varphi(\xi)e^{i(x,\xi)-i(x,\xi')} \in \mathcal{S}(\mathbb{R}^{2n}) \quad \text{for all} \quad \xi' \in \operatorname{int}\Gamma^*.$$

6.6.10.

$$F[\theta(x)] = \frac{i}{\xi + i0} = \pi\delta(\xi) + i\mathcal{P}\frac{1}{\xi}, \tag{6.12}$$

$$F[\theta(-x)] = \frac{-i}{\xi - i0} = \pi\delta(\xi) - i\mathcal{P}\frac{1}{\xi}. \tag{6.12'}$$

These formulae follow from (6.9) and from the Sochozki formulas (8.3) and (8.3′) of Sec. 1.8, for example:

$$F[\theta] = \lim_{\xi' \to +0} \int_0^\infty e^{ix(\xi + i\xi')}\,dx = \lim_{\xi' \to +0} \frac{1}{\xi + i\xi'} = \frac{i}{\xi + i0}.$$

6.6.11.

$$F[\operatorname{sgn} x] = F[\theta(x)] - F[\theta(-x)] = 2i\mathcal{P}\frac{1}{\xi}. \tag{6.13}$$

6.6.12.

$$F\left[\mathcal{P}\frac{1}{x}\right] = -2\pi F^{-1}\left[\mathcal{P}\frac{1}{x}\right] = \pi i \operatorname{sgn}\xi. \tag{6.14}$$

6.6.13. Let V^+ be a future light cone in $\mathbb{R}^{n+1}$ (see Sec. 4.4) and let $\theta_{V^+}(x)$ be its characteristic function. Then, by virtue of 6.6.9,

$$F[\theta_{V^+}] = \lim_{\xi_0'\to+0} \int_{V^+} e^{i(x,\xi)-x_0\xi_0'}\,dx$$
$$= 2^n \pi^{(n-1)/2}\Gamma\left(\frac{n+1}{2}\right)\left[-(\xi_0+i0)^2+|\xi|^2\right]^{-\frac{n+1}{2}} \qquad (6.15)$$

(for a simple method of computing this integral see Sec. 10.2).

6.6.14. *Hermite polynomials and functions.* Definitions:

$$H_n(x) = (-1)^n (n!)^{-1/2}\pi^{-1/4}2^{-n/2}e^{x^2}\frac{d^n e^{-x^2}}{dx^n}, \qquad n=0,1,\dots,$$

are *Hermite polynomials*, and

$$\mathcal{H}_n(x) = e^{-x^2/2}H_n(x), \qquad n=0,1,\dots,$$

are *Hermite functions* (the wave functions of a harmonic oscillator).

Differential equations:

$$\begin{aligned} L^-\mathcal{H}_n &= \sqrt{n}\mathcal{H}_{n-1}, \\ L^+\mathcal{H}_n &= \sqrt{n+1}\mathcal{H}_{n+1}, \\ L^+L^-\mathcal{H}_n &= n\mathcal{H}_n \end{aligned} \qquad (6.16)$$

$$\frac{dH_n(x)}{dx} = \sqrt{2n}H_{n-1}(x), \qquad n=0,1,\dots \qquad (\mathcal{H}_{-1}=H_{-1}=0), \qquad (6.17)$$

where

$$L^\pm = \frac{1}{\sqrt{2}}\left(x \mp \frac{d}{dx}\right), \qquad L^-L^+ - L^+L^- = 1. \qquad (6.18)$$

Recurrence relation:

$$\sqrt{n+1}H_{n+1}(x) = \sqrt{2}xH_n(x) - \sqrt{n}H_{n-1}(x), \qquad n=1,2,\dots. \qquad (6.19)$$

Orthonormality:

$$\int_{-\infty}^{\infty} e^{-x^2}H_n(x)H_m(x)\,dx = \int_{-\infty}^{\infty}\mathcal{H}_n(x)\mathcal{H}_m(x)\,dx = \delta_{nm}. \qquad (6.20)$$

The Fourier transform:

$$F[\mathcal{H}_n] = \sqrt{2\pi}i^n\mathcal{H}_n(\xi). \qquad (6.21)$$

We now prove (6.21), which follows from

$$H_n\left(\frac{d}{i\,d\xi}\right)e^{-\xi^2/2} = i^n\mathcal{H}_n(\xi) \qquad (6.22)$$

by virtue of (see (6.1))

$$
\begin{aligned}
F\left[e^{-x^2/2}H_n(x)\right] &= H_n\left(\frac{d}{i\,d\xi}\right)F[e^{-x^2/2}] \\
&= \sqrt{2\pi}H_n\left(\frac{d}{i\,d\xi}\right)e^{-\xi^2/2} \\
&= \sqrt{2\pi}i^n e^{-\xi^2/2}H_n(\xi) \\
&= \sqrt{2\pi}i^n \mathcal{H}_n(\xi).
\end{aligned}
$$

The equality (6.22) holds true for $n = 0$. Its truth, for $n > 0$, follows from the recurrence relations (6.19) and (6.17):

$$
\begin{aligned}
H_n\left(\frac{d}{i\,d\xi}\right)e^{-\xi^2/2} &= \sqrt{\frac{2}{n}}\frac{d}{i\,d\xi}H_{n-1}\left(\frac{d}{i\,d\xi}\right)e^{-\xi^2/2} - \sqrt{\frac{n-1}{n}}H_{n-2}\left(\frac{d}{i\,d\xi}\right)e^{-\xi^2/2} \\
&= \sqrt{\frac{2}{n}}\frac{d}{i\,d\xi}\left[i^{n-1}H_{n-1}(\xi)e^{-\xi^2/2}\right] - \sqrt{\frac{n-1}{n}}i^{n-2}H_{n-2}(\xi)e^{-\xi^2/2} \\
&= \frac{i^{n-2}}{\sqrt{n}}e^{-\xi^2/2}\left[-\sqrt{2}\xi H_{n-1}(\xi) + \sqrt{2}H'_{n-1}(\xi) - \sqrt{n-1}H_{n-2}(\xi)\right] \\
&= \frac{i^n}{\sqrt{n}}e^{-\xi^2/2}\left[\sqrt{2}\xi H_{n-1}(\xi) - 2\sqrt{n-1}H_{n-2}(\xi) - \sqrt{n-1}H_{n-2}(\xi)\right] \\
&= i^n e^{-\xi^2/2}H_n(\xi) \\
&= i^n \mathcal{H}_n(\xi).
\end{aligned}
$$

□

Smoothness: $\mathcal{H}_n \in \mathcal{S}$, and

$$
\|\mathcal{H}_n\|_p \le c_p(1+n)^{p+2}, \qquad p = 0, 1, \dots, \quad n = 0, 1, \dots. \tag{6.23}
$$

The estimate (6.23) follows from the equations (6.16) and from the formulas (6.20) and (6.21). Regarding p as even, we have

$$
\begin{aligned}
\|\mathcal{H}_p\|_p &= \sup_{x,\ 0\le\alpha\le p}\left|(1+x^2)^{p/2}\mathcal{H}_n^{(\alpha)}(x)\right| \\
&= \frac{1}{\sqrt{2\pi}}\sup_{x,\ 0\le\alpha\le p}\left|\int\left(1-\frac{d^2}{d\xi^2}\right)^{p/2}[\xi^\alpha\mathcal{H}_n(\xi)](1+\xi^2)e^{ix\xi}\frac{d\xi}{1+\xi^2}\right| \\
&\le \frac{1}{\sqrt{2\pi}}\sup_{x,\ 0\le\alpha\le p}\left|\sum_{\cdots+\alpha_s+\cdots+\alpha_m+\cdots\le 2p+2}\int c_{\dots+\alpha_s+\dots+\alpha_m+\dots}\cdots(L^+)^{\alpha_s}\cdots \right. \\
&\qquad\qquad \left. \times\cdots(L^-)^{\alpha_m}\cdots\mathcal{H}_n(\xi)\frac{e^{ix\xi}\,d\xi}{1+\xi^2}\right| \\
&\le c'_p(1+n)^{p+1}\sum_{0\le k\le 2p+2}\int|\mathcal{H}_k(\xi)|\frac{d\xi}{1+\xi^2} \\
&\le c'_p(1+n)^{p+1}\sum_{0\le k\le 2p+2}\|\mathcal{H}_k\|\sqrt{\int\frac{d\xi}{1+\xi^2}} \\
&\le c_p(1+n)^{p+2}.
\end{aligned}
$$

For odd p, the estimate (6.23) follows from (1.1) of Sec. 5.1).

Let $f \in \mathcal{S}'$. The numbers

$$a_n(f) = (f, \mathcal{H}_n), \qquad n = 0, 1, \ldots, \tag{6.24}$$

will be called *Fourier coefficients*. The formal series

$$\sum_{0 \leq n < \infty} a_n(f)\mathcal{H}_n(x) \tag{6.25}$$

will be called the *Fourier series* of the generalized function f with respect to the orthonormal system of Hermite functions $\{\mathcal{H}_n\}$.

Completeness in $\mathcal{L}^2$: *if $f \in \mathcal{L}^2$, then its Fourier series (6.25) is unique, converges in $\mathcal{L}^2$ to f, and the following Parseval–Steklov equation holds true:*

$$\|f\|^2 = \sum_{0 \leq n < \infty} |a_n(f)|^2. \tag{6.26}$$

For the function φ to belong to $\mathcal{S}$, it is necessary and sufficient that its Fourier coefficients satisfy the condition

$$\left\|(L^- L^+)^m \varphi\right\|^2 = \sum_{0 \leq n < \infty} |a_n(\varphi)|^2 n^{2m} < \infty, \qquad m = 0, 1, \ldots . \tag{6.27}$$

Then the Fourier series of φ converges to φ in $\mathcal{S}$.

Indeed, if $\varphi \in \mathcal{S}$, then $(L^- L^+)^m \varphi \in \mathcal{L}^2$ for all $m \geq 0$. Therefore, by virtue of (6.16) and (6.24),

$$\begin{aligned} a_n\left((L^- L^+)^m \varphi\right) &= \int (L^- L^+)^m \varphi(x)\mathcal{H}_n(x)\,dx \\ &= \int \varphi(x)(L^+ L^-)^m \mathcal{H}_n(x)\,dx \\ &= n^m \int \varphi(x)\mathcal{H}_n(x)\,dx = n^m a_n(\varphi), \end{aligned}$$

whence, by the Parseval–Steklov equation (6.26) follows (6.27).

Conversely, if the coefficients $\{a_n\}$ satisfy the condition

$$\sum_{0 \leq n < \infty} |a_n|^2 n^{2m} < \infty, \qquad m = 0, 1, \ldots,$$

then by (6.23) the series

$$\sum_{0 \leq n < \infty} a_n \mathcal{H}_n(x)$$

converges in $\mathcal{S}$ to some $\varphi \in \mathcal{S}$ such that $a_n = a_n(\varphi)$.

For f to belong to $\mathcal{S}'$, it is necessary and sufficient that its Fourier coefficients satisfy the following condition: there exist $p \geq 0$ and C such that

$$|a_n(f)| \leq C(1+n)^p, \qquad n = 0, 1, \ldots , \tag{6.28}$$

Here, the Fourier series of f is unique, converges to f in $\mathcal{S}'$, and the Parseval–Steklov equation holds:

$$(f, \varphi) = \sum_{0 \leq n < \infty} a_n(f) a_n(\varphi), \qquad \varphi \in \mathcal{S}. \tag{6.29}$$

Indeed, if $f \in \mathcal{S}'$ and m is the order of f (see Sec. 5.2), then by (6.24) and (6.23) the estimate (6.28) holds:

$$|a_n(f)| = \left|(f, \mathcal{H}_n)\right| \leq \|f\|_{-m} \|\mathcal{H}_n\|_m \leq c_m \|f\|_{-m} (1+n)^{m+2}.$$

Conversely, if the coefficients $\{a_n\}$ satisfy the condition (6.28), $|a_n| \le C(1+n)^p$, $n = 0, 1, \ldots$, then by virtue of (6.27) the series

$$\sum_{0\le n<\infty} a_n \mathcal{H}_n(x)$$

converges in $\mathcal{S}'$ to some $f \in \mathcal{S}'$, and the following equation holds:

$$(f, \varphi) = \sum_{0\le n<\infty} a_n a_n(\varphi), \qquad \varphi \in \mathcal{S}, \tag{6.30}$$

since

$$\left(\sum_{0\le n\le N} a_n \mathcal{H}_n, \varphi\right) = \sum_{0\le n\le N} a_n a_n(\varphi) \to (f, \varphi), \qquad N \to \infty,$$

by virtue of the completeness of the space $\mathcal{S}'$ (see Sec. 5.2). Putting $\varphi = \mathcal{H}_m$ in (6.30) and taking into account that by (6.20), $a_n(\mathcal{H}_m) = \delta_{nm}$, we get $a_n = a_n(f)$.

It remains to prove the uniqueness of the Fourier series: if $f \in \mathcal{S}'$ and $a_n(f) = 0$, $n = 0, 1, \ldots$, then $f = 0$. But this follows from (6.30). □

REMARK. Let us introduce two sequence spaces: we define convergence in them in a natural manner in accord with the estimates (6.27) and (6.28), respectively. The results that have been proved signify that the operation $f \to \{a_n(f),\ n = 0, 1, \ldots\}$ is a linear isomorphism of $\mathcal{S}$ and $\mathcal{S}'$ onto the sequence of spaces that satisfy the conditions (6.27) and (6.28) respectively. [The continuity of this operation follows from (6.27) and (6.29).]

6.6.15. *An integral representation of the Bessel function:*

$$J_\nu(x) = \frac{1}{\sqrt{\pi}\Gamma(\nu + 1/2)}\left(\frac{x}{2}\right)^\nu \int_{-1}^{1} e^{ix\xi}(1 - \xi^2)^{\nu-1/2}\, d\xi, \qquad \Re\nu > -\frac{1}{2}. \tag{6.31}$$

The Bessel function

$$J_\nu(x) = \sum_{k=0}^{\infty} \frac{(-1)^k}{k!\Gamma(k + \nu + 1)}\left(\frac{x}{2}\right)^{2k+\nu}$$

is (up to a factor) the unique solution, bounded at zero, of the Bessel equation

$$(xu')' + \left(x - \frac{\nu^2}{x}\right)u = 0.$$

By virtue of the equation

$$\begin{aligned}\frac{1}{\sqrt{\pi}\Gamma(\nu + 1/2)}\int_{-1}^{1}(1 - \xi^2)^{\nu-1/2}\, d\xi &= \frac{1}{\sqrt{\pi}\Gamma(\nu + 1/2)}\int_{0}^{1}(1 - \mu)^{-1/2}\mu^{\nu-1/2}\, d\mu \\ &= \frac{\mathrm{B}(1/2, \nu + 1/2)}{\sqrt{\pi}\Gamma(\nu + 1/2)} = \frac{\Gamma(1/2)\Gamma(\nu + 1/2)}{\sqrt{\pi}\Gamma(\nu + 1/2)\Gamma(\nu + 1)} \\ &= \frac{1}{\Gamma(\nu + 1)}\end{aligned}$$

the asymptotic behaviour, as $x \to +0$, of both sides of (6.31) is the same. And so to prove (6.31) it remains to prove that the fight-hand side of (6.31) satisfies the Bessel equation. But this is established by direct verification:

$$[xJ_\nu'(x)]' + \left(x - \frac{\nu^2}{x}\right) J_\nu(x)$$
$$= x^{\nu+1} \int_{-1}^{1} (1-\xi^2)^{\nu+1/2} e^{ix\xi}\, d\xi + (2\nu+1) i x^\nu \int_{-1}^{1} (1-\xi^2)^{\nu-1/2} e^{ix\xi} \xi\, d\xi = 0.$$

□

6.6.16. *The Hankel transform.* Let $f(|x|) \in \mathcal{L}^2$, that is, by definition,

$$'\|f\|^2 = \int_0^\infty |f(r)|^2 r^{n-1}\, dr < \infty.$$

The function

$$g(\rho) = \frac{(2\pi)^{n/2}}{\rho^{(n-2)/2}} \int_0^\infty f(r) r^{n/2} J_{(n-2)/2}(r\rho)\, dr \tag{6.32}$$

is termed a *Hankel transform of order* $(n-2)/2$ of the function $f(r)$; the integral here converges in the norm $'\|\ \|$.

The following inversion formula holds:

$$f(r) = \frac{(2\pi)^{-n/2}}{r^{(n-2)/2}} \int_0^\infty g(\rho) \rho^{n/2} J_{(n-2)/2}(r\rho)\, d\rho, \tag{6.32'}$$

and the Parseval–Steklov equation holds:

$$(2\pi)^n\, '\|f\|^2 = '\|g\|^2.$$

Special cases:

$$n = 1, \qquad g(\rho) = 2 \int_0^\infty f(r) \cos r\rho\, dr,$$

$$n = 2, \qquad g(\rho) = 2\pi \int_0^\infty f(r) r J_0(r\rho)\, dr,$$

$$n = 3, \qquad g(\rho) = \frac{4\pi}{\rho} \int_0^\infty f(r) r \sin r\rho\, dr.$$

To prove the inversion formulas (6.32) and (6.32′) and the Parseval–Steklov equation, it is sufficient to demonstrate, by the Plancherel theorem (see Sec. 6.6.3), that the right-hand sides of (6.32) and (6.32′) are the direct and inverse Fourier transforms of the functions $f(|x|)$ and $g(|\xi|)$, respectively.

Indeed, using (6.31) we have

$$F[f(|x|)] = \int f(|x|)e^{i(\xi,x)}\,dx$$
$$= \int_0^\infty f(r)r^{n-1} \int_{|x|=1} e^{ir(\xi,s)}\,ds\,dr$$
$$= \sigma_{n-1} \int_0^\infty f(r)r^{n-1} \int_0^\pi e^{ir\rho\cos\theta} \sin^{n-2}\theta\,d\theta\,dr$$
$$= \sigma_{n-1} \int_0^\infty f(r)r^{n-1} \int_{-1}^1 e^{ir\rho\mu}(1-\mu^2)^{(n-3)/2}\,d\mu\,dr$$
$$= \frac{(2\pi)^{n/2}}{\rho^{(n-2)/2}} \int_0^\infty f(r)r^{n/2} J_{(n-2)/2}(r\rho)\,dr,$$

which is what we wanted; here, σ_{n-1} is the surface area of a unit sphere in $\mathbb{R}^{n-1}$ (see Sec. 0.6). □

6.6.17. *The Fourier transform of homogeneous generalized functions.* All homogeneous generalized functions are described in Sec. 5.7.

The Fourier transform of a homogeneous generalized function f of the homogeneity degree $\alpha-1$ is a homogeneous generalized function of the homogeneity degree $-\alpha$.

Indeed, for all $t \neq 0$, we have (see (7.1) of Sec. 5.7 and (3.7) of Sec. 6.3)

$$F[f](t\xi) = F\left[f\left(\frac{x}{t}\right)\right] = |t|^{-\alpha}\pi_\nu(t)F[f](\xi).$$

By the theorem of Sec. 6.7,

$$F\left[\mathcal{P}(\pi_\nu|x|^{\alpha-1})\right](\xi) = i^\nu\Gamma_\nu(\alpha)\mathcal{P}(\pi_\nu|\xi|^{-\alpha}), \tag{6.33}$$

if $\alpha \neq -2n$, $\nu = 0$ or $\alpha \neq -2n-1$, $\nu = 1$, $n = 0, 1, \ldots$;

$$F\left[\delta^{(n)}\right] = (-i\xi)^n, \tag{6.34}$$

if $\alpha = -2n$, $\nu = 0$ or $\alpha = -2n-1$, $\nu = 1$, $n = 0, 1, \ldots$, (see (3.4) of Sec. 6.3).

The constant $\Gamma_\nu(\alpha)$ for $0 < \Re\alpha < 1$ in equality (6.33) is equal to

$$\Gamma_\nu(\alpha) = i^{-\nu}F\left[\mathcal{P}(\pi_\nu|x|^{\alpha-1})\right](1) = \int \pi_\nu(x)|x|^{\alpha-1}e^{ix}\,dx$$
$$= 2\Gamma(\alpha)\cos\frac{\pi}{2}(\alpha-\nu), \qquad \nu = 0, 1, \tag{6.35}$$

where $\Gamma_\nu(\alpha)$ is the Euler function (see Prudnikov, Brychkov, Marichev [**82**]).

The function $\Gamma_\nu(\alpha)$ is called the *gamma function* of the character $\pi_\nu(x)$ of the field $\mathbb{R}$. It can be analytically continued onto the whole plane α except simple poles $\alpha = -2n$, $\nu = 0$ or $\alpha = -2n-1$, $\nu = 1$, $n = 0, 1, \ldots$. The gamma function $\Gamma_\nu(\alpha)$ satisfies the functional relation

$$\Gamma_\nu(\alpha)\Gamma_\nu(1-\alpha) = 2\pi. \tag{6.36}$$

Relation (6.36) follows from representation (6.35) and from the appropriate relation for the Euler gamma function

$$\Gamma(\alpha)\Gamma(1-\alpha) = \frac{\pi}{\sin \pi\alpha}.$$

6.6.18. *The Fourier transform of the convolution of homogeneous generalized functions and the beta function.* For homogeneous generalized functions the product

$$\left(\pi_\nu |x|^{\alpha-1}\right)\left(\pi_\mu |x|^{\beta-1}\right) = \pi_\nu \pi_\mu |x|^{\alpha+\beta-2} \tag{6.37}$$

is defined in the domain $\Re\alpha > 0$, $\Re\beta > 0$, and the convolution

$$\begin{aligned}\left(\pi_\nu |x|^{\alpha-1}\right) * \left(\pi_\mu |x|^{\beta-1}\right) &= \int \pi_\nu(y)|y|^{\alpha-1}\pi_\mu(x-y)|x-y|^{\beta-1}\,dy \\ &= \mathrm{B}(\alpha,\nu;\beta,\mu)\pi_\nu\pi_\mu|x|^{\alpha+\beta-1}\end{aligned} \tag{6.38}$$

is defined in the domain $\Re\alpha > 0$, $\Re\beta > 0$, and $\Re(\alpha+\beta) < 1$. In (6.38) we set

$$\begin{aligned}\mathrm{B}(\alpha,\nu;\beta,\mu) &= \int \pi_\nu(y)|y|^{\alpha-1}\pi_\mu(1-y)|1-y|^{\beta-1}\,dy \\ &= \frac{1}{2\pi}\Gamma_\nu(\alpha)\Gamma_\mu(\beta)\Gamma_{\nu+\mu}(1-\alpha-\beta).\end{aligned} \tag{6.39}$$

The function $\mathrm{B}(\alpha,\nu;\beta,\mu)$ is called the *beta function* of the characters π_ν and π_μ of the field $\mathbb{R}$. (Below, we shall define the integers ν and $\mu = 0, 1$ modulo 2; $\nu, \mu \in F_2$, so $\pi_\nu(x)\pi_\mu(x) = \pi_{\nu+\mu}(x)$.)

Equalities (6.37)–(6.39) can be meromorphic continued with respect to α and β to all complex pairs $(\alpha,\beta) \in \mathbb{C}^2$ (in this case, by virtue of Sec. 6.6.17, the poles are defined uniquely). As a result, we obtain the following equalities

$$\mathcal{P}\left(\pi_\nu|x|^{\alpha-1}\right)\cdot\mathcal{P}\left(\pi_\mu|x|^{\beta-1}\right) = \mathcal{P}\left(\pi_{\nu+\mu}|x|^{\alpha+\beta-2}\right), \tag{6.37'}$$

$$\mathcal{P}\left(\pi_\nu|x|^{\alpha-1}\right) * \mathcal{P}\left(\pi_\mu|x|^{\beta-1}\right) = \mathrm{B}(\alpha,\nu;\beta,\mu)\mathcal{P}\left(\pi_{\nu+\mu}|x|^{\alpha+\beta-1}\right). \tag{6.38'}$$

Let us prove equality (6.39). To this end, we apply the formula of the Fourier transform of the convolution (see (5.5) of Sec. 6.5) to equality (6.38) and make use of equality (6.33). As a result, we obtain

$$\begin{aligned}&i^\nu\Gamma_\nu(\alpha)\mathcal{P}\left(\pi_\nu|\xi|^{-\alpha}\right)i^\mu\Gamma_\mu(\beta)\mathcal{P}\left(\pi_\mu|\xi|^{-\beta}\right) \\ &\qquad = i^{\nu+\mu}\mathrm{B}(\alpha,\nu;\beta,\mu)\Gamma_{\nu+\mu}(\alpha+\beta)\mathcal{P}\left(\pi_{\nu+\mu}|\xi|^{-\alpha-\beta}\right),\end{aligned}$$

and this, by virtue of (6.37) and (6.36), imply equality (6.39) in the domain $\Re\alpha > 0$, $\Re\beta > 0$, $\Re(\alpha+\beta) < 1$. By meromorphic continuation on (α,β) the relations (6.37)–(6.39) are valid for all $(\alpha,\beta) \in \mathbb{C}^2$ except poles. □

By virtue of (6.39) we can represent the beta function B on the variety $\alpha + \beta + \gamma = 1$, $\nu + \mu + \eta = 0$, $\alpha, \beta, \gamma \in \mathbb{C}$, $\nu, \mu, \gamma \in F_2$, in the following form symmetric with respect to the transposition of the arguments (α,ν), (β,μ), (γ,η):

$$\mathrm{B}(\alpha,\nu;\beta,\mu;\gamma,\eta) = \frac{1}{2\pi}\Gamma_\nu(\alpha)\Gamma_\mu(\beta)\Gamma_\eta(\gamma). \tag{6.40}$$

The equation $\nu + \mu + \eta = 0$ in F_2 has only four solutions: 000, 110, 101, 011. Therefore, there exists only four beta functions (of the field $\mathbb{R}$). For $\nu = \mu = \eta = 0$, the beta function

$$\mathrm{B}_0(\alpha,\beta,\gamma) = \mathrm{B}(\alpha,0;\beta,0;\gamma,0) = \frac{1}{2\pi}\Gamma_0(\alpha)\Gamma_0(\beta)\Gamma_0(\gamma), \qquad \alpha+\beta+\gamma = 1, \tag{6.41}$$

defines the crossing-symmetric amplitude of Veneziano in the quantum field theory and in the string theory (see Vladimirov, Volovich, Zelenov [**123**], Green, Schwartz, Witten [**41**]).

Let us note one more interesting representation of the beta function B_0:

$$B_0(\alpha,\beta,\gamma) = \frac{1}{2\pi}\left[B(\alpha,\beta) + B(\alpha,\gamma) + B(\beta,\gamma)\right], \qquad \alpha+\beta+\gamma = 1, \tag{6.42}$$

where $B(\alpha,\beta)$ is the Euler beta function,

$$B(\alpha,\beta) = \frac{\Gamma(\alpha)\Gamma(\beta)}{\Gamma(\alpha+\beta)}.$$

6.7. The Mellin transform. Let $\mathbb{R}^*$ denote the multiplicative group of the field $\mathbb{R}$, $\mathbb{R}^* = \mathbb{R}\setminus\{0\}$ and $d^*x = \frac{dx}{|x|}$ be the Haar measure on $\mathbb{R}^*$. We denote

$$\lambda_p(x) = \max(|x|^{-p}, |x|^p), \qquad p \geq 0.$$

Let us introduce the countable-normed space $\mathcal{S}(\mathbb{R}^*)$ of C^∞-functions $\varphi(x)$ for $x \neq 0$ with the norms

$$\|\varphi\|_p = \sup_{x,\ k\leq p} \lambda_p(x)|\varphi^{(k)}(x)|, \qquad \varphi \in \mathcal{S}(\mathbb{R}^*), \quad p = 0,1,\dots.$$

In order that $\varphi \in \mathcal{S}(\mathbb{R}^*)$, it is necessary and sufficient that its continuation at zero by zero, $\varphi(0) = 0$, belongs to $\mathcal{D}(\mathbb{R})$ and $\varphi^{(k)}(0) = 0$, $k = 0,1,\dots$.

Therefore, $\mathcal{D}(\mathbb{R}^*) \subset \mathcal{S}(\mathbb{R})$ together with the topology and, hence, $\mathcal{S}'(\mathbb{R}) \subset \mathcal{S}'(\mathbb{R}^*)$. The theory of the space $\mathcal{S}(\mathbb{R}^*)$ and its dual $\mathcal{S}'(\mathbb{R}^*)$ is completely analogous to the theory of the spaces $\mathcal{S}(\mathbb{R})$ and $\mathcal{S}'(\mathbb{R})$ (see Gel'fand and Shilov [**38**, vol. 2] and Sec. 5).

In particular, any $f \in \mathcal{S}'(\mathbb{R}^*)$ has a finite order, i.e., there exist a (smallest) integer $m_0 \geq 0$ and norms $\|f\|_{-m}$, $m \geq m_0$ such that the following inequalities hold (compare (2.3) of Sec. 5.2)

$$|(f,\varphi)| \leq \|f\|_{-m}\|\varphi\|_m, \qquad \varphi \in \mathcal{S}_m(\mathbb{R}^*), \qquad m \geq m_0, \tag{7.1}$$

and if $f_k \to 0$, $k \to \infty$ in $\mathcal{S}'(\mathbb{R}^*)$, then for some m we have $\|f_k\|_{-m} \to 0$, $k \to \infty$.

However, there are some differences. For example, the operation $\varphi(x) \to \varphi(1/x)$ is a linear isomorphism from $\mathcal{S}(\mathbb{R}^*)$ onto $\mathcal{S}(\mathbb{R}^*)$. Therefore, the operation of the *inversion* $f(1/x)$ of the generalized function $f(x)$ taken from $\mathcal{S}'(\mathbb{R}^*)$,

$$(f(1/x),\varphi) = (f(y), y^2\varphi(1/y)), \qquad \varphi \in \mathcal{S}(\mathbb{R}^*) \tag{7.2}$$

is also a linear isomorphism of $\mathcal{S}'(\mathbb{R}^*)$ onto $\mathcal{S}'(\mathbb{R}^*)$.

EXAMPLES.
$$\delta(x) = 0 \quad \text{in} \quad \mathcal{S}'(\mathbb{R}^*);$$
$$\delta(x-1) = \delta\left(\frac{1}{x} - 1\right).$$

The *Mellin transform* $M_\nu[\varphi](\alpha)$, $\nu \in F_2$ of a function $\varphi \in \mathcal{S}(\mathbb{R}^*)$ is the integrals

$$M_\nu[\varphi](\alpha) = \frac{1}{2}\int_{\mathbb{R}^*} \varphi(x)\pi_\nu(x)|x|^\alpha\, d^*x = \frac{1}{2}\int \varphi(x)\pi_\nu(x)|x|^{\alpha-1}dx, \tag{7.3}$$

where $\pi_\nu(x) = \operatorname{sgn}^\nu x$ (see Sec. 2.7). Let us note that the Mellin transform consists of two components $\{M_0[\varphi](\alpha), M_1[\varphi](\alpha)\}$, $\alpha = \sigma + i\tau$. (The usual Mellin transform is defined for even functions by formula (7.3) with $\nu = 0$ and has only one component of the form $M_0[\varphi](\alpha) = \int_0^\infty \varphi(x)|x|^{\alpha-1}\,dx$.)

THEOREM I. *The Mellin transform $M_\nu[\varphi](\alpha)$ of the function $\varphi \in S(\mathbb{R}^*)$ is an entire function of α which for any $N = 0, 1, \ldots$ satisfies the estimate*

$$|M_\nu[\varphi](\alpha)| \le C_N(\sigma)\|\varphi\|_{N+[|\sigma|]+1}|\tau|^{-N}, \qquad \nu = 0, 1, \tag{7.4}$$

the inversion formula

$$\varphi(x) = \frac{1}{2\pi i}\sum_{\nu=0,1}\pi_\nu(x)\int_{\sigma-i\infty}^{\sigma+i\infty} M_\nu[\varphi](\alpha)|x|^{-\alpha}\,d\alpha, \qquad \varphi \in S(\mathbb{R}^*), \tag{7.5}$$

is valid and the analogue of the Parseval–Steklov equality holds

$$\int \varphi(x)\psi(x)\,dx = \frac{1}{\pi i}\sum_{\nu=0,1}\int_{\sigma-i\infty}^{\sigma+i\infty} M_\nu[\varphi](\alpha)M_\nu[\psi](1-\alpha)\,d\alpha, \qquad \varphi, \psi \in S(\mathbb{R}^*), \tag{7.6}$$

where the integrals in the right-hand sides of equalities (7.5) *and* (7.6) *do not depend on σ.*

PROOF. The fact that $M_\nu[\varphi](\alpha)$ are entire functions follows from definition (7.3) if we note that $\varphi\pi_\nu \in S(\mathbb{R}^n)$ and $(\varphi\pi_\nu)^{(k)}(0) = 0$, for all $k = 0, 1, \ldots$.

Let us prove estimate (7.4). Applying the formula of integration by parts N times in integral (7.3), we obtain

$$M_\nu[\varphi](\alpha) = \frac{1}{\alpha(\alpha+1)\ldots(\alpha+N-1)}\int \left[\pi_\mu(x)\varphi(x)\right]^{(N)}|x|^{\alpha+N-1}\,dx,$$

that implies estimate (7.4):

$$|M_\nu[\varphi](\alpha)| \le |\tau|^{-N}\int \left|\varphi^{(N)}(x)\right| \lambda_{|\alpha|+N-1}(x)\,dx \le C_N(\sigma)|\tau|^{-N}\|\varphi\|_{N+[|\sigma|]+1}.$$

Now let us prove the inversion formula (7.5) for even functions $\varphi(x) = \varphi(|x|) \in S(\mathbb{R}^*)$. Denoting $\varphi(|x|) = \varphi(e^{\ln|x|}) = \psi(\ln|x|)$, $\psi \in S(\mathbb{R})$, we have

$$M_0[\varphi](\alpha) = \int_0^\infty \varphi(x)x^{\alpha-1}\,dx = \int_{-\infty}^\infty \psi(t)e^{t\alpha}\,dt, \qquad M_1[\varphi](\alpha) = 0,$$

and the right-hand side of (7.5) for $x > 0$ and $\sigma = 0$ takes the form

$$\begin{aligned}\frac{1}{2\pi i}\int_{-i\infty}^{i\infty} M_0[\varphi](\alpha)x^{-\alpha}\,d\alpha &= \frac{1}{2\pi}\int_{-\infty}^{\infty} x^{-i\tau}\int_{-\infty}^{\infty}\psi(t)e^{it\tau\ln x}\,dt\,d\tau \\ &= \frac{1}{2\pi}\int_{-\infty}^{\infty} e^{-i\tau\ln x}F[\psi](\tau)\,d\tau \\ &= F^{-1}\left[F[\psi]\right](\ln x) = \psi(\ln x) = \varphi(x),\end{aligned}$$

by virtue of formula (1.3) of Sec. 6.1 of the inversion of the Fourier transform. Since by virtue of (7.4) the estimate

$$\left|M_0[\varphi](\alpha)|x|^{-\alpha}\right| \le C_2(\sigma)|\tau|^{-2}x^{-\sigma} \qquad x > 0,$$

holds, the contour of integration $(\sigma - i\infty, \sigma + i\infty)$ in integral (7.5) can be shifted $(\sigma - i\infty, \sigma + i\infty) \to (-i\infty, i\infty)$ by any shift σ for any $x > 0$.

Similarly formula (7.5) can be proved for odd functions as well as for the arbitrary functions $\varphi = \varphi_{even} + \varphi_{odd}$ taken from $\mathcal{S}(\mathbb{R}^*)$, where

$$\varphi_{even}(x) = \frac{1}{2}[\varphi(x) + \varphi(-x)], \qquad \varphi_{odd}(x) = \frac{1}{2}[\varphi(x) - \varphi(-x)]$$

are even and odd components of the function φ.

In order to obtain formula (7.6), let us multiply equality (7.5) by $\psi(x)$ and integrate over all x. As a result, we obtain (7.6)

$$\begin{aligned}\int \varphi(x)\psi(x)\,dx &= \frac{1}{2\pi i}\sum_{\nu=0,1}\int \pi_\nu(x)\psi(x)\int_{\sigma-i\infty}^{\sigma+i\infty} M_\nu[\varphi](\alpha)|x|^{-\alpha}\,d\alpha\,dx \\ &= \frac{1}{2\pi i}\sum_{\nu=0,1}\int_{\sigma-i\infty}^{\sigma+i\infty} M_\nu[\varphi](\alpha)\int \pi_\nu(x)|x|^{-\alpha}\psi(x)\,dx\,d\alpha \\ &= \frac{1}{\pi i}\sum_{\nu=0,1}\int_{\sigma-i\infty}^{\sigma+i\infty} M_\nu[\varphi](\alpha)M_\nu[\psi](1-\alpha)\,d\alpha.\end{aligned}$$

The change of the order of integration is possible here by virtue of the Fubini theorem and estimate (7.4):

$$\left|M_\nu[\varphi](\sigma+i\tau)\pi_\nu(x)|x|^{-\alpha}\psi(x)\right| \in \mathcal{L}_1(\mathbb{R}^2), \qquad \mathbb{R}^2 = (\tau, x).$$

Theorem I is proved. □

By the *Mellin transform* $M_\nu[f](\alpha)$ of the generalized function $f \in \mathcal{S}'(\mathbb{R}^*)$ vanishing for $|x| < a$ (for $|x| < a$, respectively) we call the expressions

$$M_\nu[f](\alpha) = \frac{1}{2}\left(f(x), \eta(x)\pi_\nu(x)|x|^{\alpha-1}\right), \qquad \nu = 0, 1, \tag{7.3'}$$

where $\eta \in C^\infty$, $\eta(x) \equiv 1$, $|x| \geq a$ ($\eta(x) \equiv 1$, $|x| \leq a$, respectively). The right-hand side of (7.3′) does not depend on the auxiliary function η (for the values of α, when it exists.

THEOREM I′. *If $f \in \mathcal{S}'(\mathbb{R}^*)$, $f(x) = 0$, $|x| < a$, ($f(x) = 0$, $|x| > a$, respectively) and the m_0 is the order of f, then $M_\nu[f](\alpha)$ is a holomorphic function in the half-plane $\sigma < 1 - m_0$ ($\sigma > m_0 - 1$, respectively), satisfies the estimate*

$$|M_\nu[f](\alpha)| \leq C_m(\sigma)\|f\|_{-m}(1 + |\tau|^m), \qquad m \geq m_0, \tag{7.7}$$

and the inversion formula

$$(f,\varphi) = \frac{1}{\pi i}\sum_{\nu=0,1}\int_{\sigma-i\infty}^{\sigma+i\infty} M_\nu[f](\alpha)M_\nu[\varphi](1-\alpha)\,d\alpha, \qquad \sigma < 1 - m_0, \quad \varphi \in \mathcal{S}(\mathbb{R}^*) \tag{7.8}$$

($\sigma > m_0 - 1$, respectively) holds, where the integral in the right-hand side of (7.8) *does not depend on $\sigma < 1 - m_0$ ($\sigma > m_0 - 1$, respectively).*

PROOF. Consider the case $f(x) = 0$, $|x| < a$. The fact that the functions $M_\nu[f](\alpha)$, $\nu = 0, 1$, are holomorphic in the half-plane $\sigma < 1 - m_0$ follows from definition (7.6) and from the inclusion $\eta\pi_\nu|x|^{\alpha-1} \in \mathcal{S}_m(\mathbb{R}^*)$, by virtue of the estimate

$$|\eta(x)\pi_\nu(x)|\ |x|^{\alpha-1} \leq C\lambda_{|\alpha-1|}(x) < C\lambda_{m_0}(x).$$

Inequalities (7.7) follow from inequalities (7.1) by virtue of the estimates

$$\begin{aligned}|M_\nu[f](\alpha)| &\leq \|f\|_{-m} \left\|\eta\pi_\nu|x|^{\alpha-1}\right\|_m \\ &= \|f\|_{-m} \sup_{x,\ 0\leq k\leq m} \lambda_m(x) \left|(\eta x^{\alpha-1})^{(k)}\right| \\ &\leq C_m\|f\|_{-m} \sup_{\substack{x>b \\ 0\leq k,j\leq m}} x^{m-1+\sigma-k}|\alpha-1|\dots|\alpha-j| \\ &\leq C'_m(\sigma)\|f\|_{-m}|\alpha-m|^m \\ &\leq C_m(\sigma)\|f\|_{-m}(1+|\tau|^m).\end{aligned}$$

(Here, the number $b < a$ is such that $\eta(x) \equiv 0$ for $|x| < b$.)

Now let us prove the inversion formula (7.8). Let $\{f_k, k \to \infty\}$ be a sequence of functions $f_k \in \mathcal{S}(\mathbb{R}^*)$ converging to f in $\mathcal{S}'(\mathbb{R}^*)$, and such that $f_k(x) = 0$, $|x| < b$, $k = 1, 2, \dots$. (Such a sequence always exists, compare Sec. 5.6.2). Then $\|f - f_k\|_{m_0+1} \to 0$, $k \to \infty$ (compare Sec. 5.2).

Applying equality (7.6) to $\varphi = f_k$ and $\psi = \varphi \in \mathcal{S}(\mathbb{R}^*)$, we obtain

$$(f_k, \varphi) = \frac{1}{\pi i} \sum_{\nu=0,1} \int_{\sigma-i\infty}^{\sigma+i\infty} M_\nu[f_k](\alpha) M_\nu[\varphi](1-\alpha)\, d\alpha, \qquad k = 1, 2, \dots, \qquad \varphi \in \mathcal{S}(\mathbb{R}^*).$$

Passing in this equality to the limit as $k \to \infty$ and using estimate (7.4) and estimate (7.6) for $m = m_0 + 1$ and $\sigma < 1 - m_0$, as well as by

$$|M_\nu[f - f_k](\alpha)| \leq C_{m_0+1}(\sigma)\|f - f_k\|_{m_0+1}(1 + |\tau|^{m_0+1}),$$

we obtain equality (7.8). By virtue of the same estimates, the integral in (7.8) does not depend on $\sigma < 1 - m_0$. The case $f(x) \equiv 0$, $|x| > a$, can be considered similarly. Theorem I′ is proved. □

In the case of an arbitrary generalized function $f \in \mathcal{S}'(\mathbb{R}^*)$, we represent it in the form of the sum

$$f(x) = f_0(x) + f_\infty(x), \qquad f_0, f_\infty \in \mathcal{S}'(\mathbb{R}^*), \tag{7.9}$$

where $f_0(x) \equiv 0$, $|x| < a$ and $f_\infty \equiv 0$, $|x| > b$. (One can always make this, by setting $f_0 = (1 - \eta)f$, $f_\infty = \eta f$, where $\eta \in \mathcal{D}$, $\eta(x) \equiv 1$ $|x| < a$.)

By the *Mellin transform* of the generalized function $f \in \mathcal{S}'(\mathbb{R}^*)$ we call four functions

$$M_\nu[f_\mu](\alpha) = \frac{1}{2}\left(f_\mu(x), \pi_\nu(x)|x|^{\alpha-1}\right), \qquad \nu = 0, 1, \quad \mu = 0, \infty.$$

If the order of $f \in \mathcal{S}'(\mathbb{R}^*)$ is m_0, then the functions $M_\nu[f_\mu](\alpha)$ satisfies estimate (7.7) *and the inversion formula*

$$(f, \varphi) = \frac{1}{\pi i} \sum_{\mu=0,\infty} \sum_{\nu=0,1} \int_{\sigma_\mu - i\infty}^{\sigma_\mu + i\infty} M_\nu[f_\mu](\alpha) M_\nu[\varphi](1-\alpha)\, d\alpha, \qquad \varphi \in \mathcal{S}(\mathbb{R}^*), \tag{7.10}$$

holds; moreover, the integrals in (7.10) *does not depend on $\sigma_0 < 1 - m_0$ and $\sigma_\infty > m_0 - 1$, and the right-hand side on* (7.10) *does not depend on the representation of f as the sum* (7.9).

In particular, if f is a generalized function taken from $\mathcal{S}'(\mathbb{R}^*)$ with a compact support in $\mathbb{R}^*$, then, setting $f_0 = f$, $f_\infty = 0$, we obtain that its Mellin transform

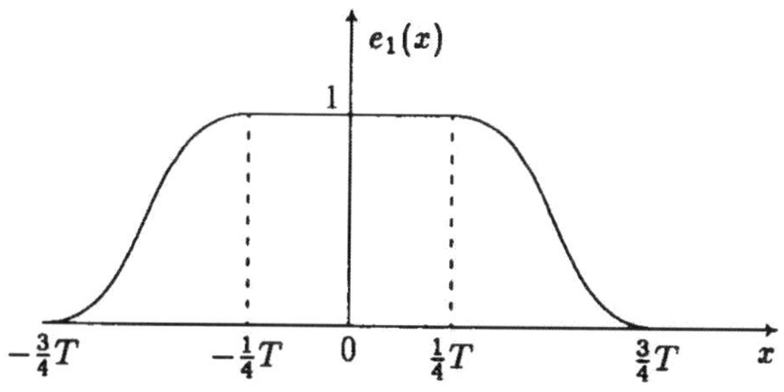

Figure 27

$M_\nu[f](\alpha)$, $\nu = 0, 1$, (7.6) is an entire function satisfying estimate (7.7), and the inversion formula (7.10) takes the form

$$(f, \varphi) = \frac{1}{\pi i} \sum_{\nu=0,1} \int_{\sigma-i\infty}^{\sigma+i\infty} M_\nu[f](\alpha) M_\nu[\varphi](1-\alpha)\, d\alpha, \qquad \varphi \in \mathcal{S}(\mathbb{R}^*). \tag{7.11}$$

EXAMPLE. $f(x) = \delta(x-1)$, $M_0[f](\alpha) = M_1[f](\alpha) = 1/2$, and formula (7.11) takes the form (compare with (7.6))

$$\varphi(1) = \frac{1}{2\pi i} \int_{\sigma-i\infty}^{\sigma+i\infty} \sum_{\nu=0,1} M_\nu[\varphi](\alpha)\, d\alpha.$$

7. Fourier Series of Periodic Generalized Functions

7.1. The definition and elementary properties of periodic generalized functions. A generalized function $f(x)$ taken from $\mathcal{D}'(\mathbb{R}^n)$ is said to be *periodic with an n-period* $T = (T_1, T_2, \ldots, T_n)$, $T_j > 0$, if it is periodic with respect to each argument x_j with period T_j, that is, if it satisfies the condition (see Sec. 1.9)

$$f(x_1, \ldots, x_j + T_j, \ldots, x_n) = f(x), \qquad j = 1, \ldots, n.$$

We use $\mathcal{D}'_T$ to denote the collection of all periodic generalized functions of an n-period T.

We now prove that for every n-period T there exists a *special partition of unity* in $\mathbb{R}^n$ (see Sec. 1.2):

$$\sum_{|k|\geq 0} e_T(x + kT) = 1, \qquad e_T \geq 0, \qquad e_T \in \mathcal{D},$$

$$\operatorname{supp} e_T \subset \left(-\tfrac{3}{4}T_1, \tfrac{3}{4}T_1\right) \times \cdots \times \left(-\tfrac{3}{4}T_n, \tfrac{3}{4}T_n\right); \tag{1.1}$$

$e_T(x)$ is an even function with respect to each variable; here we set

$$kT = (k_1 T_1, \ldots, k_n T_n).$$

Let $T > 0$. We denote by $e_T(x)$ an even function taken from $\mathcal{D}(\mathbb{R}^1)$ with the properties: $\operatorname{supp} e_T \subset \left(-\frac{3}{4}T, \frac{3}{4}T\right)$, $e_T(x) = 1$ in the neighbourhood of the interval $\left[-\frac{1}{4}T, \frac{1}{4}T\right]$, and

$$e_T(x) = 1 - e_T(x + T), \qquad x \in \left[-\tfrac{3}{4}T, -\tfrac{1}{4}T\right]$$

(Fig. 27). It is easy to see that such functions exist. Clearly, the function e_T satisfies the equation

$$\sum_{|k|\geq 0} e_T(x+kT)=1, \qquad x\in\mathbb{R}^1. \tag{1.2}$$

Setting

$$e_T(x)=e_{T_1}(x_1)\dots e_{T_n}(x_n), \tag{1.3}$$

we are convinced of the existence of the required expansion of unity.

We now introduce the generalized function

$$\delta_T(x)=\sum_{|k|\geq 0}\delta(x-kT).$$

Quite obviously, $\delta_T\in\mathcal{D}'_T\cap\mathcal{S}'$ (see Sec. 5.3).

Let us now prove the following representation: *if* $f\in\mathcal{D}'_T$, *then*

$$f=(e_Tf)*\delta_T. \tag{1.4}$$

Indeed, using (1.1) and the periodicity of $f(x)$, we have

$$\begin{aligned} f(x)&=f(x)\sum_{|k|\geq 0}e_T(x+kT)\\ &=\sum_{|k|\geq 0}f(x)e_T(x+kT)\\ &=\sum_{|k|\geq 0}f(x+kT)e_T(x+kT)\\ &=\sum_{|k|\geq 0}(e_Tf)*\delta(x+kT), \end{aligned}$$

whence, taking advantage of the continuity in $\mathcal{S}'$ of the convolution (see Sec. 5.6.1), we obtain the representation (1.4). □

From (1.4) it follows, in particular, that $\mathcal{D}'_T\subset\mathcal{S}'$. Besides, setting $f=\delta_T$ in (1.4), we obtain

$$\delta_T=(e_T\delta_T)*\delta_T. \tag{1.5}$$

Let $f\in\mathcal{D}'_T$ and $\varphi\in C^\infty\cap\mathcal{D}'_T$. We introduce the scalar product $(f,\varphi)_T$ by the rule

$$(f,\varphi)_T=(f,e_T\varphi).$$

For this definition to be proper, it is necessary to demonstrate that the right-hand side of the equation is independent of the choice of the auxiliary function $e_T(x)$ with the properties (1.1).

Indeed, let $e'_T(x)$ be another such function. Then, using the representation (1.4) and formula (6.1) of Sec. 5.6, we obtain

$$\begin{aligned}(f, e'_T\varphi) &= \big((e_T f) * \delta_T, e'_T\varphi\big)\\ &= \big(e_T(x)f(x) \times \delta_T(y), e'_T(x+y)\varphi(x+y)\big)\\ &= \big(f(x), (\delta_T(y), e_T(x)e'_T(x+y)\varphi(x+y))\big)\\ &= \left(f(x), e_T(x)\sum_{|k|\geq 0} e'_T(x-kT)\varphi(x-kT)\right)\\ &= (f, e_T\varphi),\end{aligned}$$

which is what we set out to prove. □

If $f \in \mathcal{L}^1_{\text{loc}} \cap \mathcal{D}'_T$, then

$$(f,\varphi)_T = \int_0^{T_1}\cdots\int_0^{T_n} f(x)\varphi(x)\,dx. \tag{1.6}$$

Indeed, since the scalar product $(\cdot,\cdot)_T$ is independent of the choice of the function e_T, it suffices to compute it for the concrete functions (1.3):

$$\begin{aligned}(f,\varphi)_T &= \int e_T(x)f(x)\varphi(x)\,dx\\ &= \left[\int_{-3T_1/4}^{-T_1/4} e_{T_1}(x_1) + \int_{-T_1/4}^{T_1/4} e_{T_1}(x_1) + \int_{T_1/4}^{3T_1/4} e_{T_1}(x_1)\right]\\ &\quad\times\int e_{T_2}(x_2)\ldots e_{T_n}(x_n)f(x)\varphi(x)\,dx_2\ldots dx_n\,dx_1\\ &= \int_{-T_1/2}^{T_1/2}\int e_{T_2}(x_2)\ldots e_{T_n}(x_n)f(x)\varphi(x)\,dx_2\ldots dx_n\,dx_1\\ &\quad+\left[\int_{T_1/4}^{3T_1/4} e_{T_1}(x_1) - \int_{-3T_1/4}^{-T_1/4} e_{T_1}(x_1+T_1)\right]\\ &\quad\times\int e_{T_2}(x_2)\ldots e_{T_n}(x_n)f(x)\varphi(x)\,dx_2\ldots dx_n\,dx_1\\ &= \int_{-T_1/2}^{T_1/2}\int e_{T_2}(x_2)\ldots e_{T_n}(x_n)f(x)\varphi(x)\,dx_2\ldots dx_n\,dx_1\\ &= \int_{-T_1/2}^{T_1/2}\int_{-T_2/2}^{T_2/2}\int e_{T_3}(x_3)\ldots e_{T_n}(x_n)f(x)\varphi(x)\,dx_3\ldots dx_n\,dx_2\,dx_1\\ &= \cdots = \int_{-T_1/2}^{T_1/2}\int_{-T_2/2}^{T_2/2}\cdots\int_{-T_n/2}^{T_n/2} f(x)\varphi(x)\,dx.\end{aligned}$$

In particular, the trigonometric functions

$$e^{i(k\omega,x)}, \qquad \omega = \left(\frac{2\pi}{T_1}, \dots, \frac{2\pi}{T_n}\right)$$

are periodic with n-period T and for them

$$\left(e^{i(k\omega,x)}, e^{-i(k'\omega,x)}\right)_T = \delta_{kk'} T_1 \dots T_n. \tag{1.7}$$

7.2. Fourier series of periodic generalized functions. Let $f \in \mathcal{D}'_T$. The formal series

$$f(x) \sim \sum_{|k|\geq 0} c_k(f) e^{i(k\omega,x)}, \qquad c_k(f) = \frac{(f, e^{-i(k\omega,x)})_T}{T_1 \dots T_n}, \tag{2.1}$$

is termed a *Fourier series* and the numbers $c_k(f)$ are called *Fourier coefficients* of the generalized function f.

EXAMPLE 1. If $f \in \mathcal{L}^1_{\text{loc}} \cap \mathcal{D}'_T$, then its Fourier series (2.1) turns into the classical Fourier series by virtue of (1.6).

EXAMPLE 2. The following equation in $\mathcal{S}'$ holds true:

$$\sum_{|k|\geq 0} \delta(x + kT) = \frac{1}{T_1 \dots T_n} \sum_{|k|\geq 0} e^{i(k\omega,x)}. \tag{2.2}$$

It follows from the one-dimensional formula (3.5) of Sec. 2.3 and from the continuity, in $\mathcal{S}'$, of the direct product (see Sec. 5.5). □

Let $f \in \mathcal{D}'_T$ and let m be the order of f. Then there is a number $C_m \geq 0$, not dependent on f and k, such that

$$|c_k(f)| \leq C_m \|f\|_{-m} (1 + |k|)^m. \tag{2.3}$$

Indeed, using the definition of the scalar product $(\cdot,\cdot)_T$ and fixing the auxiliary function $e_T(x)$, we obtain the estimate (2.3):

$$\begin{aligned}
|c_k(f)| &= \frac{1}{T_1 \dots T_n} \left|\left(f, e^{-i(k\omega,x)}\right)_T\right| \\
&= \frac{1}{T_1 \dots T_n} \left|\left(f, e_T e^{-i(k\omega,x)}\right)\right| \\
&\leq \frac{\|f\|_{-m}}{T_1 \dots T_n} \sup_{x, |\alpha|\leq m} \left(1 + |x|^2\right)^{m/2} \left|\partial^\alpha \left[e_T(x) e^{-i(k\omega,x)}\right]\right| \\
&\leq C' \|f\|_{-m} \sup_{x, |\alpha|\leq m} \sum_{\beta\leq\alpha} \binom{\alpha}{\beta} \left|\partial^{\alpha-\beta} e_T(x)\right| \left|(-ik\omega)^\beta\right| \\
&\leq C \|f\|_{-m} (1 + |k|)^m.
\end{aligned}$$

THEOREM. *The Fourier series of any generalized function f in $\mathcal{D}'_T$ converges to f in $\mathcal{S}'$:*

$$f(x) = \sum_{|k|\geq 0} c_k(f) e^{i(k\omega,x)}. \tag{2.4}$$

PROOF. Substituting (2.2) into the right-hand side of (1.4),

$$f = (e_T f) * \sum_{|k| \geq 0} \frac{1}{T_1 \ldots T_n} e^{i(k\omega, x)}$$
$$= \sum_{|k| \geq 0} \frac{1}{T_1 \ldots T_n} (e_T f) * e^{i(k\omega, x)},$$

and using (3.3) of Sec. 4.3 for the convolution,

$$(e_T f) * e^{i(k\omega, x)} = \left(f(y), e_T(y) e^{i(k\omega, x-y)}\right)$$
$$= T_1 \ldots T_n c_k(f) e^{i(k\omega, x)},$$

we obtain the expansion of f in the form of the Fourier series (2.4) converging in S'. The proof is complete. □

COROLLARY 1. *A generalized function f in $\mathcal{D}'_T$ is completely determined by the set of its Fourier coefficients $\{c_k(f)\}$.*

COROLLARY 2. *If $f \in \mathcal{D}'_T$ and $\varphi \in C^\infty \cap \mathcal{D}'_T$, then the following generalized Parseval-Steklov equation holds true:*

$$(f, \varphi)_T = \sum_{|k| \geq 0} c_k(f) c_k(\varphi). \tag{2.5}$$

COROLLARY 3. *The Fourier series of a generalized function f taken from $\mathcal{D}'_T$ may be differentiated termwise an infinite number of times:*

$$\partial^\alpha f(x) = \sum_{|k| \geq 0} c_k(f)(ik\omega)^\alpha e^{i(k\omega, x)}, \tag{2.6}$$

so that

$$c_k(\partial^\alpha f) = (ik\omega)^\alpha c_k(f). \tag{2.7}$$

7.3. The convolution algebra $\mathcal{D}'_T$. We introduce, on the set $\mathcal{D}'_T$, the convolution operation ⊛ via the rule

$$f \circledast g = (e_T f) * g, \qquad f, g \in \mathcal{D}'_T. \tag{3.1}$$

The convolution $f \circledast g$ does not depend on the auxiliary function e_T and is commutative, $f \circledast g = g \circledast f$.

This assertion stems from the equality

$$(e_T f) * g = (e'_T g) * f \tag{3.2}$$

that follows from the identity (1.4) and from the properties of continuity, associativity, and commutativity of the convolution $*$ (see Sec. 4.2):

$$(e_T f) * g = (e_T f) * \left((e'_T g) * \delta_T\right) = \left((e_T f) * \delta_T\right) * (e'_T g)$$
$$= f * (e'_T g) = (e'_T g) * f.$$

□

The operation $f \to f \circledast g$ is linear and continuous from $\mathcal{D}'_T$ into S' (see Sec. 5.6.1).

Finally, $f \circledast g \in \mathcal{D}'_T$.

This follows from the property of translation of a convolution (see Sec. 4.2.3):

$$(f \circledast g)(x+kT) = (e_T f) * g(x+kT)$$
$$= (e_T f) \circledast g = (f \circledast g)(x).$$

The convolution of any number of generalized functions $f_1, f_2 \dots, f_m$ taken from $\mathcal{D}'_T$ is determined in similar fashion via the rule

$$f_1 \circledast f_2 \circledast \cdots \circledast f_m = (e'_T f_1) * (e''_T f_2) * \cdots * (e_T^{(m)} f_m) * \delta_T. \tag{3.3}$$

This convolution is associative and commutative (see Sec. 4.2).

EXAMPLE 1. If f and $g \in \mathcal{L}^1_{\text{loc}} \cap \mathcal{D}'_T$, then

$$(f \circledast g)(x) = \int_0^{T_1} \cdots \int_0^{T_n} f(x-y) g(y)\, dy$$
$$= \int_0^{T_1} \cdots \int_0^{T_n} f(y) g(x-y)\, dy = (g \circledast f)(x). \tag{3.4}$$

EXAMPLE 2.

$$f \circledast e^{i(k\omega,x)} = T_1 \dots T_n c_k(f) e^{i(k\omega,x)}. \tag{3.5}$$

Indeed,

$$(f \circledast e^{i(k\omega,x)}, \varphi) = ((e_T f) * e^{i(k\omega,x)}, \varphi)$$
$$= (e_T(\xi) f(\xi) \times e^{i(k\omega,y)}, \varphi(\xi+y))$$
$$= \left(f(\xi), e_T(\xi) \int e^{i(k\omega,y)} \varphi(\xi+y)\, dy\right)$$
$$= \left(f(\xi), e_T(\xi) e^{-i(k\omega,\xi)} \int e^{i(k\omega,x)} \varphi(x)\, dx\right)$$
$$= T_1 \dots T_n c_k(f) \int e^{i(k\omega,x)} \varphi(x)\, dx.$$

□

In particular, by virtue of (1.7),

$$e^{i(k\omega,x)} \circledast e^{i(k'\omega,x)} = T_1 \dots T_n \delta_{kk'} e^{i(k\omega,x)}. \tag{3.6}$$

EXAMPLE 3. The formula (1.7) takes the form

$$f = f \circledast \delta_T, \qquad \partial^\alpha f = f \circledast \partial^\alpha \delta_T, \qquad f \in \mathcal{D}'_T \tag{3.7}$$

and, generally, if f and $g \in \mathcal{D}'_T$, then

$$c_k(f \circledast g) = T_1 \dots T_n c_k(f) c_k(g). \tag{3.8}$$

Indeed, using (3.5) we have

$$c_k(f \circledast g) = \frac{1}{T_1 \dots T_n} e^{-i(k\omega,x)} (f \circledast g) \circledast e^{i(k\omega,x)}$$
$$= \frac{1}{T_1 \dots T_n} e^{-i(k\omega,x)} (f \circledast (g \circledast e^{i(k\omega,x)}))$$
$$= c_k(g) e^{-i(k\omega,x)} (f \circledast e^{i(k\omega,x)})$$
$$= T_1 \dots T_n c_k(f) c_k(g).$$

□

From the foregoing it follows that $\mathcal{D}'_T$ forms a convolution algebra relative to the convolution operation $\circledast$ (see Sec. 4.5). The algebra $\mathcal{D}'_T$ is associative and commutative; its unit element is δ_T (see (3.7)); it contains zero divisors [see (3.6)].

What has been said in Sec. 4.9.4 holds true for equations $a \circledast u = f$ in the convolution algebra $\mathcal{D}'_T$. But here we have more precise statements.

THEOREM. *For the operator $a\circledast$, $a \in \mathcal{D}'_T$, to have an inverse $\mathcal{E}\circledast$ in $\mathcal{D}'_T$, it is necessary and sufficient that for certain $L > 0$ and m the following inequality hold:*

$$|c_k(a)| \geq L(1 + |k|)^{-m}. \tag{3.9}$$

Here, the fundamental solution $\mathcal{E}$ is unique and is expressible as a Fourier series:

$$\mathcal{E}(x) = \frac{1}{T_1^2 \dots T_n^2} \sum_{|k| \geq 0} \frac{1}{c_k(a)} e^{i(k\omega, x)}. \tag{3.10}$$

PROOF. SUFFICIENCY. By (3.9), the series (3.10) converges in $\mathcal{S}'$ and its sum $\mathcal{E} \in \mathcal{D}'_T$. We will prove that $\mathcal{E}$ satisfies the equation $a \circledast \mathcal{E} = \delta_T$. By the theorem of Sec. 7.2, it suffices to prove the equality of the Fourier coefficients:

$$c_k(a \circledast \mathcal{E}) = c_k(\delta_T) = \frac{1}{T_1 \dots T_n}.$$

But this is fulfilled by virtue of (3.8) and (3.10).

NECESSITY. Suppose, in $\mathcal{D}'_T$, there is a fundamental solution $\mathcal{E}$ of the operator $a\circledast$, $a \circledast \mathcal{E} = \delta_T$. Then it is unique (see Sec. 4.9.4) and from the equalities

$$c_k(a \circledast \mathcal{E}) = c_k(a) c_k(\mathcal{E}) T_1 \dots T_n = c_k(\delta_T) = \frac{1}{T_1 \dots T_n}$$

we derive

$$c_k(\mathcal{E}) = \frac{1}{T_1^2 \dots T_n^2 c_k(a)}. \tag{3.11}$$

Therefore the expansion (3.10) holds true. Furthermore, denoting by m the order of $\mathcal{E}$ and applying the estimate (2.3), from (3.11) we obtain the estimate (3.9):

$$|c_k(\mathcal{E})| = \frac{1}{T_1^2 \dots T_n^2} \frac{1}{|c_k(a)|} \leq C \|\mathcal{E}\|_{-m} (1 + |k|)^m.$$

The proof of the theorem is complete. □

7.4. Examples.

7.4.1. Solve the "quadratic" equation in $\mathcal{D}'_T$:

$$u \circledast u = \delta_T. \tag{4.1}$$

We have

$$c_k^2(u) = \frac{1}{T_1^2 \dots T_n^2}, \qquad c_k(u) = \pm \frac{1}{T_1 \dots T_n}.$$

Therefore equation(4.1) has a continuum of solutions:

$$u(x) = \frac{1}{T_1 \dots T_n} \sum_{|k| \geq 0} \varepsilon_k e^{i(k\omega, x)}, \qquad \varepsilon_k = \pm 1. \tag{4.2}$$

7.4.2.

$$\left(\frac{d}{dx}-\lambda\right)\mathcal{E}=\delta_T, \qquad n=1, \quad \lambda\neq ik\omega, \qquad k=0,\pm1,\dots.$$

Rewriting

$$(\delta_T'-\lambda\delta_T)\circledast\mathcal{E}=\delta_T,$$

we obtain

$$T\frac{ik\omega-\lambda}{T}c_k(\mathcal{E})=\frac{1}{T}$$

so that

$$\mathcal{E}(x)=\frac{1}{T}\sum_{|k|\geq 0}\frac{1}{ik\omega-\lambda}e^{ik\omega x}. \tag{4.3}$$

7.4.3. Let us consider the eigenvalue problem:

$$\delta_T'\circledast u=\lambda u, \qquad u\in\mathcal{D}_T',$$

$$\lambda_k=ik\omega, \qquad u_k(x)=e^{ik\omega x}, \qquad k=0,\pm1,\dots, \tag{4.4}$$

are eigenvalues and the associated eigenvalues of the operator $\delta_T'\circledast$.

7.4.4. Let $f\in\mathcal{D}_T'$, $n=1$. We consider the problem of finding the antiderivative $f^{(-1)}$ in $\mathcal{D}_T'$ (see Sec. 2.2):

$$\frac{df^{(-1)}}{dx}=f, \qquad \delta_T'\circledast f^{(-1)}=f.$$

From the equation

$$\frac{ik\omega}{T}c_k(f^{(-1)})=c_k(f)$$

it follows that *the antiderivative* $f^{(-1)}$ *exists in* $\mathcal{D}_T'$ *if and only if* $c_0(f)=0$ *and can be expressed by the Fourier series*

$$f^{(-1)}(x)=\sum_{|k|>0}\frac{c_k(f)}{ik\omega}e^{ik\omega x}+C \tag{4.5}$$

where C is an arbitrary constant.

7.4.5. *Bernoulli polynomials.* Set $f_0=1-T\delta_T$. Since $c_0(f_0)=c_0(1)-Tc_0(\delta_T)=0$, it follows that $f_0^{(-1)}$ exists in $\mathcal{D}_T'$, we choose $c_0(f_0^{(-1)})=0$, and so forth. As a result we obtain a sequence of antiderivatives $f_0^{(-m)}(x)$ in $\mathcal{D}_T'$, $m=1,2,\dots$, which are polynomials on the basic period $(0,T)$. These polynomials with leading coefficient 1 are called *Bernoulli polynomials*

$$B_m(x)=m!f_0^{(-m)}(x), \qquad m=0,1,\dots$$

Let us find their Fourier-series expansion. We have

$$(f_0^{(-m)})^{(m)}=\delta_T^{(m)}\circledast f_0^{(-m)}=f_0=1-T\delta_T$$

and therefore

$$c_k(\delta_T^{(m)}\circledast f_0^{(-m)})=(ik\omega)^m c_k(f_0^{(-m)})=1, \qquad k\neq 0.$$

Consequently,

$$B_m(x) = m! f_0^{(-m)}(x) = m! \sum_{|k|>0} \frac{1}{(ik\omega)^m} e^{ik\omega x}, \qquad 0 \le x < T. \tag{4.6}$$

The Bernoulli polynomials satisfy the recurrence relation

$$B'_m(x) = mB_{m-1}(x), \qquad B_0(x) = 1, \qquad \int_0^T B_m(x)\,dx = 0, \qquad m = 1, 2, \dots .$$

Let us write the expressions for the lower Bernoulli polynomials for $T = 1$.

$$B_0(x) = 1,$$
$$B_1(x) = x - \frac{1}{2},$$
$$B_2(x) = x^2 - x + \frac{1}{6},$$
$$B_3(x) = x^3 - \frac{3}{2}x^2 + \frac{1}{2}x, \qquad \dots .$$

The numbers $B_k = B_k(0)$ are called the *Bernoulli numbers*: $B_0 = 1$, $B_1 = -1/2$, $B_{2k+1} = 0$, $B_2 = 1/6$, $B_4 = -1/30$, ...

Let us emphasize the formula (Euler)

$$\zeta(2m) = \sum_{n=1}^{\infty} \frac{1}{n^{2m}} = (-1)^{m-1} B_{2m} 2^{2m-1} \frac{\pi^{2m}}{(2m)!}, \qquad m = 1, 2, \dots ,$$

where $\zeta(\alpha)$ is the Riemann zeta-function.

8. Positive Definite Generalized Functions

8.1. The definition and elementary properties of positive definite generalized functions. Suppose $f \in \mathcal{D}'(\mathbb{R}^n)$; a generalized function $f^*(x) = \bar{f}(-x)$ in $\mathcal{D}'(\mathbb{R}^n)$ is said to be the *$*$-Hermite conjugate of* f; if $f = f^*$, then f is said to be a *$*$-Hermite* (generalized) function.

The function $f(x)$, which is continuous in $\mathbb{R}^n$, is a *positive* function, $f \gg 0$, if for any points $x_1, \dots, x_l$ in $\mathbb{R}^n$ and for the complex numbers $z_1, \dots, z_l$ the following inequality holds true:

$$\sum_{1 \le j,k \le l} f(x_j - x_k) z_j \bar{z}_k \ge 0.$$

From this definition it follows certain properties: setting $l = 1$, we obtain $f(0) \ge 0$; also, setting $l = 2$, $x_1 = x$, $x_2 = 0$, we have

$$f(0)\big(|z_1|^2 + |z_2|^2\big) + f(x) z_1 \bar{z}_2 + f(-x) \bar{z}_1 z_2 \ge 0$$

whence it follows that f is a $*$-Hermite bounded function:

$$f = f^*, \qquad |f(x)| \le f(0). \tag{1.1}$$

Finally, replacing the integral by the limit of the sequence of the Riemann sums, we obtain the inequality

$$\int f(x - \xi)\varphi(x)\bar{\varphi}(\xi)\,dx\,d\xi \ge 0, \qquad \varphi \in \mathcal{D},$$

that is

$$(f, \varphi * \varphi^*) \geq 0, \qquad \varphi \in \mathcal{D}. \tag{1.2}$$

We take property (1.2) as the basis for defining positive definite generalized functions. A generalized function f is said to be *positive definite*, $f \gg 0$, if it satisfies the condition

$$(f, \varphi * \varphi^*) \geq 0, . \quad \varphi \in \mathcal{D}. \tag{1.3}$$

This definition immediately implies that if $f \gg 0$, then $f(-x) \gg 0$ and $\bar{f} \gg 0$ as well.

Furthermore, *for the generalized function f to be positive definite, it is necessary and sufficient that*

$$f * \alpha * \alpha^* \gg 0, \qquad \alpha \in \mathcal{E}'. \tag{1.4}$$

Indeed, if $f \gg 0$, then, using (6.4) of Sec. 4.6 and the properties of commutativity and associativity of a convolution (see Sec. 4.2), we have, for all $\alpha \in \mathcal{E}'$ and $\varphi \in \mathcal{D}$,

$$\begin{aligned}(f * \alpha * \alpha^*, \varphi * \varphi^*) &= \big(f, (\alpha * \alpha^*)(-x) * (\varphi * \varphi^*)\big) \\ &= \big(f, [\alpha(-x) * \varphi] * [\alpha^*(-x) * \varphi^*]\big) \\ &= \big(f, [\alpha(-x) * \varphi] * [\alpha(-x) * \varphi]^*\big) \geq 0,\end{aligned}$$

since $\alpha * \varphi \in \mathcal{D}$ if $\alpha \in \mathcal{E}'$ and $\varphi \in \mathcal{D}$ (see Sec. 4.2.7 and Sec. 4.6). Thus, condition (1.4) is fulfilled. Conversely, suppose the condition (1.4) is fulfilled, so that if $\alpha \in \mathcal{D}$, then $f * \alpha * \alpha^*$ is a continuous positive definite function and therefore, by (1.1), $(f * \alpha * \alpha^*)(0) \geq 0$. Now, taking advantage of formula (6.3) of Sec. 4.6, we have, for all $\alpha \in \mathcal{D}$.

$$\big(f(-y), \alpha * \alpha^*\big) = \big(f, (\alpha * \alpha^*)(-y)\big) = (f * \alpha * \alpha^*)(0) \geq 0$$

so that $f(-x) \gg 0$ and therefore $f \gg 0$. □

If $f \gg 0$, then $f = f^*$.

Indeed, from what has been proved, for all $\alpha \in \mathcal{D}$,

$$(f * \alpha * \alpha^*)^* = f^* * (\alpha * \alpha^*) = f * (\alpha * \alpha^*).$$

If in the last equation we let $\alpha \to \delta$ in $\mathcal{E}'$ [and then $\alpha^* \to \delta$ in $\mathcal{E}'$ and from formula (5.1) of Sec. 4.5 it follows that $\alpha * \alpha^* \to \delta$ in $\mathcal{E}'$ as well] and use the continuity property of a convolution (see Sec. 4.3), we obtain $f = f^*$. □

For what follows we will need the following lemma.

LEMMA. *For every integer $p \geq 0$ there is a function $\omega(x)$ with the properties:*

$$\omega \in C^{2p}; \qquad \omega(x) = 0, \quad |x| > 1;$$

$$F[\omega](\xi) \geq \frac{A}{(1 + |\xi|^2)^{p+n+1}}. \tag{1.5}$$

PROOF. Let $\chi \in \mathcal{D}$, $\chi(x) = 0$ for $|x| > 1/2$ and

$$\gamma(x) = \frac{1}{(2\pi)^n} \int \frac{e^{-i(x,\xi)}\, d\xi}{(1 + |\xi|^2)^{p+n+1}} = F^{-1}\left[\frac{1}{(1 + |\xi|^2)^{p+n+1}}\right].$$

Let us verify that the function $\omega = \gamma(\chi * \chi^*)$ has the properties (1.5). Since $\gamma \in C^{2p}$ and $\chi * \chi^* \in C^\infty$, it follows that $\omega \in C^{2p}$. Furthermore, by virtue of Sec. 4.2.7,

$\operatorname{supp}\omega \subset \operatorname{supp}\chi + \operatorname{supp}\chi^* \subset U_{1/2} + U_{1/2} = U_1$. Finally, using the formula of the Fourier transform of a convolution (see Sec. 6.5), we have

$$\begin{aligned} F[\omega](\xi) &= F\left[\gamma(\chi * \chi^*)\right] \\ &= F\left[F^{-1}\left[\frac{1}{(1+|\xi|^2)^{p+n+1}}\right]F^{-1}\left[F[\chi]F[\chi^*]\right]\right] \\ &= \frac{1}{(2\pi)^n(1+|\xi|^2)^{p+n+1}} * \left|F[\chi]\right|^2 \\ &= \frac{1}{(2\pi)^n}\int \frac{|F\chi]|^2(y)\,dy}{(1+|\xi-y|^2)^{p+n+1}} \end{aligned}$$

and therefore

$$\begin{aligned} F[\omega](\xi) &\geq \frac{1}{(2\pi)^n}\int\limits_{|y|<1} \frac{|F[\chi]|^2(y)\,dy}{(1+|\xi-y|^2)^{p+n+1}} \\ &\geq \frac{1}{(2\pi)^n[1+(|\xi|+1)^2]^{p+n+1}}\int\limits_{|y|<1} |F[\chi]|^2(y)\,dy \\ &\geq \frac{A}{(1+|\xi|^2)^{p+n+1}}. \end{aligned}$$

The proof of the lemma is complete. □

8.2. The Bochner–Schwartz theorem. Suppose $f \in \mathcal{D}'$ and $f \gg 0$. Then, in the ball $U_3 = [x : |x| < 3]$, f has a finite order $m \geq 0$ (see Sec. 1.3),

$$|(f,\varphi)| \leq K\|\varphi\|_{C^m(\overline{U}_3)}, \qquad \varphi \in C_0^m(U_3).$$

Take an integer $p \geq 0$ such that $2p \geq m$, and let ω be a function with the properties (1.5) of the lemma. Then the function $\omega * \omega^* \in C_0^m(U_2)$ and, consequently, the generalized function

$$g = f * \omega * \omega^* = f * (\omega * \omega^*) \gg 0,$$

is continuous in the ball $\overline{U}_1$.

We will now prove that g is bounded in $\mathbb{R}^n$.

Let $\alpha_n \in \mathcal{D}$, $\operatorname{supp}\alpha_n \subset U_{1/n}$, $\alpha_n \geq 0$, $\int \alpha_n\,dx = 1$, $\alpha_n \to \delta$, $n \to \infty$. The sequence of functions $g * \alpha_n * \alpha_n^*$, $n = 2, 3, \ldots$, is uniformly bounded,

$$\left|(g * \alpha_n * \alpha_n^*)(x)\right| \leq \max_{|x|\leq 2/n} |g(x)|\, \|\alpha_n * \alpha_n^*\|_{\mathcal{L}^1} \leq \max_{|x|\leq 1} |g(x)|$$

(see Sec. 4.1), and converges weakly on the set $\mathcal{D}$, which is dense in $\mathcal{L}^1$ (see Sec. 1.2). In this case, the limiting generalized function g may be identified with the function $g(x)$ taken from $\mathcal{L}^\infty$.

We now prove that $g(x)$ is the Fourier transform of a nonnegative measure with compact support on $\mathbb{R}^n$. Since g is bounded and $\mathcal{D}$ is dense in $\mathcal{S}$ (see Sec. 5.1), it follows that the inequality

$$(g, \varphi * \varphi^*) \geq 0$$

holds true for all $\varphi \in \mathcal{S}$ and, hence,

$$\left(F^{-1}[g], F[\varphi * \varphi^*]\right) = \left(F^{-1}[g], \left|F[\varphi]\right|^2\right) \geq 0, \qquad \varphi \in \mathcal{S}. \tag{2.1}$$

But the operation F is an isomorphism of $\mathcal{S}$ onto $\mathcal{S}$ (see Sec. 6.1). Therefore, the inequality (2.1) is equivalent to the inequality

$$(F^{-1}[g], |\psi|^2) \geq 0, \qquad \psi \in \mathcal{S}. \tag{2.2}$$

Now let φ be any nonnegative test function in $\mathcal{S}$ and let $\{\eta_k\}$ be a sequence of nonnegative functions taken from $\mathcal{D}$ that tend to 1 in $\mathbb{R}^n$ (see Sec. 4.1). Then

$$\left|\eta_k\sqrt{\varphi + 1/k}\right|^2 = \eta_k^2(\varphi + 1/k) \to \varphi, \qquad k \to \infty \quad \text{in} \quad \mathcal{S}$$

and, consequently, by virtue of (2.2),

$$(F^{-1}[g], \varphi) = \lim_{k\to\infty}\left(F^{-1}[g], \left|\eta_k\sqrt{\varphi + 1/k}\right|^2\right) \geq 0, \qquad \varphi \in \mathcal{S}.$$

By Theorem II of Sec. 1.7, $F^{-1}[g] = \nu$ is a nonnegative measure on $\mathbb{R}^n$ and $g = F[\nu]$. But $\nu = F^{-1}[g] \in \mathcal{S}'$. For this reason, ν is a tempered measure (see Sec. 5.3) so that for all $\varphi \in \mathcal{S}$ we have

$$(F^{-1}[g], \varphi) = \int \varphi(\xi)\nu(d\xi) = (g, F^{-1}[\varphi]). \tag{2.3}$$

Let $\{\eta_k\}$ be a nondecreasing sequence of nonnegative functions taken from $\mathcal{D}$ that tend to 1 in $\mathbb{R}^n$. Then $F^{-1}[\eta_k] \to \delta$, $k \to \infty$, on all functions bounded in $\mathbb{R}^n$ and continuous in the neighbourhood of zero. Setting $\varphi = \eta_k$ in (2.3),

$$\int \eta_k(\xi)\nu(d\xi) = (g, F^{-1}[\eta_k]),$$

passing to the limit as $k \to \infty$, and making use of the Levi theorem, we obtain

$$\int \nu(d\xi) = g(0)$$

which is precisely the assertion.

Let us prove that the equation

$$u * \omega * \omega^* = g \tag{2.4}$$

has a unique solution in the class of positive definite generalized functions from $\mathcal{D}'$ which is given by the formula

$$u = F^{-1}\left[\frac{\nu}{|F[\omega]|^2}\right]. \tag{2.5}$$

Indeed, by virtue of the inequality (1.5), the generalized function u given by (2.5) is actually the sole solution of equation (2.4) in the class $\mathcal{S}'$ by virtue of the theorem on the Fourier transform of a convolution (see Sec. 6.5):

$$F[u]\left|F[\omega]\right|^2 = F[g] = \nu. \tag{2.6}$$

It remains to prove that the solution of the homogeneous equation $u*\omega*\omega^* = 0$ in the class of generalized functions u, which can be represented in the form of a difference $u_1 - u_2$, where $u_j \gg 0$, $u_j \in \mathcal{D}'$, is trivial: $u = 0$.

Suppose such a u satisfies that equation. Then for all $\alpha \in \mathcal{D}$ the function $u * \alpha * \alpha^*$ also satisfies that equation:

$$(u * \omega * \omega^*) * \alpha * \alpha^* = 0 = (u * \alpha * \alpha^*) * \omega * \omega^*.$$

But the function $u * \alpha * \alpha^* = u_1 * \alpha * \alpha^* - u_2 * \alpha * \alpha^*$ is a difference of the continuous positive definite functions and, hence, is bounded in $\mathbb{R}^n$ (and the more so in $\mathcal{S}'$). From what has been proved, $u * \alpha * \alpha^* = 0$. Passing to the limit in this equation as $\alpha \to \delta$ in S' and using the continuity of a convolution, we obtain $u = 0$, which is what we set out to prove.

The generalized function $f \gg 0$ also satisfies equation (2.4). By virtue of the uniqueness of the solution of that equation, f coincides with the generalized function u given by (2.5). Hence, f is the inverse Fourier transform of the tempered measure $\mu = \nu |F[\omega]|^{-2}$, by the inequality (1.5).

We have thus proved the necessity of the conditions of the following theorem.

THEOREM (Bochner–Schwartz). *For a generalized function f taken from $\mathcal{D}'$ to be positive definite, it is necessary and sufficient that it be a Fourier transform of a nonnegative tempered measure, $f = F[\mu]$, $\mu \in S'$, $\mu \geq 0$.*

SUFFICIENCY. If $f = F[\mu]$, $\mu \geq 0$, $\mu \in S'$, then $(\mu, |\varphi|^2) \geq 0$ for all $\varphi \in S$, whence [compare (2.1)]

$$\left(\mu, |F[\varphi]|^2\right) = \left(F[\mu], F^{-1}[|F[\varphi]|^2]\right) = (f, \varphi * \varphi^*) \geq 0, \qquad \varphi \in \mathcal{D},$$

so that $f \gg 0$, and the theorem is proved. □

COROLLARY 1. *If $f \in \mathcal{D}'$, $f \gg 0$, then $f \in S'$.*

COROLLARY 2 (Bochner). *For a generalized function f that is continuous in the neighbourhood of zero to be positive definite, it is necessary and sufficient that it be a Fourier transform of positive measure ν with compact support in $\mathbb{R}^n$:*

$$f(x) = \int e^{i(x,\xi)} \nu(d\xi), \qquad \nu \geq 0, \qquad \int \nu(d\xi) = f(0); \tag{2.7}$$

here, $f(x)$ is a continuous function on $\mathbb{R}^n$.

COROLLARY 3. *For a generalized function f to be positive definite, it is necessary and sufficient that it be (uniquely) represented, for some integer $m \geq 0$, in the form*

$$f(x) = (1 - \Delta)^m f_0(x)$$

where $f_0(x)$ is a continuous positive definite function.

This is a consequence of the following chain of equalities:

$$\begin{aligned} f = F[\mu] &= F\left[(1 + |\xi|^2)^m \frac{\mu}{(1 + |\xi|^2)^m}\right] \\ &= (1 - \Delta)^m F[\nu] \end{aligned}$$

where the measure $\nu = \mu(1 + |\xi|^2)^{-m} \geq 0$, for sufficiently large m, may be made finite on $\mathbb{R}^n$. □

8.3. Examples.

8.3.1. Let the polynomial $P(\xi) \geq 0$. Then

$$P(-i\partial)\delta \gg 0.$$

In particular, $\delta \gg 0$.

8.3.2. If $f \gg 0$ and $g \in \mathcal{E}'$, $g \gg 0$, then $f * g \gg 0$.

Indeed, the measure $F[g] \geq 0$, $F[g] \in \theta_M$ (see Sec. 6.4) and therefore the measure $F[f * g] = F[f]F[g] \geq 0$ is tempered. □

8.3.3. If $f \gg 0$ and $F[g] \in \mathcal{E}'$, $g \gg 0$, then $gf \gg 0$.
Indeed, $g \in \theta_M$, $gf \in \mathcal{S}'$ and

$$F^{-1}[gf] = F^{-1}[g] * F^{-1}[f]$$

is a nonnegative measure taken from $\mathcal{S}'$ and, hence, tempered (see Sec. 5.3). □

8.3.4. $e^{-(Ax,x)} \gg 0$, where A is a real positive definite matrix (see Sec. 6.6.2).

8.3.5. $\dfrac{1}{|x|} \gg 0$, $n = 3$ (see Sec. 6.6.7).

8.3.6. $\pi\delta(x) \pm i\mathcal{P}\frac{1}{x} \gg 0$, $n = 1$ (see Sec. 6.6.10).

8.3.7. *For $f \in \mathcal{D}'_T$ to be positive definite, it is necessary and sufficient that its Fourier coefficients $c_k(f)$ be nonnegative. Then for all $\varphi \in C^\infty \cap \mathcal{D}'_T$ the following inequality holds:*

$$(f, \varphi \circledast \varphi^*)_T \geq 0. \tag{3.1}$$

This follows from the theorem of Sec. 7.2, by which theorem

$$F^{-1}[f] = \sum_{|k|\geq 0} c_k(f)\delta(\xi - k\omega), \tag{3.2}$$

and from the Bochner–Schwartz theorem. To prove the inequality (3.1), let us take advantage of the machinary developed in Sec. 7. Using (3.1) of Sec. 7.3 and (3.3) of Sec. 7.3, we have the chain of equalities

$$\begin{aligned}(f, \varphi \circledast \varphi^*)_T &= (e_T f, \varphi \circledast \varphi^*) \\ &= (f * \delta, e_T(\varphi \circledast \varphi^*)) \\ &= (\delta, f(-x) * e_T(\varphi \circledast \varphi^*)) \\ &= (\delta, f(-x) \circledast (\varphi \circledast \varphi^*)) \\ &= (\delta, f(-x) \circledast \varphi \circledast \varphi^*) \\ &= (\delta, f(-x) * (e_T\varphi) * (e_T\varphi^*)) \\ &= (\delta, f(-x) * [(e_T\varphi) * (e_T\varphi^*)]) \\ &= (f, (e_T\varphi) * (e_T\varphi^*)),\end{aligned}$$

so that

$$(f, \varphi \circledast \varphi^*)_T = (f, (e_T\varphi) * (e_T\varphi^*)), \tag{3.3}$$

whence follows inequality (3.1). □

9. The Laplace Transform of Tempered Generalized Functions

The fundamentals of the general theory of the Laplace transform of generalized functions were developed by Schwartz [91] and Lions [68]. However, this theory has been developed into its most complete form for the case — so important in applications of mathematical physics — of tempered generalized functions.

9.1. Definition of the Laplace transform. Let Γ be a closed convex acute cone in $\mathbb{R}^n$ with vertex at 0 (see Sec. 4.4); we put $C = \operatorname{int}\Gamma^*$ (by Lemma 1 of Sec. 4.4 the cone $C \neq \varnothing$; C is an open and convex cone). Denote by T^C a tubular domain in $\mathbb{C}^n$ with base C:

$$T^C = \mathbb{R}^n + iC = [z = x + iy : x \in \mathbb{R}^n,\ y \in C].$$

Suppose $g \in \mathcal{S}'(\Gamma+)$ (see Sec. 4.5 and 5.6). We will use the term *Laplace transform* $L[g]$ of the generalized function g for the expression

$$L[g](z) = F\left[g(\xi)e^{-(y,\xi)}\right](x), \tag{1.1}$$

where F is the Fourier transform operation.

EXAMPLE.

$$L\left[\delta(\xi - \xi_0)\right] = e^{i(z,\xi_0)}. \tag{1.2}$$

This follows from (2.6) of Sec. 6.2.

Let us now prove that for all $y \in C$

$$g(\xi)e^{-(y,\xi)} \in \mathcal{S}'$$

so that the Laplace transform $L[g]$ is a tempered generalized function with respect to x for all $y \in C$.

True enough, suppose η is any function of the class C^∞ with the following properties:

$$|\partial^\alpha \eta(\xi)| \le c_\alpha; \qquad \eta(\xi) = 1, \qquad \xi \in (\operatorname{supp} g)^\varepsilon;$$
$$\eta(\xi) = 0, \qquad \xi \notin (\operatorname{supp} g)^{2\varepsilon},$$

where ε is any positive number. Then, by what was proved in Sec. 6.6.9,

$$\eta(\xi)e^{-(y,\xi)} \in \mathcal{S}(\mathbb{R}^n) \quad \text{for all} \quad y \in C \tag{1.3}$$

and therefore, by (10.2) of Sec. 1.10,

$$g(\xi)e^{-(y,\xi)} = g(\xi)\eta(\xi)e^{-(y,\xi)} \in \mathcal{S}',$$

which is what was to be proved. □

The Laplace transform L *a linear and one-to-one operation.* This follows from the appropriate properties of the Fourier transform (see Sec. 6.2).

Let us now prove the representation

$$L[g] = \left(g(\xi), \eta(\xi)e^{i(z,\xi)}\right), \qquad z \in T^C. \tag{1.4}$$

This representation does not depend on the auxiliary function η with the above-indicated properties.

Indeed, let $y \in C$ and $\varphi \in \mathcal{S}$. Then (compare Sec. 6.6.9)

$$\begin{aligned}(L[g], \varphi) &= \left(F\left[g(\xi)e^{-(y,\xi)}\right], \varphi\right) = \left(g(\xi)e^{-(y,\xi)}, F[\varphi]\right) \\ &= \left(g(\xi), \eta(\xi)e^{-(y,\xi)} \int \varphi(x)e^{i(x,\xi)}\, dx\right) \\ &= \int \varphi(x)\left(g(\xi), \eta(\xi)e^{i(z,\xi)}\right) dx,\end{aligned}$$

whence follows (1.4). Here, we made use of formula (5.4) of Sec. 5.5, since, by virtue of (1.3),

$$\eta(\xi)e^{-(y,\xi)}\varphi(x)e^{i(x,\xi)} \in \mathcal{S}(\mathbb{R}^{2n}).$$

□

We set $f(z) = L[g]$ and will prove that *the function $f(z)$ is holomorphic in T^C and the following differentiation formula holds true:*

$$\partial^\alpha f(z) = \left((i\xi)^\alpha g(\xi), \eta(\xi)e^{i(z,\xi)}\right). \tag{1.5}$$

The proof is analogous to that of the lemma of Sec. 3.1. The continuity of the function $f(z)$ in T^C follows from the representation (1.4), from the continuity of the function $\eta(\xi)e^{i(z,\xi)}$ with respect to z in T^C in the sense of convergence in $\mathcal{S}$,

$$\eta(\xi)e^{i(z,\xi)} \to \eta(\xi)e^{i(z_0,\xi)}, \qquad z \to z_0 \quad \text{in} \quad \mathcal{S}, \quad z \in T^C, \quad z_0 \in T^C$$

(see the estimate in Sec. 6.6.9), and from the continuity of the functional g on $\mathcal{S}$,

$$f(z) = \left(g(\xi), \eta(\xi)e^{i(z,\xi)}\right) \to \left(g(\xi), \eta(\xi)e^{i(z_0,\xi)}\right) = f(z_0), \qquad z \to z_0.$$

To prove the holomorphicity of the function $f(z)$ in T^C it suffices to establish, by virtue of the familiar Hartogs theorem, the existence of all first derivatives $\frac{\partial f}{\partial z_j}$, $j = 1, \ldots, n$. Suppose $e_1 = (1, 0, \ldots, 0)$. Then for each $z \in T^C$

$$\chi_h(\xi) = \frac{1}{h}\left[\eta(\xi)e^{i(z+he_1,\xi)} - \eta(\xi)e^{i(z,\xi)}\right] \to \eta(\xi)i\xi_1 e^{i(z,\xi)},$$
$$h \to 0 \quad \text{in} \quad \mathcal{S}.$$

Therefore, from the representation (1.4) and from the linearity and the continuity of the functional g on $\mathcal{S}$ as $h \to 0$ we have

$$\begin{aligned}\frac{f(z+he_1) - f(z)}{h} &= \frac{1}{h}\left[\left(g(\xi), \eta(\xi)e^{i(z+he_1,\xi)}\right) - \left(g(\xi), \eta(\xi)e^{i(z,\xi)}\right)\right] \\ &= \left(g(\xi), \chi_h(\xi)\right) \to \left(g(\xi), \eta(\xi)i\xi_1 e^{i(z,\xi)}\right) \\ &= \left(i\xi_1 g(\xi), \eta(\xi)e^{i(z,\xi)}\right),\end{aligned}$$

so that the derivative $\frac{\partial f}{\partial z_1}$ exists and the differentiation formula (1.5) holds for $\alpha = (1, 0, \ldots, 0)$ and, hence, also for all first derivatives

$$\frac{\partial f}{\partial z_j} = \left(i\xi_j g(\xi), \eta(\xi)e^{i(z,\xi)}\right), \qquad j = 1, \ldots, n. \tag{1.6}$$

Applying this reasoning to (1.6), we see that the formulas (1.5) hold for all second derivatives, and so forth. This completes the proof. □

Our task now is to give a full description of holomorphic functions that are the Laplace transforms of generalized functions taken from the algebras $\mathcal{S}'(\Gamma+)$ and $\mathcal{S}'(\Gamma)$ (see below Sec. 12.2).

We will refer to the generalized function $g(\xi)$ of $\mathcal{S}'(\Gamma+)$, for which $f = L[g]$, as the *spectral function* of the function $f(z)$. The spectral function $g(\xi)$ is unique and, by virtue of (1.1), it is equal to

$$g(\xi) = e^{(y,\xi)}F_x^{-1}[f(x+iy)](\xi) \tag{1.7}$$

and the right-hand side of (1.7) is independent of $y \in C = \operatorname{int}\Gamma^*$.

The equality (1.7) expresses the *inverse Laplace transform*, and it can be written as

$$g(\xi) = L^{-1}[f](\xi) = e^{(y,\xi)}F_x^{-1}[f(x+iy)](\xi).$$

9.2. Properties of the Laplace transform. These properties follow from the appropriate properties of the Fourier transform (see Sec. 6.3).

9.2.1. *Differentiation of the Laplace transform:*

$$\partial^\alpha L[g] = L[(i\xi)^\alpha g], \tag{2.1}$$

which is precisely formula (1.5).

9.2.2. *The Laplace transform of a derivative:*

$$L[\partial^\alpha g] = (-iz)^\alpha L[g]. \tag{2.2}$$

It suffices to prove (2.2) for the first derivatives. We have

$$\begin{aligned} L\left[\frac{\partial g}{\partial \xi_j}\right] &= F\left[\frac{\partial g(\xi)}{\partial \xi_j} e^{-(y,\xi)}\right] \\ &= F\left[\frac{\partial}{\partial \xi_j}\left(g(\xi)e^{-(y,\xi)}\right)\right] + y_j F\left[g(\xi)e^{-(y,\xi)}\right] \\ &= (-ix_j + y_j)F\left[g(\xi)e^{-(y,\xi)}\right] \\ &= -iz_j L[g]. \end{aligned}$$

In particular, setting $g = \delta(\xi - \xi_0)$ in (2.2) and using (1.2), we obtain

$$L\left[\partial^\alpha \delta(\xi - \xi_0)\right] = (-iz)^\alpha e^{i(z,\xi_0)}. \tag{2.3}$$

9.2.3. *The translation of the Laplace transform.* If $\Im a \in C$, then

$$L\left[g(\xi)e^{i(a,\xi)}\right] = L[g](z + a). \tag{2.4}$$

Indeed,

$$\begin{aligned} L\left[g(\xi)e^{i(a,\xi)}\right] &= F\left[g(\xi)e^{i(\Re a,\xi)}e^{-(y+\Im a,\xi)}\right] \\ &= L[g](x + \Re a + iy + i\Im a) \\ &= L[g](z + a). \end{aligned}$$

□

9.2.4. *The Laplace transform of a translation:*

$$L\left[g(\xi - \xi_0)\right] = e^{i(z,\xi_0)} L[g](z) \qquad \xi_0 \in \mathbb{R}^n. \tag{2.5}$$

9.2.5. *The Laplace transform under a linear transformation of the argument:*

$$L\left[g(A\xi)\right] = \frac{1}{|\det A|} L[g]\left((A^{-1})^T z\right), \qquad z \in T^{A^T C}. \tag{2.6}$$

9.2.6. *The Laplace transform of a direct product.* If $g_1(\xi) \in \mathcal{S}'(\Gamma_1+)$ and $g_2(\eta) \in \mathcal{S}'(\Gamma_2+)$, then

$$L[g_1 \times g_2](z, \zeta) = L[g_1](z)L[g_2](\zeta), \qquad (z,\zeta) \in T^{C_1 \times C_2}. \tag{2.7}$$

9.2.7. *The Laplace transform of a convolution.* If $g \in \mathcal{S}'(\Gamma+)$ and $g_1 \in \mathcal{S}'(\Gamma+)$, then $g * g_1 \in \mathcal{S}'(\Gamma+)$ (see Sec. 5.6.2) and

$$L[g * g_1] = L[g]L[g_1]. \tag{2.8}$$

Let us prove the following formula: *if $g \in \mathcal{D}'(\Gamma+)$ and $g_1 \in \mathcal{D}'(\Gamma+)$, then for all $y \in C$*

$$\left(ge^{-(y,\xi)}\right) * \left(g_1 e^{-(y,\xi)}\right) = (g * g_1)e^{-(y,\xi)}. \tag{2.9}$$

Indeed, using the formula (5.1) of Sec. 4.5, we have, for all $\varphi \in \mathcal{D}$,

$$\begin{aligned}
\left((ge^{-(y,\xi)}) * (g_1e^{-(y,\xi)}), \varphi\right) &= \left(g(\xi)e^{-(y,\xi)} \times g_1(\xi')e^{-(y,\xi')}, \eta_1(\xi)\eta_2(\xi')\varphi(\xi+\xi')\right) \\
&= \left(g(\xi) \times g_1(\xi'), \eta_1(\xi)\eta_2(\xi')e^{-(y,\xi+\xi')}\varphi(\xi+\xi')\right) \\
&= \left(g * g_1, \varphi e^{-(y,\xi)}\right) \\
&= \left(e^{-(y,\xi)}(g * g_1), \varphi\right),
\end{aligned}$$

which is what we set out to prove.

From (2.9), for g and $g_1 \in \mathcal{S}'(\Gamma+)$, and from the formula for the Fourier transform of a convolution [see (5.1) of Sec. 6.5] there follows immediately the formula (2.8):

$$\begin{aligned}
L[g * g_1] &= F\left[(*g_1)e^{-(y,\xi)}\right] \\
&= F\left[(ge^{-(y,\xi)}) * (g_1e^{-(y,\xi)})\right] \\
&= F\left[ge^{-(y,\xi)}\right] F\left[g_1e^{-(y,\xi)}\right] \\
&= L[g]L[g_1].
\end{aligned}$$

□

REMARK. In the case of a single variable, the Laplace transform is defined differently in the operational calculus of Heaviside: if the original $g \in \mathcal{S}'([0,\infty)+)$, then its image (the Laplace transform) is the function

$$F\left[g(t)e^{-\sigma t}\right](-\omega),$$

which is holomorphic in the right half-plane $\sigma > 0$ of the complex plane $p = \sigma + i\omega$. In particular,

$$\theta(t) \leftrightarrow \int_0^\infty e^{-pt}\,dt = \frac{1}{p}.$$

However, we will adhere to the definition (1.1) in the case of $n = 1$ as well.

9.3. Examples.

9.3.1.

$$L[f_\alpha] = \frac{1}{(-iz)^\alpha}, \qquad y > 0, \tag{3.1}$$

where f_α is a generalized function from $\mathcal{S}'_+ = \mathcal{S}' \cap \mathcal{D}'_+$ that was introduced in Sec. 4.9.5. The branch of the function $(-iz)^\alpha$ in the half-plane $y > 0$ is chosen so that it is positive for $z = iy$, $y > 0$.

Let $\alpha > 0$. Then

$$L[f_\alpha](iy) = \int_0^\infty \frac{\xi^{\alpha-1}}{\Gamma(\alpha)}e^{-y\xi}\,d\xi = \frac{1}{y^\alpha\Gamma(\alpha)}\int_0^\infty u^{\alpha-1}e^{-u}\,du = \frac{1}{y^\alpha}$$

so that the functions $L[f_\alpha](z)$ and $(-iz)^{-\alpha}$, which are holomorphic in the upper half-plane, coincide on the line $z = iy$, $y > 0$. By virtue of the principle of analytic continuation, (3.1) holds for $\alpha > 0$. But if $\alpha \leq 0$, then $\alpha + m > 0$ for some integer m. Therefore, $f_\alpha = f^{(m)}_{\alpha+m}$ and, by what has been proved,

$$L[f_\alpha] = L[f^{(m)}_{\alpha+m}] = (-iz)^m L[f_{\alpha+m}]$$
$$= (-iz)^m(-iz)^{-\alpha-m} = (-iz)^{-\alpha}.$$

9.3.2.

$$L\big[\theta(\xi)\sin\omega\xi\big] = \frac{\omega}{\omega^2 - z^2},$$
$$L\big[\theta(\xi)\cos\omega\xi\big] = \frac{-iz}{\omega^2 - z^2}. \tag{3.2}$$

These follows from the equations [see 9.3.1 for $\alpha = 1$]

$$L\big[\theta(\xi)e^{i\omega\xi}\big] = \frac{1}{z+\omega}, \qquad L\big[\theta(\xi)e^{-i\omega\xi}\big] = \frac{i}{z-\omega}.$$

9.3.3. Let us prove the equation

$$L\left[\frac{\theta(\xi)\sqrt{\pi}}{\Gamma(\nu+1/2)}\left(\frac{\xi}{2}\right)^{\nu} J_\nu(\xi)\right] = (1-z^2)^{-\nu-1/2}, \qquad \nu > -1/2. \tag{3.3}$$

By 9.3.1 we have the equations

$$L\left[e^{i\xi}f_{\nu-1/2}(\xi)\right] = \left(\frac{i}{z+1}\right)^{\nu+1/2},$$
$$L\left[e^{-i\xi}f_{\nu-1/2}(\xi)\right] = \left(\frac{i}{z-1}\right)^{\nu+1/2}.$$

Therefore, using the formula for the Laplace transform of a convolution, we have

$$L\big[(e^{i\xi}f_{\nu-1/2}) * (e^{-i\xi}f_{\nu-1/2})\big] = (1-z^2)^{-\nu-1/2}.$$

But

$$(e^{i\xi}f_{\nu-1/2}) * (e^{-i\xi}f_{\nu-1/2}) = \frac{\theta(\xi)}{\Gamma^2(\nu+1/2)}\int_0^\xi e^{i(\xi-t)}e^{-it}(\xi-t)^{\nu-1/2}t^{\nu-1/2}\,dt$$
$$= \frac{\theta(\xi)e^{i\xi}\xi^{2\nu}}{\Gamma^2(\nu+1/2)}\int_0^t e^{-2i\xi u}\big[(1-u)u\big]^{\nu-1/2}\,du$$
$$= \frac{\theta(\xi)\xi^{2\nu}}{\Gamma^2(\nu+1/2)}\int_{-1}^1 e^{-i\xi v}\left(\frac{1-v^2}{4}\right)^{\nu-1/2}\frac{dv}{2}$$
$$= \frac{\theta(\xi)\sqrt{\pi}}{\Gamma(\nu+1/2)}\left(\frac{\xi}{2}\right)^{\nu} J_\nu(\xi),$$

where $u = (\nu+1)/2$, and (3.3) is proved. Here we made use of formula (6.31) of Sec. 6.6.

9.3.4. We now prove the formula

$$\sin\xi = \int_0^\xi J_0(\xi - t)J_0(t)\,dt. \tag{3.4}$$

Since the right and left members of (3.4) are odd, it suffices to prove the formula for $\xi > 0$. It is therefore sufficient to prove the convolution equation

$$\theta \sin\xi = (\theta J_0) * (\theta J_0),$$

which is equivalent, by virtue of 9.3.2 and 9.3.3, to the trivial equality

$$\frac{1}{1-z^2} = \frac{1}{\sqrt{1-z^2}}\frac{1}{\sqrt{1-z^2}}, \qquad y > 0.$$

□

9.3.5. Let us find the fundamental solution $\mathcal{E}$ of the operator $(\theta J_0)*$ in the algebra $\mathcal{S}'_+$. By 9.3.3, we have

$$L[\theta J_0] = \frac{1}{\sqrt{1-z^2}} \neq 0, \qquad y > 0.$$

Consequently,

$$L[\mathcal{E}] = \sqrt{1-z^2} = \frac{1-z^2}{\sqrt{1-z^2}}$$

whence

$$\begin{aligned}\mathcal{E}(\xi) &= \theta(\xi)J_0(\xi) + \left[\theta(\xi)J_0(\xi)\right]'' \\ &= \theta(\xi)J_0(\xi) + \delta'(\xi)J_0(\xi) + 2\delta(\xi)J_0'(\xi) + \theta(\xi)J_0''(\xi) \\ &= -\frac{\theta(\xi)J_0'(\xi)}{\xi} + \delta'(\xi),\end{aligned}$$

that is

$$\mathcal{E}(\xi) = \theta(\xi)\frac{J_1(\xi)}{\xi} + \delta'(\xi). \tag{3.5}$$

9.3.6. Let $f \in \mathcal{L}^1_{\text{loc}} \cap \mathcal{D}'_T$ (see Sec. 7), $n = 1$. Then

$$L[\theta f] = \frac{1}{1 - e^{izT}} \int_0^T f(\xi)e^{iz\xi}\,d\xi. \tag{3.6}$$

Indeed,

$$\begin{aligned}L[\theta f](z) &= \int_0^\infty f(\xi)e^{iz\xi}\,d\xi \\ &= \int_0^\infty e^{iz(t+T)}f(t+T)\,dt + \int_0^T e^{iz\xi}f(\xi)\,d\xi \\ &= e^{izT}L[\theta f] + \int_0^T e^{iz\xi}f(\xi)\,d\xi,\end{aligned}$$

whence follows (3.6). □

10. The Cauchy Kernel and the Transforms of Cauchy–Bochner and Hilbert

10.1. The space $\mathcal{H}_s$. We denote by $\mathcal{L}_s^2$ the Hilbert space consisting of all functions $g(\xi)$ with finite norm

$$\|g\|_{(s)} = \left[\int |g(\xi)|^2(1+|\xi|^2)^s\, d\xi\right]^{1/2} = \|g(\xi)(1+|\xi|^2)^{s/2}\|.$$

We denote by $\mathcal{H}_s$ the collection of all (generalized) functions $f(x)$ that are Fourier transforms of functions in $\mathcal{L}_s^2$, $f = F[g]$, with norm

$$\|f\|_s = \|F^{-1}[f]\|_{(s)} = \|g\|_{(s)}. \tag{1.1}$$

The parameter s can assume any real values.

Clearly, $\mathcal{H}_0 = \mathcal{L}^2 = \mathcal{L}_0^2$ and

$$\|g\|_{(0)} = \|g\| = (2\pi)^{-n/2}\|f\| = \|f\|_0$$

by virtue of the Parseval–Steklov equation (see Sec. 6.6.3).

From the definition of the space $\mathcal{H}_s$ we find that for $f \in \mathcal{H}_s$ it is necessary that the function f be representable as

$$f(x) = (1-\Delta)^m f_1(x), \qquad f_1 \in \mathcal{L}^2, \qquad m = 0, \quad \text{if} \quad s \geq 0;$$
$$m = 1 + \left[-\frac{s}{2}\right], \quad \text{if} \quad s < 0. \tag{1.2}$$

The space $\mathcal{H}_s$ is the Hilbert space isomorphic to $\mathcal{L}_s^2$. And

$$\mathcal{S} \subset \mathcal{H}_s \subset \mathcal{H}_{s'} \subset \mathcal{S}', \qquad s' < s,$$

where inclusion is to be understood as embedding together with the appropriate topology, $\|f\|_{s'} \leq \|f\|_s$, $f \in \mathcal{H}_s$.

Let us now prove that *$\mathcal{S}$ is dense in $\mathcal{H}_s$*.

By virtue of (1.1) it suffices to prove that $\mathcal{D}$ is dense in $\mathcal{L}_s^2$. Let $g \in \mathcal{D}_s^2$ and $\varepsilon > 0$. Then

$$\psi(\xi) = g(\xi)(1+|\xi|^2)^{s/2} \in \mathcal{L}^2.$$

But $\mathcal{D}$ is dense in $\mathcal{L}^2$ (see Sec. 1.2, Corollary 1 to Theorem II). Therefore there is a function ψ_1 in $\mathcal{D}$ such that $\|\psi - \psi_1\| < \varepsilon$. Putting

$$g_1(\xi) = \psi_1(\xi)(1+|\xi|^2)^{-s/2} \in \mathcal{D}$$

we obtain

$$\|g - g_1\|_{(s)}^2 = \int |g(\xi) - g_1(\xi)|^2(1+|\xi|^2)^s\, d\xi$$
$$= \|\psi - \psi_1\|^2 < \varepsilon^2,$$

which is what we affirmed. □

Let us now prove that

$$\mathcal{H}_s \subset \overrightarrow{C}_0^l \quad \text{if} \quad l \quad \text{is an integer}, \quad l < s - n/2.$$

REMARK. This assertion is a simple special case of the Sobolev imbedding theorem (Sobolev [97]).

To prove this, note that if $f \in \mathcal{H}_s$, then for all $|\alpha| \le l < s - n/2$

$$\xi^\alpha F^{-1}[f] \in \mathcal{L}^1$$

by the Cauchy-Bunyakovsky inequality

$$\begin{aligned}\|\xi^\alpha F^{-1}[f]\|_{\mathcal{L}^1} &= \int \left|\xi^\alpha (1+|\xi|^2)^{-s/2}(1+|\xi|^2)^{s/2} F^{-1}[f](\xi)\right| d\xi \\ &\le \left[\int |\xi|^{2|\alpha|}(1+|\xi|^2)^{-s}\, d\xi\right]^{1/2} \left\|(1+|\xi|^2)^{s/2}F^{-1}[f]\right\| \\ &= K\|F^{-1}[f]\|_{(s)} = K\|f\|_s < \infty.\end{aligned}$$

Hence,

$$\partial^\alpha f(x) = \partial^\alpha F[F^{-1}[f]] = \int (-i\xi)^\alpha F^{-1}[f](\xi) e^{i(x,\xi)}\, d\xi \in C$$

and by virtue of the Riemann–Lebesgue theorem $\partial^\alpha f(x) = o(1)$, $|x| \to \infty$ for all $|\alpha| \le l$. Which means that $f \in \overset{\circ}{C}{}^l_0$ (concerning notation see Sec. 0.5). □

Now let s be an integer ≥ 0. In that case the space $\mathcal{L}^2_s$ consists of those and only those functions $g(\xi)$ for which $\xi^\alpha g \in \mathcal{L}^2$ for all $|\alpha| \le s$. Therefore the space $\mathcal{H}_s$ consists of those and only those functions f for which the generalized derivatives $\partial^\alpha f \in \mathcal{L}^2$ for all $|\alpha| \le s$ (by the Plancherel theorem). Furthermore, as follows from the readily verifiable identity

$$\|f\|_s^2 = \|f\|_{s-1}^2 + \sum_{1\le j\le n} \|\partial_j f\|_{s-1}^2$$

the space $\mathcal{H}_s$ consists of those functions f in $\mathcal{H}_{s-1}$ for which $\partial_j f \in \mathcal{H}_{s-1}$, $j = 1,\dots,n$.

Now let us describe the conjugate to the $\mathcal{H}_s$ space. Since $\mathcal{S}$ is dense in $\mathcal{H}_s$, it follows that every continuous linear form on $\mathcal{H}_s$ is uniquely defined by its restriction on $\mathcal{S}$.

LEMMA. *If $L(f)$ is a continuous linear form on $\mathcal{H}_s$, then for some f_0 in $\mathcal{H}_{-s}$,*

$$L(f) = (f_0, f), \qquad f \in \mathcal{S}, \tag{1.3}$$

and the norm of that form is $\|f_0\|_{-s}$. Thus, $\mathcal{H}_{-s}$ is the conjugate space to $\mathcal{H}_s$.

PROOF. By hypothesis, the linear form

$$L_1(\chi) = L\left(F^{-1}[\chi(\xi)(1+|\xi|^2)^{-s/2}]\right)$$

is continuous on $\mathcal{L}^2$ and coincides with the form $L(f)$ for

$$\chi(\xi) = (1+|\xi|^2)^{1/2} F[f]. \tag{1.4}$$

Now, since the mapping $f \to \chi$ given by (1.4) is biunique and reciprocally continuous from $\mathcal{H}_s$ to $\mathcal{L}^2$, it follows that $\|L_1\| = \|L\|$. By the F. Riesz theorem there is a function $g_1 \in \mathcal{L}^2$ such that

$$L_1(\chi) = \int g_1(\xi)\chi(\xi)\, d\xi, \qquad \|g_1\| = \|L_1\|.$$

From this, if we introduce the function

$$g(\xi) = g_1(\xi)(1+|\xi|^2)^{s/2}$$

taken from $\mathcal{L}^2_{-s}$ and then set $f_0 = F[g]$, we obtain, for all f in $\mathcal{S}$, the representation (1.3):

$$\begin{aligned} L(f) &= L_1\left((1+|\xi|^2)^{s/2}F[f]\right) \\ &= \int g_1(\xi)(1+|\xi|^2)^{s/2}F[f](\xi)\,d\xi \\ &= \int g(\xi)F[f](\xi)\,d\xi \\ &= (F[g], f) = (f_0, f) \end{aligned}$$

and, besides,

$$\|L\| = \|g_1\| = \left\|g(1+|\xi|^2)^{-s/2}\right\| = \|f_0\|_{-s},$$

which is what we set out to establish. The proof of the lemma is complete. □

For $f \in \mathcal{H}_s$, it is necessary and sufficient that

$$f = g * L_s, \qquad g \in \mathcal{L}^2, \tag{1.5}$$

where the kernel L_s is given by the formula

$$L_s(x) = F^{-1}\left[(1+|\xi|^2)^{-s/2}\right]. \tag{1.6}$$

Here, the mapping $g \to f = g * L_s$ is bijective and reciprocally continuous from $\mathcal{L}^2$ onto $\mathcal{H}_s$.

Indeed, (1.5) is equivalent to (see Sec. 6.5, remark)

$$F[f] = F[g](1+|\xi|^2)^{-s/2},$$

which is what sets up the reciprocally one-to-one and reciprocally continuous correspondence between $\mathcal{L}^2$ and $\mathcal{H}_s$. □

The kernel L_s has the obvious property [by virtue of (1.6)]

$$L_s * L_\sigma = L_{s+\sigma}.$$

Explicitly, the expression is

$$L_s(x) = \begin{cases} (1-\Delta)^{-s/2}\delta(x) & \text{if } -s \text{ is even and } \geq 0, \\ \dfrac{2|x|^{s/4-n/2}}{(2\pi)^{n/2}2^{s/4}\Gamma(s/4)}K_{n/2-s/4}(|x|) & \text{if } s > n, \end{cases}$$

where K_ν is the Bessel function of an imaginary argument (see [**82**, App. II.10]).

REMARK. The convolution (1.5) is called a *Bessel potential*.

Let $f \in \mathcal{H}_s$ and $f_0 \in \mathcal{H}_\sigma$. Then the convolution $f * f_0$ exists in $\mathcal{S}'$, is expressed by the formula

$$f * f_0 = F^{-1}\left[F[f]F[f_0]\right], \tag{1.7}$$

and is continuous with respect to f and f_0 jointly: if $f \to 0$ in $\mathcal{H}_s$, and $f_0 \to 0$ in $\mathcal{H}_\sigma$, then $f * f_0 \to 0$ in $\mathcal{S}'$.

Indeed, represent f and f_0 in the form (1.2):

$$f = (1-\Delta)^m f_1, \quad f_1 \in \mathcal{L}^2, \qquad f_0 = (1-\Delta)^l f_{01}, \quad f_{01} \in \mathcal{L}^2.$$

By virtue of the formula (1.7), just proved, for the convolution $f_1 * f_{01}$ (see Sec. 6.5.2), also of the rule of differentiating a convolution (see Sec. 4.2.5), and of the properties of the Fourier transform (see Sec. 6.3), we are convinced that (1.7) holds true:

$$\begin{aligned}(1-\Delta)^{m+l}(f_1 * f_{01}) &= (1-\Delta)^m f_1 * (1-\Delta)^l f_{01} = f * f_0 \\ &= (1-\Delta)^{m+l} F^{-1}\left[F[f]F[f_0]\right] \\ &= F^{-1}\left[(1+|\xi|^2)^{m+l} F[f_1]F[f_{01}]\right] \\ &= F^{-1}\left[F\left[(1-\Delta)^m f_1\right] F\left[(1-\Delta)^l f_{01}\right]\right] \\ &= F^{-1}\left[F[f]F[f_0]\right].\end{aligned}$$

□

From the representation (1.7) follows the continuity of the convolution $f * f_0$ with respect to f and f_0 jointly.

EXAMPLE. $\mathcal{P}\frac{1}{x} * \mathcal{P}\frac{1}{x} = -\pi^2\delta(x)$.

This follows from (6.14) of Sec. 6.6:

$$\mathcal{P}\frac{1}{x} * \mathcal{P}\frac{1}{x} = F^{-1}\left[(\pi i \operatorname{sgn} \xi)^2\right] = -\pi^2 F^{-1}[1] = -\pi^2\delta(x).$$

Generalizing, we obtain the following: *if $f \in \mathcal{H}_s$, $f_0 \in \mathcal{H}_\sigma$, and $F[f_1], \dots, F[f_m]$ belong to $\mathcal{L}^\infty$, then their convolution exists in $\mathcal{S}'$ and can be represented as*

$$f * f_0 * f_1 * \dots * f_m = F^{-1}\left[F[f]F[f_0]F[f_1]\dots F[f_m]\right]. \tag{1.8}$$

Analogously, *if $f \in \mathcal{H}_s$ and $F[f_1], \dots, F[f_m]$ belong to $\mathcal{L}^\infty$, then their convolution exists in $\mathcal{H}_s$, can be represented as*

$$f * f_1 * \dots * f_m = F^{-1}\left[F[f]F[f_1]\dots F[f_m]\right], \tag{1.9}$$

and is continuous in f from $\mathcal{H}_s$ to $\mathcal{H}_s$.

If $f_0 \in \mathcal{H}_{-s}$ and $f \in \mathcal{H}_s$, $s \geq 0$, then the following formula holds:

$$f_0 * f = \left(f_0(x'), f(x-x')\right). \tag{1.10}$$

True enough, by virtue of (1.7), for all φ in $\mathcal{S}$ we have

$$\begin{aligned}(f_0 * f, \varphi) &= \left(F^{-1}\left[F[f_0]F[f]\right], \varphi\right) \\ &\left(F[f_0]F[f], \frac{1}{(2\pi)^n}\int \varphi(x)e^{-i(x,\xi)}\,dx\right) \\ &= \frac{1}{(2\pi)^n}\int \varphi(x)\left(F[f_0]F[f], e^{-i(x,\xi)}\right)\,dx,\end{aligned}$$

since $F[f_0]F[f] \in \mathcal{L}^1$. Therefore

$$\begin{aligned}
(f_0 * f)(x) &= \frac{1}{(2\pi)^n}\left(F[f_0]F[f], e^{-i(x,\xi)}\right) \\
&= \frac{1}{(2\pi)^n}\left(F[f_0], F[f]e^{-i(x,\xi)}\right) \\
&= \frac{1}{(2\pi)^n}\left(f_0, F\left[F[f]e^{-i(x,\xi)}\right]\right) \\
&= \frac{1}{(2\pi)^n}\left(f_0(x'), \int F[f](\xi)e^{-i(x-x',\xi)}\,d\xi\right) \\
&= \left(f_0(x'), F^{-1}\left[F[f]\right](x-x')\right) \\
&= \left(f_0(x'), f(x-x')\right).
\end{aligned}$$

We denote by $\mathcal{D}'_{\mathcal{L}^2}$ the inductive limit (union) of the increasing sequence of spaces $\mathcal{H}_{-s}$, $s = 0, 1, \ldots$,

$$\mathcal{D}'_{\mathcal{L}^2} = \bigcup_{s \geq 0} \mathcal{H}_{-s}.$$

By virtue of the lemma, $\mathcal{D}'_{\mathcal{L}^2}$ is a collection of continuous linear functionals on the countable-normed space $\mathcal{D}_{\mathcal{L}^2}$, which is a projective limit (intersection) of the decreasing sequence of spaces $\mathcal{H}_s$, $s = 0, 1, \ldots$,

$$\mathcal{D}_{\mathcal{L}^2} = \bigcap_{s=0}^{\infty} \mathcal{H}_s.$$

The space $\mathcal{D}_{\mathcal{L}^2}$ is an algebra with respect to the operation of ordinary multiplication (associative, commutative without unity, see Sec. 4.5); and, for all f and g in $\mathcal{D}_{\mathcal{L}^2}$,

$$\|fg\|_s \leq c_{p-s}\|f\|_p\|g\|_s, \qquad s \geq 0, \quad p > s + n/2. \tag{1.11}$$

Indeed, since $f \in \mathcal{H}_s$ and $g \in \mathcal{H}_s$ for all s, it follows, in particular, that $f \in \mathcal{L}^2$ and $g \in \mathcal{L}^2$. We put $\tilde{f} = F^{-1}[f]$ and $\tilde{g} = F^{-1}[g]$. Using the definition (1.1) of a norm in $\mathcal{H}_s$, the formula of the Fourier transform of a convolution (see Sec. 6.5), the Fubini theorem, the Cauchy–Bunyakovsky inequalities, and

$$1 + |\xi + \xi'|^2 \leq (1 + |\xi|^2)(1 + |\xi'|^2),$$

for all $s \geq 0$ and $p > s + n/2$, we obtain the inequality (1.11):

$$\begin{aligned}
\|fg\|_s^2 &= \int \left|F^{-1}[fg](\xi)\right|^2 (1+|\xi|^2)^s \, d\xi \\
&= \int (1+|\xi|^2)^s \left| \int \tilde{f}(\xi')(1+|\xi'|^2)^{p/2} \frac{\tilde{g}(\xi-\xi')\, d\xi'}{(1+|\xi'|^2)^{p/2}} \right|^2 d\xi \\
&\leq \int |\tilde{f}(\xi')|^2 (1+|\xi'|^2)^p \, d\xi' \int (1+|\xi|^2)^s \int |\tilde{g}(\xi-\xi')|^2 (1+|\xi'|^2)^{-p} \, d\xi' \, d\xi \\
&= \|f\|_p^\alpha \int |\tilde{g}(\eta)|^2 (1+|\xi'+\eta|^2)^s (1+|\xi'|^2)^{-p} \, d\eta \, d\xi' \\
&\leq \|f\|_p^2 \int |\tilde{g}(\eta)|^2 (1+|\eta|^2)^s \, d\eta \int (1+|\xi'|^2)^{s-p} \, d\xi' \\
&= \int \frac{d\xi'}{(1+|\xi'|^2)^{p-s}} \|f\|_p^2 \|g\|_s^2 .
\end{aligned}$$

□

We set

$$S_0' = \bigcup_{s \geq 0} \mathcal{L}^2_{-s} = F[\mathcal{D}'_{\mathcal{L}^2}]$$

as the inductive limit (union) of the spaces $\mathcal{L}^2_{-s}$, $s = 0, 1, \ldots$.

For f to belong to S', it is necessary and sufficient that it be representable as

$$f(x) = x^\alpha f_0(x), \qquad f_0 \in \mathcal{D}'_{\mathcal{L}^2}. \tag{1.12}$$

Sufficiency is obvious and necessity follows from the representation

$$F[f] = (i\partial)^\alpha g(\xi),$$

where g is a tempered continuous function in $\mathbb{R}^n$ (see Sec. 5.4), that is, $f_0 = F^{-1}[g] \in \mathcal{H}_s$ for some s, whence it follows the required representation (1.12). □

Let the generalized function f_0 from S' be continuously dependent, in S', on the parameter σ on the compact K, that is, $(f_\sigma, \varphi) \in C(K)$ for any $\varphi \in S$, and let μ be a finite measure on K. We introduce the generalized function $\int_K f_\sigma \mu(d\sigma)$ taken from S' by means of the equation

$$\left(\int_K f_\sigma \mu(d\sigma), \varphi \right) = \int_K (f_\sigma, \varphi) \mu(d\sigma), \qquad \varphi \in S.$$

It is easy to see that

$$F\left[\int_K f_\sigma \mu(d\sigma) \right] = \int_K F[f_\sigma] \mu(d\sigma), \tag{1.13}$$

(cf. definition of Sec. 2.7).

10.2. The Cauchy kernel $\mathcal{K}_C(z)$. Let C be a connected open cone in $\mathbb{R}^n$ with vertex at 0 and let C^* be the conjugate cone C (see Sec. 4.4). The function

$$\mathcal{K}_C(z) = \int_{C^*} e^{i(z,\xi)} \, d\xi \equiv L[\theta_{C^*}] = F[\theta_{C^*} e^{-(y,\xi)}] \tag{2.1}$$

is termed the *Cauchy kernel of a tubular region* T^C; here, $\theta_{C^*}(\xi)$ is the characteristic function of the cone C^*.

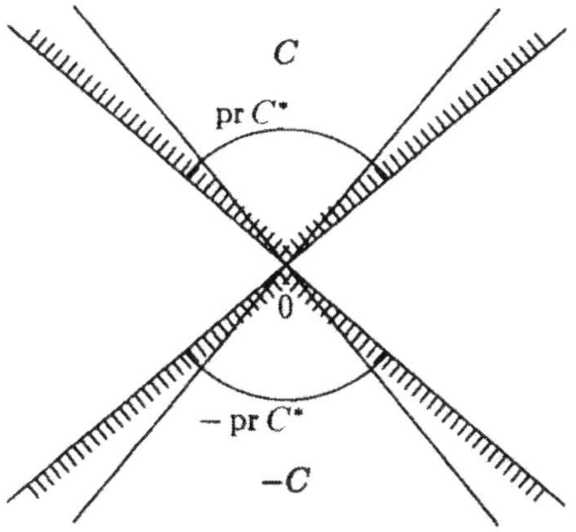

Figure 28

If the cone C is not acute, then by virtue of Lemma 1 of Sec. 4.4, $\operatorname{mes} C^* = 0$ and, hence, $\mathcal{K}_C(z) \equiv 0$. Furthermore, since $C^* = (\operatorname{ch} C)^*$, it follows that $\mathcal{K}_C(z) \equiv \mathcal{K}_{\operatorname{ch} C}(z)$. Therefore, without restricting generality, we may regard the cone C as acute and convex.

By what has been proved (see Sec. 9.1), the kernel $\mathcal{K}_C(z)$ is a holomorphic function in T^C; and, moreover, the integral in (2.1) converges uniformly with respect to z in any tubular region T^K, $K \Subset C$ (K is a compact).

We will now show that the kernel $\mathcal{K}_C(z)$ can be represented by the integral

$$\mathcal{K}_C(z) = i^n \Gamma(n) \int\limits_{\operatorname{pr} C^*} \frac{d\sigma}{(z,\sigma)^n}, \qquad z \in T^C. \tag{2.2}$$

Indeed, since $(y,\sigma) > 0$ for all $y \in C$, $\sigma \in \operatorname{pr} C^*$, it follows that the denominator of the integrand on the right of (2.2) is equal to $\left[(x,\sigma) + i(y,\sigma)\right]^n$ and does not vanish in T^C, and, consequently, the right-hand side of (2.2) is a holomorphic function in T^C. Since the kernel $\mathcal{K}_C(z)$ is also a holomorphic function in T^C, it suffices to prove (2.2) on the manifold $z = iy$, $y \in C$. But when $x = 0$ the formula (2.2) follows readily from (2.1):

$$\mathcal{K}_C(iy) = \int\limits_{C^*} e^{-(y,\xi)}\, d\xi = \int\limits_{\operatorname{pr} C^*} \int\limits_0^\infty e^{-\rho(y,\sigma)} \rho^{n-1}\, d\rho\, d\sigma$$
$$= \int\limits_{\operatorname{pr} C^*} \frac{d\sigma}{(y,\sigma)^n} \int\limits_0^\infty e^{-u} u^{n-1}\, du = i^n \Gamma(n) \int\limits_{\operatorname{pr} C^*} \frac{d\sigma}{(iy,\sigma)^n}.$$

From the representation (2.2) it follows that *the kernel* $\mathcal{K}_C(z)$ *and also the kernel* $\mathcal{K}_{-C}$ *are holomorphic in the domain*

$$D = \mathbb{C}^n \setminus \bigcup_{\sigma \in \operatorname{pr} C^*} \left[z \colon (z,\sigma) = 0\right].$$

It is easy to see that the domain D contains the tubular domains T^C and T^{-C} and also the real points of the cones C and $-C$.

The kernels $\mathcal{K}_C$ and $\mathcal{K}_{-C}$ satisfy the relations

$$\begin{aligned} \mathcal{K}_{-C}(z) &= (-1)^n \mathcal{K}_C(z) = \overline{\mathcal{K}_C(\bar z)} = \mathcal{K}_C(-z), \\ \mathcal{K}_C(\lambda z) &= \lambda^{-n} \mathcal{K}_C(z) \qquad \lambda \in \mathbb{C}^1 \setminus \{0\}, \quad z \in T^C \cup T^{-C}. \end{aligned} \tag{2.3}$$

Let us now prove the estimate

$$|\partial^\alpha \mathcal{K}_C(z)| \le M_\alpha \Delta^{-n-|\alpha|}(y), \qquad z \in T^C \cup T^{-C}, \tag{2.4}$$

where $\Delta(y) = \Delta(y, -\partial C \cup \partial C)$ is the distance from y to the boundary of the cone $-C \cup C$:

$$\Delta(y) = \inf_{\sigma \in \operatorname{pr} C^*} (\sigma, y), \qquad y \in C$$

(see Sec. 0.2 and Fig. 28).

Indeed, using the representation (2.2), we have, for $z \in T^C$, the estimate (2.4):

$$\begin{aligned}|\partial^\alpha \mathcal{K}_C(z)| &\le M_\alpha \int_{\operatorname{pr} C^*} \frac{|\sigma^\alpha|\,d\sigma}{|(z,\sigma)|^{n+|\alpha|}} \\ &\le M_\alpha \sup_{\sigma \in \operatorname{pr} C^*} (y,\sigma)^{-n-|\alpha|} = M_\alpha \Delta^{-n-|\alpha|}(y).\end{aligned}$$

The estimate (2.4) for $z \in T^{-C}$ follows from (2.4) that was proved for $z \in T^C$ and from the properties (2.3). More rigorous reasoning yields the estimate

$$|\partial^\alpha \mathcal{K}_C(z)| \le M'_\alpha \Delta^{-n+1-|\alpha|}(y)\left[|x|^2 + \Delta^2(y)\right]^{-1/2}, \qquad z \in T^C \cup T^{-C}. \tag{2.4'}$$

We now prove the estimate for all $s \ge 0$,

$$\|\partial^\alpha \mathcal{K}_C(x+iy)\|_s \le K_{s,\alpha}\left[1 + \Delta^{-s}(y)\right]\Delta^{-n/2-|\alpha|}(y), \qquad y \in -C \cup C. \tag{2.5}$$

Indeed, by (2.1) and (2.2) for $y \in C$ we have the estimate (2.5):

$$\begin{aligned}\|\partial^\alpha \mathcal{K}_C(x+iy)\|_s^2 &= \left\|F^{-1}\left[\partial^\alpha \mathcal{K}_C(x+iy)\right]\right\|_{(s)}^2 \\ &= \left\|(-i\xi)^\alpha \theta_{C^*}(\xi) e^{-(y,\xi)}\right\|_{(s)}^2 \\ &= \int_{C^*} e^{-2(y,\xi)}\left(1+|\xi|^2\right)^s |\xi^\alpha|^2\,d\xi \\ &\le \int_0^\infty (1+\rho^2)^s \rho^{n-1+2|\alpha|} \int_{\operatorname{pr} C^*} e^{-2\rho(y,\sigma)}\,d\sigma\,d\rho \\ &\le \frac{\sigma_n}{2}\int_0^\infty e^{-2\rho\Delta(y)}(1+\rho^2)^s \rho^{n-1+2|\alpha|}\,d\rho \\ &= \frac{\sigma_n}{2^{n+1+2|\alpha|}\Delta^{n+2|\alpha|}(y)}\int_0^\infty e^{-u}\left[1+\frac{u^2}{4\Delta^2(y)}\right]^s u^{n-1+2|\alpha|}\,du \\ &\le K_{s,\alpha}^2\left[1+\Delta^{-s}(y)\right]^2 \Delta^{-n-2|\alpha|}(y).\end{aligned}$$

(Here, σ_n is the surface area of a unit sphere in $\mathbb{R}^n$, see Sec. 0.6.) The case $y \in -C$ can be considered with the use of the relation (2.3). □

The kernel $\mathcal{K}_C(z)$ assumes a boundary value equal to $(\pm 1)^n F[\theta_{\pm C^*}]$, respectively,

$$\mathcal{K}_c(x+iy) \to (\pm 1)^n \mathcal{K}_C(\pm z) = (\pm 1)^n F[\theta_{\pm C^*}], \tag{2.6}$$

as $y \to 0$, $y \in \pm C$ in norm in $\mathcal{H}_s$ for arbitrary $s < -n/2$.

Indeed, by what has been proved, $\mathcal{K}_C(x+iy) \in \mathcal{H}_s$ for $y \in -C \cup C$ and for any s, while the generalized functions $F[\theta_{\pm C^*}] \in \mathcal{H}_s$ for all $s < -n/2$ (since $\theta_{\pm C^*} \in \mathcal{L}_s^2$ for $s < -n/2$). Therefore, when $s < -n/2$ and $y \in C$, $y \to 0$, we have

$$\|\mathcal{K}_C(x+iy) - F[\theta_{C^*}]\|_s^2 = \|\theta_{C^*} e^{-(y,\xi)} - \theta_{C^*}\|_{(s)}^2$$
$$= \int_{C^*} \left[e^{-(y,\xi)} - 1\right]^2 (1+|\xi|^2)^s \, d\xi \to 0.$$

But if $y \in -C$, $y \to 0$, then

$$\mathcal{K}_C(x+iy) = (-1)^n \mathcal{K}_C(-x-iy) \to (-1)^n \mathcal{K}_C(-x) = (-1)^n F[\theta_{-C^*}]$$

and the formula (2.6) is proved. □

From the formulas (2.3) and (2.6) we have the following relations for the boundary values of the kernels $\mathcal{K}_C(z)$ and $\mathcal{K}_{-C}(z)$:

$$\left.\begin{aligned} \mathcal{K}_{-C}(x) &= \overline{\mathcal{K}_C(x)} = \mathcal{K}_C(-x), && x \in \mathbb{R}^n, \\ \mathcal{K}_{-C}(x) &= (-1)^n \mathcal{K}_C(x), && x \in C \cup (-C); \end{aligned}\right\} \tag{2.7}$$

$$\left.\begin{aligned} \Re\mathcal{K}_C(x) &= \frac{1}{2} F[\theta_{C^*} + \theta_{-C^*}] \\ &= \frac{1}{2}[\mathcal{K}_C(x) + \mathcal{K}_C(-x)], \\ \Im\mathcal{K}_C(x) &= \frac{1}{2i} F[\theta_{C^*} - \theta_{-C^*}] \\ &= \frac{1}{2i}[\mathcal{K}_C(x) - \mathcal{K}_C(-x)]. \end{aligned}\right\} \tag{2.8}$$

From this, taking into account the trivial equalities

$$(\theta_{C^*} - \theta_{-C^*})^2 = (\theta_{C^*} + \theta_{-C^*})^2 = \theta_{C^*} + \theta_{-C^*},$$
$$(\theta_{C^*} - \theta_{-C^*})(\theta_{C^*} + \theta_{-C^*}) = \theta_{C^*} - \theta_{-C^*},$$

and making use of (1.9) for the convolution, we obtain the following relations between the generalized functions $\Re\mathcal{K}_C(x)$ and $\Im\mathcal{K}_C(x)$:

$$-\Im\mathcal{K}_C * \Im\mathcal{K}_C = \Re\mathcal{K}_C * \Re\mathcal{K}_C = \frac{1}{2}(2\pi)^n \Re\mathcal{K}_C, \tag{2.9}$$

$$\Im\mathcal{K}_c * \Re\mathcal{K}_C = \frac{1}{2}(2\pi)^n \Im\mathcal{K}_C, \tag{2.10}$$

Let us now calculate the real and imaginary parts of the kernel $\mathcal{K}_C(x)$. To do this, we introduce, for $k = 0, 1, \ldots$, the generalized functions

$$\delta^{(k)}[(x,\sigma)] \quad \text{and} \quad \mathcal{P}^{(k)}\frac{1}{(x,\sigma)}$$

that operates on the test functions φ in $\mathcal{S}$ via the rules

$$\left(\delta^{(k)}[(x,\sigma)], \varphi\right) = (-1)^k \int_{(x,\sigma)=0} \left(\frac{\partial}{\partial\sigma}\right)^k \varphi(x)\, dS_x,$$

$$\left(\mathcal{P}^{(k)}\frac{1}{(x,\sigma)}, \varphi\right) = (-1)^k \,\mathbf{VP} \int_{-\infty}^{\infty} \frac{1}{\lambda} \int_{(x,\sigma)=0} \frac{\partial^k}{\partial\lambda^k}\varphi(x+\lambda\sigma)\, dS_x\, d\lambda.$$

The generalized functions that have just been introduced depend continuously, in $\mathcal{S}'$, on the parameter σ on the unit sphere S_1 (in the sense of Sec. 10.1).

Let us prove the equation

$$\mathcal{K}_C(x) = \pi(-i)^{n-1} \int\limits_{\mathrm{pr}\, C^*} \delta^{(n-1)}[(x,\sigma)]\, d\sigma - (-i)^n \int\limits_{\mathrm{pr}\, C^*} \mathcal{P}^{(n-1)} \frac{1}{(x,\sigma)}\, d\sigma. \tag{2.11}$$

Using the representation (2.2), we have, for all $y \in C$ and $\varphi \in \mathcal{S}$,

$$\int \mathcal{K}_C(x+iy)\varphi(x)\, dx = i^n \Gamma(n) \int \int\limits_{\mathrm{pr}\, C^*} \frac{d\sigma}{[(x,\sigma)+i(y,\sigma)]^n} \varphi(x)\, dx$$
$$= i^n \Gamma(n) \int\limits_{\mathrm{pr}\, C^*} \int \frac{\varphi(x)\, dx}{[(x,\sigma)+i(y,\sigma)]^n} \varphi(x)\, d\sigma. \tag{2.12}$$

Now let $y \to 0$, $y \in C$. Then $0 < \varepsilon = (y,\sigma) \to 0$ for $\sigma \in \mathrm{pr}\, C^*$, and the integral

$$\int \frac{\varphi(x)\, dx}{[(x,\sigma)+i\varepsilon]^n} = \int_{-\infty}^{\infty} \frac{1}{(\lambda+i\varepsilon)^n} \int\limits_{(x,\sigma)=0} \varphi(x+\lambda\sigma)\, dS_x\, d\lambda$$
$$= \frac{1}{\Gamma(n)} \int_{-\infty}^{\infty} \frac{1}{\lambda+i\varepsilon} \frac{\partial^{n-1}}{\partial\lambda^{n-1}} \int\limits_{(x,\sigma)=0} \varphi(x+\lambda\sigma)\, dS_x\, d\lambda$$
$$= \frac{1}{\Gamma(n)} \int_{-\infty}^{\infty} \ln(\lambda+i\varepsilon) \frac{\partial^{n}}{\partial\lambda^{n}} \int\limits_{(x,\sigma)=0} \varphi(x+\lambda\sigma)\, dS_x\, d\lambda$$

is uniformly bounded with respect to (ε,σ) for all $0 < \varepsilon \le 1$ and $\sigma \in \mathrm{pr}\, C^*$; furthermore, by virtue of the Sochozki formula [see (8.3) of Sec. 1.8], that integral tends, as $\varepsilon \to +0$, to the limit

$$-\frac{i\pi}{\Gamma(n)} \int\limits_{(x,\sigma)=0} \left(\frac{\partial}{\partial\sigma}\right)^{n-1} \varphi(x)\, dS_x + \frac{1}{\Gamma(n)} \mathrm{VP} \int_{-\infty}^{\infty} \frac{1}{\lambda} \frac{\partial^{n-1}}{\partial\lambda^{n-1}} \int\limits_{(x,\sigma)=0} \varphi(x+\lambda\sigma)\, dS_x\, d\lambda,$$

that is to say, if we make use of the notation that has been introduced, then it tends to the limit

$$\left(\frac{(-1)^n i\pi}{\Gamma(n)} \delta^{(n-1)}[(x,\sigma)] + \frac{(-1)^{n-1}}{\Gamma(n)} \mathcal{P}^{(n-1)} \frac{1}{(x,\sigma)}, \varphi\right) = \left(\frac{1}{[(x,\sigma)+i0]^n}, \varphi\right).$$

Therefore, if in the integral (2.12) we pass to the limit as $y \to 0$, $y \in C$, and if we make use of the Lebesgue theorem and the limiting relation (2.6), we obtain (2.11).

In passing we also obtained the equalities

$$\mathcal{K}_C(x) = i^n \Gamma(n) \int\limits_{\operatorname{pr} C^*} \frac{d\sigma}{[(x,\sigma)+i0]^n}, \tag{2.13}$$

$$\frac{1}{[(x,\sigma)+i0]^n} = \frac{(-1)^n i\pi}{\Gamma(n)} \delta^{(n-1)}[(x,\sigma)] + \frac{(-1)^{n-1}}{\Gamma(n)} \mathcal{P}^{(n-1)} \frac{1}{(x,\sigma)}, \quad \sigma \in \operatorname{pr} C^*.$$

Finally, separating the real and imaginary parts in (2.11), we obtain the following useful formulas:

$$\Re \mathcal{K}_C(x) = \begin{cases} \pi(-1)^{\frac{n-1}{2}} \int\limits_{\operatorname{pr} C^*} \delta^{(n-1)}[(x,\sigma)]\, d\sigma & n \text{ odd}, \\ (-1)^{n/2-1} \int\limits_{\operatorname{pr} C^*} \mathcal{P}^{(n-1)} \frac{1}{(x,\sigma)}\, d\sigma & n \text{ even}, \end{cases} \tag{2.14}$$

$$\Im \mathcal{K}_C(x) = \begin{cases} (-1)^{\frac{n-1}{2}} \int\limits_{\operatorname{pr} C^*} \mathcal{P}^{(n-1)} \frac{1}{(x,\sigma)}\, d\sigma & n \text{ odd}, \\ \pi(-1)^{n/2} \int\limits_{\operatorname{pr} C^*} \delta^{(n-1)}[(x,\sigma)]\, d\sigma & n \text{ even}. \end{cases} \tag{2.15}$$

EXAMPLE 1.

$$\mathcal{K}_{\mathbb{R}^n_+}(z) = \frac{i^n}{z_1 \dots z_n} = \mathcal{K}_n(z), \qquad z \in T^{\mathbb{R}^n_+} = T^n, \tag{2.16}$$

$$\mathcal{K}_n(x) = \left[\pi\delta(x_1) + i\mathcal{P}\frac{1}{x_1}\right] \times \cdots \times \left[\pi\delta(x_n) + i\mathcal{P}\frac{1}{x_n}\right].$$

EXAMPLE 2.

$$\mathcal{K}_{V^+}(z) = 2^n \pi^{\frac{n-1}{2}} \Gamma\left(\frac{n+1}{2}\right)(-z^2)^{-\frac{n+1}{2}}, \qquad z \in T^{V^+} \tag{2.17}$$

where $z^2 = z_0^2 - z_1^2 - \cdots - z_n^2$. Let us compute the Cauchy kernel $\mathcal{K}_{V^+}(z)$. As was mentioned above, it suffices to compute it for $x = 0$; since $(V^+)^* = \overline{V^+}$ (see Sec. 4.4), it follows that

$$\mathcal{K}_{V^+}(iy) = \int\limits_{V^+} e^{-(y,\xi)}\, d\xi, \qquad y \in V^+.$$

Furthermore, by virtue of the invariance of that integral relative to the restricted Lorentz group, $L_+^\uparrow$, it suffices to compute it for $y = (y_0, 0)$, $y_0 > 0$. We have

$$\begin{aligned} \mathcal{K}_{V^+}(iy_0, 0) &= \int\limits_{V^+} e^{-y_0\xi_0}\, d\xi = \int\limits_0^\infty e^{-y_0\xi} \int\limits_{|\xi|<\xi_0} d\xi\, d\xi_0 \\ &= \frac{\sigma_n}{n} \int\limits_0^\infty e^{-y_0\xi_0} \xi_0^n\, d\xi_0 = \frac{\sigma_n}{n y_0^{n+1}} \int\limits_0^\infty e^{-u} u^n\, du \\ &= \Gamma(n)\sigma_n y_0^{-n-1} = 2^n \pi^{\frac{n-1}{2}} \Gamma\left(\frac{n+1}{2}\right)\left[-(iy_0)^2\right]^{-\frac{n+1}{2}}. \end{aligned}$$

By extending the resulting equality to all $y \in V^+$ and further onto all $z \in T^{V^+}$, we obtain (2.17). □

EXAMPLE 3.

$$\mathcal{K}_{\mathcal{P}_n}(Z) = \pi^{n(n-1)/2} i^{n^2} \frac{1! \dots (n-1)!}{(\det Z)^n}, \qquad Z \in T^{\mathcal{P}_n}, \tag{2.18}$$

where $\mathcal{P}_n$ is a cone of positive $n \times n$ matrices (see Sec. 4.4), and $T^{\mathcal{P}_n} = [Z = X + iY,\ Y = \Im Z > 0]$.

In order to compute the Cauchy kernel $\mathcal{K}_{\mathcal{P}_n}(Z)$ note that $\mathcal{P}_n^* = \overline{\mathcal{P}}_n$ and, therefore, by (2.1),

$$\mathcal{K}_{\mathcal{P}_n}(iY) = \int_{\mathcal{P}_n} e^{-\operatorname{Tr}(Y\Xi)}\, d\Xi,$$

where $d\Xi$ is the Lebesgue measure in $\mathbb{R}^{n^2}$,

$$d\Xi = d\xi_{11} \dots d\xi_{nn} \prod_{p<q} d\Re\xi_{pq}\, d\Im\xi_{pq}.$$

By virtue of the invariance of the last integral with respect to the transformations $Y \to U^{-1}YU$, where U is any unitary matrix, it suffices to compute that integral for diagonal matrices Y of the form $Y_0 = [\lambda_1, \dots, \lambda_n]$, $\lambda_j > 0$, $j = 1, \dots, n$,

$$\mathcal{K}_{\mathcal{P}_n}(iY_0) = \int_{\mathcal{P}_n} e^{-\sum_{p=1}^n \lambda_p \xi_{pp}}\, d\Xi.$$

Then the transformation $\xi_{pq} \to \frac{\xi_{pq}}{\sqrt{\lambda_p \lambda_q}}$ carries $\mathcal{P}_n$ onto itself, and its Jacobian is equal to $(\lambda_1 \dots \lambda_n)^n = (\det Y_0)^n = (\det Y)^n$. Consequently,

$$\mathcal{K}_{\mathcal{P}_n}(iY_0) = \mathcal{K}_{\mathcal{P}_n}(iY) = (\det Y)^{-n} \int_{\mathcal{P}_n} e^{-\operatorname{Tr}\Xi}\, d\Xi.$$

The last constant has been computed (see, for example, Bochner [6])

$$\int_{\mathcal{P}_n} e^{-\operatorname{Tr}\Xi}\, d\Xi = \pi^{n(n-1)/2} 1! \cdots (n-1)!.$$

Therefore

$$\mathcal{K}_{\mathcal{P}_n}(iY) = i^{n^2} \pi^{n(n-1)/2} 1! \cdots (n-1)! [\det(iY)]^{-n}, \qquad Y \in \mathcal{P}_n.$$

Extending this relation to all $Z \in T^{\mathcal{P}_n}$, we obtain the formula (2.18). □

10.3. The Cauchy–Bochner transform. Suppose $f \in \mathcal{H}_s$. The function

$$f(z) = \frac{1}{(2\pi)^n} (f(x'), \mathcal{K}_C(z - x')), \qquad z \in T^C \cup T^{-C}, \tag{3.1}$$

is called the *Cauchy-Bochner transform* (*integral*). It is assumed here that the cone C is convex and acute.

Since $\mathcal{K}_C(x + iy) \in \mathcal{H}_s$ far all s and $y \in -C \cup C$ (see Sec. 10.2), it follows that by (1.10) the right-hand side of (3.1) may be rewritten in the form of a convolution:

$$f(z) = \frac{1}{(2\pi)^n} f(x') * \mathcal{K}_C(x' + iy), \qquad z \in T^C \cup T^{-C}. \tag{3.2}$$

EXAMPLE. When $n = 1$, $C = (0, \infty)$ and $f \in \mathcal{L}^2$, the Cauchy–Bochner integral (3.1) turns into the classical Cauchy integral:

$$f(z) = \frac{1}{2\pi i} \int_{-\infty}^{\infty} \frac{f(x')}{x' - z} \, dx'.$$

The function $f(z)$ is holomorphic in $T^C \cup T^{-C}$, and

$$\partial^\alpha f(z) = \frac{1}{(2\pi)^n} (f(x'), \partial^\alpha \mathcal{K}_C(z - x')), \tag{3.3}$$

$$|\partial^\alpha f(z)| \le \frac{K_{|s|,\alpha}}{(2\pi)^n} \|f\|_s [1 + \Delta^s(y)] \Delta^{-n/2-|\alpha|}(y), \tag{3.4}$$

where the numbers $K_{|s|,\alpha}$ are the same as in the estimate (2.5).

The holomorphy of the function $f(z)$ in $T^C \cup T^{-C}$ and the differentiation formula (3.3) follow directly from the facts that the Cauchy kernel $\mathcal{K}_C(z)$ is a holomorphic function in $T^C \cup T^{-C}$ and $\mathcal{K}_C(x + iy) \in \mathcal{H}_s$ for all s and $y \in -C \cup C$ (see Sec. 10.2). The estimate (3.4) follows from (3.3) and also from the lemma of Sec. 10.1 and from (2.5):

$$\begin{aligned} |\partial^\alpha f(z)| &\le \frac{1}{(2\pi)^n} \|f\|_s \|\partial^\alpha \mathcal{K}_C(z - x')\|_{-s} \\ &= \frac{\|f\|_s}{(2\pi)^n} \|\partial^\alpha \mathcal{K}_C(x + iy)\|_{-s} \\ &\le \frac{K_{|s|,\alpha}}{(2\pi)^n} \|f\|_s [1 + \Delta^s(y)] \Delta^{-n/2-|\alpha|}(y). \end{aligned}$$

Now let us prove, for all s and $y \in -C \cup C$, the estimates

$$\|\partial^\alpha f(x + iy)\|_s \le N_\alpha \|f\|_s \Delta^{-|\alpha|}(y), \tag{3.5}$$

$$\|\partial^\alpha f(x + iy)\|_{s-|\alpha|} \le \|f\|_s. \tag{3.6}$$

Indeed, from the representation (3.2) and from the definition of the kernel $\mathcal{K}_C(z)$ it follows that

$$\begin{aligned} F^{-1}[\partial^\alpha f(x + iy)] &= \frac{1}{(2\pi)^n} F^{-1}[f * \partial^\alpha \mathcal{K}_C] \\ &= F^{-1}[f] F^{-1}[\partial^\alpha \mathcal{K}_C] \\ &= F^{-1}[f] (i\xi)^\alpha F^{-1}[\mathcal{K}_C] \\ &= (i\xi)^\alpha F^{-1}[f](\xi) e^{-(y,\xi)} \theta_{C^*}(\xi). \end{aligned} \tag{3.7}$$

Therefore

$$\|\partial^\alpha f(x+iy)\|_s^2 = \int_{C^*} \left|F^{-1}[f](\xi)\right|^2 |\xi^\alpha|^2 e^{-2(y,\xi)}((1+|\xi|)^s\, d\xi$$
$$\leq \|f\|_s^2 \sup_{\xi\in C^*} |\xi|^{2|\alpha|} e^{-2(y,\xi)}$$
$$= \|f\|_s^2 \sup_{\rho\geq 0} \rho^{2|\alpha|} \sup_{\sigma\in \operatorname{pr} C^*} e^{-2\rho(y,\xi)}$$
$$= \|f\|_s^2 \sup_{\rho\geq 0} \rho^{2|\alpha|} e^{-2\rho\Delta(y)}$$
$$= \|f\|_s^2 2^{-2|\alpha|}\Delta^{-2|\alpha|}(y) \sup_{u\geq 0} u^{2|\alpha|}e^{-u},$$

which is what yields the estimate (3.5). The estimate (3.6) is derived in similar fashion but more simply. □

As $y \to 0$ for $y \in \pm C$, the function $f(z)$ assumes in $\mathcal{H}_s$, in norm, the boundary values $f_\pm(x)$, which are respectively equal to

$$f_+ = \frac{1}{(2\pi)^n} f * \mathcal{K}_C, \qquad f_- = \frac{(-1)^n}{(2\pi)^n} f * \overline{\mathcal{K}}_C, \tag{3.8}$$

Indeed, taking into account (3.7), we have, for $y \in C$,

$$F^{-1}[f(x+iy) - f_+] = F^{-1}[f](\xi)[e^{-(y,\xi)} - 1]\theta_{C^*}(\xi).$$

Therefore, when $y \to 0$, $y \in C$, we obtain

$$\|f(x+iy) - f_+(x)\|_s^2 = \int_{C^*} \left|F^{-1}[f](\xi)\right|^2 [e^{-(y,\xi)} - 1]^2 (1+|\xi|^2)^s\, d\xi \to 0,$$

which is what we set out to prove. The case of $y \to 0$, $y \in -C$, is considered in similar fashion with use made of the formulas (2.6) and (2.7). □

10.4. The Hilbert transform. Suppose $f \in \mathcal{H}_s$ for some s. The *Hilbert transform* f_1 of the generalized function f is the convolution

$$f_1 = -\frac{2}{(2\pi)^n} f * \Im\mathcal{K}_C. \tag{4.1}$$

Applying the Fourier transform to (4.1) and using (2.8), we obtain

$$F[f_1] = -i(\theta_{C^*} - \theta_{-C^*})F[f], \tag{4.2}$$

whence it follows that

$$f_1 \in \mathcal{H}_s \quad \text{and} \quad \operatorname{supp} F[f_1] \subset -C^* \cup C^*. \tag{4.3}$$

If $f \in \mathcal{H}_s$, then the conditions

(1) $$\operatorname{supp} F[f] \subset -C^* \cup C^*, \tag{4.4}$$

(2) $$f = \frac{2}{(2\pi)^n} f_1 * \Im\mathcal{K}_C, \tag{4.5}$$

(3) $$f = \frac{2}{(2\pi)^n} f * \Re\mathcal{K}_C, \tag{4.6}$$

are equivalent.

Indeed, from (1) → (2), by virtue of (4.2) and (2.8),

$$F[f] = i(\theta_{C^*} - \theta_{-C^*})F[f_1].$$

From (2) → (3), by virtue of (4.1) and (2.9),

$$\begin{aligned} f &= \frac{2}{(2\pi)^n} f_1 * \Im\mathcal{K}_C = -\frac{4}{(2\pi)^{2n}}(f * \Im\mathcal{K}_C) * \Im\mathcal{K}_C \\ &= -\frac{4}{(2\pi)^{2n}} f * (\Im\mathcal{K}_C * \Im\mathcal{K}_C) = \frac{2}{(2\pi)^n} f * \Re\mathcal{K}_C \end{aligned}$$

and from the associativity of the convolution (see Sec. 10.1). Finally, from (3) → (1), by (2.8),

$$F[f] = (\theta_{C^*} + \theta_{-C^*})F[f].$$

□

We will say that the generalized functions f and f_1 in $\mathcal{H}_s$ form a *pair of Hilbert transforms* if they satisfy the relations (4.1) and (4.5):

$$\begin{aligned} f_1 &= -\frac{2}{(2\pi)^n} f * \Im\mathcal{K}_C, \\ f &= \frac{2}{(2\pi)^n} f_1 * \Im\mathcal{K}_C. \end{aligned} \tag{4.7}$$

EXAMPLE. When $n = 1$ (see Sec. 10.2),

$$\mathcal{K}_C(z) = \frac{1}{2}, \qquad \Re\mathcal{K}_C(x) = \pi\delta(x), \qquad \Im\mathcal{K}_C(x) = \mathcal{P}\frac{1}{x},$$

the formulas (4.7) take the form

$$\begin{aligned} f_1 &= -\frac{1}{\pi} f * \mathcal{P}\frac{1}{x}, \\ f &= \frac{1}{\pi} f_1 * \mathcal{P}\frac{1}{x}, \end{aligned} \tag{4.8}$$

and the relation (4.6) turns into the identity $f = f$. When $f \in \mathcal{L}^2$, the formulas (4.8) turn into the classical Hilbert transform formulas.

REMARK 1. We note here the difference between the cases $n = 1$ and $n \geq 2$: for $n = 1$, the condition (4.4) is absent because $-C^* \cup C^* = \mathbb{R}^1$, whereas for $n \geq 2$ that condition is essential.

REMARK 2. The results of this subsection were obtained by Beltrami and Wohlers [4] ($n = 1$) and Vladimirov [107] ($n \geq 2$).

10.5. Holomorphic functions of the class $\mathcal{H}_a^{(s)}(C)$. Suppose C is a convex acute open cone, $a \geq 0$, and let s be a real number. Denote by $H_a^{(s)}(C)$ the Banach space consisting of functions $f(z)$ holomorphic in T^C with norm

$$\|f\|_a^{(s)} = \sup_{y \in C} e^{-a|y|}\|f(x+iy)\|_s. \tag{5.1}$$

We write $H_0^{(s)}(C) = H^{(s)}(C)$ and $\| \ \|_0^{(s)} = \| \ \|^{(s)}$.

LEMMA. *Let a function $f(z)$ be holomorphic in T^C and let it satisfy the following condition of growth: for any $\varepsilon > 0$ there is a number $M(\varepsilon)$ such that*

$$\|f(x+iy)\|_s \leq M(\varepsilon)e^{(a+\varepsilon)|y|}\left[1+\Delta^{-\gamma}(y)\right], \qquad y \in C, \tag{5.2}$$

for certain s, $a \geq 0$ and $\gamma \geq 0$ (all dependent only on f). Then $f(z)$ is the Laplace transform of the function g in $\mathcal{L}^2_{s'}(C^* + \bar{U}_a)$, where $s' = s$ if $\gamma = 0$, $s' < s - \gamma$ if $\gamma > 0$; here the following estimates hold true:

$$\|g\|_{(s)} \leq 2 \inf_{0<\varepsilon\leq 1} M(\varepsilon), \quad \textit{if} \quad \gamma = 0; \tag{5.3}$$

and

$$\|g\|_{(s')} \leq \sqrt{\frac{1+\gamma}{s-s'-\gamma}}\, e^{2+a} \inf_{0<\varepsilon\leq 1} M(\varepsilon) \inf_{\sigma \in \operatorname{pr} C}\left[1+\Delta^{-\gamma}(\sigma)\right], \quad \textit{if}\ \gamma > 0,\ s' < s-\gamma. \tag{5.3'}$$

PROOF. We introduce the generalized function $g_y(\xi)$, taken from $\mathcal{D}'(\mathbb{R}^n \times C)$, via the formula

$$g_y(\xi) = e^{(y,\xi)}F_x^{-1}\left[f(x+iy)\right](\xi, y). \tag{5.4}$$

Here, F_x^{-1} signifies the Fourier transform with respect to the variables x (see Sec. 6.2). We will prove that $g_y(\xi)$ does not depend on $y \in C$. Indeed, differentiating (5.4) with respect to y_j and using the Cauchy–Riemann conditions, we have

$$\begin{aligned}\frac{\partial g_y(\xi)}{\partial y_j} &= \xi_j e^{(y,\xi)}F_x^{-1}\left[f(x+iy)\right] + e^{(y,\xi)}F_x^{-1}\left[\frac{\partial f(x+iy)}{\partial y_j}\right] \\ &= e^{(\xi,y)}\left\{\xi_j F_x^{-1}\left[f(x+iy)\right] + iF_x^{-1}\left[\frac{\partial}{\partial y_j}f(x+iy)\right]\right\} \\ &= e^{(\xi,y)}F_x^{-1}\left[f(x+iy)\right](\xi_j + i^2\xi_j) = 0, \qquad j = 1, \ldots, n,\end{aligned}$$

whence, by virtue of the criterion of Sec. 3.3, we conclude that $g_y(\xi)$ does not depend on $y \in C$, $g_y(\xi) = g(\xi)$. And then from (5.4) it follows that $g(\xi)e^{-(y,\xi)} \in \mathcal{S}'$ for all $y \in C$ and

$$f(x+iy) = F\left[g(\xi)e^{-(y,\xi)}\right], \qquad z \in T^C. \tag{5.5}$$

Furthermore, by the hypothesis $f(x+iy) \in \mathcal{H}_s$ for every $y \in C$ so that, by (5.5), $g(\xi)e^{-(y,\xi)}$ is a function in $\mathcal{L}^2_s$ and

$$\left\|g(\xi)e^{-(y,\xi)}\right\|^2_{(s)} = \left\|f(x+iy)\right\|^2_{(s)}, \qquad y \in C.$$

From this, by (5.2), for all $\varepsilon > 0$, we derive the inequality

$$\int \left|g(\xi)\right|^2 e^{-2(y,\xi)}\left(1+|\xi|^2\right)^s d\xi \leq M^2(\varepsilon)e^{2(a+\varepsilon)|y|}\left[1+\Delta^{-\gamma}(y)\right]^2, \qquad y \in C. \tag{5.6}$$

We will now prove that $g(\xi) = 0$ almost everywhere outside $C^* + \bar{U}_a$. Let $\xi_0 \notin C^* + \bar{U}_a$. By Lemma 3 of Sec. 4.4

$$C^* + \bar{U}_a = [\xi : \mu_C(\xi) \leq a]$$

so that

$$\mu_C(\xi_0) = -\inf_{y \in \operatorname{pr} C}(\xi_0, y) > a.$$

Therefore there is a point $y_0 \in \operatorname{pr} C$ such that $(\xi_0, y_0) < -a - \varkappa$ for certain $\varkappa > 0$ (see Fig. 23). This inequality also holds, in continuity, in a sufficiently small neighbourhood $|\xi - \xi_0| < \delta$. Therefore, putting $y = ty_0$ in (5.6), we obtain, for all $t > 0$, the inequality

$$e^{2t(a+\varkappa)} \int_{|\xi-\xi_0|<\delta} |g(\xi)|^2(1+|\xi|^2)^s \, d\xi \le \int_{|\xi-\xi_0|<\delta} |g(\xi)|^2 e^{-2(y,\xi)}(1+|\xi|^2)^s \, d\xi$$
$$\le M^2(\varepsilon)e^{2t(a+\varepsilon)}\left[1 + t^{-\gamma}\Delta^{-\gamma}(y_0)\right]^2,$$

which is possible (assuming $\varepsilon < \varkappa$) only if $g(\xi) = 0$ almost everywhere in $|\xi - \xi_0| < \delta$. Since ξ_0 is an arbitrary point outside $C^* + \bar{U}_a$, it follows that $g(\xi) = 0$ almost everywhere outside $C^* + \bar{U}_a$ so that $\operatorname{supp} g \subset C^* + \bar{U}_a$.

In (5.6), set $y = t\sigma$, $t > 0$. We then get

$$t^{2\gamma} \int_{C^*+\bar{U}_a} |g(\xi)|^2 e^{-2t(\sigma,\xi)}(1+|\xi|^2)^s \, d\xi \le M^2(\varepsilon)e^{2t(a+\varepsilon)}\left[t^\gamma + \frac{1}{\Delta^\gamma(\sigma)}\right]^2,$$
$$t > 0. \quad (5.7)$$

Let $\gamma = 0$. Passing to the limit in (5.7) as $t \to +0$ and using the Fatou lemma, we obtain

$$\int_{C^*+\bar{U}_a} |g(\xi)|^2(1+|\xi|^2)^s \, d\xi \le 4M^2(\varepsilon), \qquad \varepsilon > 0,$$

whence follows the inequality (5.3).

Now let $\gamma > 0$. Take into account the inequality $(\sigma, \xi) \le |\xi|$, divide the inequality (5.7) through by $t^{1-2\delta}$, where δ is an arbitrary number $0 < \delta \le 1$, integrate the resulting inequality with respect to t on $(0, 1)$, and take advantage of the Fubini theorem. Assuming $0 < \varepsilon \le 1$, we then obtain the inequality

$$\int_{C^*+\bar{U}_a} |g(\xi)|^2(1+|\xi|^2)^s \int_0^1 t^{2(\gamma+\delta)-1}e^{-2t|\xi|} \, dt \, d\xi$$
$$\le \frac{1}{2\delta}e^{2(a+1)}M^2(\varepsilon)\left[1+\Delta^{-\gamma}(\sigma)\right]^2. \quad (5.8)$$

Now, taking into account the estimate

$$\int_0^1 t^{2(\gamma+\delta)-1}e^{-2t|\xi|} \, dt \ge \min\left(1, |\xi|^{-2(\gamma+\delta)}\right) \int_0^1 u^{2\gamma+1}e^{-2u} \, du$$
$$\ge \frac{e^{-2}}{2(\gamma+1)}(1+|\xi|^2)^{-\gamma-\delta},$$

we derive from (5.8) the estimate

$$\int_{C^*+\bar{U}_a} |g(\xi)|^2(1+|\xi|^2)^{s-\gamma-\delta} \, d\xi \le \frac{e^{2(a+2)}}{\delta}(\gamma+1)M^2(\varepsilon)\left[1+\Delta^{-\gamma}(\sigma)\right]^2,$$

whence it follows that $g \in \mathcal{L}^2_{s'}(C^* + \bar{U}_a)$ for all $s' = s - \gamma - \delta < s - \gamma$ and the estimate (5.3′) holds true. Finally, from (5.4) it follows that $f = L[g]$. The proof of the lemma is complete. □

THEOREM. *For the function $f(z)$ to belong to the class $H_a^{(s)}(C)$, it is necessary and sufficient that its spectral function $g(\xi)$ belong to the class $\mathcal{L}_s^2(C^* + \bar{U}_a)$.*

Here, the following equalities hold:

$$\|f\|_a^{(s)} = \|f\|_s = \|g\|_{(s)}, \tag{5.9}$$

where $f_+(x)$ is the boundary value in $\mathcal{H}_s$ of the function $f(z)$ as $y \to 0$, $y \in C$, and $f_+ = F[g]$.

PROOF. NECESSITY. Let $f \in H_a^{(s)}(C)$. Then from the lemma [for $\gamma = 0$ and $M(\varepsilon)$ independent of ε] it follows that $f(z) = L[g]$, where $g \in \mathcal{L}_s^2(C^* + \bar{U}_a)$.

SUFFICIENCY. Let $f(z) = L[g]$, where $g \in \mathcal{L}_s^2(C^* + \bar{U}_a)$. By what has been proved (see Sec. 9.1), the function $f(z)$ is holomorphic in $T^{\operatorname{int} C^{**}} = T^C$ and it is given by the integral

$$f(z) = \int_{C^*+\bar{U}_a} g(\xi)e^{i(z,\xi)}\, d\xi = F\left[g(\xi)e^{-(y,\xi)}\right], \qquad z \in T^C. \tag{5.10}$$

Let us prove that $f \in H_a^{(s)}(C)$. Using the relations of the norms in the spaces $H_a^{(s)}(C)$, $\mathcal{H}_s$ and $\mathcal{L}_s^2$, we obtain from (5.10)

$$\begin{aligned}\|f\|_a^{(s)^2} &= \sup_{y\in C} e^{-2a|y|}\|f(x+iy)\|_s^2\\ &= \sup_{y\in C} e^{-2a|y|}\left\|g(\xi)e^{-(y,\xi)}\right\|_{(s)}^2\\ &= \sup_{y\in C} e^{-2a|y|}\int_{C^*+\bar{U}_a} |g(\xi)|^2 e^{-2(y,\xi)}\left(1+|\xi|^2\right)^s d\xi\\ &= \int |g(\xi)|^2\left(1+|\xi|^2\right)^s d\xi = \|g\|_{(s)}^2.\end{aligned}$$

That is,

$$\|f\|_a^{(s)} = \|g\|_{(s)}. \tag{5.11}$$

We now prove that the function $f(z)$ assumes, when $y \to 0$, $y \in C$, a (unique) boundary value in $\mathcal{H}_s$ equal to $f_+ = F[g]$. This follows from the limiting relation

$$\begin{aligned}\|f(x+iy) - F[g]\|_s^2 &= \|L[g](x+iy) - F[g](x)\|_s^2\\ &= \int_{C^*+\bar{U}_a} |g(\xi)|^2\left[e^{-(y,\xi)}-1\right]^2\left(1+|\xi|^2\right)^s d\xi \to 0, \qquad y\to 0, \quad y \in C.\end{aligned}$$

Thus, $\|g\|_{(s)} = \|f_+\|_s$, which together with (5.11) yields (5.9). The proof is complete. □

COROLLARY 1. *The spaces $H_a^{(s)}(C)$ and $\mathcal{L}_s^2(C^* + \bar{U}_a)$ are (linearly) isomorphic and isometric, and the isomorphism is realized via the Laplace transformation $g \to L[g] = f$.*

COROLLARY 2. *Any function $f(z)$ in $H_a^{(s)}(C)$ has, for $y \to 0$, $y \in C$, a (unique) boundary value $f_+(x)$ in $\mathcal{H}_s$ and the correspondence $f \to f_+$ is isometric.*

REMARK. The theorem on the existence of boundary values in $\mathcal{D}'_{L^p}$ was proved by a different method by Tillmann [103] and Luszczki and Zieleźny [70] ($n = 1$).

COROLLARY 3 (an analogue of Liouville's theorem). *It the cone C is not acute and $f \in H_0^{(s)}(C)$, then $f(z) \equiv 0$.*

True enough, by Lemma 1 of Sec. 4.4, $\operatorname{mes} C^* = 0$. In that case, as follows from the proof of the Lemma, the function $g(\xi) = 0$ almost everywhere in $\mathbb{R}^n$ so that $f(z) = L[g] \equiv 0$. □

10.6. The generalized Cauchy–Bochner representation. Here we continue the investigation started in Subsection 10.5 when $a = 0$.

THEOREM I. *For a function $f(z)$ to belong to $H^{(s)}(C)$, it is necessary and sufficient that it possess the generalized Cauchy–Bochner integral representation*

$$\frac{1}{(2\pi)^n}(f_+(x'), \mathcal{K}_C(z - x')) = \begin{cases} f(z), & z \in T^C \\ 0, & z \in T^{-C}, \end{cases} \tag{6.1}$$

where $f_+(x)$ is a boundary value in $\mathcal{H}_s$ of the function $f(z)$ as $y \to 0$, $y \in C$.

PROOF. NECESSITY. Let $f \in H^{(s)}(C)$. By the theorem of Sec. 10.5, $f_{(s)}$ is the Laplace transform of the function g in $\mathcal{L}_s^2(C^*)$ so that

$$f(z) = F\big[g(\xi)e^{-(y,\xi)}\theta_{C^*}(\xi)\big], \qquad z \in T^C,$$
$$0 = F\big[g(\xi)e^{-(y,\xi)}\theta_{-C^*}(\xi)\big], \qquad z \in T^{-C}.$$

From this fact, using the definition of the kernel $\mathcal{K}_C(z)$ [see (2.1)] and using (1.7) and (1.9) for the convolution, we obtain the representation (6.1):

$$f(z) = \frac{1}{(2\pi)^n}F[g] * \mathcal{K}_C = \frac{1}{(2\pi)^n}(f_+(x'), \mathcal{K}_C(z - x')), \qquad z \in T^C,$$
$$0 = \frac{1}{(2\pi)^n}F[g] * \mathcal{K}_{-C} = \frac{(-1)^n}{(2\pi)^n}(f_+(x'), \mathcal{K}_C(z - x')), \qquad z \in T^{-C}.$$

Here we made use of one of the equalities of (2.3): $\mathcal{K}_{-C}(z) = (-1)^n\mathcal{K}_C(z)$, and also used the relation $f_+ = F[g]$.

SUFFICIENCY. Suppose $f(z)$ has the representation (6.1). Then $f \in H^{(s)}(C)$ (see Sec. 10.3). Theorem I is proved. □

REMARK. For $s = 0$, Theorem I becomes the Bochner theorem [6]; for $n = 1$, Theorem I was obtained by Beltrami and Wohlers [4]; for arbitrary n and s see Vladimirov [107].

THEOREM II. *The following statements are equivalent:*

(1) *f_+ is a boundary value in $\mathcal{H}_s$ of some function taken from $H^{(s)}(C)$;*
(2) *f_+ belongs to $\mathcal{H}_s$ and satisfies the relations*

$$\Re f_+ = -\frac{2}{(2\pi)^n}\Im f_+ * \Im\mathcal{K}_C,$$
$$\Im f_+ = \frac{2}{(2\pi)^n}\Re f_+ * \Im\mathcal{K}_C. \tag{6.2}$$

That is, $\Re f_+$ and $\Im f_+$ form a pair of Hilbert transforms;
(3) *f_+ belongs to $\mathcal{H}_s$ and $\operatorname{supp} F^{-1}[f_+] \subset C^*$.*

PROOF. (1) → (2). Let $f_+(x)$ be a boundary value in $\mathcal{H}_s$ of a function $f(z)$ taken from $H^{(s)}(C)$. Then, by Theorem I, $f_+ \in \mathcal{H}_s$ and for $f(z)$ the generalized Cauchy–Bochner representation (6.1) holds true, from which follow, by (3.8), the relations

$$\begin{aligned} f_+ &= \frac{1}{(2\pi)^n} f_+ * \mathcal{K}_C, \\ 0 &= \frac{1}{(2\pi)^n} f_+ * \overline{\mathcal{K}}_C. \end{aligned} \tag{6.3}$$

From this we obtain relations (6.2) by separating the real and imaginary parts.

(2) → (3). Let f_+ in $\mathcal{H}_s$ satisfy the relations (6.2). Then it will also satisfy (6.3). Applying the inverse Fourier transform to the first of the relations (6.3) and making use of (2.6), we obtain

$$F^{-1}[f_+] = F^{-1}[f_+]F^{-1}[\mathcal{K}_C] = F^{-1}[f_+]\theta_{C^*}(\xi),$$

whence it follows that $\operatorname{supp} F^{-1}[f_+] \subset C^*$.

(3) → (1). If $f_+ \in \mathcal{H}_s$ and $\operatorname{supp} F^{-1}[f_+] \subset C^*$, then $F^{-1}[f_+] \in \mathcal{L}^2_s(C^*)$. By the theorem of Sec. 10.5, the function $f(z) = L[g] \in H^{(s)}(C)$ and the boundary value of it in $\mathcal{H}_s$ is equal to $F[g] = f_+$.

Theorem II is proved. □

REMARK. For $n = 1$, $C = (0, \infty) = \mathbb{R}^1_+$, the formulas (6.2) take on the following form:

$$\begin{aligned} \Re f_+ &= -\frac{1}{\pi}\Im f_+ * \mathcal{P}\frac{1}{x}, \\ \Im f_+ &= \frac{1}{\pi}\Re f_+ * \mathcal{P}\frac{1}{x}. \end{aligned} \tag{6.4}$$

In physics the formulas (6.4) are called dispersion relations (without subtraction). It is natural to regard (6.2) as a generalization of the dispersion relations to the multidimensional case with causality with respect to an arbitrary convex acute closed cone C^*.

11. Poisson Kernel and Poisson Transform

11.1. The definition and properties of the Poisson kernel. Let C be a convex acute open cone in $\mathbb{R}^n$ (with vertex at 0). The function

$$\mathcal{P}_C(x, y) = \frac{|\mathcal{K}_C(x+iy)|^2}{(2\pi)^n \mathcal{K}_C(2iy)} = \frac{|\mathcal{K}_C(x+iy)|^2}{\pi^n \mathcal{K}_C(iy)}, \qquad (x, y) \in T^C \tag{1.1}$$

is termed the *Poisson kernel* of the tubular domain T^C. Here, $\mathcal{K}_C$ is the Cauchy kernel (see Sec. 10.2).

EXAMPLE 1 (see (2.16) of Sec. 10.2).

$$\mathcal{P}_{\mathbb{R}^n_+}(x, y) = \frac{y_1 \dots y_n}{\pi^n |z_1|^2 \dots |z_n|^2} \equiv \mathcal{P}_n(x, y). \tag{1.2}$$

EXAMPLE 2 (see (2.17) of Sec. 10.2).

$$\mathcal{P}_{V^+}(x, y) = \frac{2^n \Gamma(\frac{n+1}{2})}{\pi^{\frac{n+3}{2}}} \frac{(y^2)^{\frac{n+1}{2}}}{|(x+iy)^2|^{n+1}}. \tag{1.3}$$

The following is a list of the properties of the Poisson kernel $\mathcal{P}_C$ that follow from the corresponding properties of the Cauchy kernel $\mathcal{K}_C$ (see Sec. 10.2):

11.1.1.

$$0 \le \mathcal{P}_C(x,y) = \mathcal{P}_C(-x,y) \in C^\infty(T^C) \tag{1.4}$$

follows from the holomorphicity of the kernel $\mathcal{K}_C(z)$ in T^C and from the fact that $\mathcal{K}_C(2iy) > 0$, $y \in C$.

11.1.2.

$$\int \mathcal{P}_C(x,y)\,dx = 1, \qquad y \in C \tag{1.5}$$

follows from the Parseval–Steklov equation applied to (2.1) of Sec. 10.2:

$$\int |\mathcal{K}_C(x+iy)|^2\,dx = (2\pi)^n \int_{C^*} e^{-2(y,\xi)}\,d\xi = (2\pi)^n \mathcal{K}_C(2iy), \qquad y \in C.$$

11.1.3.

$$\mathcal{P}_C(x,y) \le \frac{\mathcal{K}_C^2(iy)}{(2\pi)^n \mathcal{K}_C(2iy)} = \frac{1}{\pi^n}\mathcal{K}_C(iy), \qquad (x,y) \in T^C \tag{1.6}$$

follows from the estimate

$$|\mathcal{K}_C(x,iy)| \le \int_{C^*} e^{-(y,\xi)}\,d\xi = \mathcal{K}_C(iy).$$

11.1.4.

$$\left\|\mathcal{P}_C(x,y)\right\|_{\mathcal{L}^p} \le \pi^{-n(1-1/p)}\mathcal{K}_C^{1-\frac{1}{p}}(iy), \qquad y \in C, \qquad 1 \le p \le \infty \tag{1.7}$$

follows from (1.4)–(1.6) by virtue of

$$\begin{aligned}\|\mathcal{P}_C(x,y)\|_{\mathcal{L}^p}^p &= \int \mathcal{P}_C^p(x,y)\,dx\\ &\le \sup_x \mathcal{P}_C^{p-1}(x,y)\int \mathcal{P}_C(x,y)\,dx\\ &\le \frac{\mathcal{K}_C^{2p-2}(iy)}{(2\pi)^{n(p-1)}\mathcal{K}_C^{p-1}(2iy)}\\ &= \pi^{-n(p-1)}\mathcal{K}_C^{p-1}(iy).\end{aligned}$$

11.1.5.

$$0 < F_x\big[\mathcal{P}_C(x,y)\big](\xi) = \frac{[\theta_{C^*}(\xi)e^{-(y,\xi)}] * [\theta_{C^*}(-\xi)e^{(y,\xi)}]}{\mathcal{K}_C(2iy)} \le e^{-|(y,\xi)|}, \qquad \xi \in \mathbb{R}^n, \quad y \in C \tag{1.8}$$

follows from the Fourier transform formula of a convolution and from the fact that

$$F_x^{-1}\big[\mathcal{K}_C(z)\big] = \theta_{C^*}(\xi)e^{-(y,\xi)} \in \mathcal{L}^1,$$
$$F_x^{-1}\big[\overline{\mathcal{K}}_C(z)\big] = \theta_{C^*}(-\xi)e^{(y,\xi)} \in \mathcal{L}^1.$$

11.1.6.

$$\mathcal{P}_C(x,y) \gg 0$$

is a continuous positive definite function for all $y \in C$ (see Sec. 8.2); it follows from (1.8) and from the fact that

$$\big[\theta_{C^*}(\xi)e^{-(y,\xi)}\big] * \big[\theta_{C^*}(-\xi)e^{(y,\xi)}\big] \in \mathcal{L}^1, \qquad y \in C.$$

11.1.7.

$$\left[\theta_{C^*}(\xi)e^{-(y,\xi)}\right] * \left[\theta_{C^*}(-\xi)e^{(y,\xi)}\right] \gg 0 \tag{1.9}$$

is a continuous positive definite function for all $y \in C$; it follows from (1.4) and (1.8).

11.1.8.

$$F_x\left[\mathcal{P}_C(x,y)\right](\xi) = e^{-|(y,\xi)|}, \qquad \xi \in -C^* \cup C^*, \qquad y \in C \tag{1.10}$$

follows from (1.8) by virtue of the following manipulations for $\xi \in C^*$:

$$\begin{aligned} F_x\left[\mathcal{P}_C(x,y)\right](\xi) &= \frac{1}{\mathcal{K}_C(2iy)} \int\limits_{\substack{-\xi' \in C^* \\ \xi-\xi' \in C^*}} e^{-(y,\xi-\xi')+(y,\xi')}\, d\xi' \\ &= \frac{1}{\mathcal{K}_C(2iy)} e^{-(y,\xi)} \int\limits_{-C^*} e^{2(y,\xi')}\, d\xi' = e^{-(y,\xi)}. \end{aligned}$$

But if $\xi \in -C^*$, then the equation being proved remains true because of the evenness of the kernel $\mathcal{P}_C(x,y)$ with respect to x.

11.1.9.

$$|\partial_x^\alpha \mathcal{P}_C(x,y)| \le M'_\alpha |y|^n \Delta^{-2n-|\alpha|}(y), \qquad (x,y) \in T^C \tag{1.11}$$

follows from the estimates (2.4) of Sec. 10.2 and from the estimate

$$\mathcal{K}_C(2iy) = \int\limits_{C^*} e^{-2(y,\xi)}\, d\xi \ge \int\limits_{C^*} e^{-2|y|\,|\xi|}\, d\xi = \frac{\varkappa}{|y|^n}, \qquad \varkappa > 0. \tag{1.12}$$

11.1.10.

$$\begin{gathered} \|\partial_x^\alpha \mathcal{P}_C(x,y)\|_s \le K_{s,\alpha,p}\left[1+\Delta^{-s}(y)\right]\left[1+\Delta^{-p}(y)\right]\Delta^{-n-|\alpha|}(y)|y|^n, \\ y \in C, \qquad s \ge 0, \qquad p > s + n/2, \end{gathered} \tag{1.13}$$

so that $\mathcal{P}_C(x,y) \in \mathcal{H}_s$ for all s and $y \in C$.

This follows from the inequalities (1.11) and (2.5) of Sec. 10.2 and from the estimate (1.12):

$$\begin{aligned} \|\partial_x^\alpha \mathcal{P}_C(x,y)\|_s &= \frac{1}{(2\pi)^n \mathcal{K}_C(2iy)} \left\|\partial^\alpha \mathcal{K}_C(x+iy)\overline{\mathcal{K}_C(x+iy)}\right\|_s \\ &\le \frac{|y|^n}{(2\pi)^n \varkappa} \left\| \sum_\beta \binom{\alpha}{\beta} \partial^\beta \mathcal{K}_C(x+iy) \overline{\partial^{\alpha-\beta}\mathcal{K}_C(x+iy)} \right\|_s \\ &\le \frac{|y|^n}{(2\pi)^n \varkappa} c_{p-s} \sum_\beta \binom{\alpha}{\beta} \|\partial^\beta \mathcal{K}_C(x+iy)\|_s \|\partial^{\alpha-\beta}\mathcal{K}_C(x+iy)\|_p \\ &\le \frac{c_{p-s}}{(2\pi)^n \varkappa} \sum_\beta \binom{\alpha}{\beta} K_{s,\beta} K_{p,\alpha-\beta}\left[1+\Delta^{-s}(y)\right] \\ &\quad \times \left[1+\Delta^{-p}(y)\right]\Delta^{-n-|\alpha|}(y)|y|^n. \end{aligned}$$

□

11.2. The Poisson transform and Poisson representation. Let $f \in \mathcal{H}_s$ for some s, $-\infty < s < \infty$. We call the convolution [see (1.10) of Sec. 10.1]

$$\begin{aligned}\mathcal{F}(x,y) &= f(x) * \mathcal{P}_C(x,y) \\ &= (f(x'), \mathcal{P}_C(x-x',y)), \qquad (x,y) \in T^C, \end{aligned} \tag{2.1}$$

the *Poisson transform* (or *integral*).

By virtue of Subsec. 11.1.10, the Poisson integral exists for every $y \in C$ and is continuous operation from $\mathcal{H}_s$ to $\mathcal{H}_s$.

EXAMPLE. If $f \in \mathcal{L}^2 = \mathcal{H}_0$, then the Poisson integral becomes the classical Poisson integral:

$$\mathcal{F}(x,y) = \int f(x')\mathcal{P}_C(x-x',y)\,dx'.$$

The following is a partial list of the properties of the Poisson integral.

11.2.1.

$$\mathcal{F}(x,y) \in C^\infty(T^C). \tag{2.2}$$

This follows from (1.4) and from (1.13).

11.2.2.

$$\|\mathcal{F}(x,y)\|_s^2 \le \|f\|_s^2, \qquad y \in C. \tag{2.3}$$

This follows from (1.8) by virtue of the following manipulations:

$$\|\mathcal{F}(x,y)\|_s^2 = \|F_x^{-1}[\mathcal{F}(x,y)]\|_{(s)}^2 = \|F[f]F_x^{-1}[\mathcal{P}_C(x,y)]\|_{(s)}^2 \le \|f\|_s^2.$$

11.2.3.

THEOREM (generalized Poisson representation). *For $f(x)$ to belong to $H^{(s)}(C)$, it is necessary and sufficient that it be uniquely represented as the Poisson integral*

$$f(z) = (\chi(x'), \mathcal{P}_C(x-x',y)), \qquad z \in T^C, \tag{2.4}$$

where $\chi \in \mathcal{H}_s$ and $\operatorname{supp} F^{-1}[\chi] \subset C^*$; here, $\chi = f_+$ where $f_+(x)$ is the boundary value in $\mathcal{H}_s$ of the function $f(z)$ as $y \to 0$, $y \in C$.

PROOF. NECESSITY. Since $f \in H^{(s)}(C)$, it follows, by the theorem of Sec. 10.5, that there is a function $g \in \mathcal{L}_s^2(C^*)$ such that $f_+ = F[g] \in \mathcal{H}_s$ and

$$f(z) = F\left[g(\xi)e^{-(y,\xi)}\right](x), \qquad z \in T^C. \tag{2.5}$$

From this, using (1.10), we obtain for the function $f(z)$ the generalized Poisson representation (2.4):

$$\begin{aligned} f(z) &= F\left[g(\xi)F_x\left[\mathcal{P}_C(x,y)\right](\xi)\right] = F[g](x) * \mathcal{P}_C(x,y) \\ &= f_+(x) * \mathcal{P}_C(x,y) = (f_+(x'), \mathcal{P}_C(x-x',y)), \qquad z \in T^C. \end{aligned}$$

The generalized Poisson representation (2.4) is unique since, by (1.8),

$$F_x^{-1}\left[\mathcal{P}_C(x,y)\right](\xi) \neq 0, \qquad \xi \in \mathbb{R}^n, \quad y \in C.$$

SUFFICIENCY. Suppose a generalized function χ is such that $g = F^{-1}[\chi] \in \mathcal{L}_s^2(C^*)$. Then by the theorem of Sec. 10.5 the function $f(z)$ defined by (2.5) belongs to $H^{(s)}(C)$ and, by what has been proved, can be represented by the integral (2.4) with $\chi = F[g] = f_+$. This completes the proof of the theorem. □

COROLLARY 1. *Under the hypothesis of the theorem, we get*

$$\begin{aligned} \Re f(z) &= (\Re f_+(x'), \mathcal{P}(x - x', y)), \\ \Im f(z) &= (\Im f_+(x'), \mathcal{P}(x - x', y)). \end{aligned} \tag{2.6}$$

COROLLARY 2. *If $f(x)$ is a real generalized function in $\mathcal{H}_s$ and* $\operatorname{supp} F[f] \subset -C^* \cup C^*$, *then the function*

$$u(x, y) = (f(x'), \mathcal{P}_C(x - x', y)) \tag{2.7}$$

is a real part of some function of the class $H^{(s)}(C)$ and assumes, in the sense of $\mathcal{H}_s$ as $y \to 0$, $y \in C$, the value of $f(x)$.

Indeed, putting

$$f_+(x) = F\big[\theta_{C^*}(\xi) F^{-1}[f](\xi)\big](x),$$

we obtain

$$f_+ \in \mathcal{H}_s, \qquad \operatorname{supp} F^{-1}[f_+] \subset C^* \quad \text{and} \quad f = 2\Re f_+,$$

so that

$$u(x, y) = 2\Re\,(f_+(x'), \mathcal{P}_C(x - x', y)).$$

From this and from the theorem follow the required assertions. □

EXAMPLE. The function $\mathcal{K}_C(z + iy')$ belongs to the class $H^{(s)}(C)$ for all $y' \in C$ and s [see estimate (2.5) of Sec. 10 in which $\Delta(y + y') \geq \Delta(y')$, $y \in C$]. Suppose C' is an arbitrary (convex open) subcone of the cone C, $C' \subset C$. Applying (2.4) to the function $\mathcal{K}_C(z + iy')$ of the class $H^{(s)}(C')$, we obtain

$$\begin{gathered} \mathcal{K}_C(z + iy') = \int \mathcal{K}_C(x' + iy') \mathcal{P}_{C'}(x - x', y)\, dx', \\ (x, y) \in T^{C'}, \qquad y' \in C. \end{gathered} \tag{2.8}$$

From this, using the Cauchy–Bunyakowsky inequality and (1.5), we obtain the following inequality:

$$\begin{aligned} |\mathcal{K}_C(z + iy')|^2 &\leq \int |\mathcal{K}_C(x' + iy')|^2 \mathcal{P}_{C'}(x - x', y)\, dx' \int \mathcal{P}_{C'}(x - x', y)\, dx' \\ &= \int \mathcal{P}_{C'}(x - x', y) |\mathcal{K}_C(x' + iy')|^2\, dx'. \end{aligned} \tag{2.9}$$

In terms of the Poisson kernel (1.1), the inequality (2.9) takes the form

$$\begin{gathered} \mathcal{P}_C(x, y + y') \leq \frac{\mathcal{K}_C(iy')}{\mathcal{K}_C(iy + iy')} \int \mathcal{P}_{C'}(x - x', y) \mathcal{P}_C(x', y')\, dx', \\ (x, y) \in T^{C'}, \qquad y' \in C. \end{gathered} \tag{2.10}$$

In particular, for $C' = C$, $y' = y$ the formula (2.10) assumes the form

$$\mathcal{P}_C(x, 2y) \leq 2^n \int \mathcal{P}_C(x - x', y) \mathcal{P}_C(x', y)\, dx', \qquad (x, y) \in T^C. \tag{2.11}$$

Here we made use of the property of homogeneity (of degree $-n$) of the kernel $\mathcal{K}_C$ [see (2.3) of Sec. 10.2].

11.3. Boundary values of the Poisson integral.

11.3.1.

$$\int \mathcal{P}_C(x,y)\varphi(x)\,dx \to \varphi(0), \qquad y \to 0, \qquad y \in C, \tag{3.1}$$

for any function $\varphi \in \mathcal{L}^\infty$ continuous at 0.

By virtue of (1.5), it suffices, when proving this assertion, to establish the following limiting relation: for any $\delta > 0$

$$\int\limits_{|x|>\delta} \mathcal{P}_C(x,y)\,dx \to 0, \qquad y \to 0, y \in C. \tag{3.2}$$

Let us construct an auxiliary function $\omega(x)$ with the properties:

(1) ω is a real continuous function in $\mathbb{R}^n$, $\omega(x) \to 0$, $|x| \to \infty$;

(2) $\omega(0) = 1$, $|\omega(x)| < 1$, $x \neq 0$;

(3) $\int \mathcal{P}_C(x,y)\omega(x)\,dx \to 1$, $y \to 0$, $y \in C$.

Suppose $\eta \in \mathcal{D}(\operatorname{int} C^*)$, $\eta \geq 0$, $\int \eta(\xi)\,d\xi = 1$ ($\operatorname{int} C^* \neq \varnothing$ since C is an acute cone; see Sec. 4.4). The function

$$\omega'(x) = \Re \int \eta(\xi)e^{i(x,\xi)}\,d\xi = \int \eta(\xi)\cos(x,\xi)\,d\xi$$

possesses the required properties (1) to (3).

Indeed, property (1) follows from the Riemann–Lebesgue theorem. Let us prove property (2). It is clear that $\omega(0) = 1$ and $|\omega(x)| \leq 1$. Suppose $\omega(x_0) = \pm 1$, $x_0 \neq 0$, that is, $1 = \pm \int \eta(\xi)\cos(x_0,\xi)\,d\xi$; but this contradicts the hypothesis $\int \eta(\xi)\,d\xi = 1$. Property (3) follows from the corollary to the theorem of Sec. 11.2 by virtue of which

$$\Re \int \eta(\xi)e^{i(z,\xi)}\,d\xi = \int \mathcal{P}_C(x - x', y)\omega(x')\,dx', \qquad z \in T^C.$$

Putting $x = 0$ here, then taking into account (1.4) and passing to the limit as $y \to 0$, $y \in C$, we obtain relation (3).

Suppose $\delta > 0$. By the properties (1) and (2) there exists a number $\varepsilon > 0$ such that $|\omega(x)| \leq 1 - \varepsilon$, $|x| \geq \delta$. From this fact, taking into account property (3), we obtain

$$\begin{aligned}
1 &= \lim_{\substack{y\to 0,\\ y\in C}} \left[\int\limits_{|x|\leq\delta} \mathcal{P}_C(x,y)\omega(x)\,dx + \int\limits_{|x|>\delta} \mathcal{P}_C(x,y)\omega(x)\,dx \right] \\
&\leq \lim_{\substack{y\to 0,\\ y\in C}} \left[\int\limits_{|x|\leq\delta} \mathcal{P}_C(x,y)\,dx + (1-\varepsilon) \int\limits_{|x|>\delta} \mathcal{P}_C(x,y)\,dx \right] \\
&\leq \lim_{\substack{y\to 0,\\ y\in C}} \left[1 - \varepsilon \int\limits_{|x|>\delta} \mathcal{P}_C(x,y)\,dx \right],
\end{aligned}$$

which completes the proof of relation (3.2) and thus the relation (3.1). □

11.3.2. *If $f \in \mathcal{H}_s$, then its Poisson integral*

$$\mathcal{F}(x,y) \to f(x), \qquad y \to 0, \qquad y \in C \quad \text{in} \quad \mathcal{H}_s. \tag{3.3}$$

Indeed, by (1.8) and (3.1)

$$\left|F_x[\mathcal{P}_C(x,y)](\xi) - 1\right|^2 \le 4, \qquad \xi \in \mathbb{R}^n, \qquad y \in C;$$

$$\left|F_x[\mathcal{P}_C(x,y)](\xi) - 1\right|^2 = \left|\int \mathcal{P}_C(x,y)e^{-i(x,\xi)}\,dx - 1\right|^2 \to 0, \qquad y \to 0, \qquad y \in C.$$

For this reason [compare (2.3)], by the Lebesgue theorem, as $y \to 0$, $y \in C$, we have

$$\|\mathcal{F}(x,y) - f(x)\|_s^2 = \|F_x^{-1}[\mathcal{F}(x,y) - f(x)]\|_{(s)}^2$$
$$= \int \left|F_x[\mathcal{P}_C(x,y)](\xi) - 1\right|^2 \left|F^{-1}[f](\xi)\right|^2 (1+|\xi|^2)^s\,d\xi \to 0,$$

which is what we set out to prove. □

11.3.3. $\mathcal{P}_C(x,y) \to \delta(x)$, $y \to 0$, $y \in C$ in $\mathcal{H}_s$, $s < -n/2$.

This follows from (3.3) since $\delta \in \mathcal{H}_s$ for all $s < -n/2$ and

$$\mathcal{P}_C(x,y) = \delta(x) * \mathcal{P}_C(x,y) \to \delta(x), \qquad y \to 0, \qquad y \in C \quad \text{in} \quad \mathcal{H}_s.$$

□

11.3.4. In the case of an n-hedral cone (see Sec. 4.4)

$$C = [y\colon (y,e_1) > 0, \dots, (y,e_n) > 0]$$

the limiting relation (3.1) admits of extension to a more general class of functions $\varphi(s)$, namely: *if $\varphi(x)$ is continuous at 0 and such that the integral*

$$\int \mathcal{P}_C(x,y)|\varphi(x)|\,dx \le K, \qquad y \in C, \quad |y| < a \tag{3.4}$$

is bounded, then

$$\int \mathcal{P}_C(x,y)\varphi(x)\,dx \to \varphi(0), \qquad y \to 0, \qquad y \in C. \tag{3.5}$$

Indeed, since the (nonsingular) linear mapping

$$z \to Tz = [(z,e_1), \dots, (z,e_n)]$$

carries the domain T^C onto the domain T^n [see (2.16) of Sec. 10.2], it follows that it suffices to prove the assertion for the cone $\mathbb{R}^n_+$ and the kernel (see (1.2))

$$\mathcal{P}_n(x,y) = \prod_{1 \le j \le n} \frac{y_j}{\pi} \frac{1}{x_j^2 + y_j^2}.$$

By virtue of (3.1), we only need to prove that for all $\delta > 0$

$$\int_{|x|>\delta} \mathcal{P}_n(x,y)|\varphi(x)|\,dx \to 0, \qquad y \to 0, \qquad y \in \mathbb{R}^n_+. \tag{3.6}$$

But the limiting relation (3.6) follows from the estimate

$$\frac{1}{x_k^2 + y_k^2} \le \frac{2}{x_k + \delta^2 a^2/n}, \qquad |x_k| > \frac{\delta a}{\sqrt{n}}$$

and from the estimate (3.4) by virtue of the following chain of inequalities for $|y| < a\sqrt{1-\delta^2/n}$, $y \in \mathbb{R}^n_+$ $(a < 1,\ \delta < \sqrt{n})$:

$$\begin{aligned}
\int\limits_{|x|>\delta} \mathcal{P}_n(x,y)|\varphi(x)|\,dx &\le \sum_{1\le k\le n} \frac{y_1\cdots y_n}{\pi^n} \int\limits_{|x_k|>\delta/\sqrt{n}} \frac{|\varphi(x)|\,dx}{(x_1^2+y_1^2)\dots(x_n^2+y_n^2)} \\
&\le 2\sum_{1\le k\le n} \frac{y_1\cdots y_n}{\pi^n} \int \frac{\varphi(x)|\,dx}{(x_1^2+y_1^2)\dots(x_k^2+\delta^2a^2/n)\dots(x_n^2+y_n^2)} \\
&\le 2\frac{\sqrt{n}}{\delta a} \sum_{1\le k\le n} y_k \int \mathcal{P}_n(x, y_1, \dots, \delta a/\sqrt{n}, \dots, y_n)|\varphi(x)|\,dx \\
&\le 2\frac{\sqrt{n}}{\delta a} K \sum_{1\le k\le n} y_k .
\end{aligned}$$

□

11.3.5. The following holds for an n-hedral cone C [see 11.3.4]: *if $P(x)$ is a polynomial and the integral*

$$\int |P(x)|\mathcal{P}_C(x,y)\,dx < \infty$$

for some $y \in C$, then $P(x) = \text{const}$.

Indeed, as in 11.3.4, it suffices to prove this assertion for the cone $\mathbb{R}^n_+$ and the kernel $\mathcal{P}_n(x,y)$. For $n = 1$ it is readily demonstrated by induction on the degree of the polynomial P. Then we apply induction on n: let

$$P(x) = x_n^m P_m(\tilde{x}) + \dots + x_n P_1(\tilde{x}) + P_0(\tilde{x}),$$
$$\tilde{x} = (x_1, \dots, x_{n-1}).$$

Then by the Fubini theorem the integral

$$\int\limits_{-\infty}^{\infty} |x_n^m P_m(\tilde{x}) + \dots + P_0(\tilde{x})|\,\mathcal{P}_1(x_n, y_n)\,dx_n < \infty$$

for almost all $\tilde{x} \in \mathbb{R}^{n-1}$ and therefore, by what has been proved, $P(x) = P_o(\tilde{x})$, and so forth. □

REMARK 1. The question arises as to whether the assertions of 11.3.4 and 11.3.5 hold true for an arbitrary acute (convex) cone C. The assertion 11.3.5 has been proved for the cone $V^+(n = 4$ by Vladimirov [**114**, (II)].

REMARK 2. For $s = 0$, that is, in $\mathcal{H}_s = \mathcal{L}^2$ (and in $\mathcal{L}^p$, $1 \le p \le \infty$), the theory was given (by use of a different method) in Stein and Weiss [**99**].

12. Algebras of Holomorphic Functions

In this section we give an internal description of the Laplace transform of generalized functions from the algebras $\mathcal{S}'(C^*+)$ and $\mathcal{S}'(C^*)$ in a manner similar to that in Sec. 10.5 for functions from $\mathcal{L}^2_s(C^* + \overline{U}_a)$.

12.1. The definition of the $H_+(C)$ and $H(C)$ algebras. Let C be a connected open cone with vertex at 0. Denote by $H_a^{(\alpha,\beta)}(C)$, $\alpha \geq \beta$, $\beta \geq 0$, the set of all functions $f(z)$ that are holomorphic in T^C and that satisfy the following growth condition:

$$|f(z)| \leq Me^{a|y|}(1+|z|^2)^{\alpha/2}\left[1+\Delta^{-\beta}(y)\right], \qquad z \in T^C. \tag{1.1}$$

we introduce the convergence (topology) in $H_a^{(\alpha,\beta)}(C)$ in accordance with the estimate (1.1) by means of the norm

$$\|f\|_a^{(\alpha,\beta)} = \sup_{z\in T^C} \frac{|f(z)|e^{-a|y|}}{(1+|z|^2)^{\alpha/2}\left[1+\Delta^{-\beta}(y)\right]}.$$

The spaces $H_a^{(\alpha,\beta)}(C)$ are Banach spaces and

$$H_a^{(\alpha,\beta)}(C) \subset H_{a'}^{(\alpha',\beta')}(C), \qquad \alpha' \geq \alpha, \qquad \beta' \geq \beta, \qquad a' \geq a, \tag{1.2}$$

with the inclusion (1.2) to be understood together with the appropriate topology, by virtue of the obvious inequality

$$\|f\|_{a'}^{(\alpha',\beta')} \leq 2\|f\|_a^{(\alpha,\beta)}. \tag{1.3}$$

We set

$$H_a(C) = \bigcup_{\alpha\geq 0,\ \beta\geq 0} H_a^{(\alpha,\beta)}(C), \qquad H_+(C) = \bigcup_{a\geq 0} H_a(C),$$

$$H_a^0(C) = \bigcup_s H_a^{(s)}(C), \qquad H_+^0(C) = \bigcup_{a\geq 0} H_a^0(C).$$

The set $H_+(C)$ forms an algebra of functions that are holomorphic in T^C and that satisfy the estimate (1.1) for certain $a \geq 0$, $\alpha \geq 0$ and $\beta \geq 0$ relative to the operation of ordinary multiplication. This algebra is associative, commutative, contains a unit element but does not contain divisors of zero. Furthermore, $H_0(C) = H(C)$ is a subalgebra of the algebra $H_+(C)$ and contains the unit element. We endow the spaces $H_a(C)$, $H_+(C)$, $H_a^0(C)$ and $H_+^0(C)$ with a topology of the inductive limit (union) of the increasing sequence of spaces $H_a^{(\alpha,\beta)}(C)$, $H_a(C)$, $H_a^{(s)}(C)$ and $H_a^0(C)$, respectively (see Diedonné and Schwartz **[17]**, Bourbaki **[11]**). In what follows, we will drop the index 0 for $a = 0$.

12.2. Isomorphism of the algebras $\mathcal{S}'(C^*+) \sim H_+(C)$ and $\mathcal{S}'(C^*) \sim H(C)$. Let C be an acute convex cone. By Lemma 1 of Sec. 4.4 $\operatorname{int} C^* \neq \varnothing$. We choose an (arbitrary) basis $e_1, \dots, e_n$ in $\mathbb{R}^n$ such that $e_j \in \operatorname{pr}\operatorname{int} C^*$, $j = 1, \dots, n$. Then we construct the polynomial

$$l(z) = (e_1, z)\dots(e_n, z).$$

We will say that $l(z)$ is an *admissible polynomial for the cone* C. Since $(e_j, y) > 0$ for all $y \in C$, it follows that

$$l(z) = \left[(e_1, x) + i(e_1, y)\right]\dots\left[(e_n, x) + i(e_n, y)\right] \neq 0, \qquad z \in T^C. \tag{2.1}$$

We now convince ourselves that the following lemma holds true:

LEMMA. *Let a function $f(z)$ be holomorphic in T^C and let it satisfy the following growth condition: for any number $\varepsilon > 0$ there is a number $M(\varepsilon)$ such that*

$$|f(x+iy)| \leq M(\varepsilon)e^{(a+\varepsilon)|y|}(1+|z|^2)^{\alpha/2}\left[1+\Delta^{-\beta}(y)\right], \qquad z \in T^C \tag{2.2}$$

for certain $a \geq 0$, $\alpha \geq 0$ and $\beta \geq 0$ (that depend solely on f). Then $f(z)$ can, for $\delta > \alpha + n/2$, be represented in the form

$$f(z) = l^\delta(z) f_\delta(z), \qquad f_\delta \in H_a^{(s)}(C), \qquad s < -\beta - n(\delta - 1/2), \tag{2.3}$$

$$\|f_\delta\|_a^{(s)} \leq K_{s,\delta} \inf_{0<\varepsilon\leq 1} M(\varepsilon) \inf_{\sigma \in \operatorname{pr} C} \left[1 + \Delta^{-\beta - n(\delta-1/2)}(\sigma)\right]. \tag{2.4}$$

Here, $l(z)$ is any admissible polynomial for the cone C.

PROOF. By (2.1) the function

$$f_\delta(z) = f(z) l^{-\delta}(z)$$

is holomorphic in T^C. Since $(e_j, y) \geq \sigma|y|$, $j = 1, \ldots, n$, $y \in C$, for some $\sigma > 0$ [see (4.1) of Sec. 4.4], it follows that

$$\begin{aligned}
|(l, z)|^2 &= \left[(e_1, x)^2 + (e_1, y)^2\right] \ldots \left[(e_n, x)^2 + (e_n, y)^2\right] \\
&\geq \left[(e_1, x)^2 + \sigma^2|y|^2\right] \ldots \left[(e_n, x)^2 + \sigma^2|y|^2\right] \\
&\geq (\sigma|y|)^{2n-2}\left[(e_1, x)^2 + \cdots + (e_n, x)^2 + \sigma^2|y|^2\right], \qquad z \in T^C.
\end{aligned} \tag{2.5}$$

Since the vectors $e_1, \ldots, e_n$ are linearly independent, there is a number $b > 0$ such that

$$(e_1, x)^2 + \cdots + (e_n, x)^2 \geq b^2|x|^2.$$

From this, continuing the estimates (2.5), we obtain

$$|l(x+iy)|^2 \geq (\sigma|y|)^{2n-2}\left[b^2|x|^2 + \sigma^2|y|^2\right], \qquad z \in T^C.$$

Taking into account the estimate thus obtained and the estimate (2.2), we have, for all $z \in T^C$,

$$\begin{aligned}
|f_\delta(x+iy)|^2 &= |f(x+iy)|^2 |l(x+iy)|^{-2\delta} \\
&\leq M^2(\varepsilon) e^{2(a+\varepsilon)|y|} \frac{(1 + |x+iy|^2)^\alpha \left[1 + \Delta^{-\beta}(y)\right]^2}{(\sigma|y|)^{2\delta(n-1)}\left[b^2|x|^2 + \sigma^2|y|^2\right]^\delta} \\
&\leq K_1^2 M^2(\varepsilon) e^{2(a+\varepsilon)|y|} \frac{\left[1 + \Delta^\beta(y)\right]^2 (1 + |x|^2 + |y|^2)^\alpha}{\Delta^{2\beta}(y)|y|^{2\delta(n-1)}\left(|x|^2 + |y|^2\right)^\delta} \\
&\leq K_1^2 M^2(\varepsilon) e^{2(a+\varepsilon)|y|} \frac{\left[1 + \Delta^\beta(y)\right]^2 \left[1 + |x|^2 + \Delta^2(y)\right]^\alpha}{\Delta^{2\beta + 2\delta(n-1)}(y)\left[|x|^2 + \Delta^2(y)\right]^\delta}.
\end{aligned}$$

Here, we also took into account that $\Delta(y) \leq |y|$ and $\delta > \alpha$. Therefore

$$\begin{aligned}
\|f_\delta(x+iy)\|^2 &\leq K_1^2 M^2(\varepsilon) e^{2(a+\varepsilon)|y|} \frac{\left[1 + \Delta^\beta(y)\right]^2}{\Delta^{2\beta+2\delta(n-1)}(y)} \int \frac{\left[1 + |x|^2 + \Delta^2(y)\right]^\alpha}{\left[|x|^2 + \Delta^2(y)\right]^\delta} dx \\
&\leq K_1^2 M^2(\varepsilon) e^{2(a+\varepsilon)|y|} \frac{\left[1 + \Delta^\beta(y)\right]^2}{\Delta^{2\beta+n(2\delta-1)}(y)} \int \frac{\left[1 + \Delta^2(y)(1 + |\xi|^2)\right]^\alpha}{(1 + |\xi|^2)^\delta} d\xi.
\end{aligned}$$

By virtue of the choice of the number δ, $2\delta - 2\alpha > n$, but then $2\alpha \leq n(2\delta - 1)$ and, continuing our estimates, we obtain

$$\|f_\delta(x+iy)\|_0^2 \leq K_2^2 M^2(\varepsilon) e^{2(a+\varepsilon)|y|} \left[1 + \Delta^{-\beta - n(\delta-1/2)}(y)\right]^2, \qquad y \in C, \tag{2.6}$$

where the number K_2 depends solely on α, β, δ and the admissible polynomial l. The estimate (2.6) shows that the function f_δ satisfies the conditions of the lemma of Sec. 10.5 for $s = 0$ and $\gamma = \beta + n(\delta - 1/2)$. Therefore $f_\delta = L[g_\delta]$, where $g_\delta \in \mathcal{L}_s^2(C^* + \overline{U}_a)$, $s < -\beta - n(\delta - 1/2)$, and satisfies the estimate (see (5.3′) of Sec. 10.5)

$$\|g_\delta\|_{(s)} \le K_2 e^{2+\epsilon} \sqrt{\frac{\beta + n(\delta - 1/2) + 1}{-s - \beta - n(\delta - 1/2)}} \inf_{0 \le s \le 1} M(\epsilon) \inf_{\sigma \in \operatorname{pr} C} [1 + \Delta^{-\gamma}(\sigma)].$$

By the theorem of Sec. 10.5, $f_\delta = L[g_\delta] \in H_a^{(s)}(C)$ and satisfies the estimate (2.4) with a certain $K_{s,\delta}$. The proof of the lemma is complete. □

THEOREM. *The following statements are equivalent:*

(1) *f belong to $H_a(C)$;*
(2) *f can be represented in the form*

$$f(z) = l^\delta(z) f_\delta(z), \qquad f_\delta = L[g_\delta] \in H_a^{(s)}(C) \tag{2.7}$$

for all admissible polynomials $l(z)$ for the cone C for all $s < s_0$ and for all $\delta > \delta_0 \ge n/2$ (s_0 and δ_0 depend solely on f);
(3) *f possesses the spectral function g taken from $S'(C^* + \overline{U}_a)$.*

Here the following operations are continuous: $f \to f_\delta \to g_\delta \to g \to f$.

PROOF. (1) → (2). Let $f \in H_a^{(\alpha,\beta)}(C)$. From the lemma, for $M(\epsilon) = \|f\|_a^{(\alpha,\beta)}$, it follows that for $\delta > \alpha + n/2$, $f(z)$ can be represented as (2.7), and the function $f_\delta \in H_a^{(s)}(C)$, $s < -\beta - n(\delta - 1/2)$, and satisfies the estimate

$$\|f_\delta\|_a^{(s)} \le K_{s,\delta} \|f\|_a^{(\alpha,\beta)} \inf_{\sigma \in \operatorname{pr} C} [1 + \Delta^{-\beta - n(\delta - 1/2)}(\sigma)], \tag{2.8}$$

with some $K_{s,\delta}$. The estimate (2.8) is what signifies that the operation $f \to f_\delta$ is continuous from $H_a^{(\alpha,\beta)}(C)$ to $H_a^{(s)}(C)$.

(2) → (3). Suppose $f(z)$ can be represented as (2.7). Assuming $\delta > \delta_0 \ge n/2$ to be integer and using the theorem of Sec. 10.5 and property 9.2.2 of Sec. 9.2, we conclude that the spectral function g of the function f can be represented as

$$g(\xi) = l^\delta(-i\partial) g_\delta(\xi), \qquad g_\delta \in \mathcal{L}_s^2(C^* + \overline{U}_a), \tag{2.9}$$

that is, $g \in S'(C^* + \overline{U}_a)$.

(3) → (1). Let $f = L[g]$, where $g \in S'(C^* + \overline{U}_a)$. Then $f(z)$ is a holomorphic function in $T^{\operatorname{int} C^{**}} = T^C$ and can be represented in the form (see Sec. 9.1)

$$f(z) = (g(\xi), \eta(\xi) e^{i(z,\xi)}), \qquad z \in T^C,$$

where $\eta \in C^\infty$; $\eta(\xi) = 1$, $\xi \in (C^* + \overline{U}_a)^{\epsilon/2}$; $\eta(\xi) = 0$, $\xi \notin (C^* + \overline{U}_a)^\epsilon$; $|\partial^\alpha \eta(\xi)| \le c_\alpha(\epsilon)$; ϵ is an arbitrary number, $0 < \epsilon \le 1$. Since $g \in S'$, it follows, by the Schwartz theorem (see Sec. 5.2), that it is of finite order m. Furthermore, by what was proved in Sec. 9.1, $\eta(\xi) e^{i(z,\xi)} \in S$ for all $z \in T^C$. Hence, for all $z \in T^C$ the following estimates hold true:

$$
\begin{aligned}
|f(z)| &\le \|g\|_{-m}\|\eta(\xi)e^{i(z,\xi)}\|_m \\
&= \|g\|_{-m} \sup_{\substack{\xi \\ |\alpha|\le m}} (1+|\xi|^2)^{m/2}|\partial^\alpha[\eta(\xi)e^{i(z,\xi)}]| \\
&\le \|g\|_{-m} \sup_{\substack{\xi \\ |\alpha|\le m}} (1+|\xi|^2)^{m/2}\sum_{\beta\le\alpha}\binom{\alpha}{\beta}e^{-(y,\xi)}|z^\beta|\,|\partial^{\alpha-\beta}\eta(\xi)| \\
&\le K_m'(\varepsilon)\|g\|_{-m}(1+|z|^2)^{m/2}\sup_{\xi\in(C^*+\overline{U}_a)^\varepsilon}(1+|\xi|^2)^{m/2}e^{-(y,\xi)} \\
&\le K_m'(\varepsilon)\|g\|_{-m}(1+|z|^2)^{m/2}\sup_{\substack{\xi_1\in C^*, \\ |\xi_2|\le a+\varepsilon}}(1+|\xi_1+\xi_2|^2)^{m/2}e^{-(y,\xi_1)-(y,\xi_2)} \\
&\le K_m''(\varepsilon)\|g\|_{-m}e^{(a+\varepsilon)|y|}(1+|z|^2)^{m/2}\sup_{\xi_1\in C^*}(1+|\xi_1|^2)^{m/2}e^{-(y,\xi_1)} \\
&\le K_m''(\varepsilon)\|g\|_{-m}e^{(a+\varepsilon)|y|}(1+|z|^2)^{m/2}\sup_{\rho\ge 0}(1+\rho^2)^{m/2}e^{-\Delta(y)\rho} \\
&\le K_m''(\varepsilon)\|g\|_{-m}e^{(a+\varepsilon)|y|}(1+|z|^2)^{m/2}\sup_{t\ge 0}\left[1+\frac{t^2}{\Delta^2(y)}\right]^{m/2}e^{-t},
\end{aligned}
$$

that is

$$|f(z)| \le K_m(\varepsilon)\|g\|_{-m}e^{(a+\varepsilon)|y|}(1+|z|^2)^{m/2}[1+\Delta^{-m}(y)], \qquad z\in T^C.$$

Thus, the function $f(z)$ satisfies the conditions of the lemma with $\alpha=\beta=m$ and $M(\varepsilon)=K_m(\varepsilon)\|g\|_{-m}$. In this case, when $\delta>m+n/2$, it can be represented as (2.7), where $f_\delta\in H_a^{(s)}(C)$ for $s<-m-n(\delta-1/2)<0$, and it satisfies the estimate

$$\|f_\delta\|_a^{(s)} \le K_{s,\delta}'\|g\|_{-m}\inf_{0<\varepsilon\le 1}K_m(\varepsilon)\inf_{\sigma\in \mathrm{pr}\, C}[1+\delta^{-m-n(\delta-1/2)}(\sigma)]$$
$$= K_{s,\delta}\|g\|_{-m}, \tag{2.10}$$

By the theorem of Sec. 10.5, the function $f_\delta(z)$ is the Laplace transform of the function g_δ taken from $\mathcal{L}_s^2(C^*+\overline{U}_a)$,

$$f_\delta(z)=\int_{C^*+U_a} g_\delta(\xi)e^{i(z,\xi)}d\xi, \qquad z\in T^C, \tag{2.11}$$

which, by virtue of (2.10), satisfies the inequality

$$\|g_\delta\|_{(s)}=\|f\|_a^{(s)}\le K_{s,\delta}\|g\|_{-m}. \tag{2.12}$$

Applying to the integral (2.11) the Cauchy–Bunyakovsky inequality and using the definition of a norm in the space $\mathcal{L}_s^2$, we get, for all $z\in T^C$,

$$|f_\delta(z)|\le \int_{C^*+U_a}|g_\delta(\xi)|(1+|\xi|^2)^{s/2}e^{-(y,\xi)}(1+|\xi|^2)^{-s/2}d\xi$$

$$\le \|g_\delta\|_{(s)}\left[\int_{C^*+U_a}(1+|\xi|^2)^{-s}e^{-2(y,\xi)}d\xi\right]^{1/2} \tag{2.13}$$

Before continuing the estimate (2.13), we take note of the inequality

$$-(y,\xi) \le -\Delta(y)(|\xi|-a)\theta(|\xi|-a) + a|y|, \qquad y \in C, \quad \xi \in C^* + \overline{U}_a. \tag{2.14}$$

Indeed, if $\xi = \xi_1 + \xi_2$, $\xi_1 \in C^*$, $|\xi_2| \le a$, then

$$-(y,\xi) = -(y,\xi_1) - (y,\xi_2) \le -\Delta(y)|\xi_1| + a|y| \le \begin{cases} a|y| & \text{if } |\xi| < a, \\ -\Delta(y)(|\xi|-a) + a|y| & \text{if } |\xi \ge a. \end{cases}$$

Taking into account the inequality (2.14), we continue the estimate (2.13):

$$|f_\delta(z)|^2 \le \|g_\delta\|^2_{(s)} \int_{C^*+U_a} (1+|\xi|^2)^{-s} e^{-2\Delta(y)(|\xi|-a)\theta(|\xi|-a)+2a|y|}\, d\xi$$

$$\le \|g_\delta\|^2_{(s)} e^{2a|y|} \left[\int_{|\xi|<a} (1+|\xi|^2)^{-s}\, d\xi \int_{|\xi|>a} (1+|\xi|^2)^{-s} e^{-2\Delta(y)(|\xi|-a)}\, d\xi \right]$$

$$= \|g_\delta\|^2_{(s)} e^{2a|y|} \left\{ M_s'(a) + \sigma_n \int_0^\infty [1+(r+a)^2]^{-s} (r+a)^{n-1} e^{-2\Delta(y)r}\, dr \right\}$$

$$= \|g_\delta\|^2_{(s)} e^{-2a|y|} \left\{ M_s'(a) + \sigma_n \int_0^\infty \left[1 + \left(\frac{u}{\Delta(y)} + a\right)^2\right]^{-s} \left(\frac{u}{\Delta(y)} + a\right)^{n-1} e^{-2u} \frac{du}{\Delta(y)} \right\}.$$

This is to say, for some $M_s(a)$,

$$|f_\delta(z)| \le M_s(a) \|g_\delta\|_{(s)} e^{a|y|} [1 + \Delta^{s-n/2}(y)], \qquad z \in T^C.$$

Whence, taking into account the estimate (2.12), we obtain from the representation (2.7)

$$|f(z)| = |I^\delta(z)|\, |f_\delta(z)| \le M_s(a)|z|^{n\delta} \|g_\delta\|_{(s)} e^{a|y|} [1 + \Delta^{s-n/2}(y)]$$

$$\le M_s(a) K_{s,\delta} \|g\|_{-m} e^{a|y|} (1+|z|^2)^{n\delta/2} [1 + \Delta^{s-n/2}(y)], \qquad z \in T^C,$$

so that $f \in H_a^{(n\delta, s-n/2)}(C)$ and the operation $g \to f$ is continuous from $S'(C^* + \overline{U}_a)$ to $H_a(C)$.

It remains to note that the operation $f_\delta \to g_\delta$ is continuous from $H_a^0(C)$ to $S_0'(C^* + \overline{U}_a)$ (by the theorem of Sec. 10.5)[2], and the operation $g_\delta \to g$ is continuous from $S_0'(C^* + \overline{U}_a)$ to $S'(C^* + \overline{U}_a)$ [by (2.9)]. The proof of the theorem is complete. □

COROLLARY 1. *The algebras $H_+(C)$ and $S'(C^*+)$ and also their subalgebras $H(C)$ and $S'(C^*)$ are isomorphic, and that isomorphism is accomplished via the Laplace transformation.*

---

[2]The definition of the space S_0' see in Sec. 10.1.

COROLLARY 2. *For $g \in \mathcal{S}'(C^* + \overline{U}_a)$, it is necessary and sufficient that for any admissible polynomial for the cone C and for any integer $\delta \geq \delta_0(g)$ it is representable in the form*

$$g(\xi) = l^\delta(-i\partial)g_\delta(\xi), \qquad g_\delta \in \mathcal{S}'_0(C^* + \overline{U}_a), \tag{2.15}$$

the operation $g \to g_\delta$ being continuous from $\mathcal{S}'(C^* + \overline{U}_a)$ to $\mathcal{S}'_0(C^* + \overline{U}_a)$.

COROLLARY 3. *The operation $f \to \partial^\alpha f$ is continuous in $H_a(C)$.*

This follows from the continuity of the operations

$$f \to g \to (i\xi)^\alpha g \to \partial^\alpha f.$$

REMARK. These results have been proved by Vladimirov **[108]** by a different method.

COROLLARY 4. *Any function $f(z)$ in $H_+(c)$ has a (unique) boundary value $f_+(x)$ as $y \to 0$, $y \in C$ in $\mathcal{S}'$, which value is equal to $F[g] = f_+$, and the operation $f \to f_+$ is continuous from $H_+(C)$ to $\mathcal{S}'$.*

REMARK. The theorem on the existence, in $\mathcal{S}'$, of boundary values of functions taken from the algebra $H(C)$ has been proved by Vladimirov **[109, 112]** and Tillmann **[102]**. The proof given here is taken from Vladimirov **[110]**. More general conceptions of boundary values of holomorphic functions have been considered in the works of Köthe **[64]** ($\mathcal{D}'$), Sato **[87]** (hyperfunctions), Komatsu **[59]** (ultradistributions), and Martineau **[74]**, Zharinov **[129]** (Fourier-ultrahyperfunctions).

12.3. The Paley–Wiener–Schwartz theorem and its generalizations.

THEOREM (Paley–Wiener–Schwartz). *For a function $f(z)$ to be entire and to satisfy the conditions of growth: for any $\varepsilon > 0$ there is a number $M(\varepsilon)$ such that*

$$|f(z)| \leq M(\varepsilon)e^{(a+\varepsilon)|y|}(1 + |z|^2)^{\alpha/2}, \qquad z \in \mathbb{C}^n, \tag{3.1}$$

for certain $a \geq 0$ and $\alpha \geq 0$ (that are dependent on f), it is necessary and sufficient that its spectral function g belong to $\mathcal{S}'(\overline{U}_a)$. Here, $f(z)$ satisfies the growth condition

$$|f(z)| \leq Me^{a|y|}(1 + |z|^2)^{\alpha'/2}, \qquad z \in \mathbb{C}^n, \tag{3.2}$$

for certain M and $\alpha' \geq \alpha$.

PROOF. NECESSITY. Suppose $f(z)$ is an entire function satisfying the growth condition (3.1). Then by the theorem of Sec. 12.2, $f(z)$ is the Laplace transform of the generalized function g in $\mathcal{S}'(C^* + \overline{U}_a)$ for any convex cone C. Hence

$$\operatorname{supp} g \subset \bigcap_C (C^* + \overline{U}_a) = \overline{U}_a,$$

so that $g \in \mathcal{E}'(\overline{U}_a)$.

SUFFICIENCY. Suppose $f(z)$ is the Laplace transform of the generalized function g taken from $\mathcal{E}'(\overline{U}_a)$.

Let us cover $\mathbb{R}^n \setminus \{0\}$ with a finite number of convex acute open cones C_j, $j = 1, \ldots, N$. Then $g \in \mathcal{S}'(C_j^* + \overline{U}_a)$, $j = 1, \ldots, N$. By the theorem of Sec. 12.2 in each T^{C_j}, $f(z)$ satisfies an estimate of the type (3.2),

$$|f(z)| \leq M_j e^{a|y|}(1 + |z|^2)^{\alpha_j/2}, \qquad z \in T^{C_j}, \quad j = 1, \ldots, N. \tag{3.3}$$

Setting $M = \max_j M_j$ and $\alpha' = \max_j \alpha_j$, we obtain from (3.3) the estimate (3.2) in $\mathbb{C}^n$. The theorem is proved. □

COROLLARY. *For a function $f(z)$ to be entire and to satisfy the growth condition,*

$$|f(z)| \le K_N e^{a|y|}(1+|z|^2)^{-N}, \qquad \xi \in \mathbb{C}^n, \tag{3.4}$$

it is necessary and sufficient, for all $N \ge 0$, that its spectral function φ belong to $\mathcal{D}(\overline{U}_a)$. Here,

$$K_N = \int_{|\xi|<a} |(1-\Delta)^N \varphi(\xi)|\, d\xi. \tag{3.5}$$

This follows from the Paley–Wiener–Schwartz theorem and from the estimate

$$|L[\varphi](z)| = (1+|z|^2)^{-N} \left| \int_{|\xi|<a} e^{i(z,\xi)}(1-\Delta)^N \varphi(\xi)\, d\xi \right|$$
$$\le (1+|z|^2)^{-N} e^{a|y|} \int_{|\xi|<a} |(1-\Delta)^N \varphi(\xi)|\, d\xi, \qquad \varphi \in \mathcal{D}, \quad \operatorname{supp}\varphi \subset \overline{U}_a,$$

where N is an integer ≥ 0. □

12.4. The space $H_a(C)$ is the projective limit of the spaces $H_{a'}(C')$. Suppose C' is a convex open cone, compact in the cone C, and $a' > a \ge 0$. We denote by $\Delta'(y)$ the distance from the point y to the boundary of the cone C'. Then $\Delta'(y) < \Delta(y)$, $y \in C'$, and therefore the following inequality holds:

$$\sup_{z \in T^{C'}} e^{-(a+\varepsilon)|y|} \frac{|f(z)|}{(1+|z|^2)^{\alpha/2}[1+\Delta'^{-\beta}(y)]}$$
$$\le \sup_{z \in T^C} e^{-a|y|} \frac{|f(z)|}{(1+|z|^2)^{\alpha/2}[1+\Delta^{-\beta}(y)]}.$$

That is, (the norm $\|\ \|'$ corresponds to the cone C')

$$\|f\|'^{(\alpha,\beta)}_{a+\varepsilon} \le \|f\|^{(\alpha,\beta)}_a, \tag{4.1}$$

whence we conclude that

$$H_a(C) \subset H_{a'}(C'), \tag{4.2}$$

and this imbedding is continuous.

We introduce the intersection of spaces

$$\bigcap_{C' \Subset C,\ a' > a} H_{a'}(C')$$

with convergence $f_k \to 0$, $k \to \infty$, if $f_k \to 0$, $k \to \infty$, in each of the $H_{a'}(C')$. In other words, we equip this intersection with a topology of the projective limit (of the intersection) of a decreasing sequence of the spaces $H_{a'}(C')$, $a' \to a + 0$, $C' \to C$, $C' \Subset C$.

We have *the equality*

$$H_a(c) = \bigcap_{C' \Subset C,\ a'>a} H_{a'}(C') \tag{4.3}$$

which holds true together with the corresponding topology.

Indeed, the truth of the imbedding

$$H_a(c) \subset \bigcap_{C' \Subset C,\ a'>a} H_{a'}(C') \tag{4.4}$$

has already been proved by (4.2). We now prove the inverse imbedding

$$\bigcap_{C' \Subset C,\ a'>a} H_{a'}(C') \subset H_a(c). \tag{4.5}$$

Let $f \in H_{a'}(C')$ for all $C' \Subset C$ and $a' > a$. By the theorem of Sec. 12.2 $f = L[g]$, where $g \in \mathcal{S}'(C'^* + \overline{U}_{a'})$. Noting that

$$\bigcap_{C' \Subset C,\ a'>a} (C'^* + \overline{U}_{a'}) = C^* + \overline{U}_a,$$

we conclude that $g \in \mathcal{S}'(C^* + \overline{U}_a)$ and, hence, $f(z) \in H_a(C)$. Furthermore, the operation $f \to g$ is continuous from $H_{a'}(C')$ to $\mathcal{S}'(C'^* + \overline{U}_{a'})$. But

$$\bigcap_{C' \Subset C,\ a'>a} \mathcal{S}'(C'^* + \overline{U}_{a'}) = \mathcal{S}'(C^* + \overline{U}_a)$$

and this equality is continuous in both directions. Finally, the operation $g \to f$ is continuous from $\mathcal{S}'(C^* + \overline{U}_a)$ to $H_a(C)$. And this means that the imbedding $f \to f$ is continuous from

$$\bigcap_{C' \Subset C,\ a'>a} H_{a'}(C') \quad \text{to} \quad H_a(C).$$

The imbedding (4.5) together with the imbedding (4.4) is what yields (4.3), which is what we set out to prove. □

The equality (4.3) gives a different definition of functions of the class $H_a(C)$, which definition is convenient for applications.

For a function $f(z)$ that is holomorphic in T^C to belong to $H_a(C)$, it is necessary and sufficient that, for any arbitrary cone $C' \Subset C$ and an arbitrary number $\varepsilon > 0$, there exist numbers $\alpha' \geq 0$, $\beta' \geq 0$, and $M' > 0$ such that

$$|f(z)| \leq M' e^{(a+\varepsilon)|y|} \frac{(1+|z|^2)^{\alpha'/2}}{|y|^{\beta'}}, \qquad z \in T^{C'}. \tag{4.6}$$

Indeed, if $f \in H_a(C)$, then $f \in H_a^{(\alpha,\beta)}(C)$ for certain $\alpha \geq 0$ and $\beta \geq 0$ such that f satisfies the inequality (1.1). Let $C' \Subset C$. By Lemma 1 of Sec. 4.4 there is a number $\varkappa > 0$ such that

$$\Delta(y) = \inf_{\sigma \in \operatorname{pr} C^*} (\sigma, y) \geq \varkappa |y|, \qquad y \in C'.$$

From this and from the inequality (1.1) it follows the inequality (4.6) for $\varepsilon = 0$, $\beta' = \beta$ and for certain $\alpha' \geq \alpha$ and $M'(C') \geq M$.

Conversely, if $f(z)$ is holomorphic in T^C and, for arbitrary $C' \Subset C$ and $\varepsilon > 0$, satisfies the estimate (4.6), then, taking into account that $\Delta'(y) \leq |y|$, where $\Delta'(y)$ is the distance from y to $\partial C'$, we obtain $f \in H_{a+\varepsilon}(C')$, whence, by (4.3), it follows that $f \in H_a(C)$. □

12.5. The Schwartz representation. Suppose an acute (convex open) cone C is such that the Cauchy kernel $\mathcal{K}_C(z) \neq 0$ in the tube $T^C = \mathbb{R}^n + iC$. Such cones C will be called *regular*[3]. For example, the cones $\mathbb{R}^n_+$ and V^+ are regular (see Sec. 13.5 below).

LEMMA. *If a cone C is regular, then its Cauchy kernel $\mathcal{K}_C(z)$ is a divisor of the unity in the algebra $H(C)$, i.e., $\mathcal{K}_C^{-1}(z) \in H(C)$.*

PROOF. Since $\mathcal{K}_C(z) \neq 0$ in T^C, to prove the lemma it is sufficient to establish the following estimate (see Sec. 12.4): for any cone $C' \Subset C$ there exist nonnegative numbers μ, α, and β such that

$$|\mathcal{K}_C^{-1}(z)| \leq M \frac{1+|z|^\alpha}{|y|^\beta}, \qquad z \in T^{C'}.$$

However, this estimate follows immediately from representation (2.2) in Sec. 10.2 of the kernel $\mathcal{K}_C$

$$|\mathcal{K}_C(z)| = \Gamma(n)|z|^{-n} \left| \int_{\mathrm{pr}\, C^*} \frac{d\sigma}{[(p,\sigma)+i(q,\sigma)]^n} \right|, \qquad p = \frac{x}{|z|}, \quad q = \frac{y}{|z|},$$

if we note that the function

$$\left| \int_{\mathrm{pr}\, C^*} \frac{d\sigma}{[(p,\sigma)+i(q,\sigma)]^n} \right|$$

is positive and continuous on the compact

$$[(p,q) \in \mathbb{R}^{2n} : |p|^2 + |q|^2 = 1, \ q \in C'],$$

hence, it is bounded from below by a positive number $\sigma = \sigma(C')$. This and the previous equality imply the estimate

$$|\mathcal{K}_C(z)| \geq \sigma \Gamma(n) |z|^{-n}.$$

Setting $\alpha = n$ and $\beta = 0$, we obtain the statement of the lemma. □

The *Schwartz kernel* of the region T^C, where C is a regular cone, relative to the point $z^0 = x^0 + iy^0 \in T^C$ is the function

$$\mathcal{S}_C(z; z^0) = \frac{2\mathcal{K}_C(z)\mathcal{K}_C(-\overline{z^0})}{(2\pi)^n \mathcal{K}_C(z-\overline{z^0})} - \mathcal{P}_C(x^0, y^0), \qquad z \in T^C. \tag{5.1}$$

We note some properties of the Schwartz kernel.

12.5.1.

$$\mathcal{S}_C(z; z) = \mathcal{P}_C(x, y), \qquad z \in T^C. \tag{5.2}$$

This property follows from (5.1) when $z^0 = z$, from the definition of the Poisson kernel (1.1) of Sec. 11.1, and from the property (2.3) of Sec. 10.2 of the Cauchy kernel.

---

[3]For $n = 1, 2, 3$ all acute cones are regular; for $n \geq 4$ it is not the case (Danilov [14]); homogeneous cones of positivity are regular (Rothaus [86]).

12.5.2.

$$\int \mathcal{S}_C(z - x'; z^0 - x')\, dx' = 1, \qquad z \in T^C, \quad z^0 \in T^C. \tag{5.3}$$

This property follows from the Parseval–Steklov equation applied to (2.1) of Sec. 10.2,

$$\begin{aligned}\int \mathcal{K}_C(z - x')\mathcal{K}_C(-\overline{z^0} + x')\, dx' &= \int \mathcal{K}_C(z - x')\overline{\mathcal{K}_C(z^0 - x')}\, dx' \\ &= (2\pi)^n \int_{C^*} e^{i(z - \bar{z}^0, \xi)}\, d\xi \\ &= (2\pi)^n \mathcal{K}_C(z - \overline{z^0}),\end{aligned}$$

and from the property (1.5), Sec. 11.1, of the Poisson kernel. □

12.5.3.

$$|\mathcal{S}_C(z; z^0)| \le \frac{\mathcal{K}_C(2iy)}{|\mathcal{K}_C(z - \overline{z^0})|}\mathcal{P}_C(x, y) + \left[\frac{\mathcal{K}_C(2iy^0)}{|\mathcal{K}_C(z - \overline{z^0})|} + 1\right]\mathcal{P}_C(x^0, y^0), \tag{5.4}$$
$$z \in T^C, \qquad z^0 \in T^C.$$

This property follows from the definitions of the Schwartz and the Poisson kernels and from the estimate $2|ab| \le |a|^2 + |b|^2$. □

EXAMPLE 1 (see (2.16) of Sec. 10.2 and (1.2) of Sec. 11.1).

$$\mathcal{S}_{\mathbb{R}^n_+}(z; z^0) = \mathcal{S}_n(z; z^0) = \frac{2i^n}{(2\pi)^n}\left(\frac{1}{z_1} - \frac{1}{\overline{z^0}_1}\right)\cdots\left(\frac{1}{z_n} - \frac{1}{\overline{z^0}_n}\right) - \mathcal{P}_n(x^0, y^0).$$

In particular, for $n = 1$, $C = (0, \infty)$,

$$\mathcal{S}_1(z; z^0) = \frac{i}{\pi}\left(\frac{1}{z} - \frac{x^0}{|z^0|^2}\right), \qquad \Re\mathcal{S}_1(z; z^0) = \mathcal{P}_1(x, y).$$

EXAMPLE 2 (see (2.17) of Sec. 10.2 and (1.3) of Sec. 11.1).

$$\mathcal{S}_{V^+}(z; z^0) = \frac{\Gamma(\frac{n+1}{2})\left[-(z - \overline{z^0})^2\right]^{(n+1)/2}}{\pi^{(n+3)/2}(-z^2)^{(n+1)/2}\left[-(\overline{z^0})^2\right]^{(n+1)/2}} - \mathcal{P}_{V^+}(x^0, y^0).$$

Let the boundary value $f_+(x)$ of a function $f(x)$ of the class $H(C)$ (see Sec. 12.2) satisfy the condition

$$f_+(x)\mathcal{K}_C(x - \overline{z^0}) \in \mathcal{H}_s \tag{5.5}$$

for some s and for all $z^0 \in T^C$. Then the generalized function (5.5) is the boundary value in $\mathcal{S}'$ of the function $f(z)\mathcal{K}_C(z - \overline{z^0})$ of the class $H(C)$ and therefore the support of its inverse Fourier transform is contained in the cone C^*. By Theorem II of Sec. 10.6, the function $f(z)\mathcal{K}_C(z - \overline{z^0})$ belongs to the class $H^{(s)}(C)$ and its boundary value in $\mathcal{H}_s$ is equal to $f_+(x)\mathcal{K}_C(x - \overline{z^0})$ since $\mathcal{K}_C(x + iy) \in \theta_M$ for all $y \in C$ [see Sec. 5.3 and Sec. 10.2, estimate (2.4)]. Applying Theorem I of Sec. 10.6

to the function $f(z)\mathcal{K}_C(z-\overline{z^0})$, we obtain

$$f(z)\mathcal{K}_C(z-\overline{z^0}) = \frac{1}{(2\pi)^n}(f_+(x')\mathcal{K}_C(x'-\overline{z^0}), \mathcal{K}_C(z-x')), \quad z \in T^C, \quad z^0 \in T^C. \tag{5.6}$$

Putting $z^0 = z$ in (5.6) and taking into account (5.2) for the function $f(z)$, we derive the generalized Poisson representation

$$f(z) = (f_+(x'), \mathcal{P}_C(x-x', y)), \qquad z \in T^C. \tag{5.7}$$

Then, interchanging z and z^0 in (5.6), we obtain

$$f(z^0)\mathcal{K}_C(z^0-\bar{z}) = \frac{1}{(2\pi)^n}\left(f_+(x')\mathcal{K}_C(x'-\bar{z}), \mathcal{K}_C(z^0-x')\right),$$

whence, passing to the complex conjugate, we derive

$$\bar{f}(z^0)\mathcal{K}_C(z-\overline{z^0}) = \frac{1}{(2\pi)^n}\left(\bar{f}_+(x')\mathcal{K}_C(z-x'), \mathcal{K}_C(x'-\overline{z^0})\right). \tag{5.8}$$

Subtracting (5.8) from (5.6), we get the relation

$$\mathcal{K}_C(z-\overline{z^0})[f(z)-\bar{f}(z^0)] = \frac{2i}{(2\pi)^n}\left(\Im f_+(x'), \mathcal{K}_C(z-x')\mathcal{K}(x'-\overline{z^0})\right), \quad z \in T^C, \quad z^0 \in T^C. \tag{5.9}$$

Suppose C is a regular cone so that $\mathcal{K}_C(z) \neq 0$, $z \in T^C$. Divide (5.9) by $\mathcal{K}_C(z-\overline{z^0})$ and, in (5.9), in accordance with formula (5.7), make the substitution

$$\Im f(z^0) = (\Im f_+(x'), \mathcal{P}_C(x^0-x', y^0)), \qquad z^0 \in T^C. \tag{5.10}$$

As a result we obtain the representation

$$f(z) = i\left(\Im f_+(x'), \frac{2}{(2\pi)^n}\mathcal{K}_C(z-x')\mathcal{K}_C(x'-\overline{z^0}) - \mathcal{P}_C(x^0-x', y^0)\right) + \Re f(z^0)$$

or, using the definition (5.1) of the Schwartz kernel,

$$f(z) = i(\Im f_+(x'), \mathcal{S}_C(z-x'; z^0-x')) + \Re f(z^0), \quad z \in T^C, \quad z^0 \in T^C. \tag{5.11}$$

Formula (5.11) is called the *generalized Schwartz representation*.

This completes the proof of the following theorem.

THEOREM. *If C is an acute cone, then any function $f(z)$ of the class $H(C)$ that satisfies the condition* (5.5) *can be represented in terms of its boundary value f_+ by the Poisson integral* (5.7) *and can also be represented in terms of the imaginary part of its boundary value by the formula* (5.9). *And if, besides, the cone C is regular, then for any such function $f(z)$ the generalized Schwartz representation* (5.11) *holds true.*

12.6. A generalization of the Phragmén–Lindelöf theorem. The Phragmén–Lindelöf theorem in the theory of holomorphic functions is defined as any generalization of the maximum principle to the case of unbounded domains or to more general (than continuous) boundary values. Here we give one such generalization of the maximum principle that will be used later on in Sec. 21.1.

THEOREM. *If the boundary value $f_+(x)$ of a function $f(z)$ of the class $H(C)$, where C is an acute cone, is bounded: $|f_+(x)| \le M$, $x \in \mathbb{R}^n$, then we also have $|f(z)| \le M$, $z \in T^C$; what is more, for $f(z)$ we have the generalized Poisson representation*

$$f(z) = \int \mathcal{P}_C(x - x', y) f_+(x')\, dx', \qquad z \in T^C. \tag{6.1}$$

REMARK. For $n = 1$ this theorem was proved by Nevanlinna [**78**].

PROOF. Since $\mathcal{K}_C(x + iy) \in \mathcal{L}^2$ for all $y \in C$ (see Sec. 10.2), it follows that $f_+(x)\mathcal{K}_C(x - \overline{z^0}) \in \mathcal{L}^2$ for all $z^0 \in T^C$ and, hence, the condition (5.5) is fulfilled for $s = 0$. By the theorem of Sec. 12.5, for the function $f(z)$ the Poisson representation (6.1) holds; from this and from the property (1.5) of Sec. 11.1, of the kernel $\mathcal{P}_C$ follows the estimate

$$|f(z)| \le M \int \mathcal{P}_C(x - x', y)\, dx' = M, \qquad z \in T^C,$$

which completes the proof of the theorem. □

13. Equations in Convolution Algebras

Let Γ be a closed convex acute solid cone in $\mathbb{R}^n$ (with vertex at 0). Then the sets of tempered generalized functions $\mathcal{S}'(\Gamma+)$ and $\mathcal{S}'(\Gamma)$ form convolution algebras [$\mathcal{S}'(\Gamma)$ is a subalgebra of $\mathcal{S}'(\Gamma+)$] (see Sec. 5.6.2) that are isomorphic to the algebras of the holomorphic functions $H_+(C)$ and $H(C)$, respectively, where $C = \operatorname{int}\Gamma^*$, and the isomorphism is accomplished by the operation of the Laplace transform (see Sec. 12.2).

13.1. Divisors of unity in the $H_+(C)$ and $H(C)$ algebras. As was shown in Sec. 4.9.4 the solvability of the equation

$$a * u = f \qquad a \text{ and } f \in \mathcal{S}'(\Gamma+),$$

in the convolution algebra $\mathcal{S}'(\Gamma+)$ reduces to the existence of a fundamental solution $\mathcal{E}$ (the kernel of the inverse operator $a^{-1}*$) of the convolution operator $a*$,

$$a * \mathcal{E} = \delta, \tag{1.1}$$

in the same algebra $\mathcal{S}'(\Gamma+)$. The equation (1.1) is equivalent to the algebraic equation

$$L[a]f = 1 \tag{1.2}$$

in the algebra $H_+(C)$ with respect to the unknown function $f(z) = L[\mathcal{E}]$. Therefore the question of the existence of a fundamental solution of the operator $a*$ in the algebra $\mathcal{S}'(\Gamma+)$ reduces to the question of the possibility of dividing unity by the function $f_0(z) = L[a]$ in the $H_+(C)$ algebra. In other words, the question reduces to studying the divisors of unity in the $H_+(C)$ algebra: if $f \in H_+(C)$, then we want to know under what conditions $1/f \in H_+(C)$.

The necessary condition for this, $f(z) \neq 0$, $z \in T^C$, is not a sufficient condition, as will be seen by the following simple example: $f(z) = e^{-i/z} \in H(0,\infty)$ since $|f(z)| = e^{-y/|z|^2} \leq 1$. However, $1/f \notin H_+(0,\infty)$ since

$$\left|\frac{1}{f(z)}\right| = e^{\frac{y}{|z|^2}} > e^{\frac{1}{y}\left(1-\frac{x^2}{y^2}\right)}.$$

EXAMPLE. If $H(z) \neq 0$ is holomorphic and homogeneous in T^C, then $H^{-1}(z) \in H(C)$.

The proof is similar to the one given for the Cauchy kernel (see the lemma in Sec. 12.5). □

We first note that the study of divisors of unity in the $H_+(C)$ algebra reduces to studying the divisors of unity in its subalgebra, the $H(C)$ algebra. Indeed, any function $f(z)$ in $H_+(C)$ [that is, $f \in H_a^{(\alpha,\beta)}(C)$ for certain $a \geq 0$, $\alpha \geq 0$ and $\beta \geq 0$] can be represented in the form

$$f(z) = e^{-i(z,e)} f_e(z), \qquad f_e \in H(C), \tag{1.3}$$

where e is an arbitrary point in $\operatorname{int}\Gamma$ such that $(y,e) \geq a|y|$ for all $y \in C$ (by Lemma 1 of Sec. 4.4, such points exist). Indeed,

$$|f_e(z)| = |e^{i(z,e)} f(z)| \leq \|f\|_a^{(\alpha,\beta)} (1+|z|^2)^{\alpha/2} [1+\Delta^{-\beta}(y)], \qquad z \in T^C,$$

so that $f_e \in H^{(\alpha,\beta)}(C)$.

From the result of Sec. 12.4 we have the following theorem.

THEOREM. *For $f \in H(C)$ to be a divisor of unity in the $H(C)$ algebra, it is necessary and sufficient that, for any cone $C' \Subset C$ and any number $\varepsilon > 0$, there exist numbers $\alpha' \geq 0$, $\beta' \geq 0$ and $M' > 0$ such that*

$$|f(z)| \geq M' e^{-\varepsilon|y|} (1+|z|^2)^{-\alpha'/2} |y|^{\beta'}, \qquad z \in T^{C'}. \tag{1.4}$$

The condition (1.4) is hard to verify. We now point to several sufficient criteria for the divisibility of unity in the $H(C)$ algebra that follow from the theorem that has just been proved.

13.2. On division by a polynomial in the $H(C)$ algebra.

THEOREM. *Suppose $P(z) \neq 0$ is a polynomial, and a function $f(z)$ is holomorphic in T^C and $Pf \in H(C)$. Then $f \in H(C)$ and the operation $f \to Pf$ has a continuous inverse in $H(C)$.*

COROLLARY. *If the polynomial $P(z)$ does not vanish in T^C, then $\frac{1}{P} \in H(C)$.*

Indeed, in that case, $\frac{1}{P(z)}$ is a holomorphic function in T^C and $P\frac{1}{P} = 1 \in H(C)$.

PROOF OF THE THEOREM. To prove this theorem we take advantage of the following result obtained by Hörmander [see inequality (2.3) of Sec. 15.2 for $p = 0$]:

For a given polynomial $P(z) \not\equiv 0$ there are numbers $m \geq 0$ (an integer) and $K > 0$ such that for any $\varphi \in C^m(\mathbb{R}^{2n})$ the following estimate holds true:

$$|\varphi(x,y)| \leq K \sup_{\substack{(x,y) \\ |\gamma|\leq m}} (1+|z|^2)^{m/2} \left|\partial^\gamma_{(x,y)}[P(z)\varphi(x,y)]\right|. \tag{2.1}$$

By hypothesis $Pf \in H(C)$. By Corollary 3 (see Sec. 12.2), $\partial^\gamma(Pf) \in H(C)$. Therefore there will be numbers $\alpha_0 \geq 0$ and $\beta_0 \geq 1$ such that $\partial^\gamma(Pf) \in H^{(\alpha,\beta)}(C)$, $|\gamma| \leq m$, $\alpha \geq \alpha_0$, $\beta \geq \beta_0$.

Let C' be an open convex cone, $C' \Subset C$. Then there will be an (open convex) cone C'' such that $C' \Subset C'' \Subset C$. Let us construct a function $\eta(\sigma)$ of the class $C^\infty(S_1)$ that is equal to 1 on $\operatorname{pr} C'$ and equal to 0 outside $\operatorname{pr} C''$. (From the lemma of Sec. 1.2 it follows that such functions do exist.) If in the inequality (2.1) we put

$$\varphi(x,y) = \frac{f(z)\eta(\frac{y}{|y|})|y|^m}{(1+|z|^2)^{m+\alpha/2}(1+|y|^{-\beta})} \in C^m(\mathbb{R}^{2n}),$$

for all $z \in T^{C'}$ we obtain

$$\frac{|f(z)|\,|y|^m}{(1+|z|^2)^{m+\alpha/2}(1+|y|^{-\beta})}$$

$$\leq K \sup_{\substack{(x,y)\\ |\gamma|\leq m}} (1+|z|^2)^{m/2} \left|\partial^\gamma_{(x,y)}\left[\frac{P(z)f(z)\eta(\frac{y}{|y|})|y|^{\beta+m}}{(1+|z|^2)^{m+\alpha/2}(1+|y|^\beta)}\right]\right|$$

$$\leq K_1(C') \sup_{\substack{(x,y)\in T^{C''}\\ |\gamma|\leq m}} \frac{(|y|^\beta + |y|^{\beta+m})\,|\partial^\gamma[P(z)f(z)]|}{(1+|z|^2)^{m+\alpha/2}(1+|y|^\beta)}$$

$$\leq K_2(C') \sup_{\substack{(x,y)\in T^{C''}\\ |\gamma|\leq m}} \frac{|\partial^\gamma[P(z)f(z)]|}{(1+|z|^2)^{\alpha/2}(1+|y|^{-\beta})}.$$

Now, taking into account that

$$\Delta'(y) \leq |y|, \quad y \in C'; \qquad \Delta'(y) \geq \sigma|y|, \quad y \in C'', \tag{2.2}$$

where $\Delta(y)$ and $\Delta'(y)$ are the distances from the point y to ∂C and $\partial C'$, respectively, let us continue our estimates

$$\frac{|f(z)|}{(1+|z|^2)^{m+\alpha/2}[1+(\Delta')^{-\beta-m}(y)]} \leq K_3(C') \sup_{\substack{(x,y)\in T^C\\ |\gamma|\leq m}} \left|\frac{\partial^\gamma[P(z)f(z)]}{(1+|z|^2)^{\alpha/2}[1+\Delta^{-\beta}(y)]}\right|,$$

whence, by (2.2) (the norm $\|\ \|'$ corresponds to the cone C')

$$\|f\|'^{(2m+\alpha,m+\beta)} \leq K_3(C') \max_{|\gamma|\leq m} \|\partial^\gamma(Pf)\|^{(\alpha,\beta)}. \tag{2.3}$$

Now, by Corollary 3 (see Sec. 12.2), the operation $Pf \to \partial^\alpha(Pf)$ is continuous in $H(C)$ so that for certain $M_1 > 0$, $\alpha' \geq 0$, and $\beta' \geq 0$ the following estimates hold:

$$\|\partial^\gamma(Pf)\|^{(\alpha,\beta)} \leq M_1\|Pf\|^{(\alpha',\beta')}, \qquad |\gamma| \leq m.$$

Taking into account these estimates, let us rewrite the inequality (2.3) as

$$\|f\|'^{(2m+\alpha,m+\beta)} \leq M(C')\|Pf\|^{(\alpha',\beta')}. \tag{2.4}$$

The estimate (2.4) shows (see Sec. 12.4) that $f \in H(C)$ and the operation $f \to Pf$ has a continuous inverse in $H(C)$. The proof is complete. □

REMARK. This theorem was proved in Bogolyubov and Vladimirov **[10]**. It resembles the theorem of Hörmander **[47]** on the division of a tempered generalized function by a polynomial.

13.3. Estimates for holomorphic functions with nonnegative imaginary part in T^C.

THEOREM. *Suppose a function $f(z)$ is holomorphic in T^C and $\Im f(z) \geq 0$, $z \in T^C$. Then it satisfies the following estimate: for any cone $C' \Subset C$ there is a number $M(C')$ such that*

$$|f(z)| \leq M(C')\frac{1+|z|^2}{|y|}, \qquad z \in T^{C'}. \tag{3.1}$$

That is, $f \in H^{(2,1)}(C')$ for all $C' \Subset C$ so that $f \in H(C)$.

COROLLARY. *If under the hypothesis of the theorem $f(z) \not\equiv 0$ in T^C, then $1/f \in H(C)$.*

We now prove the corollary. If $\Im f(z) > 0$ in T^C, then the function $\frac{1}{f(z)}$ is holomorphic in T^C and

$$\Im\frac{1}{f(z)} = \frac{-\Im f(z)}{|f|^2} < 0.$$

By the theorem, $1/f \in H(C)$,. But if $\Im f(z) \geq 0$ vanishes at some point in the domain T^C, then, by virtue of the maximum principle for harmonic functions, $\Im f(z) \equiv 0$ in T^C so that $f(z) = \text{const} \neq 0$ and therefore, trivially, $1/f \in H(C)$. □

REMARK. The estimate (3.1) was obtained in Vladimirov [**111**].

To prove the theorem, let us first prove three lemmas.

LEMMA 1. *If a function $f(z)$ is holomorphic and $\Im f(z) \geq 0$ in the unit polycircle $S^n = [z: |z_1| < 1, \ldots, |z_n| < 1]$, then*

$$\Im f(0)\frac{1-\max_{1\leq j\leq n}|z_j|}{1+\max_{1\leq j\leq n}|z_j|} \leq |f(z) - \Re f(0)| \leq \Im f(0)\frac{1+\max_{1\leq j\leq n}|z_j|}{1-\max_{1\leq j\leq n}|z_j|}, \qquad z \in S^n. \tag{3.2}$$

In particular,

$$|f(z)| \leq \frac{2|f(0)|}{1-\max_{1\leq j\leq n}|z_j|}, \qquad z \in S^n. \tag{3.3}$$

PROOF. If $\Im f(z) \equiv 0$, then, as we have seen, $f(z) = \text{const}$ and the estimate (3.2) is trivially fulfilled. We can therefore assume that $\Im f(z) > 0$, $z \in S^n$. Let us fix an arbitrary point $z \in S^n$ and put $\rho = \max_{1\leq j\leq n}|z_j|$ so that $0 \leq \rho < 1$. We consider the function $\varphi(\lambda) = f\left(\lambda\frac{z}{\rho}\right)$ that is holomorphic and $\Im\varphi(\lambda) > 0$ in the circle $|\lambda| < 1$. The function

$$\psi(\lambda) = \frac{\varphi(\lambda) - \Re\varphi(0) - i\Im\varphi(0)}{\varphi(\lambda) - \Re\varphi(0) + i\Im\varphi(0)}$$

is holomorphic and $|\psi(\lambda)| < 1$ in the circle $|\lambda| < 1$; what is more, $\psi(0) = 0$. By the Schwartz lemma, $|\psi(\lambda)| \leq |\lambda|$ and therefore

$$\Im\varphi(0)\frac{1-|\lambda|}{1+|\lambda|} \leq |\varphi(\lambda) - \Re\varphi(0)| = \left|i\Im\varphi(0)\frac{1+\psi(\lambda)}{1-\psi(\lambda)}\right|$$
$$\leq \Im\varphi(0)\frac{1+|\lambda|}{1-|\lambda|}, \qquad |\lambda| < 1, \tag{3.4}$$

and so

$$\left|\varphi(\lambda)\right| \leq \left|\Re\varphi(0)\right| + \Im\varphi(0)\frac{1+|\lambda|}{1-|\lambda|} \leq \frac{2|\varphi(0)|}{1-|\lambda|}, \qquad |\lambda| < 1. \tag{3.5}$$

If in the estimates (3.4) and (3.5) we put $\lambda = \rho < 1$, we obtain the estimates (3.2) and (3.3). And that is the end of the proof of Lemma 1. □

LEMMA 2. *If a function $f(z)$ is holomorphic and $\Im f(z) \geq 0$ in $T^n = [z\colon y_1 > 0, \ldots, y_n > 0]$, then*

$$\left|f(z)\right| \leq \sqrt{2}\left|f(\mathrm{i})\right| \max_{1\leq j\leq n} \frac{1+|z_j|^2}{y_j}, \qquad z \in T^n \tag{3.6}$$

where $\mathrm{i} = (i, i, \ldots, i)$.

PROOF. The holomorphic mapping

$$w_j = \frac{z_j - i}{z_j + i}, \qquad z_j = i\frac{1+w_j}{1-w_j}, \qquad j = 1, \ldots, n$$

transforms the tubular domain T^n on S^n, and the function $f(z)$ into the function

$$\varphi(w) = f\left(i\frac{1+w_1}{1-w_1}, \ldots, i\frac{1+w_n}{1-w_n}\right)$$

which is holomorphic, and $\Im\varphi(w) \geq 0$ in S^n. Applying (3.3) to $\varphi(w)$, we obtain

$$\left|\varphi(w)\right| \leq \frac{2|\varphi(0)|}{1 - \max_{1\leq j\leq n}|w_j|}, \qquad w \in S^n.$$

From this, it we pass to the old variables, we obtain the estimate

$$\left|f(z)\right| \leq \frac{2|f(\mathrm{i})|}{1 - \max_{1\leq j\leq n}\left|\frac{z_j - i}{z_j + i}\right|}, \qquad z \in T^n. \tag{3.7}$$

Let us prove the inequality

$$\left|\frac{z-i}{z+i}\right| < 1 - \frac{\sqrt{2}y}{1+|z|^2}, \qquad y > 0. \tag{3.8}$$

Putting

$$\alpha = \frac{1+x^2}{y} + y \geq 2,$$

we reduce inequality (3.8) to the equivalent inequality

$$2\alpha^2 - (\alpha+2)(\sqrt{2}\alpha - 1) > 0, \qquad \alpha \geq 2.$$

Now the latter inequality does indeed occur: it holds for $\alpha = 2$, and the derivative of the left-hand side is greater than zero:

$$(4 - 2\sqrt{2})\alpha + 1 - 2\sqrt{2} > 0, \qquad \alpha \geq 2.$$

If we take into account the inequality (3.8), we obtain from (3.7) the inequality (3.6). Lemma 2 is proved. □

LEMMA 3. *Suppose the function $f(z)$ is holomorphic and $\Im f(z) \geq 0$ in T^C, where C is an n-hedral acute cone,*

$$C = [y\colon (e_j, y) > 0,\ j = 1, \ldots, n] \qquad (|e_j| = 1).$$

Then $f(z)$ satisfies the estimate

$$|f(z)| \leq \sqrt{2}|f(T^{-1}\mathrm{i})|\frac{1+|z|^2}{\Delta(y)}, \qquad z \in T^C, \tag{3.9}$$

where T is a linear transformation,

$$y \to Ty = [(e_1, y), \ldots, (e_n, y)].$$

PROOF. Since the cone C is not empty, the vectors $e_1, \ldots, e_n$ are linearly independent and, hence, the matrix T^{-1} exists. The biholomorphic mapping

$$w = Tz, \qquad z = T^{-1}w$$

transforms the domain T^C into the domain T^n, and the function $f(z)$ into the function $f(T^{-1}w)$, which is holomorphic, and $\Im f(T^{-1}w) \geq 0$ in T^n. Applying the estimate (3.6) to that function, we obtain

$$|f(T^{-1}w)| \leq \sqrt{2}|f(T^{-1}\mathrm{i})| \max_{1\leq j\leq n} \frac{1+|w_j|^2}{\Im w_j}, \qquad w \in T^n.$$

From this, passing to the variables z, we derive the estimate

$$|f(z)| \leq \sqrt{2}|f(T^{-1}\mathrm{i})| \max_{1\leq j\leq n} \frac{1+|(e_j, z)|^2}{(e_j, y)}, \qquad z \in T^C,$$

from which, and also from the relations

$$|(e_j, z)|^2 \leq |e_j|^2|z|^2 = |z|^2; \qquad \Delta(y) = \max_{1\leq j\leq n}(e_j, y), \qquad y \in C,$$

follows the inequality (3.9). The proof of Lemma 3 is complete. □

PROOF OF THE THEOREM. Let $C' \Subset C$. Cover the cone $\overline{C'}$ with a finite number of n-hedral open cones $C_k \Subset C$, $k = 1, \ldots, N$, and in each cone C_k choose a cone $C'_k \Subset C_k$ so that the cones $C'_1, \ldots, C'_N$ still cover the cone $\overline{C'}$. In each domain T^{C_k} the estimate (3.9) holds true:

$$|f(z)| \leq \sqrt{2}|f(T_k^{-1}\mathrm{i})|\frac{1+|z|^2}{\Delta_k(y)}, \qquad k = 1, \ldots, N, \tag{3.10}$$

where $\Delta_k(y)$ and T_k have the same meaning relative to the cone C_k as $\Delta(y)$ and T do relative to the cone C. Furthermore, since $C'_k \Subset C_k$, it follows that there exist numbers σ_k such that $\Delta_k(y) \geq \sigma_k|y|$ for all $y \in C'_k$ (see Lemma 1 of Sec. 4.4). Taking into consideration this inequality, we obtain from (3.10) the estimates

$$|f(z)| \leq \frac{\sqrt{2}}{\sigma_k}|f(T_k^{-1}\mathrm{i})|\frac{1+|z|^2}{|y|}, \qquad z \in T^{C'_k}, \quad k = 1, \ldots, N,$$

whence follows the estimate (3.1) in $T^{C'}$ for

$$M(C') = \max_{1\leq k\leq N} \frac{\sqrt{2}}{\sigma_k}|f(T_k^{-1}\mathrm{i})|.$$

The theorem is proved. □

13.4. Divisors of unity in the algebra $W(C)$. Denote by $\dot{\mathbb{R}}^n$ the fact that the point at infinity is adjoined to $\mathbb{R}^n$. Denote by $W(C)$ the Banach algebra consisting of functions holomorphic in T^C that are Laplace transforms of generalized functions of the form $\lambda\delta(\xi) + g(\xi)$, where λ is an arbitrary number and g is an arbitrary function in $\mathcal{L}^1(C^*)$. (The set of such generalized functions forms a convolution algebra; see Sec. 4.1.) The algebra $W(C)$ is called a *Wiener algebra*. Thus, any element $f \in W(C)$ can be represented as

$$f(z) = \lambda + \int_{C^*} g(\xi)e^{i(z,\xi)}\,d\xi,$$

$$\|f\|_{W(C)} = |\lambda| + \int_{C^*} |g(\xi)|\,d\xi.$$

Here, the function $f(z)$ is continuous in $\dot{\bar{T}}^C$, the closure of T^C in $\dot{\mathbb{R}}^{2n}$, and the following inequality holds:

$$\|f\|_0^{(0,0)} = \frac{1}{2}\sup_{z\in T^C}|f(z)| \le \frac{1}{2}\|f\|_{W(C)},$$

so that the $W(C)$ algebra is a subalgebra of the $H(C)$ algebra and the embedding of $W(C)$ in $H(C)$ is continuous.

If $f \in W(C)$, and $f(z) \neq 0$ in $T^C \cup \dot{\mathbb{R}}^n$, then $1/f \in W(C)$.

Indeed, in that case

$$f^+(x) = \lambda + \int_{C^*} g(\xi)e^{i(x,\xi)}\,d\xi \neq 0, \qquad x \in \mathbb{R}^n, \quad \lambda \neq 0, \quad g \in \mathcal{L}^1.$$

By Wiener's theorem (see Wiener [125])

$$\frac{1}{f^+(x)} = \frac{1}{\lambda} + \int g_1(\xi)e^{i(x,\xi)}d\xi, \qquad g_1 \in \mathcal{L}^1. \tag{4.1}$$

Furthermore, since $f(z) \neq 0$, $z \in T^C \cup \dot{\mathbb{R}}^n$, it follows that for all $C' \Subset C$

$$\inf_{z\in T^{C'}}|f(z)| > 0.$$

By the theorem of Sec. 13.1 (for $\alpha' = \beta' = \varepsilon = 0$), $1/f \in H(C)$ so that $1/f = L[g]$, $g \in \mathcal{S}'(C^*)$. Therefore $1/f^+ = F[g]$. Comparing that equality with (4.1), we obtain

$$g(\xi) = \frac{1}{\lambda}\delta(\xi) + g_1(\xi)$$

so that $\operatorname{supp} g_1 \subset C^*$ and therefore $g_1 \in \mathcal{L}^1(C^*)$. But then, by (4.1),

$$\frac{1}{f(z)} = \frac{1}{\lambda} + \int_{C^*} g_1(\xi)e^{i(z,\xi)}\,d\xi \in W(C).$$

□

13.5. Example. The Cauchy kernel $\mathcal{K}_{V^+}(z)$ [see (2.17) Sec. 10.2] is a divisor of unity in the algebra $H(V^+)$,

$$\frac{1}{\mathcal{K}_{V^+}(z)} = \frac{1}{2^n\pi^{(n-1)/2}\Gamma\left(\frac{n+1}{2}\right)}(-z^2)^{(n+1)/2} \in H(V^+),$$

hence, the cone V^+ is regular (see Sec. 12.5).

This follows from the corollary to the theorem of Sec. 13.2 and from the fact that the polynomial

$$z^2 = (x+iy)^2 = x^2 - y^2 + 2i(x,y) \neq 0, \qquad z \in T^{V^+}.$$

Indeed, we would otherwise have

$$x^2 = y^2, \qquad x_0 y_0 = (\mathbf{x}, \mathbf{y}), \qquad y^2 > 0, \quad y_0 > 0,$$

that is

$$0 < y^2 = x_0^2 - \mathbf{x}^2 = \frac{(\mathbf{x},\mathbf{y})^2}{y_0^2} - |\mathbf{x}|^2 \leq \frac{|\mathbf{x}|^2}{y_0^2}(|\mathbf{y}|^2 - y_0^2) = -\frac{|\mathbf{x}|^2}{y_0^2} y^2 < 0,$$

which is a contradiction. □

Thus, in the algebra $\mathcal{S}'(\overline{V}^+)$ there is an inverse operator of $\theta_{\overline{V}^+}*$. What is more, it is possible to define arbitrary real powers $\theta^\alpha_{\overline{V}^+}*$ of that convolution operator by putting

$$L[\theta^\alpha_{\overline{V}^+}*] = \mathcal{K}^\alpha_{V^+}(z), \qquad z \in T^{V^+}. \tag{5.1}$$

For the sake of definiteness, we choose that branch of the holomorphic function $\mathcal{K}^\alpha_{V^+}(z)$ that is positive for $z = iy$ [see (2.2) of Sec. 10.2]. From (5.1) it follows that

$$\theta^\alpha_{\overline{V}^+} * \theta^\beta_{\overline{V}^+} = \theta^{\alpha+\beta}_{\overline{V}^+}, \qquad -\infty < \alpha, \beta < \infty. \tag{5.2}$$

The powers $\Box^\alpha$ of the d'Alembert operator are defined in similar fashion:

$$\Box = \Box\delta * = \frac{\partial^2}{\partial\xi_0^2} - \frac{\partial^2}{\partial\xi_1^2} - \cdots - \frac{\partial^2}{\partial\xi_n^2} = \frac{\partial^2}{\partial\xi_0^2} - \Delta.$$

We have

$$L[\Box] = -z^2 = c_n \mathcal{K}_{V^+}^{-\frac{2}{n+1}}(z), \qquad c_n = 2^{\frac{2n}{n+1}} \pi^{\frac{n-1}{n+1}} \Gamma^{\frac{2}{n+1}}\left(\frac{n+1}{2}\right).$$

Therefore, setting

$$L[\Box^\alpha] = c_n^\alpha \mathcal{K}_{V^+}^{-\frac{2\alpha}{n+1}}(z),$$

we obtain

$$\Box^\alpha = c_n^\alpha \theta_{\overline{V}^+}^{-\frac{2\alpha}{n+1}} *, \qquad \Box = c_n \theta_{\overline{V}^+}^{-\frac{2}{n+1}} * \tag{5.3}$$

by virtue of (5.1). By (5.2) the following relation holds true:

$$\Box^\alpha \Box^\beta = \Box^{\alpha+\beta}, \qquad -\infty < \alpha, \beta < \infty. \tag{5.4}$$

In particular, for $\alpha = -1$, we have, from (5.3),

$$\Box^{-1} = \frac{1}{c_n} \theta_{\overline{V}^+}^{\frac{2}{n+1}} * = \frac{1}{c_n}\left(\theta_{\overline{V}^+}^{-\frac{n-1}{n+1}} * \theta_{\overline{V}^+}\right) * = c_n^{-\frac{n+1}{2}} \Box^{\frac{n-1}{2}} \theta_{\overline{V}^+} *.$$

That is

$$\mathcal{E}(\xi) = \frac{1}{2^n \pi^{\frac{n-1}{2}} \Gamma\left(\frac{n+1}{2}\right)} \Box^{\frac{n-1}{2}} \theta_{\overline{V}^+}(\xi). \tag{5.5}$$

From this, for $n = 3$, we obtain a known result:

$$\mathcal{E}(\xi) = \frac{1}{8\pi} \Box \theta_{\overline{V}^+}(\xi) \tag{5.6}$$

for the fundamental solution of a three-dimensional wave operator.

REMARK 1. Fractional powers of the operator $\square$ were introduced in a different manner by M. Riesz [**85**].

REMARK 2. In similar fashion we can introduce fractional and negative powers of the operator $\theta_{C^*}*$ in the $H(C)$ algebra for any regular cone C (see Sec. 14.1 and also Vladimirov, Drozhzhinov and Zavialov [**24**, §§ 2–8]:

$$L[\theta^\alpha_{C^*}] = \mathcal{K}^\alpha_C(z), \qquad z \in T^C, \quad -\infty < \alpha < \infty.$$

14. Tauberian Theorems for Generalized functions

Theorems connecting the asymptotic behaviour of a (generalized) function in a neighbourhood of zero with the asymptotic behaviour of its integral transform (Fourier, Laplace, Mellin or others) at infinity are called *Tauberian*. The theorems inverse to Tauberian ones are called *Abelian*.

The presentation of this section follows, in general, the book by Vladimirov, Drozhzhinov and Zavialov [**122**] (see also Vladimirov [**119**] and Drozhzhinov and Zavialov [**24**]).

14.1. Preliminary results. Always below Γ is a closed convex acute solid cone and $C = \operatorname{int}\Gamma^*$ is a regular cone with the vertex at 0 (see Sec. 12.5); $\mathcal{K}_C(z)$ is the Cauchy kernel of the tube domain $T^C = \mathbb{R}^n + iC$ and

$$\theta^\alpha_\Gamma(\xi) = L^{-1}[\mathcal{K}^\alpha_C], \qquad -\infty < \alpha < \infty,$$

is the convolution group of generalized functions from $\mathcal{S}'(\Gamma)$ (associative, commutative, with unity and without divisors of zero, see Sec. 13.5).

We list here some properties of θ^α_Γ:

a) $\theta^\alpha_\Gamma * \theta^\beta_\Gamma = \theta^{\alpha+\beta}_\Gamma$, $-\infty < \alpha, \beta < \infty$;

b) $\theta^\alpha_\Gamma(t\xi) = t^{n(\alpha-1)}\theta^\alpha_\Gamma(\xi)$, $t > 0$;

c) for any $m = 0, 1, \ldots$ there exists N such that

$$\theta^\alpha_\Gamma \in C^m(\mathbb{R}^n), \qquad \alpha > N,$$

follows from estimate (2.4′) of Sec. 10.2 (see also Lemma 1);

d) $|\partial^m\theta^\alpha_\Gamma(\xi)| \le C|\xi|^{n(\alpha-1)-m}$, $\xi \in \mathbb{R}^n$, $\alpha > N$,
follows from the definition of θ^α_Γ and from its homogeneity (see b)).

DEFINITION. Let $f \in \mathcal{S}'(\Gamma)$ (see Sec. 4.5). By the *primitive* of order α, $f^{(-\alpha)}(\xi)$, with respect to the cone Γ we call the convolution

$$f^{(-\alpha)} = \theta^\alpha_\Gamma * f. \tag{1.1}$$

EXAMPLE. For $n = 1$, $\Gamma = [0,\infty)$, $C = (0,\infty)$, $\mathcal{K}_C(z) = \frac{i}{z}$,

$$\theta^m_\Gamma(z) = \underbrace{\theta * \theta * \cdots * \theta}_{m \text{ times}} = *\theta^m, \qquad m = 0, 1, \ldots .$$
$$\theta^{-m}_C(\xi) = \delta^{(m)}(\xi),$$

Therefore, the operator $\theta^\alpha_\Gamma *$ for $\alpha > 0$ is the (fractional) antiderivative of order α; for $\alpha = 0$ it is the identity operator; for $\alpha < 0$ it is the (fractional) derivative (see 4.9.5).

LEMMA 1. *If $f \in \mathcal{S}'(\Gamma)$, then its primitive $f^{(-\alpha)}$ for all sufficiently large $\alpha > N$ is continuous in $\mathbb{R}^n$, the following representation is valid*

$$f^{(-\alpha)}(\xi) = (f(\xi'), \eta(\xi')\theta^\alpha_\Gamma(\xi - \xi')), \tag{1.2}$$

and the inequality

$$|f^{(-\alpha)}(\xi)| \le C\|f\|_{-m}|\xi|^r \tag{1.3}$$

holds for some $C > 0$, $r \ge 0$, *where* m *is the order of* f. *(In* (1.2) η *is an arbitrary* C^∞*-function,* $|\partial^\beta \eta(\xi)| \le C_\beta$, $\xi \in \mathbb{R}^n$, *which is equal to 1 in* Γ^ε *and equal to 0 outside* $\Gamma^{2\varepsilon}$, $\varepsilon > 0$ *is arbitrary.)*

PROOF. Let m denote the order of $f \in \mathcal{S}'(\Gamma)$ (see Sec. 5.2). By virtue of the reasoning above, $\theta_\Gamma^\alpha \in C^m(\mathbb{R}^n)$ for all sufficiently large $\alpha > N$ and $\operatorname{supp}\theta_\Gamma^\alpha \subset \Gamma$. Using the standard reasoning (cf. Sec. 5.6.2), one can deduce representation (1.2) from this fact and from representation (6.2) of Sec. 5.6. Representation (1.2) implies the continuity of $f^{(-\alpha)}(\xi)$ in $\mathbb{R}^n$ and inequality (1.3) if we note that the set of the functions

$$\{\xi' \to \eta(\xi')\theta_\Gamma^\alpha(\xi - \xi'),\ \xi \in \mathbb{R}^n\}$$

is continuous in $\mathcal{S}_m$ with respect to ξ and make use of inequality (2.3) of Sec. 5.2 and estimate d)

$$|f^{(-\alpha)}(\xi)| \le \|f\|_{-m}\,\|\eta(\xi')\theta_\Gamma^\alpha(\xi - \xi')\|_m \le C\|f\|_{-m}|\xi|^{n(\alpha-1)+m}.$$

The lemma is proved. □

DEFINITIONS. 1. A function $f(\xi)$ has an *asymptotic* $g(\xi)$ of order α in the cone Γ as $|\xi| \to \infty$ if, for any $\xi \in \operatorname{int}\Gamma$, there exists the limit

$$\lim_{|\xi|\to\infty} |\xi|^{-\alpha} f(\xi) = g\left(\frac{\xi}{|\xi|}\right) \tag{1.4}$$

and there exist constants M and R such that

$$|\xi|^{-\alpha}|f(\xi)| \le M, \qquad |\xi| > R, \qquad \xi \in \operatorname{int}\Gamma. \tag{1.5}$$

2. A generalized function $f(\xi)$ taken from $\mathcal{S}'(\Gamma)$ has a *quasiasymptotic* $g(\xi)$ of order α at ∞ if

$$k^{-\alpha} f(k\xi) \to g(\xi), \qquad k \to \infty \quad \text{in} \quad \mathcal{S}'. \tag{1.6}$$

2′. A generalized function $\tilde{f}(\xi)$ taken from $\mathcal{S}'$ has a *quasiasymptotic* $\tilde{g}(\xi)$ of order α at 0 if

$$\rho^\alpha \tilde{f}(\rho x) \to \tilde{g}(x), \qquad \rho \to +0 \quad \text{in} \quad \mathcal{S}'.$$

3. A function $\tilde{f}(z)$ holomorphic in T^C has an *asymptotic* $h(z)$ of order α at 0 in T^C if

(i)

$$\lim_{\rho\to+0} \rho^\alpha f(\rho z) = h(z), \qquad z \in T^C, \tag{1.7}$$

(ii) there exist numbers M, a and b such that

$$\rho^\alpha|f(\rho z)| \le M\frac{1+|z|^a}{\Delta_C^b(y)}, \qquad 0 < \rho \le 1, \qquad z \in T^C. \tag{1.8}$$

Definition 2 implies that the quasiasymptotic g of order α at ∞ (if exists) belongs to $\mathcal{S}'(\Gamma)$ and is a homogeneous generalized function of degree α (cf. Sec. 5.7),

$$g(t\xi) = t^\alpha g(\xi), \qquad t > 0.$$

Its primitive $g^{(-N)}$ is a homogeneous generalized function from $\mathcal{S}'(\Gamma)$ of the homogeneity degree $\alpha + nN$.

EXAMPLE. $\delta(\xi)$ has the quasiasymptotic $\delta(\xi)$ of order $\alpha = -n$ at ∞.

In order that $f \in \mathcal{S}'(\Gamma)$ has the quasiasymptotic g of order α at ∞ it is necessary and sufficient that its Fourier transform $\tilde{f}$ has the quasiasymptotic $\tilde{g}$ of order $\alpha+n$ at 0.

This assertion follows from Definitions 2 and 2', from equality (3.7) of Sec. 6.3.5:

$$F\left[k^{-\alpha}f(k\xi)\right] = \rho^{\alpha+n}\tilde{f}(\rho x), \qquad \rho = \frac{1}{k} > 0,$$

and from the continuity of the operation of the Fourier transform in $\mathcal{S}'$. □

In particular, if $f \in \mathcal{S}'(\Gamma)$ has a quasiasymptotic g of order α at ∞, then $f^{(-N)}$, $-\infty < N < \infty$, also has the quasiasymptotic $g^{(-N)}$ of order $\alpha + nN$ at ∞.

The assertion follows from Definition 2, from the equality

$$\widetilde{f^{(-N)}}(x) = \mathcal{K}_C^N(x)\tilde{f}(x),$$

and from the homogeneity of the kernel $\mathcal{K}_C(x)$ (see Sec. 10.2). □

If a function has the ordinary asymptotic, then it also has the quasiasymptotic of the same order. More exactly, the following lemma is valid.

LEMMA 2. *If a function $f(\xi)$ taken from $\mathcal{S}'(\Gamma)$ has the asymptotic $g(\xi)$ of order $\alpha > -n$ in the cone Γ as $|\xi| \to \infty$, then f also has the quasiasymptotic g of the same order α at ∞.*

PROOF. It follows from (1.4) and (1.5) that

$$k^{-\alpha}f(k\xi) \to |\xi|^\alpha g\left(\frac{\xi}{|\xi|}\right) = g(\xi), \quad k \to \infty \quad \text{almost everywhere in} \quad \mathbb{R}^n$$

(we assume that g is continued by zero onto the whole $\mathbb{R}^n$) and, moreover,

$$\left|k^{-\alpha}f(k\xi)\right| \le M|\xi|^\alpha, \qquad |\xi| > R/k, \quad \xi \in \mathbb{R}^n.$$

Let $\varphi \in \mathcal{S}$. Then

$$(k^{-\alpha}f(k\xi), \varphi) = k^{-\alpha}\int f(k\xi)\varphi(\xi)\,d\xi = \int\limits_{|\xi|>R/k} k^{-\alpha}f(k\xi)\varphi(\xi)\,d\xi$$

$$+\,k^{-\alpha}\int\limits_{|\xi|<R/k} f(k\xi)\varphi(\xi)\,d\xi \to \int g(\xi)\varphi(\xi)\,d\xi = (g,\varphi), \qquad k \to \infty,$$

since one can pass to the limit under the integral sign in the first summand (by the Lebesgue theorem), and the second summand, which is equal to

$$k^{-n-\alpha}\int\limits_{|\xi|<1} f(\xi)\varphi(\xi/k)\,d\xi,$$

tends to zero as $k \to \infty$, if $n + \alpha > 0$. The limiting relation obtained proves that f has the quasiasymptotic g of order α at ∞. □

THEOREM. *For $f \in \mathcal{S}'(\Gamma)$ to have the quasiasymptotic g of order α at ∞ it is necessary and sufficient that there exists $N > -1 - \alpha/n$ such that the function $f^{(-N)}(\xi)$ has the asymptotic $g^{(-N)}(\xi)$ of order $\alpha + nN$ in the cone Γ and*

$$|\xi|^{-\alpha-nN}f^{(-N)}(\xi) \overset{\frac{\xi}{|\xi|}\in \operatorname{pr}\Gamma}{\Longrightarrow} g^{(-N)}\left(\frac{\xi}{|\xi|}\right), \qquad |\xi| \to \infty, \tag{1.9}$$

and the function $g^{(-N)}(\xi)$ is continuous in $\mathbb{R}^n$ with the support in Γ, $g^{(-N)} \in C_0(\Gamma)$.

PROOF. The sufficiency follows from Lemma 2 and from the preceding criterion.

Let us prove the necessity. Since by assumption the sequence $\{k^{-\alpha} f(k\xi), 1 \leq k \to \infty\}$ converges in S', it converges (and is bounded) in some space S'_m (see Sec. 5.2). Lemma 1 of Sec. 4.4 implies that the set of functions

$$\{\xi' \to \eta(\xi')\theta_\Gamma^N(e - \xi'),\ |e| = 1\} \tag{1.10}$$

is bounded in S_m for sufficiently large $N > -1 - \alpha/n$. From this fact, applying formula (1.2) and using property b) as well as the equality

$$\eta(k\xi)f(k\xi) = \eta(\xi)f(k\xi),$$

as $k \to \infty$, we obtain the following chain of equalities

$$\begin{aligned} k^{-\alpha-nN} f^{(-N)}(ke) &= (k^{-\alpha-nN} f(\xi'), \eta(\xi')\theta_\Gamma^N(ke - \xi')) \\ &= (k^{-\alpha-n} f(\xi'), \eta(\xi')\theta_\Gamma^N(e - \xi'/k)) \\ &= (k^{-\alpha} f(k\xi), \eta(\xi)\theta_\Gamma^N(e - \xi)) \\ &\overset{|e|=1}{\Longrightarrow} (g(\xi), \eta(\xi)\theta_\Gamma^N(e - \xi)) = g^{(-N)}(e) \end{aligned} \tag{1.11}$$

and, moreover, by virtue of (1.3), the inequality

$$|\xi|^{-\alpha-nN} \left| f^{(-N)}(\xi) \right| \leq C \sup_{k \geq 1} \|k^{-\alpha} f(k\xi)\|_{-m} \leq M, \qquad |\xi| \geq 1.$$

Thus, by Definition 1, there exists the asymptotic $g^{(-N)}(\xi)$ of the function $f^{(-N)}(\xi)$ of order $\alpha + nN$ in the cone Γ and $g^{(-N)} \in C$, $\operatorname{supp} g^{(-N)} \subset \Gamma$; hence, $g^{(-N)} \in C_0(\Gamma)$. The theorem is proved. □

Let us note the following remark useful for applications.

REMARK. By virtue of the following lemma, it is sufficient to verify the existence of the quasiasymptotic on the test functions from $\mathcal{D}$.

LEMMA 3. *Let $f \in \mathcal{D}'(\Gamma)$ and*

$$k^{-\alpha} f(k\xi) \to g(\xi), \qquad k \to \infty \quad \text{in} \quad \mathcal{D}'.$$

Then $f \in S'(\Gamma)$ and f has the quasiasymptotic $g \in S'(\Gamma)$ of order α at ∞.

PROOF. Let us come back to the proof of the necessity in the preceding theorem. The supports of all functions (1.10) are contained in a ball U (see Sec. 4.4). Let m be the order of $f \in \mathcal{D}'(\Gamma)$ in U (see Sec. 1.3) and $N > -1 - \frac{\alpha}{n}$ be such that functions (1.10) belong to the space $C^m(\overline{U})$, are bounded and continuous with respect to e in this space. It follows from (1.11) that the sequence of continuous functions

$$k^{-\alpha-nN} f^{(-N)}(ke), \qquad k \to \infty,$$

converges uniformly with respect to e, $|e| = 1$, to the continuous function $g^{(-N)}(e)$. This fact together with the following inequality (see (3.1) of Sec. 1.3)

$$|\xi|^{-\alpha-nN} \left| f^{(-N)}(\xi) \right| \leq \sup_{k \geq 1,\ |e|=1} \|k^{-\alpha} f(k\xi)\|_{C^{m'}(\overline{U})} \|\eta(\xi)\theta_\Gamma^N(e - \xi)\|_{C^m(\overline{U})} \leq M, \tag{1.12}$$

for $|\xi| > 1$, shows that $f^{(-N)}$ has the asymptotic $g^{(-N)}$ of order $\alpha+nN$ in the cone Γ and $f^{(-N)} \in S'(\Gamma)$; therefore, $f \in S'(\Gamma)$. By Lemma 2, $f^{(-N)}$ has the quasiasymptotic $g^{(-N)}$ of order $\alpha+nN$. Then $f = f^{(-N)(N)}$ has the quasiasymptotic $g^{(-N)(N)} = g$ of order α at ∞.

The lemma is proved. □

COROLLARY. *Let $f \in \mathcal{D}'(\Gamma)$. If the set of generalized functions*

$$\{k^{-\alpha} f(k\xi),\ k \geq 1\}$$

is bounded in $\mathcal{D}'$, then it is also bounded in $S'(\Gamma)$.

It follows from inequality (1.12). □

14.2. General Tauberian theorem. Let $f \in S'(\Gamma)$. Its Laplace transform

$$\tilde{f}(z) = L[f] = \left(f(\xi), \eta(\xi)e^{i(z,\xi)}\right), \tag{2.1}$$

where the auxiliary function η satisfies the conditions of Lemma 1 of Sec. 14.1, belongs to the $H(C)$-algebra of functions holomorphic in T^C and satisfying the growth condition

$$\left|\tilde{f}(z)\right| \leq M\frac{1+|z|^c}{\Delta_C^d(y)}, \qquad z = x+iy = T^C, \tag{2.2}$$

for some M, c, d. (Estimate (2.2) follows immediately from inequality (1.1) of Sec. 12.1 for $a = 0$.)

Let us prove the following General Tauberian theorem.

THEOREM. *In order that $f \in S'(\Gamma)$ to have the quasiasymptotic g of order α at ∞ it is necessary that $\tilde{f}(z)$ has the asymptotic $h(z)$ of order $\alpha+n$ at 0 in T^C, i.e.,*

$$1) \qquad \lim_{\rho\to+0} \rho^{\alpha+n}\tilde{f}(\rho z) = h(z), \qquad z \in T^C \tag{2.3}$$

$$2) \qquad \rho^{\alpha+n}\left|f(\rho z)\right| \leq M\frac{1+|z|^a}{\Delta_C^b(y)}, \qquad 0<\rho\leq 1, \quad z\in T^C, \tag{2.4}$$

and it is sufficient that the following conditions hold:

a) *there exists a solid subcone $C' \subset C$ such that $\tilde{f}(iy)$ has an asymptotic $h(iy)$ of order $\alpha+n$ at 0 in the cone C',*

$$\lim_{\rho\to 0} \rho^{\alpha+n}\tilde{f}(i\rho y) = h(iy), \qquad y \in C'; \tag{2.5}$$

b) *there exist numbers M, q, and $\beta \in [0,1)$ and a vector $e \in C$ such that*

$$\rho^{\alpha+n}\left|\tilde{f}(\rho x + i\rho\lambda e)\right| \leq M\lambda^{-q}, \qquad 0<\rho\leq 1, \quad 0<\lambda\leq 1, \quad |x|\leq\lambda^{\beta}. \tag{2.6}$$

In this case the equalities

$$h(z) = L[g] = \tilde{g}(z), \qquad z \in T^C, \tag{2.7}$$

$$\mathcal{K}_C^N(z)h(z) = \Gamma(\alpha+n+nN)\int\limits_{\mathrm{pr}\,\Gamma} \frac{g^{(-N)}(\sigma)\,d\sigma}{(-iz,\sigma)^{\alpha+n+nN}}, \qquad z\in T^C, \tag{2.8}$$

hold for all sufficiently large N.

PROOF. NECESSITY. If $f(\xi)$ has the quasiasymptotic of order α at ∞, then conditions (i) and (ii) (see Definition 3) follow, as before, from the relations

$$\rho^{\alpha+n}\tilde{f}(\rho z) = \rho^{\alpha+n}\left(f(\xi), \eta(\xi)e^{i(\xi,\rho z)}\right) = \left(\rho^{\alpha}f(\xi'/\rho), \eta(\xi')e^{i(\xi',z)}\right)$$
$$\to \left(g(\xi'), \eta(\xi')e^{i(\xi',z)}\right) = \tilde{g}(z), \qquad \rho \to 0, \quad z \in T^C, \tag{2.9}$$
$$\rho^{\alpha+n}\left|\tilde{f}(\rho z)\right| \le \sup_{0<\rho\le 1} \|\rho^{\alpha}f(\xi/\rho)\|_{-m}\left\|\eta(\xi)e^{i(\xi,z)}\right\|_m$$
$$\le M\frac{1+|z|^a}{\Delta_C^b(y)}, \qquad 0<\rho\le 1, \quad z \in T^C, \tag{2.10}$$

where m is the order of the set of generalized functions from S'

$$\{\rho^{\alpha}f(\xi/\rho),\ 0<\rho\le 1\} = \{k^{-\alpha}f(k\xi),\ k \ge 1\}. \tag{2.11}$$

SUFFICIENCY. Let the conditions a) and b) hold. First, we prove that the set of generalized functions (2.11) is bounded in some space S'_m. By virtue of the corollary of Lemma 3, it is sufficient to verify this fact on the test functions φ from $\mathcal{D}$. Fix an arbitrary $e \in C$. Then for all $k \ge 1$ we have

$$(k^{-\alpha}f(k\xi), \varphi) = \left(k^{-\alpha-n}\tilde{f}(x/k), \tilde{\varphi}(x)\right)$$
$$= k^{-\alpha-n}\int \tilde{f}(x/k + ie/k)\tilde{\varphi}(x+i\epsilon)\,dx. \tag{2.12}$$

(When shifting the contour of integration, we use estimates (3.4) of Sec. 12.3 and (2.2).)

Let us prove the estimate

$$\left|k^{-\alpha-n}\tilde{f}(x/k+ie/k)\right| \le K(1+|x|^s), \qquad k\ge 1, \quad x \in \mathbb{R}^n, \tag{2.13}$$

where K and s do not depend on k.

For $|x| \le 1$ estimate (2.13) follows immediately from estimate (2.6) for $\lambda = 1$. Under the condition

$$1 < |x|^{1/(1-\beta)} \le k$$

estimate (2.13) also follows from (2.6). Indeed, denoting

$$\rho = k^{-1}|x|^{1/(1-\beta)}, \qquad \lambda = |x|^{-1/(1-\beta)},$$

we have

$$0<\rho\le 1, \qquad 0<\lambda<1, \qquad |x|\,|x|^{-1/(1-\beta)} = |x|^{-\beta/(1-\beta)} = \lambda^{\beta},$$

and estimate (2.6) implies estimate (2.13):

$$k^{-\alpha-n}\left|\tilde{f}\left(x/k+ie/k\right)\right| = |x|^{-\frac{\alpha+n}{1-\beta}}\rho^{\alpha+n}\left|\tilde{f}\left(\rho x|x|^{-1/(1-\beta)} + i\rho|x|^{-1/(1-\beta)}e\right)\right|$$
$$\le M|x|^{-\frac{\alpha+n}{1-\beta}}|x|^{\frac{a}{1-\beta}}$$
$$= M|x|^{\frac{a-\alpha-n}{1-\beta}}.$$

Finally, for $k < |x|^{1/(1-\beta)}$ inequality (2.13) follows from estimate (2.2)

$$k^{-\alpha-n}\left|\tilde{f}\left(x/k+i\rho/k\right)\right| \le Mk^{-\alpha-n-bn}\left(1+\frac{\sqrt{1+|x|^2}}{k}\right)^c\Delta_C^{-d}(e) \le K\left(1+|x|^s\right).$$

It follows immediately from estimates (2.13), (2.10) and from (2.11) that the set of the numbers

$$\{k^{-\alpha}(f(k\xi),\varphi),\ k\geq 1\}$$

is bounded for any $\varphi\in\mathcal{D}$; hence, the sequence (2.11) is bounded in some $\mathcal{S}'_m$. Condition a) implies that sequence (2.11) converges as $k\to\infty$ on the functions

$$\left\{\eta(\xi)e^{-(y,\xi)},y\in C'\right\} \tag{2.14}$$

to the function $h(iy)$, by virtue of (2.9)

$$\rho^{\alpha+n}\tilde{f}(i\rho y)=\left(k^{-\alpha}f(k\xi),\eta(\xi)e^{-(y,\xi)}\right),\qquad \rho\to 0,\quad y\in C'. \tag{2.15}$$

Prove now that the linear hull of functions (2.14) is dense in the set of functions $\{\psi=\eta\varphi,\ \varphi\in\mathcal{S}\}$. Indeed, if $g_1\in\mathcal{S}'$ vanishes on functions (2.14), then

$$\tilde{g}_1(iy)=\left(g_1(\xi),\eta(\xi)e^{-(y,\xi)}\right)=0,\qquad y\in C'.$$

By the uniqueness theorem for holomorphic functions, we deduce that $\tilde{g}_1(z)=0$, $z\in T^C$; hence, $g_1=0$. This fact and the Hahn–Banach theorem imply that the sequence of functionals

$$\{\eta(\xi)k^{-\alpha}f(k\xi)=k^{-\alpha}f(k\xi),\ 1\leq k<\infty\}$$

is bounded in $\mathcal{S}'_m$ and converges in $\mathcal{S}'$ to a generalized function, say, to $g\in\mathcal{S}'(\Gamma)$. From (2.15) we deduce that

$$h(iy)=\left(g(\xi),\eta(\xi)e^{-(y,\xi)}\right)=\tilde{g}(iy).$$

Thus, $f\in\mathcal{S}'(\Gamma)$ has the quasiasymptotic g of order α at ∞ and equalities (2.7) hold.

Prove equality (2.8). Since g is a homogeneous generalized function of degree α, we have that $\tilde{g}(x)$ is a homogeneous function of degree $-\alpha-n$. By the theorem of Sec. 14.1, for all sufficiently large N, we make sure of the validity of equality (2.8):

$$\begin{aligned}
\mathcal{K}_C^N(z)\tilde{g}(z)&=\left(g^{(-N)}(\xi),\eta(\xi)e^{i(z,\xi)}\right)\\
&=\int_\Gamma|\xi|^{\alpha+nN}g^{(-N)}\left(\frac{\xi}{|\xi|}\right)e^{i(z,\xi)}\,d\xi\\
&=\int_{\mathrm{pr}\,\Gamma}g^{(-N)}(\sigma)\int_0^\infty\rho^{\alpha+nN+n-1}e^{i\rho(z,\sigma)}\,d\rho\,d\sigma\\
&=\Gamma(\alpha+nN+n)\int_{\mathrm{pr}\,\Gamma}\frac{g^{(-N)}(\sigma)\,d\sigma}{(-iz,\sigma)^{\alpha+n+nN}}.
\end{aligned}$$

The theorem is proved. □

14.3. One-dimensional Tauberian theorems. For $n = 1$, we have $\Gamma = [0, \infty)$, $C = (0, \infty)$, $T^C = T^1$, $\mathcal{K}_C(z) = iz^{-1}$. The General Tauberian theorem of Sec. 14.2 is simplified and takes a more specific form.

First, we prove a lemma which gives a description of homogeneous generalized functions from $\mathcal{S}'_+$.

LEMMA. *If $g \in \mathcal{S}'_+$ is a homogeneous generalized function of degree α, then*

$$g(\xi) = Cf_{\alpha+1}(\xi), \tag{3.1}$$

where C is some constant and f_α is the kernel of the Riemann–Liouville operator (see 4.9.5).

PROOF. Let $\alpha = 0$. Reasoning as in the proof of the theorem in Sec. 5.7, we deduce that $g(\xi) = C\theta(\xi)$ for some constant C, where θ is the Heaviside function. The case $\alpha \neq 0$ can be reduced to the considered one: the generalized function $g^{(\alpha)} = g * f_{-\alpha} \in \mathcal{S}'_+$ of the homogeneity degree 0 and therefore $g^{(\alpha)} = C\theta = Cf_1$. This implies representation (3.1):

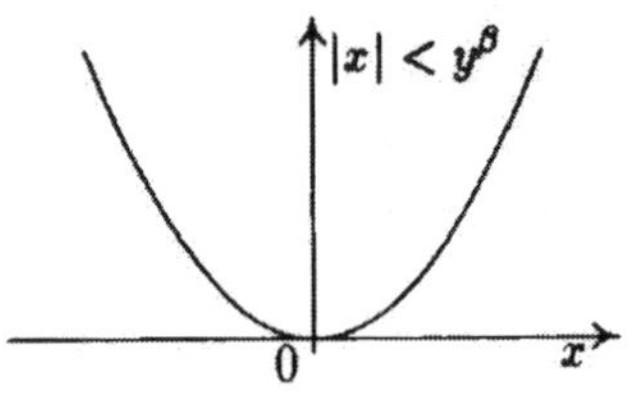

Figure 29

$$g = g * \delta = g * (f_{-\alpha} * f_\alpha) = (g * f_{-\alpha}) * f_\alpha$$
$$= g^{(\alpha)} * f_\alpha = Cf_1 * f_\alpha = Cf_{\alpha+1}$$

(see (9.10) of Sec. 4.9.5). The lemma is proved. □

REMARK. For $\alpha > -1$ we have

$$f_{\alpha+1}(\xi) = \frac{\theta(\xi)\xi^\alpha}{\Gamma(\alpha+1)};$$

for $\alpha = -1$ we have $f_0(\xi) = \delta(\xi)$.

THEOREM. *For $f \in \mathcal{S}'_+$ to have a quasiasymptotic $Cf_{\alpha+1}$ of order α at ∞ the following conditions are necessary:*

(1) $\lim_{\rho\to+0} \rho^{\alpha+1}\tilde{f}(\rho z) = C(iz)^{-\alpha-1}$, $z \in T^1$;

(2) $\rho^{\alpha+1}\left|\tilde{f}(\rho z)\right| \leq M\frac{(1+|z|)^a}{y^b}$, $0 < \rho \leq 1$, $z \in T^1$,

and the following conditions are sufficient:

a) $\lim_{y\to+0} y^{1+\alpha}\tilde{f}(iy) = C$;

b) *there exist numbers M, q, r_0, and $\beta \in [0,1)$ such that*

$$r^{\alpha+1}\left|\tilde{f}(re^{i\varphi})\right| \leq M\sin^{-q}\varphi, \qquad 0 < r \leq r_0, \quad |x| \leq y^\beta.$$

In this case, for all sufficiently large q the function $f^{(-q)}(\xi)$ is continuous with respect to $\xi > 0$ and has the asymptotic

$$\lim_{\xi\to+\infty} \frac{f^{(-q)}(\xi)}{\xi^{q+\alpha+1}} = \frac{C}{\Gamma(\alpha+1+q)}. \tag{3.2}$$

(Fig. 29 depicts the domain $|x| < y^\beta$ in the half-plane $y > 0$.)

The proof of the theorem follows from the General Tauberian theorem of Sec. 14.2. □

14.4. Tauberian and Abelian theorems for nonnegative measures. In this case the General Tauberian theorem of Sec. 14.2 is simplified, namely, condition b) can be omitted (and then Condition (2) should be automatically fulfilled).

Let $\mu(d\xi)$ be a nonnegative measure with the support in the cone Γ (see Sec. 1.7). Its primitive $\mu^{(-1)}(\xi) = \mu * \theta_\Gamma$ can be almost everywhere in $\mathbb{R}^n$ represented by the integral

$$\mu^{(-1)}(\xi) = \int_{\Delta(\xi)} \mu(d\xi'), \tag{4.1}$$

where $\Delta(\xi) = \Gamma \cap (\xi - \Gamma)$ (see Fig. 30).

Its Laplace transform $\tilde{\mu}(z)$ can be expressed by the integral

$$\tilde{\mu}(z) = \int_\Gamma e^{i(z,\xi)}\mu(d\xi), \qquad z \in T^C. \tag{4.2}$$

LEMMA. *If $\mu \in \mathcal{D}'(\Gamma)$ is a nonnegative homogeneous measure, then its primitive $\mu^{(-1)}(\xi)$ is continuous in* $\operatorname{int}\Gamma$.

PROOF. Let $\xi_n \to \xi$, $n \to \infty$, $\xi \in \operatorname{int}\Gamma$. Then for any $\varepsilon > 0$ there exists a $\delta > 0$ such that

$$0 < \mu^{(-1)}((1+\delta)\xi) - \mu^{(-1)}((1-\delta)\xi) < \delta, \tag{4.3}$$

$$\mu^{(-1)}((1-\delta)\xi) \le \mu^{(-1)}(\xi) \le \mu^{(-1)}((1+\delta)\xi), \tag{4.4}$$

by virtue of homogeneity and monotonicity of the function $\mu^{(-1)}(\xi)$ with respect to Γ.

Then, starting from some number N, for $n > N$ the inclusions $\xi_n \in \operatorname{int}\Gamma$ are valid and the inequalities

$$\Delta((1-\delta)\xi) \le \Delta(\xi_n) < \Delta((1+\delta)\xi)$$

hold, which imply the inequalities

$$\mu^{(-1)}((1-\delta)\xi) \le \mu^{(-1)}(\xi_n) \le \mu^{(-1)}((1+\delta)\xi).$$

Comparing these inequalities with (4.3) and (4.4), we obtain

$$\left|\mu^{(-1)}(\xi_n) - \mu^{(-1)}(\xi)\right| < \delta, \qquad n > N,$$

which is what we set out to prove. □

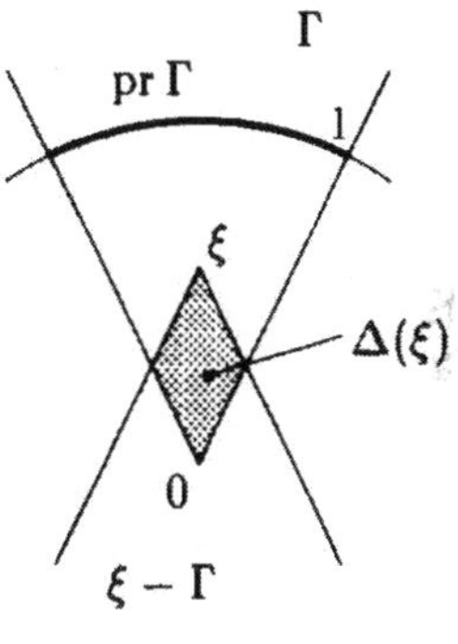

Figure 30

The General Tauberian theorem of Sec. 14.2 immediately implies the following theorem.

THEOREM. *For a nonnegative measure $\mu(d\xi)$ from $\mathcal{S}'(\Gamma)$ to have a quasiasymptotic g of order α at ∞ it is necessary that the following conditions hold*

(1) $\lim\limits_{\rho\to+0} \rho^{\alpha+n}\tilde{\mu}(\rho z) = h(z)$, $z \in T^C$,

(2) $\rho^{\alpha+n}\,|\tilde{\mu}(\rho z)| \le M\dfrac{1+|z|^a}{\Delta_C^b(y)}$, $0 < \rho \le 1$, $z \in T^C$,

and it is sufficient that there exists a solid subcone $C' \subset C$ such that

(a)

$$\lim_{\rho \to +0} \rho^{\alpha+n} \tilde{\mu}(i\rho y) = h(iy), \qquad y \in C'. \tag{4.5}$$

In this case $\tilde{g}(z) = h(z)$ and $g(\xi)$ is a nonnegative homogeneous measure of degree α with support in Γ which satisfies relations (2.8) for $N = 1, 2, \ldots$.

In order to prove the theorem it is sufficient to note that condition b) (see (2.6)) of the General Tauberian theorem always holds for the functions $\tilde{\mu}(z)$

$$\begin{aligned}
\rho^{\alpha+n} |\tilde{\mu}(\rho x + i\rho\lambda e)| &\le \rho^{\alpha+n} \int_{\Gamma} e^{-\rho\lambda(e,\xi)} \mu(d\xi) \\
&= \rho^{\alpha+n} h(i\rho\lambda e) \\
&= \lambda^{-\alpha-n} (\rho\lambda)^{\alpha+n} h(i\rho\lambda e) \\
&\le \lambda^{-\alpha-n} \sup_{0<\rho\le 1} \rho^{\alpha+n} h(i\rho e) \\
&= M\lambda^{-\alpha-n}, \qquad 0 < \rho \le 1, \qquad 0 < \lambda \le 1, \quad x \in \mathbb{R}^n.
\end{aligned}$$

□

REMARK 1. The primitive $\mu^{(-1)}$ of the measure μ has the asymptotic $g^{(-1)}$ of order $\alpha + n$ in the cone Γ. One can prove that

$$|\xi|^{-\alpha-n} \mu^{(-1)}(\xi) \overset{\frac{\xi}{|\xi|} \in \operatorname{pr} \Gamma'}{\Longrightarrow} g^{(-1)}\left(\frac{\xi}{|\xi|}\right), \qquad |\xi| \to \infty, \tag{4.6}$$

for any $\Gamma' \Subset \Gamma$.

REMARK 2. For $n = 1$, the theorem contains the classical Tauberian theorem and the Abelian theorems of Hardy and Littlewood. In this case,

$$h(iy) = h(i) y^{-\alpha-1}, \qquad g(\xi) = C f_{\alpha+1}(\xi)$$

and relations (2.8) for $N = 1$ takes the form

$$C = h(i). \tag{4.7}$$

14.5. Tauberian theorems for holomorphic functions of bounded argument. Consider the second case when conditions 2) and b) in General Tauberian theorem of Sec. 14.2 are skipped (they are fulfilled automatically). This case concerns the functions $\tilde{f}(z)$ of a bounded argument in T^C, i.e.,

$$\tilde{f}(z) \ne 0, \qquad \left|\arg \tilde{f}(z)\right| < M, \qquad z \in T^C. \tag{5.1}$$

This case includes, in particular, the functions with the positive imaginary (or real) part, i.e., the functions of the class $H_+(T^C)$ (see Sec. 13.3 and Secs. 18, 19).

THEOREM. *Let $f \in \mathcal{S}'(\Gamma)$ and $\tilde{f}(z)$ have a bounded argument in T^C. For f to have a quasiasymptotic g of order α at ∞, it is necessary and sufficient that there exists a solid subcone $C' \subset C$ such that*

$$\rho^{\alpha+n} \tilde{f}(i\rho y) \to h(iy), \qquad \rho \to +0, \qquad y \in C'. \tag{5.2}$$

In this case for any $z \in T^C$ there exists the limit

2) $\lim_{\rho \to +0} \rho^{\alpha+n} \tilde{f}(\rho z) = h(z)$

and the following estimate holds: there exists a number K independent of ρ, λ and x such that for any vector $e \in \operatorname{pr} C$ the inequality holds

b) $\rho^{\alpha+n}\left|\tilde{f}(\rho x + i\rho\lambda e)\right| \leq K\lambda^{-\frac{Mn}{\pi}},\ 0 < \rho \leq 1,\ 0 < \lambda \leq 1,\ |x| \leq 1.$

PROOF. By virtue of the General Tauberian theorem of Sec.14.2, it is sufficient to prove estimate (2.4) (more exactly, estimate b)) under condition (5.2). If the function $\tilde{f}(z)$ has a bounded argument in T^C, then, by virtue of (5.1), the function

$$f_+(z) = e^{i\frac{\pi}{2}}\tilde{f}^{\frac{\pi}{M}}(z) = \left|\tilde{f}(z)\right|^{\frac{\pi}{M}} \exp\left(i\frac{\pi}{M}\arg\tilde{f}(z) + i\frac{\pi}{2}\right) \tag{5.3}$$

has the nonnegative imaginary part and the representation

$$\tilde{f}(z) = e^{-i\frac{M}{2}} f_+^{\frac{M}{\pi}}(z), \qquad f_+ \in H_+(T^C), \tag{5.4}$$

holds and, by virtue of (5.2),

$$\lim_{\rho\to+0} \rho^{\pi(\alpha+n)/M} f_+(i\rho y) = \lim_{\rho\to+0} e^{i\pi/2}\left[\rho^{\alpha+n}\tilde{f}(i\rho y)\right]^{\pi/M}$$
$$= e^{i\pi/2}h^{\pi/M}(iy), \qquad y \in C'. \tag{5.5}$$

Applying now inequality (3.1) of Sec. 17.3 to the function $f_+(z)$ for $z = \rho x + i\rho\lambda e$, $z^0 = i\rho e$, $0 < \rho \leq 1$, $|x| \leq 1$, $e \in \operatorname{pr} C$, we obtain

$$|\mathcal{K}_C[\rho x + (\lambda+1)i\rho e]|^2 \left|f_+(\rho x + i\rho\lambda e) - \bar{f}_+(i\rho e)\right|^2$$
$$\leq 4\mathcal{K}_C(2i\rho\lambda e)\mathcal{K}_C(2i\rho e)\Im f_+(\rho x + i\rho\lambda e)\Im f_+(i\rho e).$$

Dividing this inequality by ρ^{-2n}, we rewrite it in the form

$$|\mathcal{K}_C[x + (\lambda+1)ie]|^2 \left|f_+(\rho x + i\rho\lambda e) - \bar{f}_+(i\rho e)\right|^2$$
$$\leq 2^{2-2n}\lambda^{-n}\mathcal{K}_C^2(ie)\Im f_+(\rho x + i\rho\lambda e)\Im f_+(i\rho e). \tag{5.6}$$

By assumption, the cone Γ is regular; hence, $|\mathcal{K}_C(z)| > 0$, $z \in T^C$. Therefore, the continuous positive function $|\mathcal{K}_C[x + i(\lambda+1)e]|$ is bounded from below on the compact $(\lambda, x) = (0 \leq \lambda \leq 1,\ |x| \leq 1)$ by some number $\sigma > 0$; hence,

$$|\mathcal{K}_C[x + i(\lambda+1)e]| \geq \sigma, \qquad 0 < \lambda \leq 1, \qquad |x| \leq 1.$$

This fact and inequality (5.6) imply the inequality

$$\left|f_+(\rho x + i\rho\lambda e)\right|^2 \leq 2\sigma^{-2}\left|f_+(\rho x + i\rho\lambda e) - \bar{f}_+(i\rho e)\right|^2 + 2\sigma^{-2}\left|f_+(i\rho e)\right|^2$$
$$\leq 2\sigma^{-2}A\lambda^{-n}\left|f_+(\rho x + i\rho\lambda e)\right|\ \left|f_+(i\rho e)\right| + 2\sigma^{-2}\left|f_+(i\rho e)\right|^2, \tag{5.7}$$

where we use the notation

$$A = 2^{2-2n}\mathcal{K}_C^2(ie).$$

Multiplying inequality (5.7) by $\rho^{\frac{2\pi}{M}(\alpha+n)}$ and denoting

$$B = \sup_{0<\rho\leq 1} \rho^{\frac{\pi}{M}(\alpha+n)}\left|f_+(i\rho e)\right|$$

(by virtue of (5.5), $B < \infty$), we rewrite it in the form

$$\rho^{\frac{2\pi}{M}(\alpha+n)}\left|f_+(\rho x + i\rho\lambda e)\right|^2 \leq 2\sigma^{-2}AB\lambda^{-n}\rho^{\frac{\pi}{m}(\alpha+n)}\left|f_+(\rho x + i\rho\lambda e)\right| + 2B^2 e^{-2},$$

or, denoting $C = 2B\sigma^{-2}(A+1)$, we obtain

$$\rho^{\frac{\pi}{M}(\alpha+n)}\left|f_+(\rho x + i\rho\lambda e)\right| \leq C\lambda^{-n}, \qquad 0 < \rho \leq 1, \qquad 0 < \lambda \leq 1, \qquad |x| \leq 1.$$

This and (5.4) imply estimate b):

$$\rho^{\alpha+n}\left|\tilde{f}(\rho x + i\lambda\rho e)\right| = \left[\rho^{\frac{r}{M}(\alpha+n)}\left|f_+(\rho x + i\rho\lambda e)\right|\right]^{\frac{M}{r}} \le C^{\frac{M}{r}}\lambda^{-\frac{Mn}{r}},$$

which proves the theorem. □

The Tauberian theory presented in this section remains valid if we replace the scale (automodelling) function k^α of order α by a *regularly varying* function $\rho(k)$. A continuous positive function $\rho(k)$, $k \in (0, \infty)$ is called *regularly varying*, if for any $k > 0$ there exists the limit

$$\lim_{t\to\infty}\frac{\rho(tk)}{\rho(t)} = C(k),$$

and the convergence is uniform with respect to k on any compact of the semi-axis $(0, \infty)$. One can easily see that $C(k)C(k_1) = C(kk_1)$; hence, $C(k) = k^\alpha$ for some real α. The number α is called the order of automodellity of the regularly varying function $\rho(k)$.

Let us give some examples of the regularly varying functions of order α:

$$k^\alpha, \quad k^\alpha \ln^\beta(1+k), \quad k^\alpha \ln\ln(k+e), \quad k^\alpha(2+\sin\sqrt{k}).$$

Concerning the regularly varying functions see Seneta [92].

At present, the Tauberian theory of generalized functions is developing in other directions. The quasiasymptotic on the orbits of one-parametric groups of transformations preserving a cone and relating theorems of the Keldysh type and the comparison theorems are being investigated. Essential progress is achieved concerning the extension of theorems of the Wiener type on the generalized functions. All these results can be found in the book by Vladimirov *at al* [122] and in recent papers by Drozhzhinov and Zavialov [26, 27, 28].

CHAPTER 3

SOME APPLICATIONS IN MATHEMATICAL PHYSICS

15. Differential Operators with Constant Coefficients

The theory of generalized functions has exerted a strong influence on the development of the theory of linear differential equations. First to be mentioned here are the fundamental works of L. Gårding, L. Hörmander, B. Malgrange, I.M. Gel'fand, L. Ehrenpreis of the 1950s devoted to the general theory of linear partial differential equations irrespective of their type. The results of these studies are summarized in the Analysis of Linear Partial Differential Operators in four volumes by Hörmander [**51**] (1985). Big advances have been made in the theory of the so-called pseudodifferential operators [a generalization of differential and integral (singular) operators][1].

15.1. Fundamental solutions in $\mathcal{D}'$. One of the basic and most profound results is the proof of the existence of a fundamental solution $\mathcal{E}(x)$ in $\mathcal{D}'$ of any linear differential operator $P(\partial) \not\equiv 0$ with constant coefficients (see Sec. 4.9.3), that is,

$$P(\partial)\mathcal{E}(x) = \delta(x) \tag{1.1}$$

where

$$P(\partial) = \sum_{|\alpha| \le m} a_\alpha \partial^\alpha, \qquad \sum_{|\alpha|=m} |a_\alpha| \ne 0, \tag{1.2}$$

is a differential operator of the mth order.

This result was first obtained independently by L. Ehrenpreis [**31**] (1954) and B. Malgrange [**73**] (1953).

Before proceeding to the proof of the existence of a fundamental solution, we will first prove two lemmas on polynomials.

LEMMA 1. *If*

$$P(\xi) = \sum_{|\alpha| \le m} a_\alpha \xi^\alpha, \qquad \sum_{|\alpha|=m} |a_\alpha| \ne 0,$$

is an arbitrary polynomial of degree $m \ge 1$, then there exists a nonsingular linear real transformation of coordinates

$$\xi = C\xi', \qquad \det C \ne 0, \qquad C = (c_{kl}),$$

that transforms the polynomial P to the form

$$\tilde{P}(\xi') = a{\xi_1'}^m + \sum_{0 \le k \le m-1} \tilde{P}_k(\xi_2', \ldots, \xi_n'){\xi_1'}^k, \qquad a \ne 0.$$

[1]See Hörmander [**48**]-[**50**], Kohn and Nirenberg [**56, 57**] and Plamenevskii [**81**].

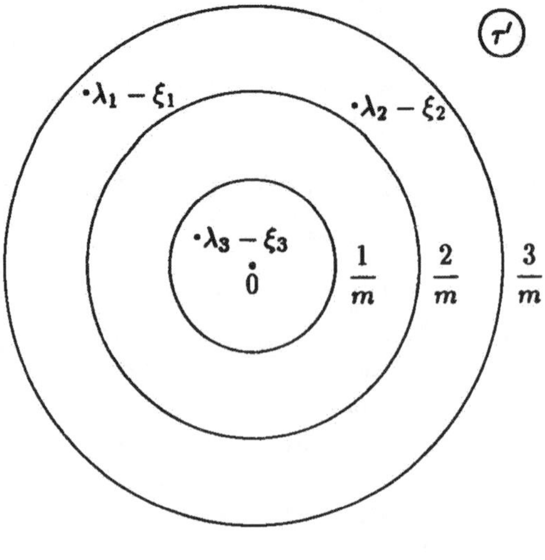

Figure 31

PROOF. The coefficient of $\xi_1'^m$ in the polynomial $\tilde{P}(\xi') = P(C\xi')$ is equal to

$$\sum_{|\alpha|=m} a_\alpha c_{11}^{\alpha_1} c_{21}^{\alpha_2} \ldots c_{n1}^{\alpha_n}. \tag{1.3}$$

Since $\sum_{|\alpha|=m} |a_\alpha| \neq 0$, we can choose n real numbers $c_{11}, c_{21}, \ldots, c_{n1}$ so that the expression (1.3) is not zero; we then have $\sum_{1\leq k\leq n} |c_{k1}| \neq 0$. The remaining numbers c_{kl} are chosen to be arbitrary real numbers so that $\det C \neq 0$. The proof of Lemma 1 is complete. □

LEMMA 2. *Suppose*

$$P(\xi) = a\xi_1^m + \sum_{0\leq k\leq m-1} P_k(\xi_2,\ldots,\xi_n)\xi_1^k, \qquad a \neq 0, \tag{1.4}$$

is a polynomial. Then there is a constant $\varkappa$, depending solely on m, such that for every point $\xi \in \mathbb{R}^n$ there is an integer k, $0 \leq k \leq m$, such that the following inequality holds:

$$\left|P\left(\xi_1 + i\tau\frac{k}{m}, \xi_2, \ldots, \xi_n\right)\right| \geq a\varkappa, \qquad |\tau| = 1. \tag{1.5}$$

PROOF. Fix $\xi \in \mathbb{R}^n$. Expand the polynomial $P(z, \xi_2, \ldots, \xi_n)$ into factors involving z:

$$P(z, \xi_2, \ldots, \xi_n) = a(z - \lambda_1)\ldots(z - \lambda_m)$$

so that $\lambda_j = \lambda_j(\xi_2, \ldots, \xi_n)$, $j = 1, 2, \ldots, m$, and

$$P\left(\xi_1 + i\tau\frac{k}{m}, \xi_2, \ldots, \xi_n\right) = a\left(\xi_1 - \lambda_1 + i\tau\frac{k}{m}\right)\ldots\left(\xi_1 - \lambda_m + i\tau\frac{k}{m}\right). \tag{1.6}$$

Using the "box" principle, we conclude that among the $m+1$ circles $|\tau'| = \frac{j}{m}$, $j = 0, 1, \ldots, m$, there is at least one, $|\tau| = \frac{k}{m}$, distant from m points $\lambda_1 - \xi_1, \ldots, \lambda_m - \xi_1$ at least by $\frac{1}{2m}$ (Fig. 31). From this and from (1.6) it follows the inequality (1.5) for $\varkappa = \left(\frac{1}{2m}\right)^m$, which completes the proof of Lemma 2. □

THEOREM (Malgrange–Ehrenpreis). *Every differential operator with constant coefficients $P(\partial) \not\equiv 0$ has a fundamental solution in $\mathcal{D}'$.*

PROOF. Since a nonsingular linear real transformation carries $\mathcal{D}'$ onto $\mathcal{D}'$ (see Sec. 1.10), then by virtue of Lemma 1 it suffices to prove the theorem for the case where the polynomial $P(i\xi)$ is of the form (1.4).

Suppose $f_0, f_1, \ldots, f_m$ are measurable nonnegative functions specified on $\mathbb{R}^n$ and such that $\sum_{0\le k\le m} f_k(\xi) \equiv 1$, $\xi \in \mathbb{R}^n$, and $f_k(\xi) = 0$ for those ξ for which

$$\min_{|\tau|=1} \left| P\left(i\xi_1 - \tau\frac{k}{m}, i\xi_2, \ldots, i\xi_n\right)\right| < a\varkappa \tag{1.7}$$

(by Lemma 2, such functions and an $\varkappa > 0$ exist).

We now determine the generalized function $\mathcal{E}$ by putting, for all $\varphi \in \mathcal{D}$,

$$(\mathcal{E}, \varphi) = \frac{1}{(2\pi)^n} \sum_{0\le k\le m} \int f_k(\xi) \frac{1}{2\pi i} \int_{|\tau|=1} \frac{L[\varphi]\left(\xi_1 + i\tau\frac{k}{m}, \xi_2, \ldots, \xi_n\right)}{P\left(i\xi_1 - \tau\frac{k}{m}, i\xi_2, \ldots, i\xi_n\right)} \frac{d\tau}{\tau}\, d\xi, \tag{1.8}$$

where $L[\varphi]$ is the Laplace transform of the function φ (see Sec. 9.1). We will prove that the expression on the right of (1.8) exists and determines a linear and continuous functional on $\mathcal{D}$, that is, $\mathcal{E} \in \mathcal{D}'$. But this assertion follows from the following estimates:

$$\begin{aligned} |(\mathcal{E}, \varphi)| &\le \frac{1}{(2\pi)^n} \sum_{0\le k\le m} \int f_k(\xi) \frac{\max_{|\tau|=1} \left|L[\varphi]\left(\xi_1 + i\tau\frac{k}{m}, \xi_2, \ldots, \xi_n\right)\right|}{\min_{|\tau|=1}\left|P\left(i\xi_1 - \tau\frac{k}{m}, i\xi_2, \ldots, i\xi_n\right)\right|}\, d\xi \\ &\le \frac{1}{(2\pi)^n a\varkappa} \sum_{0\le k\le m} \max e^{|\Re\tau|\frac{k}{m}} \int \frac{d\xi}{\left(1 + \left|\xi_1 + \frac{i\tau k}{m}\right|^2 + \xi_2^2 + \cdots + \xi_n^2\right)^N} \\ &\quad \times \int_{|x|<R} \left|(1-\Delta)^N \varphi(x)\right| dx. \end{aligned}$$

That is to say, from the estimate

$$|(\mathcal{E}, \varphi)| \le K_n \int_{|x|<R} \left|(1-\Delta)^N \varphi(x)\right| dx \tag{1.9}$$

which holds for all integers $N > n/2$ and for all $\varphi \in \mathcal{D}(U_R)$. In deriving the estimate (1.9), we made use of the estimates (3.4) to (3.5) of Sec. 12.3 for the entire functions $L[\varphi](\zeta)$ and also the estimate (1.7) and the properties of the functions $\{f_k\}$.

It remains to verify that the constructed generalized function $\mathcal{E}$ in $\mathcal{D}'$ satisfies the equation (1.1). Using (1.8), for all $\varphi \in \mathcal{D}$ we have

$$
\begin{aligned}
(P(\partial)\mathcal{E}, \varphi) &= (\mathcal{E}, P(-\partial)\varphi) \\
&= \frac{1}{(2\pi)^n} \sum_{0 \le k \le m} \int f_k(\xi) \frac{1}{2\pi i} \int_{|\tau|=1} \frac{L[P(-\partial)\varphi]\left(\xi_1 + i\tau\frac{k}{m}, \xi_2, \dots, \xi_n\right)}{P\left(i\xi_1 - \tau\frac{k}{m}, i\xi_2, \dots, i\xi_n\right)} \frac{d\tau}{\tau}\, d\xi \\
&= \frac{1}{(2\pi)^n} \sum_{0 \le k \le m} \int f_k(\xi) \frac{1}{2\pi i} \int_{|\tau|=1} L[\varphi]\left(\xi_1 + i\tau\frac{k}{m}, \xi_2, \dots, \xi_n\right) \frac{d\tau}{\tau}\, d\xi \\
&= \frac{1}{(2\pi)^n} \sum_{0 \le k \le m} \int f_k(\xi) F[\varphi](\xi)\, d\xi \\
&= \frac{1}{(2\pi)^n} \int F[\varphi](\xi)\, d\xi = \varphi(0) = (\delta, \varphi),
\end{aligned}
$$

which is what was required, and the theorem is proved. □

Having the fundamental solution $\mathcal{E}$ of the operator $P(\partial)$, we can construct a solution u in $\mathcal{D}'$ of the equation

$$P(\partial)u = f, \qquad f \in \mathcal{D}' \tag{1.10}$$

in the form of a convolution

$$u = \mathcal{E} * f \tag{1.11}$$

for those f in $\mathcal{D}'$ for which this convolution exists in $\mathcal{D}'$ (see Sec. 4.9.3). Thus, by choosing various fundamental solutions, it is possible to obtain various classes of right-hand members for which the equation (1.10) is solvable in the form of the convolution (1.11).

The convolution $V = \mathcal{E} * f$ is called the *potential* of the operator $P(\partial)$ with the density f.

The solution of equation (1.10) *is unique in the class of the generalized functions from* $\mathcal{D}'$ *for which the convolution with* $\mathcal{E}$ *exists.*

Indeed, if $u \in \mathcal{D}'$ is a solution of the homogeneous equation (1.10), $P(\partial)u = 0$, then, using the properties of the convolution (see Sec. 4.2), we obtain

$$u = u * \delta = u * P(\partial)\mathcal{E} = P(\partial)u * \mathcal{E} = 0 * \mathcal{E} = 0,$$

what is required. □

15.2. Tempered fundamental solutions. In Subsection 15.1 we established the fact that every nonzero differential operator with constant coefficients has at least one fundamental solution in $\mathcal{D}'$. The question arises — one that is important in applications — of how to find a fundamental solution with the required properties of growth, of support, of smoothness, and so forth. A convenient tool in this respect is the method of Fourier transforms. However, the Fourier transform technique that was developed in Sec. 6 is applicable to tempered generalized functions. For this reason, in constructing a fundamental solution by the method of Fourier transforms we confine ourselves from the very start to the class $\mathcal{S}'$.

The equation (1.1) in the class $\mathcal{S}'$ is equivalent to the algebraic equation (see Sec. 6.3.2)

$$P(-i\xi)\tilde{\mathcal{E}}(\xi) = 1 \tag{2.1}$$

with respect to the Fourier transform $F[\mathcal{E}] = \tilde{\mathcal{E}}$. Thus, the problem of seeking a tempered fundamental solution turns out to be a special case of the more general problem of "dividing" a tempered generalized function by a polynomial, that is, of the problem of finding a solution u in S' of the equation

$$P(\xi)u = f \tag{2.2}$$

where $P \not\equiv 0$ is a polynomial and f is a specified generalized function in S'. The solvability of the problem of "division" was proved in 1958 independently by Hörmander [**47**] and Łojasiewicz [**69**].

The proof is based on the following lemma.

LEMMA (Hörmander). *The mapping*

$$\varphi \to P\varphi, \qquad \varphi \in S,$$

has a continuous inverse in S; in other words, for every integer $p \geq 0$ there are numbers $K_p > 0$ and an integer $p' = p'(p) \geq p$ such that the following inequality holds:

$$\|\varphi\|_p \leq K_p \|P\varphi\|_{p'}, \qquad \varphi \in C^{p'}(\mathbb{R}^n). \tag{2.3}$$

The existing proofs of this lemma are extremely complicated. We confine ourselves here to the proof of only the case $n = 1$.

First we will prove (2.3) for the case $P(\xi) = \xi$. Setting $\psi = \xi\varphi$, we have

$$\varphi = \frac{\psi}{\xi}, \qquad |\varphi(\xi)| \leq \begin{cases} \max_{|\xi|\leq 1} |\psi'(\xi)|, & |\xi| \leq 1, \\ |\psi(\xi)|, & |\xi| > 1; \end{cases}$$

$$\varphi' = \frac{\psi'}{\xi} - \frac{\psi}{\xi^2}, \qquad |\varphi'(\xi)| \leq \begin{cases} \frac{3}{2} \max_{|\xi|\leq 1} |\psi''(\xi)|, & |\xi| \leq 1, \\ |\psi'(\xi)| + |\psi(\xi)|, & |\xi| > 1; \end{cases}$$

and so forth. Consequently,

$$\begin{aligned} \|\varphi\|_p &= \sup_{\substack{\xi \\ |\alpha|\leq p}} (1 + |\xi|^2)^{p/2} |\varphi^{(\alpha)}(\xi)| \\ &\leq K_p \sup_{\substack{\xi \\ |\alpha|\leq p+1}} (1 + |\xi|^2)^{p/2} |\psi^{(\alpha)}(\xi)| \\ &\leq K_p \|\xi\varphi\|_{p+1}, \end{aligned}$$

which is what was required ($p' = p + 1$).

From the fact that the inequality (2.3) holds for $P = \xi$ follows that it also holds for $P = \xi - \xi_0$ and, hence, for all polynomials $P = a\prod_{1\leq k\leq m}(\xi - \xi_0)$:

$$\|\varphi\|_p \leq K_p^{(1)} \|(\xi - \xi_1)\varphi\|_{p+1} \leq K_p^{(2)} \|(\xi - \xi_1)(\xi - \xi_2)\varphi\|_{p+2} \leq$$
$$\cdots \leq K_p^{(m)} \|(\xi - \xi_1)(\xi - \xi_2)\dots(\xi - \xi_m)\varphi\|_{p+m} = K_p \|P\varphi\|_{p+m}.$$

□

Using the Hörmander lemma, we will prove that *the equation* (2.2) *is always solvable in S'*.

Indeed, consider the linear functional

$$P\varphi \to (f, \varphi)$$

defined on a linear subspace $[\psi: \psi = P\varphi,\ \varphi \in \mathcal{S}]$ of the space $\mathcal{S}$. By the Hörmander lemma, this functional is continuous; if $P\varphi_k \to 0$, $k \to \infty$ in $\mathcal{S}$, then $\varphi_k \to 0$, $k \to \infty$ in $\mathcal{S}$ and therefore $(f, \varphi_k) \to 0$, $k \to \infty$. By the Hahn–Banach theorem, there exists a (linear) continuous extension u, $\varphi \to (u, \varphi)$, of that functional on the whole of $\mathcal{S}$ so that $u \in \mathcal{S}'$ and $(u, P\varphi) = (f, \varphi)$. And this means that the functional u satisfies equation (2.2).

Passing to the Fourier transform, we obtain that the following theorem holds.

THEOREM (Hörmander–Lojasiewicz). *The equation*

$$P(\partial)u = f, \tag{2.4}$$

where $P(\partial) \not\equiv 0$, is solvable in $\mathcal{S}'$ for all $f \in \mathcal{S}'$.

COROLLARY. *Every nonzero linear differential operator with constant coefficients has a tempered fundamental solution.*

15.3. A descent method. Let us consider the linear differential equation with constant coefficients in the space $\mathbb{R}^{n+1}$ of the variables $(x, t) = (x_1, \ldots, x_n, t)$,

$$P(\partial, \partial_0)u = f(x) \times \delta(t), \qquad f \in \mathcal{D}'(\mathbb{R}^n), \tag{3.1}$$

where

$$\partial_0 = \frac{\partial}{\partial t}, \qquad P(\partial, \partial_0) = \sum_{1 \le q \le p} P_q(\partial)\partial_0^q + P_0(\partial),$$

and $P_q(\partial)$ are differential operators with respect to the variables x.

We will say that a generalized function $u(x, t)$ taken from $\mathcal{D}'(\mathbb{R}^{n+1})$ *admits of an extension to functions of the form* $\varphi(x)1(t)$, where $\varphi \in \mathcal{D}(\mathbb{R}^n)$, if no matter what the sequence of test functions $\eta_k(x, t)$, $k = 1, 2, \ldots$, in $\mathcal{D}(\mathbb{R}^{n+1})$, which sequence converges to 1 in $\mathbb{R}^{n+1}$ (see Sec. 4.1), there exists a limit

$$\lim_{k \to \infty} (u, \varphi(x)\eta_k(x, t)) = (u, \varphi(x)1(t)), \qquad \varphi \in \mathcal{D}(\mathbb{R}^n); \tag{3.2}$$

in fact that limit is independent of the sequence $\{\eta_k\}$.

We denote the functional (3.2) by u_0,

$$(u_0, \varphi) = (u, \varphi(x)1(t)) = \lim_{k \to \infty} (u, \varphi(x)\eta_k(x, t)), \qquad \varphi \in \mathcal{D}. \tag{3.3}$$

Clearly, for any k the functional $(u, \varphi(x)\eta_k(x, t))$ is linear and continuous on $\mathcal{D}$, that is, it belongs to $\mathcal{D}'$. Therefore, by the theorem on the completeness of the space $\mathcal{D}'$ (see Sec. 1.4) the functional u_0 as well belongs to $\mathcal{D}'$: $u_0 \in \mathcal{D}'$.

We will call the generalized function $u_0(x)$ the *generalized integral with respect to t* of the generalized function $u(x, t)$.

We now give a criterion of existence for a generalized integral with respect to t.

THEOREM I. *In order that for u in $\mathcal{D}'(\mathbb{R}^{n+1})$ there exist u_0 — a generalized integral with respect to t — it is necessary and sufficient that there exist a convolution $u * [\delta(x) \times 1(t)]$. Here, the following equation holds:*

$$u * [\delta(x) \times 1(t)] = u_0(x) \times 1(t). \tag{3.4}$$

We prove sufficiency. Suppose there exists a convolution $u * [\delta(x) \times 1(t)]$. Then there exists a generalized function u_0 in $\mathcal{D}'$ such that (3.4) holds (see Sec. 4.2.3 and Sec. 3.3).

We will prove that u_0 is a generalized integral with respect to t for $u(x,t)$. Suppose $\{\xi_k(x)\}$, $\{\eta_k(x,t)\}$, and $\{\chi_l(t)\}$ are sequences of test functions in $\mathcal{D}(\mathbb{R}^n)$, $\mathcal{D}(\mathbb{R}^{n+1})$, and $\mathcal{D}(\mathbb{R}^1)$, which sequences converge to 1 in $\mathbb{R}^n$, $\mathbb{R}^{n+1}$, and $\mathbb{R}^1$ respectively. Suppose $\varphi \in \mathcal{D}(\mathbb{R}^n)$. Then there is a number N such that $\xi_k\varphi = \varphi$ for all $k \geq N$. Furthermore, for every k there is a number i_k such that

$$\eta_k(x,t)\int \chi_{i_k}(t')\omega_\varepsilon(t+t')\,dt' = \eta_k(x,t) \tag{3.5}$$

where ω_ε is the "cap" (see Sec. 1.2). Indeed, if $(x,t) \in \operatorname{supp}\eta_k \subset [(x,t)\colon |t| < R_k]$, then, choosing the number i_k so that $\chi_{i_k}(t) = 1$ for $|t| \leq R_k + \varepsilon$, we obtain

$$\begin{aligned}\int \chi_{i_k}(t')\omega_\varepsilon(t+t')\,dt' &= \int\limits_{|t'|<\varepsilon} \omega_\varepsilon(t')\chi_{i_k}(t'-t)\,dt' \\ &= \int \omega_\varepsilon(t')\,dt' = 1.\end{aligned}$$

Now, making use of (3.4) and (3.5) and also of the definitions of a generalized integral with respect to t (3.3) and of a convolution (see Sec. 4.1), and also noting that the sequence

$$\xi_k(x)\xi_k(x')\eta_k(x,t)\chi_{i_k}(t'), \qquad k = 1,2,\ldots,$$

of test functions in $\mathcal{D}(\mathbb{R}^{2n+2})$ converges to 1 in $\mathbb{R}^{2n+2}$, we have, for all $\varphi \in \mathcal{D}(\mathbb{R}^n)$,

$$\begin{aligned}&\lim_{k\to\infty}(u, \varphi(x)\eta_k(x,t)) \\ &= \lim_{k\to\infty}\left(u(x,t), \xi_k(x)\varphi(x)\eta_k(x,t)\int \chi_{i_k}(t')\omega_\varepsilon(t+t')\,dt'\right) \\ &= \lim_{k\to\infty}(u(x,t)\times\delta(x')\times 1(t'), \xi_k(x)\xi_k(x')\eta_k(x,t)\chi_{i_k}(t')\varphi(x+x')\omega_\varepsilon(t+t')) \\ &= (u * [\delta(x)\times 1(t)], \varphi(x)\omega_\varepsilon(t)) \\ &= (u_0(x)\times 1(t), \varphi(x)\omega_\varepsilon(t)) \\ &= \left(u_0(x), \varphi(x)\int \omega_\varepsilon(t)\,dt\right) \\ &= (u_0, \varphi),\end{aligned}$$

which is what we set out to prove.

We now prove necessity. Suppose for u there exists u_0, which is a generalized integral with respect to t. Suppose $\xi_k(x,t;x',t')$, $k = 1,2,\ldots$, is a sequence of test functions in $\mathcal{D}(\mathbb{R}^{2n+2})$ that converges to 1 in $\mathbb{R}^{2n+2}$. Let $\varphi \in \mathcal{D}(\mathbb{R}^{n+1})$. Then for every compact $K \subset \mathbb{R}^{n+1}$ there is a number N such that for all $k \geq N$

$$\int \xi_k(x,t;0,t')\varphi(x,t+t')\,dt' = \int \varphi(x,t+t')\,dt' = \int \varphi(x,t')\,dt', \qquad (x,t) \in K.$$

Consequently, there exists a sequence $\eta_k(x,t)$ of functions taken from $\mathcal{D}(\mathbb{R}^{n+1})$, which sequence converges to 1 in $\mathbb{R}^{n+1}$ and is such that the sequence of functions

$$\begin{aligned}\chi_k(x,t) = &\int \xi_k(x,t;0,t')\varphi(x,t+t')\,dt' \\ &- \eta_k(x,t)\int \varphi(x,t')\,dt' + \eta_k(x,t), \qquad k = 1,2,\ldots,\end{aligned} \tag{3.6}$$

in $\mathcal{D}(\mathbb{R}^{n+1})$ converges to 1 in $\mathbb{R}^{n+1}$. Let $\varphi_0(x)$ be a function in $\mathcal{D}(\mathbb{R}^n)$ equal to 1 on $\operatorname{supp}\varphi(x,t)$ so that $\varphi = \varphi_0\varphi$. Then, using (3.6) and the definitions of a generalized integral in t and of a convolution, we have

$$\begin{aligned}
(u_0(x) \times 1(t), \varphi) &= \left(u_0(x), \int \varphi(x,t)\,dt\right) \\
&= \left(u_0(x), \varphi_0(x) \int \varphi(x,t)\,dt\right) \\
&= \lim_{k\to\infty} \left(u(x,t), \varphi_0(x)\eta_k(x,t) \int \varphi(x,t)\,dt\right) \\
&= -\lim_{k\to\infty} (u(x,t), \varphi_0(x)\chi_k(x,t)) + \lim_{k\to\infty} (u(x,t), \varphi_0(x)\eta_k(x,t)) \\
&\quad + \lim_{k\to\infty} \left(u(x,t), \varphi_0(x) \int \xi_k(x,t;0,t')\varphi(x,t+t')\,dt'\right) \\
&= -(u_0, \varphi_0) + (u_0, \varphi_0) + \lim_{k\to\infty} (u(x,t) \times [\delta(x') \times 1(t')], \\
&\quad \xi_k(x,t;x',t')\varphi_0(x+x')\varphi(x+x',t+t')) \\
&= (u * [\delta(x) \times 1(t)], \varphi_0\varphi) \\
&= (u * [\delta(x) \times 1(t)], \varphi),
\end{aligned}$$

which is what we set out to prove. □

COROLLARY 1. *Suppose the function $u(x,t)$ is measurable and $\int |u(x,t)|\,dt \in \mathcal{L}^1_{\text{loc}}$. Then its generalized integral in t exists in $\mathcal{L}^1_{\text{loc}}$ and can be represented by the classical integral*

$$u_0(x) = \int u(x,t)\,dt. \tag{3.7}$$

REMARK. The formula (3.7) shows that a generalized integral in t is an extension of the classical concept of an integral in t to generalized functions.

COROLLARY 2. *If $u = f(x) \times \delta(t)$, where $f \in \mathcal{D}'$, then $u_0 = f$.*

THEOREM II. *If the solution u in $\mathcal{D}'(\mathbb{R}^{n+1})$ of the equation* (3.1) *possesses u_0 (a generalized integral in t), then u_0 satisfies the equation*

$$P_0(\partial)u_0 = f(x). \tag{3.8}$$

PROOF. Let $\eta_k(x,t)$, $k = 1, 2, \ldots$, be a sequence of functions in $\mathcal{D}(\mathbb{R}^{n+1})$ that converges to 1 in $\mathbb{R}^{n+1}$. Then for $q = 1, 2, \ldots$, the sequence of functions $\eta_k + \partial_0^q \eta_k$, $k = 1, 2, \ldots$, also converges to 1 in $\mathbb{R}^{n+1}$ and, hence, for all φ in $\mathcal{D}(\mathbb{R}^n)$,

$$\begin{aligned}
&\lim_{k\to\infty} (u, \varphi(x)\partial_0^q \eta_k(x,t)) \\
&\quad = \lim_{k\to\infty} (u, \varphi(x)[\eta_k(x,t) + \partial_0^q \eta_k(x,t)]) - \lim_{k\to\infty} (u, \varphi(x)\eta_k(x,t)) \\
&\quad = (u_0, \varphi) - (u_0, \varphi) = 0.
\end{aligned}$$

Taking into account the resulting equation, we verify that u_0 satisfies (3.8):

$$\begin{aligned}(P_0(\partial)u_0,\varphi) &= (u_0, P_0(-\partial)\varphi)\\ &= \lim_{k\to\infty}(u, P_0(-\partial)\varphi(x)\eta_k(x,t))\\ &= \lim_{k\to\infty}\left(u, P_0(-\partial)\varphi(x)\eta_k(x,t) + \sum_{1\le q\le p}(-1)^q P_q(-\partial)\varphi(x)\partial_0^q\eta_k(x,t)\right)\\ &= \lim_{k\to\infty}(u, P(-\partial,-\partial_0)\varphi(x)\eta_k(x,t))\\ &= \lim_{k\to\infty}(P(\partial,\partial_0)u, \varphi(x)\eta_k(x,t))\\ &= \lim_{k\to\infty}(f(x)\times\delta(t), \varphi(x)\eta_k(x,t))\\ &= \lim_{k\to\infty}(f(x), \varphi(x)\eta_k(x,0)) = (f,\varphi).\end{aligned}$$

The theorem is proved. □

The foregoing method of obtaining a solution $u_0(x)$ of the equation (3.8) in n variables in terms of the solution $u(x,t)$ of equation (3.1) in $n+1$ variables is termed the *method of descent with respect to the variable t*.

The descent method is particularly convenient for the construction of fundamental solutions. Applying Theorem II for $f = \delta(x)$, we obtain the following corollary.

COROLLARY. *If a fundamental solution $\mathcal{E}(x,t)$ of the operator $P(\partial,\partial_0)$ possesses $\mathcal{E}_0(x)$ (a general integral in t), then $\mathcal{E}_0$ is a fundamental solution of the operator $P_0(\partial)$.*

The fundamental solution $\mathcal{E}_0$ satisfies the relation

$$\mathcal{E}_0(x)\times 1(t) = \mathcal{E} * \left[\delta(x)\times 1(t)\right]. \tag{3.9}$$

The physical meaning of (3.9) consists in the fact that $\mathcal{E}_0(x)$ is a perturbation (independent of t) of a source $\delta(x)\times 1(t)$ concentrated along the axis t (compare Sec. 4.9.3).

15.4. Examples.

15.4.1. Particular solutions of the equation $\xi u = 1$ are the generalized functions

$$\frac{1}{\xi+i0}, \qquad \frac{1}{\xi-i0}, \qquad \mathcal{P}\frac{1}{\xi}$$

which, by virtue of the Sochozki formulae (8.3) and (8.3′) of Sec. 1.8, differ by the expression const $\delta(\xi)$, which is a general solution of the homogeneous equation $\xi u = 0$ (see Sec. 2.6).

15.4.2. If a polynomial $P(\xi)$ does not have real zeros, then the function $\frac{1}{P(\xi)}$ belongs to θ_M and is the unique solution of the equation $P(\xi)u = 1$.

This assertion follows from the following lemma.

LEMMA. *If a polynomial $P(\xi)\neq 0$, $\xi\in\mathbb{R}^n$, there are constants $C>0$ and ν such that the following inequality holds true:*

$$|P(\xi)| \ge C(1+|\xi|^2)^{-\nu}, \qquad \xi\in\mathbb{R}^n. \tag{4.1}$$

Proof. It suffices to prove the estimate (4.1) for $|\xi| > 1$. To do this, perform an inversion transformation (Fig. 32):

$$\xi^* = \frac{\xi}{|\xi|^2}, \qquad \xi = \frac{\xi^*}{|\xi^*|^2}, \qquad |\xi|\,|\xi^*| = 1.$$

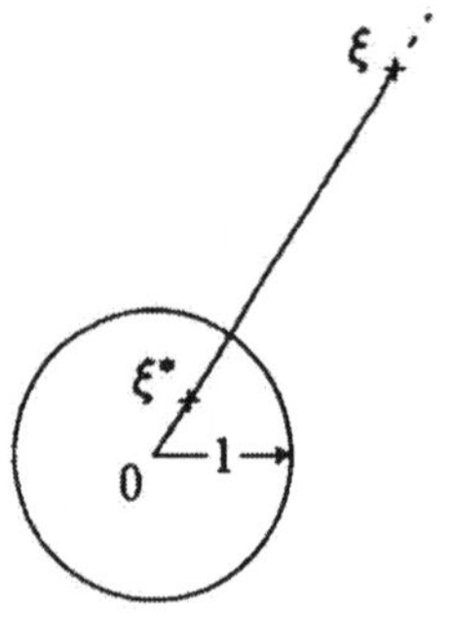

Figure 32

Suppose m is the degree of P. The polynomial

$$P^*(\xi^*) = |\xi^*|^{2m} P\left(\frac{\xi^*}{|\xi^*|^2}\right) \tag{4.2}$$

may have a unique zero in $\mathbb{R}^n$: $\xi^* = 0$. Therefore there exist numbers $C_1 > 0$ and $\mu \geq 0$ such that

$$|P^*(\xi^*)| \geq C_1 |\xi^*|^\mu, \qquad |\xi^*| < 1,$$

and so, by (4.2),

$$|P(\xi)| \geq C_1 |\xi^*|^{\mu - 2m} = C_1 |\xi|^{2m-\mu}, \qquad |\xi| > 1.$$

The proof of the lemma is complete. □

15.4.3. *The equation*

$$\xi_1 u(\xi) = f(\xi)$$

is solvable for any f in S' and its general solution is of the form

$$(u, \varphi) = (f, \psi) + \big(\delta(\xi_1) \times u_1(\xi_2, \ldots, \xi_n), \varphi\big), \qquad \varphi \in S, \tag{4.3}$$

where u_1 is an arbitrary generalized function in $S'(\mathbb{R}^{n-1})$,

$$\psi(\xi) = \frac{1}{\xi_1}\left[\varphi(\xi) - \eta(\xi_1)\varphi(0, \xi_2, \ldots, \xi_n)\right]$$

where $\eta(\xi_1)$ is an arbitrary function in $\mathcal{D}$ equal to 1 in the neighbourhood of 0.

The proof of this assertion is similar to that for the space $\mathcal{D}'$ (see Sec. 3.3). □

15.4.4. The function $\mathcal{E}(t) = \theta(t)Z(t)$, where $Z(t)$ is a solution of the homogeneous differential equation (compare Sec. 4.9.6)

$$P\left(\frac{d}{dt}\right)Z = Z^{(m)} + a_1 Z^{(m-1)} + \cdots + a_m Z = 0$$

that satisfies the conditions

$$Z(0) = Z'(0) = \cdots = Z^{(m-2)}(0) = 0, \qquad Z^{(m-1)}(0) = 1,$$

is a fundamental solution of the operator $P(\frac{d}{dt})$.

Indeed, using (3.1) of Sec. 2.3, we obtain

$$\mathcal{E}'(t) = \theta(t)Z'(t), \ldots, \mathcal{E}^{(m-1)}(t) = \theta(t)Z^{(m-1)}(t),$$

$$\mathcal{E}^{(m)}(t) = \delta(t) + \theta(t)Z^{(m)}(t),$$

whence

$$P\left(\frac{d}{dt}\right)\mathcal{E}(t) = \theta(t)P\left(\frac{d}{dt}\right)Z(t) + \delta(t) = \delta(t),$$

which completes the proof. □

In particular, the function

$$\mathcal{E}(t) = \theta(t)e^{-at} \tag{4.4}$$

is a fundamental solution of the operator $\frac{d}{dt} + a$.

15.4.5. *A fundamental solution of the heat conduction operator,*

$$\frac{\partial \mathcal{E}}{\partial t} - a^2\Delta\mathcal{E} = \delta(x,t). \tag{4.5}$$

Applying the Fourier transform F_x (see Sec. 6.2) to (4.5) and using the formulae (3.8) and (3.9) of Sec. 6.3,

$$F_x[\delta(x,t)] = F_x[\delta(x) \times \delta(t)] = F[\delta](\xi) \times \delta(t) = 1(\xi) \times \delta(t),$$

$$F_x\left[\frac{\partial \mathcal{E}}{\partial t}\right] = \frac{\partial}{\partial t}F_x[\mathcal{E}], \qquad F_x[\Delta\mathcal{E}] = -|\xi|^2 F_x[\mathcal{E}],$$

we obtain, for the generalized function $\bar{\mathcal{E}}(\xi,t) = F_x[\mathcal{E}]$, the equation

$$\frac{\partial \bar{\mathcal{E}}}{\partial t} + a^2|\xi|^2\bar{\mathcal{E}} = 1(\xi) \times \delta(t). \tag{4.6}$$

Taking into account (4.4) with $a^2|\xi|^2$ substituted for a, we conclude that the solution in $\mathcal{S}'$ of the equation (4.6) is the function

$$\bar{\mathcal{E}}(\xi,t) = \theta(t)e^{-a^2|\xi|^2 t}.$$

From this, using the inverse Fourier transform F_ξ^{-1} and (6.2) of Sec. 6.6, we obtain

$$F_\xi^{-1}[\bar{\mathcal{E}}] = \frac{\theta(t)}{(2\pi)^n}\int e^{-a^2|\xi|^2 t - i(\xi_0,x)}\,d\xi = \frac{\theta(t)}{(2a\sqrt{\pi t})^n}e^{-\frac{|x|^2}{4a^2 t}}.$$

That is,

$$\mathcal{E}(x,t) = \frac{\theta(t)}{(2a\sqrt{\pi t})^n}e^{-\frac{|x|^2}{4a^2 t}}. \tag{4.7}$$

□

15.4.6. *Fundamental solution of the wave operator* □. It is demonstrated in Sec. 13.5 that the (generalized) function

$$\mathcal{E}_n(x) = \frac{1}{2^n\pi^{\frac{n-2}{2}}\Gamma\left(\frac{n+1}{2}\right)}\Box^{\frac{n-1}{2}}\left[\theta(x_0)\theta(x^2)\right], \qquad x = (x_0,\mathbf{x}), \tag{4.8}$$

where $\theta(x_0)\theta(x^2) = \theta_{\overline{V}^+}(x)$ is the characteristic function of the closed future light cone $\overline{V}^+ = [x\colon x_0 \geq |\mathbf{x}|]$, (see Sec. 4.4) is a fundamental solution of the wave operator □. Putting $n = 1$ in (4.8), we have

$$\mathcal{E}_1(x) = \frac{1}{2}\theta(x_0)\theta(x^2). \tag{4.9}$$

We will prove, for $n \geq 2$, the equality

$$\Box[\theta(x_0)\theta(x^2)] = 2(n-1)\theta(x_0)\delta(x^2), \tag{4.10}$$

where the generalized function $\theta(x_0)\delta(x^2)$ is given by

$$(\theta(x_0)\delta(x^2),\varphi) = \frac{1}{2}\int_0^\infty \frac{1}{x_0}\int_{|x|=x_0}\varphi\,dS_{\mathbf{x}}\,dx_0$$

$$= \frac{1}{2}\int_{\mathbb{R}^n}\frac{\varphi(\mathbf{x},|\mathbf{x}|)}{|\mathbf{x}|}\,d\mathbf{x}, \qquad \varphi \in \mathcal{D}(\mathbb{R}^{n+1}). \tag{4.11}$$

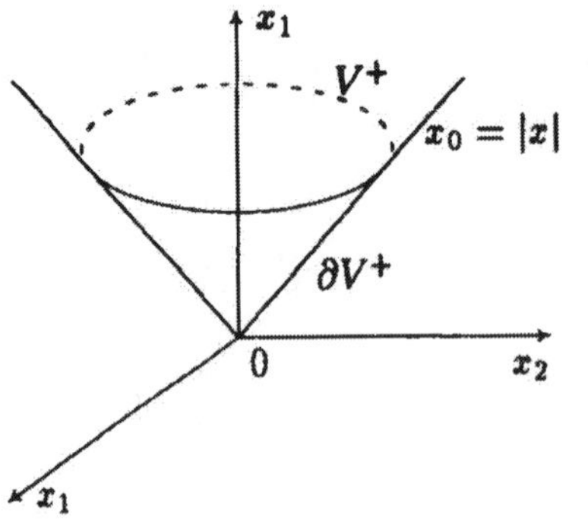

Figure 33

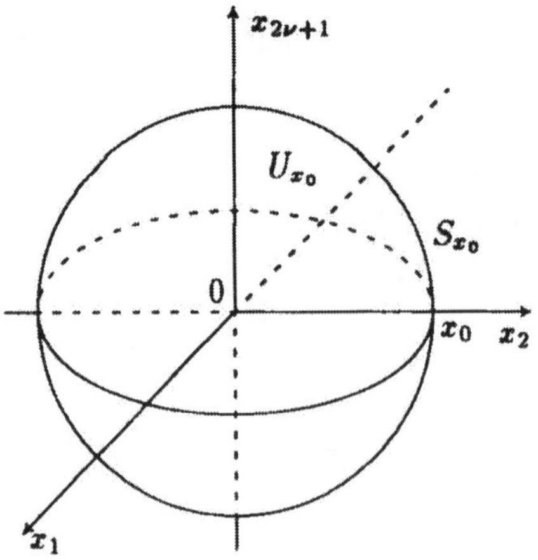

Figure 34

Using the technique of differential forms and the Stokes theorem (see, for example, Vladimirov [**105**]), for all $\varphi \in \mathcal{D}(\mathbb{R}^{n+1})$ we have

$$\begin{aligned}(\Box[\theta(x_0)\theta(x^2)], \varphi) &= (\theta(x_0)\theta(x^2), \Box\varphi) \\ &= \int_{V^+} \Box\varphi(x)\, dx_0 \wedge dx_1 \wedge \cdots \wedge dx_n \\ &= \int_{V^+} d\Big(\frac{\partial\varphi}{\partial x_0} dx_1 \wedge \cdots \wedge dx_n + \frac{\partial\varphi}{\partial x_1} dx_0 \wedge dx_2 \wedge \cdots \wedge dx_n + \\ &\qquad \cdots + (-1)^{n-1}\frac{\partial\varphi}{\partial x_n} dx_0 \wedge dx_1 \wedge \cdots \wedge dx_{n-1}\Big) \\ &= \int_{\partial V^+} \frac{\partial\varphi}{\partial x_0} dx_1 \wedge \cdots \wedge dx_n + \frac{\partial\varphi}{\partial x_1} dx_0 \wedge dx_2 \wedge \cdots \wedge dx_n + \\ &\qquad \cdots + (-1)^{n-1}\frac{\partial\varphi}{\partial x_n} dx_0 \wedge dx_1 \wedge \cdots \wedge dx_{n-1}.\end{aligned}$$

But on $\partial V^+ \setminus \{0\}$ we have, by virtue of the equation $x_0^2 = |\mathbf{x}|^2$, the relation

$$x_0\, dx_0 = x_1\, dx_1 + \cdots + x_n\, dx_n.$$

Therefore, continuing our chain of equations, we obtain

$$\begin{aligned}(\Box[\theta(x_0)\theta(x^2)], \varphi) = \int_{\partial V^+} d\Big\{\frac{\varphi}{x_0}\big[x_1\, dx_2 \wedge \cdots \wedge dx_n + \\ \cdots + (-1)^{n-1} x_n\, dx_1 \wedge \cdots \wedge dx_{n-1}\big]\Big\} \\ - 2(n-1)\int_{\partial V^+} \varphi \frac{dx_1 \wedge \cdots \wedge dx_n}{2x_0}. \qquad (4.12)\end{aligned}$$

By the Stokes theorem and by virtue of the fact that φ has a compact support, the first integral in the right-hand member of (4.12) is zero. In the second integral, integration is performed along the outer side of the surface ∂V^+ (Fig. 33) so that $dx_1 \wedge \cdots \wedge dx_n = -d\mathbf{x}$ and therefore

$$(\Box[\theta(x_0)\theta(x^2)], \varphi) = 2(n-1)\int\limits_{\mathbb{R}^n} \frac{\varphi(\mathbf{x}, |\mathbf{x}|)}{|\mathbf{x}|}\, d\mathbf{x},$$

whence, by (4.11), follows (4.10). □

Putting $n = 2\nu + 1$ in (4.8) and using (4.10), we obtain

$$\mathcal{E}_{2\nu+1}(x) = \frac{1}{2^{2\nu-1}\pi^\nu \Gamma(\nu)} \Box^{\nu-1}[\theta(x_0)\delta(x^2)]. \tag{4.13}$$

In particular, for $\nu = 1$ we derive the following from (4.13):

$$\mathcal{E}_3(x) = \frac{1}{2\pi}\theta(x_0)\delta(x^2). \tag{4.14}$$

To find a formula similar to (4.13) for $n = 2\nu$, we take advantage of the descent method with respect to the variable x_{n+1} (see Sec. 15.3). To do this, it must be shown that $\mathcal{E}_{2\nu+1}(x, x_{2\nu+1})$, $x = (x_0, x_1, \ldots, x_{2\nu})$ possesses a generalized integral with respect to variable $x_{2\nu+1}$. Let the sequence $\eta_k(x, x_{2\nu+1})$, $k = 1, 2, \ldots$, of test functions in $\mathcal{D}(\mathbb{R}^{n+2})$ converge to 1 in $\mathbb{R}^{n+2}$. Then, using (4.13) and (4.11), we have, for all $\varphi \in \mathcal{D}(\mathbb{R}^{n+2})$,

$$\lim_{k\to\infty} (\mathcal{E}_{2\nu+1}(x, x_{2\nu+1}), \varphi(x)\eta_k(x, x_{2\nu+1})) = \frac{1}{2^{2\nu-1}\pi^\nu\Gamma(\nu)}$$

$$\times \lim_{k\to\infty} \frac{1}{2}\int_0^\infty \frac{1}{x_0} \int\limits_{|\mathbf{x}|^2 + x_{2\nu+1}^2 = x_0^2} \Box^{\nu-1}[\varphi(x)\eta_k(x, x_{2\nu+1})]\, dS_{(x, x_{2\nu+1})}\, dx_0$$

$$= \frac{1}{2^{2\nu}\pi^\nu\Gamma(\nu)} \int_0^\infty \frac{1}{x_0} \int\limits_{|\mathbf{x}|^2 + x_{2\nu+1}^2 = x_0^2} \Box^{\nu-1}\varphi(x)\, dS_{(x, x_{2\nu+1})}\, dx_0$$

Transform the last integral. Since $\Box^{\nu-1}\varphi(x)$ does not depend on $x_{2\nu+1}$, then, by replacing the surface integral over the sphere $S_{x_0} = [(x, x_{2\nu+1}) : |\mathbf{x}|^2 + x_{2\nu+1}^2 = x_0^2]$ with twice the integral over the ball $|\mathbf{x}| < x_0$ (Fig. 34), we get

$$(\mathcal{E}_{2\nu}, \varphi) = \frac{1}{2^{2\nu-1}\pi^\nu\Gamma(\nu)} \int_0^\infty \int\limits_{|\mathbf{x}| < x_0} \frac{\Box^{\nu-1}\varphi(x)}{\sqrt{x^2}}\, d\mathbf{x}\, dx_0.$$

That is

$$\mathcal{E}_{2\nu}(x) = \frac{1}{2^{2\nu-1}\pi^\nu\Gamma(\nu)} \Box^{\nu-1}\left[\theta(x_0)x_+^2\right]^{-1/2}, \tag{4.15}$$

where

$$\left[\theta(x_0)x_+^2\right]^{-1/2} = \begin{cases} (x^2)^{-1/2}, & \text{if } x_0 \geq |\mathbf{x}|, \\ 0, & \text{if } x_0 < |\mathbf{x}|. \end{cases} \tag{4.16}$$

Putting $\nu = 1$ in (4.15), we have

$$\mathcal{E}_2(x) = \frac{1}{2\pi\sqrt{\theta(x_0)x_+^2}}. \tag{4.17}$$

15.4.7. *Fundamental solution of the Laplace operator* Δ. In Sec. 2.3.8, it was demonstrated that the functions

$$\begin{aligned} \mathcal{E}_n(x) &= -\frac{1}{(n-2)\sigma_n|x|^{n-2}}, \qquad n \geq 3; \\ \mathcal{E}_2(x) &= \frac{1}{2\pi}\ln|x| \end{aligned} \tag{4.18}$$

are a fundamental solution of the Laplace operator.

Let us compute $\mathcal{E}_n$ by the method of the Fourier transform. We have

$$-|\xi|^2 F[\mathcal{E}_n] = 1.$$

Let $n = 2$. The generalized function $-Pf\frac{1}{|\xi|^2}$ defined in Sec. 6.6.8 satisfies that equation, and its Fourier transform is equal to $2\pi \ln|x| + 2\pi C_0$, where C_0 is some constant. Therefore

$$F^{-1}\left[-Pf\frac{1}{|\xi|^2}\right] = \frac{1}{4\pi^2}F\left[-Pf\frac{1}{|\xi|^2}\right] = \frac{1}{2\pi}\ln|x| + \frac{C_0}{2\pi}.$$

Since the constant satisfies the homogeneous Laplace equation, then by dropping the term $\frac{C_0}{2\pi}$ we see that $\mathcal{E}_2$ may be chosen equal to $\frac{1}{2\pi}\ln|x|$.

Now let $n = 3$. In this case, the function $-|\xi|^2$ is locally integrable in $\mathbb{R}^3$ and is tempered, and therefore, in accordance with Sec. 15.2,

$$\mathcal{E}_n(x) = -F^{-1}\left[\frac{1}{|\xi|^2}\right] = -\frac{1}{(2\pi)^n}F\left[\frac{1}{|\xi|^2}\right],$$

whence, using (6.7) of Sec. 6.6, we obtain (4.18) for $n = 3$. The computation of $\mathcal{E}_n(x)$ is similar for $n > 3$ as well. □

It is particularly simple to construct $\mathcal{E}_n(x)$ for $n \geq 3$ by the descent method with respect to the variable t (see Sec. 15.3) from the fundamental solution of the heat conduction operator or the wave operator. For example, by using (3.7), we obtain from (4.7) for $a = 1$ the formula (4.18) for $n \geq 3$:

$$\begin{aligned} \mathcal{E}_n(x) &= -\int \mathcal{E}(x,t)\,dt \\ &= -\int_0^\infty \frac{1}{(2\sqrt{\pi t})^n} e^{-\frac{|x|^2}{4t}}\,dt \\ &= \frac{-|x|^{-n+2}}{4\pi^{n/2}} \int_0^\infty e^{-u} u^{n/2-2}\,du \\ &= -\Gamma(n/2-1)\frac{|x|^{-n+2}}{4\pi^{n/2}} \\ &= -\frac{1}{(n-2)\sigma_n|x|^{n-2}}. \end{aligned}$$

Computation is analogous in the case of the fundamental solution $\mathcal{E}_{n,k}(x)$ of the iterated Laplace operator Δ^k for $2k < n$:

$$\mathcal{E}_{n,k}(x) = \frac{(-1)^k \Gamma(n/2-k)}{2^{2k}\pi^{n/2}(k-1)!}|x|^{2k-n}. \tag{4.19}$$

15.4.8. *Fundamental solution of the Cauchy-Riemann operator,*

$$\frac{\partial}{\partial \bar{z}}\mathcal{E} = \delta(x,y), \qquad \frac{\partial}{\partial \bar{z}} = \frac{1}{2}\left(\frac{\partial}{\partial x} + i\frac{\partial}{\partial y}\right). \tag{4.20}$$

Applying the operator $\frac{\partial}{\partial x} - i\frac{\partial}{\partial y}$ to the equation (4.20), we obtain

$$\frac{1}{2}\Delta\mathcal{E} = \left(\frac{\partial}{\partial x} - i\frac{\partial}{\partial y}\right)\delta,$$

whence, using formulae (1.11) and (4.18), we have, for $n = 2$,

$$\mathcal{E} = 2\mathcal{E}_2 * \left(\frac{\partial}{\partial x} - i\frac{\partial}{\partial y}\right)\delta = \frac{1}{\pi}\left(\frac{\partial}{\partial x} - i\frac{\partial}{\partial y}\right)\ln\sqrt{x^2+y^2},$$

that is,

$$\mathcal{E}(x,y) = \frac{1}{\pi z}, \qquad z = x + iy. \tag{4.21}$$

15.4.9. *Fundamental solution of the transfer operator,*

$$\frac{1}{v}\frac{\partial \mathcal{E}_s}{\partial t} + (s, \nabla\mathcal{E}_s) + \alpha\mathcal{E}_s = \delta(x,t), \qquad |s| = 1, \qquad v > 0, \qquad ,\alpha \geq 0. \tag{4.22}$$

Applying the Fourier transform F_x to (4.22), we obtain, for the generalized function $F_x[\mathcal{E}_s] = \tilde{\mathcal{E}}_s(\xi,t)$, the equation

$$\frac{1}{v}\frac{\partial \tilde{\mathcal{E}}_s}{\partial t} + [\alpha - i(s,\xi)]\tilde{\mathcal{E}}_s = 1(\xi) \times \delta(t). \tag{4.23}$$

From this, using (4.4), we conclude that the solution, in S', of the equation (4.23) is the function $\tilde{\mathcal{E}}_s(\xi,t) = v\theta(t)e^{[i(s,\xi)-\alpha]vt}$. Now applying the inverse Fourier transform F_ξ^{-1} and using the formula (2.6) of Sec. 6.2, we obtain, for $x_0 = vts$, the fundamental solution of the transfer operator

$$\mathcal{E}_s(x,t) = v\theta(t)e^{-\alpha tv}\delta(x - vts). \tag{4.24}$$

To compute the fundamental solution $\mathcal{E}_s^0(x)$ of the stationary transfer operator

$$(s, \nabla\mathcal{E}_s^0) + \alpha\mathcal{E}_s^0 = \delta(x) \tag{4.25}$$

let us take advantage of the descent method with respect to the variable t (see Sec. 15.3):

$$(\mathcal{E}_s, \varphi(x)1(t)) = v \int_0^\infty e^{-\alpha v t} (\delta(x - vts), \varphi)\, dt$$

$$= v \int_0^\infty e^{-\alpha v t} \varphi(vts)\, dt$$

$$= v \int_0^\infty e^{-\alpha u} \varphi(us)\, du$$

$$= \left(\frac{e^{-\alpha|x|}}{|x|^2} \delta\left(s - \frac{x}{|x|}\right), \varphi \right),$$

so that

$$\mathcal{E}_s^0(x) = \frac{e^{-\alpha|x|}}{|x|^2} \delta\left(s - \frac{x}{|x|}\right). \tag{4.26}$$

15.4.10. *Fundamental solution of the Schrödinger operator,*

$$i\frac{\partial \mathcal{E}}{\partial t} + \frac{1}{2m}\Delta\mathcal{E} = \delta(x, t). \tag{4.27}$$

Applying the Fourier transform F_x to (4.27), we obtain, for the generalized function $F_x[\mathcal{E}] = \tilde{\mathcal{E}}(\xi, t)$, the equation [compare item 15.4.5]

$$i\frac{\partial \tilde{\mathcal{E}}}{\partial t} - \frac{1}{2m}|\xi|^2\tilde{\mathcal{E}} = 1(\xi) \times \delta(t),$$

whence, by (4.4), we have

$$\tilde{\mathcal{E}}(\xi, t) = -i\theta(t) \exp\left(-\frac{i}{2m}|\xi|^2 t\right)$$

$$= -i \lim_{\varepsilon \to +0} \theta(t) \left(\frac{m}{m + i\varepsilon}\right)^{n/2} \exp\left(-\frac{i}{2(m + i\varepsilon)}|\xi|^2 t\right).$$

Using the continuity, in S', of the operator of the Fourier transform F_ξ^{-1} and applying formula (4.7) for $a^2 = \frac{i}{2(m+i\varepsilon)}$, $\varepsilon > 0$, we obtain

$$\mathcal{E}(x, t) = -i\theta(t) \left(\frac{m}{2\pi t}\right)^{n/2} \exp\left(i\left[\frac{|x|^2}{2t}(m + i0) - \frac{\pi n}{4}\right]\right). \tag{4.28}$$

In particular, for $n = 1$ we have

$$\mathcal{E}(x, t) = -\frac{1 + i}{\sqrt{2}}\theta(t)\sqrt{\frac{m}{2\pi t}} e^{i\frac{m}{2t}x^2}. \tag{4.29}$$

15.5. A comparison of differential operators. Let $P(\partial)$ be a differential operator with constant coefficients of order m defined by (1.2). We set

$$\tilde{P}^2\xi = \sum_{|\beta|\le m} \left|P^{(\beta)}(i\xi)\right|^2$$

$$P^{(\beta)}(\xi) = \partial^\beta P(\xi) = \beta! \sum_{\beta\le\alpha} a_\alpha \binom{\alpha}{\beta} \xi^{\alpha-\beta}.$$

The Leibniz formula takes the form

$$P(\partial)(fg) = \sum_{|\beta|\le m} \frac{\partial^\beta f}{\beta!} P^{(\beta)}(\partial) g. \tag{5.1}$$

It can be verified directly:

$$\begin{aligned} P(\partial)(fg) &= \sum_{|\alpha|\le m} a_\alpha \partial^\alpha (fg) \\ &= \sum_{|\alpha|\le m} a_\alpha \sum_{\beta\le\alpha} \binom{\alpha}{\beta} \partial^\beta f \partial^{\alpha-\beta} g \\ &= \sum_{|\beta|\le m} \partial^\beta f \sum_{\beta\le\alpha} a_\alpha \binom{\alpha}{\beta} \partial^{\alpha-\beta} g \\ &= \sum_{|\beta|\le m} \frac{\partial^\beta f}{\beta!} P^{(\beta)}(\partial) g. \end{aligned}$$

We now prove the *Hörmander inequality*: if $\mathcal{O}$ is a bounded open set, then for any α there is a number $C_\alpha = C_\alpha(P, \mathcal{O})$ such that

$$\left\|P^{(\alpha)}(\partial)\varphi\right\| \le C_\alpha \left\|P(\partial)\varphi\right\|, \qquad \varphi \in \mathcal{D}(\mathcal{O}). \tag{5.2}$$

Here, $\|\ \| = \|\ \|_{\mathcal{L}^2(\mathcal{O})}$ (see Sec. 0.3).

PROOF. It suffices to prove the inequality (5.2) for $|\alpha| = 1$ and apply the induction process. We provide it for $\alpha = (1, 0, \dots, 0)$. Suppose $\varphi \in \mathcal{D}(\mathcal{O})$. Then by (5.1)

$$P(\partial)(x_1\varphi) = x_1 P(\partial)\varphi + P^{(1)}(\partial)\varphi, \qquad P^{(1)}(\xi) = \frac{\partial P(\xi)}{\partial \xi_1}. \tag{5.3}$$

Forming the scalar product of (5.3) on the right by $P^{(1)*}(\partial)\varphi = \overline{P^{(1)}(-\partial)}\varphi$ (compare Sec. 8.1) and taking into account that the operators $P(\partial)$ and $P^{(1)}(\partial)$ commute, we obtain

$$\left(P^{(1)}(\partial)(x_1\varphi), P^*(\partial)\varphi\right) = \left(x_1 P(\partial)\varphi, P^{(1)*}(\partial)\varphi\right) + \left\|P^{(1)}(\partial)\varphi\right\|^2. \tag{5.4}$$

But, by (5.3),

$$P^{(1)}(\partial)(x_1\varphi) = x_1 P^{(1)}(\partial)\varphi + P^{(2)}(\partial)\varphi, \qquad P^{(2)}(\xi) = \frac{\partial P^{(1)}(\xi)}{\partial \xi_1}.$$

Substituting the resulting expression into (5.4), we have

$$\begin{aligned}\left\|P^{(1)}(\partial)\varphi\right\|^2 &= \langle x_1 P^{(1)}(\partial)\varphi, P^*(\partial)\varphi\rangle + \langle P^{(2)}(\partial)\varphi, P^*(\partial)\varphi\rangle \\ &\quad - \langle x_1 P(\partial)\varphi, P^{(1)*}(\partial)\varphi\rangle \\ &\le \left\|x_1 P^{(1)}(\partial)\varphi\right\| \|P^*(\partial)\varphi\| + \left\|P^{(2)}(\partial)\varphi\right\| \|P^*(\partial)\varphi\| \\ &\quad + \|x_1 P(\partial)\varphi\| \left\|P^{(1)*}(\partial)\varphi\right\|.\end{aligned}$$

From this, putting $R_1 = \sup_{x\in\mathcal{O}}\left[|x_1|, 1\right]$ and noting that

$$\left\|P(\partial)\varphi\right\| = \left\|P^*(\partial)\varphi\right\|$$

we obtain the inequality

$$\left\|P^{(1)}(\partial)\varphi\right\|^2 =\le 2R_1\left[\left\|P^{(1)}(\partial)\varphi\right\| + \left\|P^{(2)}(\partial)\varphi\right\|\right]\left\|P(\partial)\varphi\right\|. \tag{5.5}$$

Suppose the inequality (5.2) is proved for all polynomials of degree $< m$. (If the degree of P is 0, then it is trivial.) Then there is a number C_1 such that

$$\left\|P^{(2)}(\partial)\varphi\right\| \le C_1\left\|P^{(1)}(\partial)\varphi\right\|, \qquad \varphi\in\mathcal{D}(\mathcal{O}).$$

Substituting this inequality into the inequality (5.5) and cancelling $\left\|P^{(1)}(\partial)\varphi\right\|$, we obtain that (5.2) holds true for $\alpha = (1, 0, \ldots, 0)$ with $C_\alpha = 2R_1(1 + C_1)$. □

From the Hörmander inequality follows the corollary: *if $\mathcal{O}$ is an open bounded set and $P(\partial) \not\equiv 0$, then*

$$P(\partial)\mathcal{L}^2(\mathcal{O}) \supset \mathcal{L}^2(\mathcal{O}). \tag{5.6}$$

To prove the inclusion (5.6) it is necessary to establish the existence, in $\mathcal{L}^2(\mathcal{O})$, of the solution of the equation

$$P(\partial)u = f, \qquad f\in\mathcal{L}^2(\mathcal{O}),$$

that is, of the equation

$$\langle u, P^*(\partial)\varphi\rangle = \langle f, \varphi\rangle, \qquad \varphi\in\mathcal{D}(\mathcal{O}). \tag{5.7}$$

The equation (5.7) defines an antilinear form on the functions $P^*(\partial)\mathcal{D}(\mathcal{O})$, which form is continuous in the norm $\mathcal{L}^2(\mathcal{O})$, by virtue of the inequality (5.2),

$$\|\varphi\| \le C\|P^*(\partial)\varphi\|, \qquad \varphi\in\mathcal{D}(\mathcal{O}).$$

This form, by the Hahn–Banach theorem, may be extended antilinearly and continuously onto all $\mathcal{L}^2(\mathcal{O})$, which form, by the Riesz theorem, is what determines the required solution $u(x)$ in $\mathcal{L}^2(\mathcal{O})$.

Note that we have again obtained the existence of a fundamental solution of the operator $P(\partial) \not\equiv 0$ in $\mathcal{D}'(\mathcal{O})$ for any open bounded set $\mathcal{O}$. To prove this, it suffices to represent $\delta = \partial^\alpha f$, $f\in\mathcal{L}^2(\mathcal{O})$, and take advantage of the inclusion (5.6).

We will say that the operator $P(\partial)$ is *stronger* than the operator $Q(\partial)$ in an open bounded set $\mathcal{O}$, and we write: $Q < P$ in $\mathcal{O}$ if there is a constant $K = K(P, Q, \mathcal{O})$ such that

$$\left\|Q(\partial)\varphi\right\| \le K\left\|P(\partial)\varphi\right\|, \qquad \varphi\in\mathcal{D}(\mathcal{O}). \tag{5.8}$$

THEOREM. *The following statements are equivalent:*

(1) *the operator P is stronger than the operator Q in $\mathcal{O}$;*

(2) *there exists a constant C such that*

$$|Q(i\xi)| \le C\tilde{P}(\xi), \qquad \xi \in \mathbb{R}^n. \tag{5.9}$$

PROOF. (1) → (2). Let $\varphi \in \mathcal{D}(\mathcal{O})$, $\varphi \not\equiv 0$. Applying the inequality (5.8) with $e^{i(x,\xi)}\varphi(x)$ substituted for $\varphi(x)$, we obtain

$$\|Q(\partial)e^{i(x,\xi)}\varphi\|^2 \le K^2\|P(\partial)e^{i(x,\xi)}\varphi\|^2, \qquad \xi \in \mathbb{R}^n, \tag{5.10}$$

where K does not depend on ξ. Noting that

$$P(\partial)e^{i(x,\xi)} = P(i\xi)e^{i(x,\xi)}$$

and using the Leibniz formula (5.1), from (5.10) we obtain

$$\left\|\sum_\alpha Q^{(\alpha)}(i\xi)\frac{\partial^\alpha\varphi}{\alpha!}\right\|^2 \le K^2\left\|\sum_\alpha P^{(\alpha)}(i\xi)\frac{\partial^\alpha\varphi}{\alpha!}\right\|^2. \tag{5.11}$$

Suppose m is the larger of the orders of the operators Q and P. We will prove that the quadratic form

$$\left\|\sum_{|\alpha|\le m}\frac{\partial^\alpha\varphi}{\alpha!}\lambda_\alpha\right\|^2$$

of the variables $\{\lambda_\alpha\}$ is positive definite. Indeed, if for $\{\lambda_\alpha\} \ne 0$ that form vanishes, then

$$\sum_{|\alpha|\le m}\frac{\lambda_\alpha}{\alpha!}\partial^\alpha\varphi(x) = 0, \qquad x \in \mathbb{R}^n,$$

Applying the Laplace transform to the resulting equation,

$$\sum_{|\alpha|\le m}\frac{\lambda_\alpha}{\alpha!}(-iz)^\alpha L[\varphi](z) = 0, \qquad z \in \mathbb{C}^n,$$

and taking into account that $L[\varphi](z)$ is an entire function (see Sec. 12.3), we conclude that $L[\varphi] \equiv 0$, whence, $\varphi \equiv 0$, which cannot be.

Thus, there exists a constant $\sigma > 0$, not dependent on $\{\lambda_\alpha\}$, such that

$$\frac{1}{\sigma}\sum_{\|\alpha|\le m}|\lambda_\alpha|^2 \le \left\|\sum_{|\alpha|\le m}\frac{\partial^\alpha\varphi}{\alpha!}\lambda_\alpha\right\|^2 \le \sigma\sum_{\|\alpha|\le m}|\lambda_\alpha|^2. \tag{5.12}$$

Applying the first of the inequalities of (5.12) to $\lambda_\alpha = Q^{(\alpha)}(i\xi)$, the second to $\lambda_\alpha = P^{(\alpha)}(i\xi)$, and taking into account the inequality (5.11), we obtain the inequality

$$\tilde{Q}^2(\xi) \le (\sigma K)^2\tilde{P}^2(\xi) \tag{5.13}$$

whence follows the inequality (5.9) for $C = \sigma K$.

(2) → (1). Let $\varphi \in \mathcal{D}(\mathcal{O})$. Multiplying inequality (5.9) by $|F^{-1}[\varphi](\xi)|^2$ and applying the Parseval–Steklov equation (see Sec. 6.6.3), we obtain

$$\|Q(i\xi)F^{-1}[\varphi](\xi)\|^2 = \|F^{-1}[Q(\partial)\varphi]\|^2 = \frac{1}{(2\pi)^n}\|Q(\partial)\varphi\|^2$$

$$\le \frac{C}{(2\pi)^n}\sum_{|\alpha|\le m}\left\|P^{(\alpha)}(\partial)\varphi\right\|^2. \tag{5.14}$$

Applying to the right-hand side of (5.14) the Hörmander inequality (5.2), we see that inequality (5.8) holds for a certain K (dependent on $\mathcal{O}$). And this means that $Q < P$ in $\mathcal{O}$. The proof of the theorem is complete. □

COROLLARY 1. *If $Q < P$ in some $\mathcal{O}$, then $Q < P$ in any open bounded set.*

COROLLARY 2. *The inequality* (5.9) *is equivalent to the inequality*

$$\tilde{Q}(\xi) \le C_1 \tilde{P}(\xi), \qquad \xi \in \mathbb{R}^n.$$

15.6. Elliptic and hypoelliptic operators. An operator $P(\partial)$ is said to be *elliptic* (or, respectively, *hypoelliptic*) if it possesses a (real) analytic (respectively, C^∞) fundamental solution $\mathcal{E}(x)$ for $x \ne 0$. Every elliptic operator is hypoelliptic.

EXAMPLES. The Laplace and Cauchy–Riemann operators are elliptic (see Sec. 15.4.7 and 15.4.8); the heat conduction operator is hypoelliptic (see Sec. 15.4.5).

THEOREM I. *For the operator $P(\partial)$ to be hypoelliptic, it is necessary and sufficient that for any open set $\mathcal{O}$ every solution $u(x)$ in $\mathcal{D}'(\mathcal{O})$ of the equation $P(\partial)u = f$, where $f \in C^\infty(\mathcal{O})$, belong to $C^\infty(\mathcal{O})$.*

PROOF. Sufficiency is obvious. We will prove necessity. Let $\mathcal{E}(x)$ be a fundamental solution of the class $C^\infty(\mathbb{R}^n \setminus \{0\})$ of the operator $P(\partial)$ and $u \in \mathcal{D}'(\mathcal{O})$ a solution in $\mathcal{O}$ of the equation $P(\partial)u = f$. Suppose $\mathcal{O}'$ is an arbitrary open set, compact in $\mathcal{O}$, and $\eta \in \mathcal{D}(\mathcal{O})$, $\eta(x) = 1$, $x \in \mathcal{O}'$. The generalized function ηu is of compact support and satisfies the equation [see (5.1)]

$$P(\partial)(\eta u) = \eta f + f_1,$$

where $\eta f \in C^\infty$, ηf is of compact support, $f_1 \in \mathcal{D}'$ and $\operatorname{supp} f_1 \subset \operatorname{supp} \eta \setminus \mathcal{O}'$. Therefore (see Sec. 15.1)

$$\eta u = \mathcal{E} * (\eta f) + \mathcal{E} * f_1$$

whence it follows that $u \in C^\infty(\mathcal{O}')$ (see Sec. 4.3). Since $\mathcal{O}' \Subset \mathcal{O}$ is arbitrary, it follows that $u \in C^\infty(\mathcal{O})$. Theorem I is proved. □

The algebraic conditions for hypoellipticity may be indicated: *for an operator $P(\partial)$ to be hypoelliptic, it is necessary and sufficient that for all α, $|\alpha| \ge 1$,*

$$\frac{P^{(\alpha)}(-i\xi)}{P(-i\xi)} \to 0, \qquad |\xi| \to \infty. \tag{6.1}$$

This result was obtained by Hörmander [46].

The proof of the following theorem is similar to that of Theorem I.

THEOREM I′. *For an operator $P(\partial)$ to be elliptic, it is necessary and sufficient that, for any $\mathcal{O}$, every solution $u(x)$ in $\mathcal{D}'(\mathcal{O})$ of the equation $P(\partial)u = 0$ be (real) analytic in $\mathcal{O}$.*

The algebraic condition for ellipticity: *for an operator $P(\partial)$ to be elliptic, it is necessary and sufficient that its principal part*

$$P_m(\partial) = \sum_{|\alpha|=m} a_\alpha \partial^\alpha$$

satisfies the condition $P_m(\xi) \ne 0$, $\xi \ne 0$.

This result was obtained by Petrovskii **[80]** (1939), for classical solutions, by Weyl **[124]** (1940) for generalized solutions for the Laplace operator, and by Hörmander (1955) (see **[46]**) in the general case.

We now prove a theorem on wiping out isolated singularities of harmonic functions.

A generalized function $u(x)$ in $\mathcal{D}'(G)$ is said to be *harmonic* in a domain G if it satisfies the Laplace equation $\Delta u = 0$ in G (and then $u \in C^\infty$ via Theorem I).

THEOREM II. *Let* $0 \in G$. *If the function* u *is harmonic in the domain* $G \setminus \{0\}$ *and*

$$\begin{aligned} u(x) &= o(|x|^{-n+2}), \qquad n \geq 3; \\ u(x) &= o(\ln |x|), \qquad n = 2, \end{aligned} \qquad |x| \to 0, \tag{6.2}$$

then u *is harmonic in* G.

PROOF. Let $U_R \Subset G$. We introduce the function $\tilde{u}(x)$, which is equal to $u(x)$ in $\overline{U}_R$ and 0 outside $\overline{U}_R$. This function is integrable on $\mathbb{R}^n$ and by (3.7′) of Sec. 2.3 it is the generalized function in $\mathcal{D}'(\mathbb{R}^n)$ such that

$$\Delta \tilde{u} + \frac{\partial u}{\partial \mathbf{n}} \delta_{S_R} + \frac{\partial}{\partial \mathbf{n}}(u \delta_{S_R}) = 0 \quad \text{in} \quad \mathbb{R}^n \setminus \{0\}.$$

From this, by the theorem of Sec. 2.6, we have

$$\Delta \tilde{u} = -\frac{\partial u}{\partial \mathbf{n}} \delta_{S_R} - \frac{\partial}{\partial \mathbf{n}}(u \delta_{S_R}) + \sum_{|\alpha| \leq m} c_\alpha \partial^\alpha \delta \quad \text{in} \quad \mathbb{R}^n. \tag{6.3}$$

Since $\tilde{u}$ is of compact support, by using (1.11), we obtain, from (6.3),

$$\begin{aligned} \tilde{u} &= \mathcal{E}_n * \Delta \tilde{u} \\ &= -\mathcal{E}_n * \frac{\partial u}{\partial \mathbf{n}} \delta_{S_R} - \mathcal{E}_n * \frac{\partial}{\partial \mathbf{n}}(u \delta_{S_R} + \sum_{|\alpha| \leq m} c_\alpha \mathcal{E}_n * \partial^\alpha \delta \\ &= V_n^{(0)} + V_n^{(1)} + \sum_{|\alpha| \leq m} c_\alpha \partial^\alpha \mathcal{E}_n \end{aligned} \tag{6.4}$$

where

$$\begin{aligned} \mathcal{E}_n(x) &= k_n |x|^{-n+2}, \qquad n \geq 3; \\ \mathcal{E}_2(x) &= \frac{1}{2\pi} \ln |x| \end{aligned}$$

is the fundamental solution of the Laplace operator, and $V_n^{(0)}(x)$ and $V_n^{(1)}(x)$ are surface potentials of a simple and double layer on the sphere S_R (see Sec. 4.9). From the representation (6.4) for $|x| < R$ and from the condition (6.2) it follows that $c_\alpha = 0$, so that

$$u(x) = V_n^{(0)}(x) + V_n^{(1)}(x), \qquad |x| < R,$$

whence it follows that $u(x)$ is a harmonic function in the ball U_R. The proof of the theorem is complete. □

15.7. Hyperbolic operators. Let C be a convex open cone in $\mathbb{R}^n$ with vertex at 0. The operator $P(\partial)$ is said to be *hyperbolic relative to the cone C* if it satisfies the condition: there is a point $y_0 \in \mathbb{R}^n$ such that

$$P(y_0 - iz) \neq 0 \quad \text{for all} \quad z \in T^C. \tag{7.1}$$

THEOREM. *For the operator $P(\partial)$ to be hyperbolic relative to a cone C, it is necessary and sufficient that it have a (unique) fundamental solution $\mathcal{E}(x)$ in the algebra $\mathcal{D}'(C^*)$, which solution can be represented as*

$$\mathcal{E}(x) = e^{(y_0,x)}\mathcal{E}_0(x), \qquad \mathcal{E}_0 \in \mathcal{S}'(C^*), \tag{7.2}$$

where the point $y_0 \in \mathbb{R}^n$ is defined in (7.1).

PROOF. NECESSITY. If the operator $P(\partial)$ is hyperbolic relative to the cone C, then the polynomial $P(y_0 - iz)$ does not vanish in the tubular domain T^C. Therefore $1/P(y_0 - iz) \in H(C)$ (see Sec. 13.2) so that

$$\frac{1}{P\big[-i(iy_0 + z)\big]} = L[\mathcal{E}_0](z), \qquad \mathcal{E}_0 \in \mathcal{S}'(C^*). \tag{7.3}$$

Setting $\zeta = z + iy_0$, we obtain (see Sec. 9.2.3)

$$\frac{1}{P(-i\zeta)} = L[\mathcal{E}_0](\zeta - iy_0) = L\big[\mathcal{E}_0(x)e^{(y_0,x)}\big],$$

whence follows the representation (7.2).

SUFFICIENCY. If the operator $P(\partial)$ has a fundamental solution of the form of (7.2), then the function $L[\mathcal{E}_0](z)$ is holomorphic in T^C and, hence, by virtue of (7.3) the polynomial $P(y_0 - iz)$ does not vanish in T^C since the operator $P(\partial)$ is hyperbolic relative to the cone C.

The uniqueness of a fundamental solution in the algebra $\mathcal{D}'(C^*)$ was proved in Sec. 4.9.4.

The theorem is proved. □

EXAMPLE 1. The wave operator □ is hyperbolic relative to the future light cone V^+, and (see Sec. 13.5)

$$\Box(-iz) = -z_0^2 + z_1^2 + \cdots + z_n^2 \neq \quad \text{in} \quad T^{V^+}.$$

EXAMPLE 2. The differential operator $P(\frac{d}{dt})$ (see Sec. 15.4.4) is hyperbolic relative to the cone $(0, \infty)$.

REMARK. The cone C is a connected component of the open set (see Hörmander [46])

$$\big[y \colon P_m(y) \neq 0\big].$$

15.8. The sweeping principle. Let $P(\partial)$ be a differential operator with constant coefficients of order m that is defined by equality (1.2) in Sec. 15.1 and $\mathcal{E}(x)$ be its fundamental solution. Suppose that $\mathcal{O}$ is an open set in $\mathbb{R}^n$, $\partial\mathcal{O}$ is its boundary and $\theta_{\mathcal{O}}(x)$ is its characteristic function (see Sec. 0.2). Let $u \in \mathcal{D}'(\mathcal{O})$ be the solution of the homogeneous equation $P(\partial) = 0$ in $\mathcal{O}$ which is representable in $\mathcal{O}$ in the form of the potential $V = \mathcal{E} * f$ with the density $f \in \mathcal{D}'$.

DEFINITION. If there exists a generalized function $h \in \mathcal{D}'$, $\operatorname{supp} h \subset \partial\mathcal{O}$, such that

$$\mathcal{E} * h = \begin{cases} u = \mathcal{E} * f, & x \in \mathcal{O}, \\ 0, & x \in \mathbb{R}^n \setminus \overline{\mathcal{O}}, \end{cases} \tag{8.1}$$

then we say that the *sweeping* of the density f on $\partial\mathcal{O}$ occurs.

THEOREM. *Let $\mathcal{O}$ be a bounded open set in $\mathbb{R}^n$. If the potential $V = \mathcal{E} * f$, $f \in \mathcal{D}'$, exists in $\mathcal{D}'$, satisfies the homogeneous equation $P(\partial)V = 0$ in $\mathcal{O}$ and there exists the generalized function $\theta_{\mathcal{O}}(x)V(x)$ in $\mathcal{D}'$,[2] then the sweeping of the density f on $\partial\mathcal{O}$ occurs.*

PROOF. Write

$$h = P(\partial)[\theta_{\mathcal{O}} V], \qquad h \in \mathcal{D}'. \tag{8.2}$$

By the hypothesis of the theorem, $\operatorname{supp} h \subset \partial\mathcal{O}$. Since the convolution $\mathcal{E}$ and $\theta_{\mathcal{O}} V$ exists in $\mathcal{D}'$ ($\mathcal{O}$ is bounded! see Sec. 4.3), by virtue of (8.2) $\theta_{\mathcal{O}} V$ can be represented as the potential (see Sec. 15.1)

$$\theta_{\mathcal{O}}(x)V(x) = \mathcal{E} * h, \qquad x \in \mathbb{R}^n. \tag{8.3}$$

The restriction of equality (8.3) onto $\mathcal{O}$ yields $\mathcal{E} * h = V = u$, and its restriction onto $\mathbb{R}^n \setminus \overline{\mathcal{O}}$ yields $\mathcal{E} * h = 0$, and equalities (8.1) are proven. The theorem is proved. ☐

EXAMPLE. Let $\mathcal{O} = G$ be a bounded domain in $\mathbb{R}^n$ with a piecewise smooth boundary $\partial G = S$ and the potential

$$V = \frac{1}{|x|^{n-2}} * f, \qquad f \in \mathcal{D}', \qquad \operatorname{supp} f \subset \mathbb{R}^n \setminus G$$

exist in $\mathcal{D}' \cap C^1(\bar{G})$. In this case formula (3.7′) of 2.3.7 yields an explicit expression for h,

$$h(x) = -\frac{\partial V}{\partial n}\delta_S(x) - \frac{\partial}{\partial n}(V\delta_S)(x). \tag{8.4}$$

16. The Cauchy Problem

16.1. The generalized Cauchy problem for a hyperbolic equation. Let S be a C-like surface of the class C^∞ and let S_+ be a domain lying above S (see Sec. 4.5); $P(\partial)$ is a hyperbolic operator relative to the cone C of order m.

We consider the classical Cauchy problem

$$P(\partial)u = f(x), \qquad x \in S_+, \tag{1.1}$$

$$\left.\frac{\partial^k u}{\partial n^k}\right|_S = u_k(x), \qquad x \in S, \qquad k = 0, 1, \ldots, m-1, \tag{1.2}$$

that is, the problem of finding a function $u \in C^m(S_+) \cap C^{m-1}(\overline{S}_+)$ that satisfies the equation (1.1) in S_+ and the condition (1.2) on S. For solvability of the Cauchy problem (1.1)–(1.2) it is necessary that $f \in C(S_+)$ and $u_k \in C^{m-k-1}(S)$.

Suppose that the classical solution u of the problem (1.1)–(1.2) exists and $f \in C(\overline{S}_+)$. We extend the functions f and u by zero onto S_- and denote the

[2] Concerning the multiplication of a generalized function by the characteristic function of an open set see Sec. 1.10.

extended functions by $\tilde{f}$ and $\tilde{u}$ respectively. Then the function $\tilde{u}(x)$ satisfies the following differential equations over the entire space $\mathbb{R}^n$:

$$P(\partial)\tilde{u} = \tilde{f}(x) + \sum_{0\le k\le m-1} \frac{\partial^k}{\partial \mathbf{n}^k}(v_k\delta_S)(x), \tag{1.3}$$

where $v_k\delta_S$ is the density of a simple layer on S with surface density v_k (see Sec. 1.7), uniquely defined by the functions $\{u_j\}$, by the surface S, and by the operator $P(\partial)$. The generalized function $\frac{\partial^k}{\partial \mathbf{n}^k}(v_k\delta_S)$ acts on the test functions φ via the rule (compare Sec. 2.3.2)

$$\left(\frac{\partial^k}{\partial \mathbf{n}^k}(v_k\delta_S), \varphi\right) = (-1)^k \int_S v_k(x)\frac{\partial^k \varphi(x)}{\partial \mathbf{n}^k}\, dS.$$

Let us prove (1.3). Using equation (1.1) and the condition (1.2), we have, for all $\varphi \in \mathcal{D}$,

$$\begin{aligned}(P(\partial)\tilde{u}, \varphi) &= (\tilde{u}, P(-\partial)\varphi)\\ &= \int_{S_+} u(x)P(-\partial)\varphi(x)\, dx\\ &= \int_{S_+} P(\partial)u(x)\varphi(x)\, dx + \sum_{0\le k\le m-1} (-1)^k \int_S v_k(x)\frac{\partial^k\varphi(x)}{\partial \mathbf{n}^k} dS\\ &= \int \tilde{f}(x)\varphi(x)\, dx + \sum_{0\le k\le m-1}\left(\frac{\partial^k}{\partial \mathbf{n}^k}(v_k\delta_s), \varphi\right),\end{aligned}$$

which is equivalent to (1.3). □

EXAMPLE 1. For the Cauchy problem for the ordinary differential equation

$$P\left(\frac{d}{dt}\right)u \equiv u^{(m)} + a_1u^{(m-1)} + \cdots + a_mu = f(t), \tag{1.4}$$

$$u^{(k)}(0) = u_k, \qquad k = 0, 1, \ldots, m-1, \tag{1.5}$$

the equation (1.3) takes the form (see Sec. 2.3.3)

$$P\left(\frac{d}{dt}\right)\tilde{u} = \tilde{f}(t) + \sum_{0\le k\le m-1} v_k\delta^{(k)}(t) \tag{1.6}$$

where

$$v_k = \sum_{0\le j\le m-1-k} a_{m-1-j}u_j \qquad (a_0 = 1). \tag{1.7}$$

Indeed, using formula (3.1) of Sec. 2.3 and the initial conditions (1.5), we have

$$\tilde{u}_{\text{cl}}^{(k)}(t) + \sum_{0\le j\le k-1} u_j\delta^{(k-j-1)}(t), \qquad k = 1, 2, \ldots, n.$$

From this and from (1.4) it follows equation (1.6):

$$P\left(\frac{d}{dt}\right)\tilde{u} = P_{\text{cl}}\left(\frac{d}{dt}\right)\tilde{u}(t) + u_0\delta^{(m-1)}(t) + (a_1u_0 + u_1)\delta^{(m-2)}(t) + \cdots + (a_{m-1}u_0 + \cdots + a_1u_{m-2} + u_{m-1})\delta(t)$$
$$= \tilde{f}(t) + \sum_{0\le k\le m-1} v_k\delta^{(k)}(t),$$

where the numbers v_k are defined by equations (1.7). □

EXAMPLE 2. For the Cauchy problem for the wave equation

$$\Box u = f(x), \qquad x = (x_0, \mathbf{x}), \tag{1.8}$$

$$u\Big|_{x_0=0} = u_0(\mathbf{x}), \qquad \frac{\partial u}{\partial x_0}\Big|_{x_0=0} = u_1(\mathbf{x}), \tag{1.9}$$

the equation (1.3) takes the form

$$\Box\tilde{u} = \tilde{f}(x) + u_0(\mathbf{x}) \times \delta'(x_0) + u_1(\mathbf{x}) \times \delta(x_0). \tag{1.10}$$

Thus, the classical solution $u(x)$ of the Cauchy problem (1.1)–(1.2), being a continuation on S_- via zero, satisfies equation (1.3) in $\mathbb{R}^n$. Here the initial conditions (1.2) play the role of sources concentrated on the surface S (as the sum of densities of layers of different orders). For example, for the wave equation, by virtue of (1.10), which is the sum of the densities of a simple layer and a double layer on the plane $x_0 = 0$, i.e., an instantly operating source (for $x_0 = 0$).

Now we can generalize the classical Cauchy problem for a hyperbolic operator $P(\partial)$ in the following manner. Suppose the generalized function $F \in \mathcal{D}'(\overline{S}_+)$. We use the term *generalized Cauchy problem* for the operator $P(\partial)$ with the source F to describe the problem of finding, in $\mathbb{R}^n$, a generalized solution u in $\mathcal{D}'(\overline{S}_+)$ of the equation

$$P(\partial)u = F(x). \tag{1.11}$$

By the foregoing, all solutions of the classical Cauchy problem are contained among the solutions of the generalized Cauchy problem.

The following theorem holds true:

THEOREM. *A solution of the generalized Cauchy problem exists uniquely and is expressed by the formula*

$$u = \mathcal{E} * F, \tag{1.12}$$

where $\mathcal{E}$ is a fundamental solution of the operator $P(\partial)$ in $\mathcal{D}'(C^*)$. This solution depends continuously on F in the sense of convergence in the space $\mathcal{D}'(\overline{S}_+)$.

To prove the theorem, it is necessary to take advantage of the results of Sec. 4.5, for $\Gamma = C^*$ and $K = \varnothing$, and also of Sec. 4.9.4. Here the operation $F \to \mathcal{E} * F$ is continuous from $\mathcal{D}'(\overline{S}_+)$ to $\mathcal{D}'(\overline{S}_+)$. □

REMARK. The foregoing is carried over in obvious fashion to hyperbolic systems. Let A be an $N \times N$ matrix whose elements are differential operators with constant coefficients. The operator $A(\partial)$ is said to be hyperbolic with respect to the cone C if the operator $\det A(\partial)$ is hyperbolic relative to C.

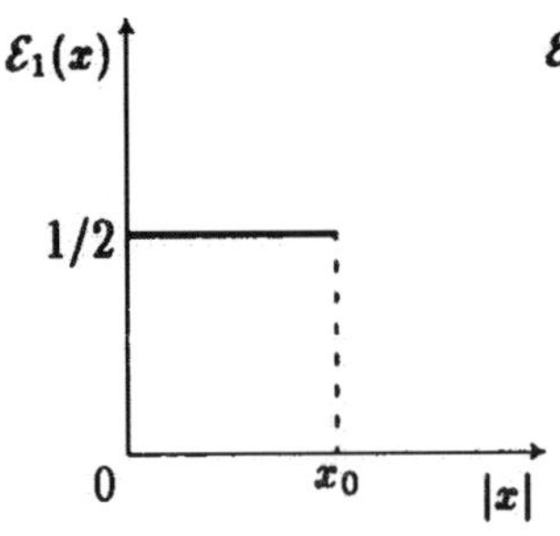

Figure 35

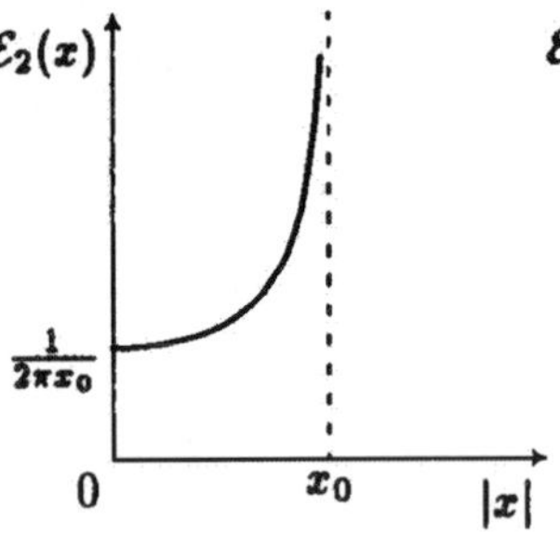

Figure 36

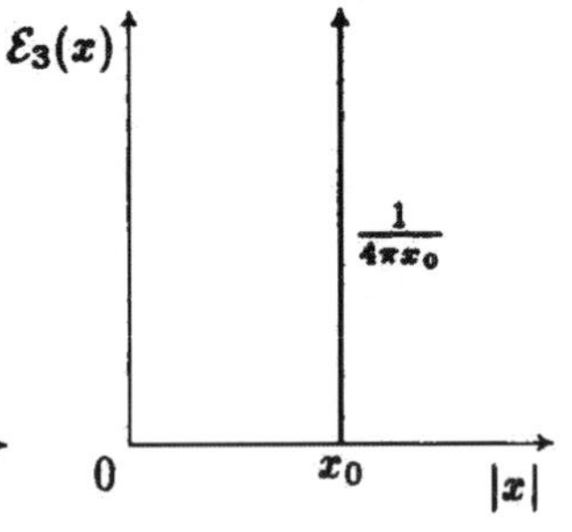

Figure 37

EXAMPLE 1′. The formula (1.12) for solving the classical Cauchy problem (1.4)–(1.5) takes the form

$$u(t) = \int_0^t f(\tau)Z(t-\tau)\,d\tau + \sum_{0\le k\le m-1} v_k Z^{(k)}(t), \tag{1.13}$$

where $Z(t)$ is a solution of the homogeneous equation $P(\frac{d}{dt})Z = 0$ that satisfies the conditions $Z^{(k)}(0) = 0$, $0 \le k \le m-2$, $Z^{(m-1)}(0) = 1$, and the numbers v_k are given by the equations (1.7).

To obtain formula (1.13) compute the convolution of the fundamental solution $\mathcal{E}(t) = \theta(t)Z(t)$ of the operator $P(\frac{d}{dt})$ (see Sec. 15.4.4) with the right-hand side of (1.6):

$$\begin{aligned}\tilde{u} &= \tilde{f} * \mathcal{E} + \sum_{0\le k\le m-1} v_k\delta^{(k)} * \mathcal{E}\\ &= \int \tilde{f}(\tau)\mathcal{E}(t-\tau)\,d\tau + \sum_{0\le k\le m-1} v_k\mathcal{E}^{(k)}(t)\\ &= \int_0^t f(\tau)Z(t-\tau)\,d\tau + \theta(t)\sum_{0\le k\le m-1} v_k Z^{(k)}(t).\end{aligned}$$

Here we took into account the equations (see Sec. 15.4.4)

$$\mathcal{E}^{(k)}(t) = \left[\theta(t)Z(t)\right]^{(k)} = \theta(t)Z^{(k)}(t), \qquad k = 0, 1, \ldots, m-1.$$

16.2. Wave potential. The (generalized) functions defined by equations (4.9), (4.13) and (4.15) of Sec. 15.4 constitute the fundamental solution of the wave operator:

$$\mathcal{E}_n(x) = \begin{cases} \dfrac{1}{2^{2\nu-1}\pi^\nu\Gamma(\nu)}\square^{\nu-1}\left[\theta(x_0)\delta(x^2)\right], & n = 2\nu+1,\ \nu \ge 1;\\ \dfrac{1}{2^{2\nu-1}\pi^\nu\Gamma(\nu)}\square^{\nu-1}\left[\theta(x_0)x_+^2\right]^{-1/2}, & n = 2\nu;\\ \dfrac{1}{2}\theta(x_0)\theta(x^2), & n = 1.\end{cases} \tag{2.1}$$

The supports of $\mathcal{E}_1$ and $\mathcal{E}_{2\nu}$ coincide with $\overline{V^+}$ and the support of $\mathcal{E}_{2\nu+1}$, $\nu \ge 1$, with $\partial\overline{V^+}$. These peculiarities of structure of the support of the fundamental solution are what determine the difference in the nature of wave propagation in odd-dimensional ($n \geq 3$) and even-dimensional spaces and on the straight line. Figures 35 to 37 depict schematically the graphs of the fundamental solutions $\mathcal{E}_n(x)$ with respect to $|\mathbf{x}|$ for a fixed $x_0 > 0$.

LEMMA. *The fundamental solution $\mathcal{E}_n(x)$ belongs to the class $C^\infty([0,\infty))$ with respect to x_0 and its restriction $\mathcal{E}_{nx_0}(\mathbf{x})$ for $x_0 > 0$ possesses the support S_{x_0} (for odd $n \geq 3$), $\overline{U}_{x_0}$ (for even n or $n = 1$), and satisfies the limiting relations, as $x_0 \to +0$,*

$$\mathcal{E}_{nx_0}(\mathbf{x}) \to 0, \qquad \frac{\partial \mathcal{E}_{nx_0}(\mathbf{x})}{\partial x_0} \to \delta(x), \qquad \frac{\partial^2 \mathcal{E}_{nx_0}(\mathbf{x})}{\partial x_0^2} \to 0 \quad \text{in} \quad \mathcal{D}'(\mathbb{R}^n). \tag{2.2}$$

PROOF. Let $n = 2\nu + 1 \geq 3$ be odd. We will prove that, for $x_0 > 0$,

$$\mathcal{E}_{n\varphi}(x) = \frac{1}{2^{2\nu}\pi^\nu\Gamma(\nu)} \sum_{0\leq\alpha\leq\nu-1} (-1)^{\nu-1-\alpha}\binom{\nu-1}{\alpha}$$
$$\times \frac{d^{2\alpha}}{dx_0^{2\alpha}}\left[x_0^{2\nu-1}\int_{|s|=1}(\Delta^{\nu-1-\alpha}\varphi)(x_0 s)\,ds\right], \qquad \varphi \in \mathcal{D}(\mathbb{R}^n). \tag{2.3}$$

Indeed, for all $\varphi \in \mathcal{D}(\mathbb{R}^n)$ and $\psi \in \mathcal{D}(x_0 > 0)$ we have, by virtue of (2.1) and (4.11) of Sec. 15.4,

$$\begin{aligned}
(\mathcal{E}_{n\varphi}, \psi) &= (\mathcal{E}_n, \varphi(\mathbf{x})\psi(x_0)) \\
&= \frac{1}{2^{2\nu-1}\pi^\nu\Gamma(\nu)}\left(\Box^{\nu-1}\theta[(x_0)\delta(x^2)], \varphi(\mathbf{x})\psi(x_0)\right) \\
&= \frac{1}{2^{2\nu-1}\pi^\nu\Gamma(\nu)}\Bigg(\theta(x_0)\delta(x^2), \\
&\qquad \sum_{0\leq\alpha\leq\nu-1}(-1)^{\nu-1-\alpha}\binom{\nu-1}{\alpha}\psi^{(2\alpha)}\Delta^{\nu-1-\alpha}\varphi(\mathbf{x})\Bigg) \\
&= \frac{1}{2^{2\nu}\pi^\nu\Gamma(\nu)}\sum_{0\leq\alpha\leq\nu-1}(-1)^{\nu-1-\alpha}\binom{\nu-1}{\alpha} \\
&\qquad \times \int_0^\infty \frac{\psi^{(2\alpha)}(x_0)}{x_0}\int_{|\mathbf{x}|=x_0}\Delta^{\nu-1-\alpha}\varphi(\mathbf{x})\,dS_{\mathbf{x}}\,dx_0 \\
&= \frac{1}{2^{2\nu}\pi^\nu\Gamma(\nu)}\sum_{0\leq\alpha\leq\nu-1}(-1)^{\nu-1-\alpha}\binom{\nu-1}{\alpha} \\
&\qquad \times \int_0^\infty \psi^{(2\alpha)}(x_0)x_0^{2\nu-1}\int_{|s|=1}(\Delta^{\nu-1-\alpha}\varphi)(x_0 s)\,ds\,dx_0,
\end{aligned}$$

whence follows formula (2.3) (see the notations and techniques developed in Sec. 3.4). From (2.3) it follows that $\mathcal{E}_n \in C^\infty([0,\infty))$ with respect to x_0, $\operatorname{supp}\mathcal{E}_{nx_0} = S_{x_0}$ and, as $x_0 \to +0$,

$$(\mathcal{E}_{nx_0}, \varphi) = \mathcal{E}_{n\varphi}(x_0) \to 0,$$

$$\left(\frac{\partial \mathcal{E}_{nx_0}}{\partial x_0}, \varphi\right) = \mathcal{E}'_{n\varphi}(x_0)$$

$$= \frac{1}{2^{2\nu}\pi^\nu\Gamma(\nu)} \sum_{0\le\alpha\le\nu-1} (-1)^{\nu-1-\alpha}\binom{\nu-1}{\alpha}$$

$$\times \frac{d^{2\alpha+1}}{dx_0^{2\alpha+1}}\left[x_0^{2\nu-1}\int_{|s|=1}(\Delta^{\nu-1-\alpha}\varphi)(x_0 s)\,ds\right]$$

$$\to \frac{(2\nu-1)!}{2^{2\nu}\pi^\nu\Gamma(\nu)}\int_{|s|=1}\varphi(0)\,ds$$

$$= \frac{\Gamma(2\nu)}{2^{2\nu}\pi^\nu\Gamma(\nu)}\sigma_{2\nu+1}\varphi(0)$$

$$= \frac{2\Gamma(2\nu)\pi^{\nu+1/2}}{2^{2\nu}\pi^\nu\Gamma(\nu)\Gamma(\nu+1/2)}\varphi(0) = \varphi(0) = (\delta,\varphi),$$

$$\left(\frac{\partial^2 \mathcal{E}_{nx_0}}{\partial x_0^2}, \varphi\right) = \mathcal{E}''_{n\varphi}(x_0)$$

$$= \frac{1}{2^{2\nu}\pi^\nu\Gamma(\nu)} \sum_{0\le\alpha\le\nu-1} (-1)^{\nu-1-\alpha}\binom{\nu-1}{\alpha}$$

$$\times \frac{d^{2\alpha+2}}{dx_0^{2\alpha+2}}\left[x_0^{2\nu-1}\int_{|s|=1}(\Delta^{\nu-1-\alpha}\varphi)(x_0 s)\,ds\right] \to 0.$$

In this last relation we took advantage of the fact that the function

$$\int_{|s|=1}\varphi(x_0 s)\,ds = \int_{|s|=1}\varphi(-x_0 s)\,ds$$

is even, infinitely differentiable, and therefore its first derivative with respect to x_0 at zero is equal to zero.

Thus the limiting relations (2.3) have been proved for odd $n \ge 3$. For even $n = 2\nu$ the proof is analogous: the simplest thing is to take advantage of the descent method with respect to the variable $x_{2\nu+1}$ (see Sec. 15.4.6). For $n = 1$ the proof is trivial. The proof of the lemma is complete. □

EXAMPLE.

$$\mathcal{E}_{2\nu+1x_0}(\mathbf{x}) = \frac{1}{2^{2\nu}\pi^\nu\Gamma(\nu)}\Box^{\nu-1}\frac{1}{x_0}\delta_{S_{x_0}}(\mathbf{x}), \qquad x_0 > 0, \tag{2.4}$$

where $\delta_{S_{x_0}}(\mathbf{x})$ is a simple layer on the sphere $|\mathbf{x}| = x_0$ (see Sec. 1.7).

Suppose $F \in \mathcal{D}'(\overline{\mathbb{R}}^1_+ \times \mathbb{R}^n)$. The convolution $V_n = F * \mathcal{E}_n$ that exists in $\mathcal{D}'(\overline{\mathbb{R}}^1_+ \times \mathbb{R}^n)$ (see Sec. 4.5) is termed a *wave potential* wiht density F. The wave potential V_n depends continuously on F in the sense of convergence in $\mathcal{D}'(\overline{\mathbb{R}}^1_+ \times \mathbb{R}^n)$. Finally, that potential satisfies the wave equation $\Box V_n = F$ (see Sec. 15.1).

The other properties of the potential V_n are substantially dependent on the properties of the density F.

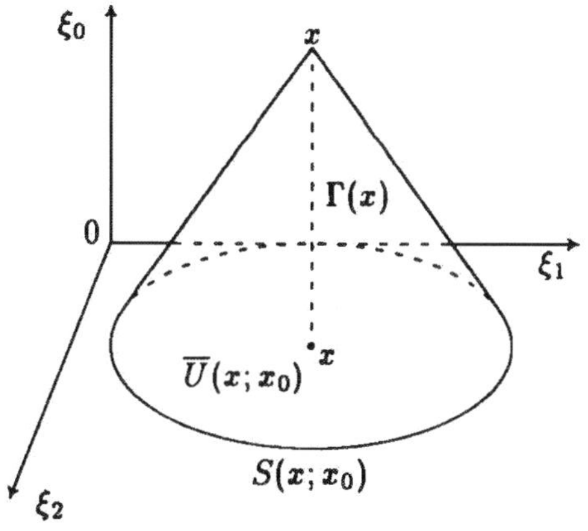

Figure 38

If $F = f \in \mathcal{L}^1_{\rm loc}(\overline{\mathbb{R}}^1_+ \times \mathbb{R}^n)$, *then the wave potential* V_n *is given by the formulae*

$$V_{2\nu+1}(x) = \frac{1}{2^{2\nu}\pi^\nu\Gamma(\nu)}\Box^{\nu-1}\int\limits_{|\mathbf{x}-\boldsymbol{\xi}|<x_0} \frac{f(x_0 - |\mathbf{x}-\boldsymbol{\xi}|, \boldsymbol{\xi})}{|\mathbf{x}-\boldsymbol{\xi}|}\,d\boldsymbol{\xi}, \qquad \nu \geq 1,$$

$$V_{2\nu}(x) = \frac{1}{2^{2\nu-1}\pi^\nu\Gamma(\nu)}\Box^{\nu-1}\int\limits_0^{x_0}\int\limits_{|\mathbf{x}-\boldsymbol{\xi}|<x_0-\xi_0} \frac{f(\xi)\,d\boldsymbol{\xi}\,d\xi_0}{\sqrt{|\mathbf{x}-\boldsymbol{\xi}|^2}},$$

$$V_1(x) = \frac{1}{2}\int\limits_0^{x_0}\int\limits_{x_1-x_0+\xi_0}^{x_1+x_0-\xi_0} f(\xi)\,d\xi_1\,d\xi_0. \tag{2.5}$$

Suppose $n = 2\nu + 1 \geq 3$. Using the representation (5.1) of Sec. 4.5 for the convolution $f * \mathcal{E}_n$, we have, from (2.1) and (4.11) of Sec. 15.4 for all $\varphi \in \mathcal{D}(\mathbb{R}^{n+1})$,

$$\begin{aligned}
(V_n, \varphi) &= (f * \mathcal{E}_n, \varphi)\\
&= c_\nu\big(f * \Box^{\nu-1}[\theta(x_0)\delta(x^2)], \varphi\big)\\
&= c_\nu\,\big(\Box^{\nu-1}(f * \theta(x_0)\delta(x^2)), \varphi\big)\\
&= c_\nu\,\big(f * \theta(x_0)\delta(x^2), \Box^{\nu-1}\varphi\big)\\
&= c_\nu\,\big(f(\xi) \times \theta(y_0)\delta(y^2), \eta(\xi)\eta_1(y)\Box^{\nu-1}\varphi(\xi + y)\big)\\
&= c_\nu\left(\theta(y_0)\delta(y^2), \eta_1(y)\int f(\xi)\Box^{\nu-1}\varphi(\xi + y)\,d\xi\right)\\
&= c_\nu\left(\theta(y_0)\delta(y^2), \eta_1(y)\int f(x - y)\Box^{\nu-1}\varphi(x)\,dx\right)\\
&= \frac{c_\nu}{2}\int\left[\int f(x_0 - |\mathbf{y}|, \mathbf{x} - \mathbf{y})\Box^{\nu-1}\varphi(x)\,dx\right]\frac{d\mathbf{y}}{|\mathbf{y}|}\\
&= \frac{c_\nu}{2}\int \Box^{\nu-1}\varphi(x)\int\frac{f(x_0 - |\mathbf{x}-\boldsymbol{\xi}|, \boldsymbol{\xi})}{|\mathbf{x}-\boldsymbol{\xi}|}\,d\boldsymbol{\xi}\,dx,
\end{aligned}$$

whence follows the first of the formulae of (2.5); here, $c_\nu 2^{2\nu-1}\pi^\nu\Gamma(\nu) = 1$. Similarly, and more simply, we can prove the other formulae of (2.5).

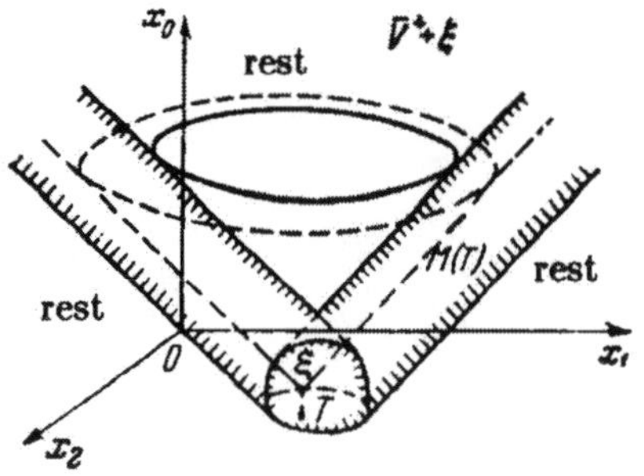

Figure 39

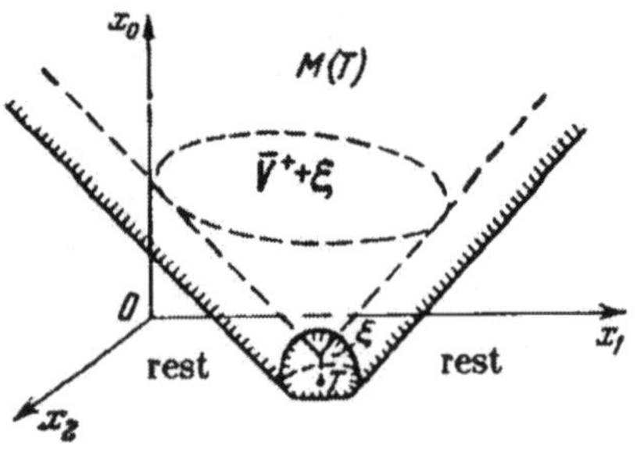

Figure 40

REMARK. From (2.5), for $n = 2\nu + 1 \geq 3$, it follows that the potential $V_{2\nu+1}(x)$ at the point $\mathbf{x}$ at time $t = x_0 > 0$ is completely specified by the values of the source $f(\xi)$ on the lateral surface of the cone (Fig. 38)

$$\Gamma(x) = \left[\xi : \xi_0 - x_0 \leq -|\mathbf{x} - \xi|,\ \xi_0 \geq 0\right].$$

That is, by the values of the source $f(x_0 - |\mathbf{x} - \xi|, \xi)$ in the ball $\overline{U}(\mathbf{x}; x_0)$, which values are taken at early times $\xi_0 = x_0 - |\mathbf{x} - \xi|$; and the delay time $|\mathbf{x} - \xi|$ is the time required for the perturbation to move from point ξ to point $\mathbf{x}$.[3] On the other hand, from (2.5), for $n = 2\nu$, it follows that the value of the potential $V_{2\nu}(x)$ is completely determined by the values of the source $f(\xi)$ on the cone $\Gamma(x)$ itself.

Let the source f be concentrated on a closed set $T \subset \mathbb{R}^{n+1}$. By the foregoing, a perturbation, for odd $n \geq 3$, is propagated from T onto the set $M(T)$, which is the union of boundaries of the future light cones $V^+ + \xi$ when their vertices ξ run through T (Fig. 39). For even n, the perturbation is propagated onto a union of the closed cones themselves $\overline{V^+} + \xi$, $\xi \in T$ (Fig. 40). The set $M(T)$ obtained in this fashion is called the *influence* region of the set T. It is clear that outside $M(T)$ we have a *region of rest*.

16.3. Surface wave potentials. Suppose the density $F = u_1(\mathbf{x}) \times \delta(x_0)$ or $F = u_0(\mathbf{x}) \times \delta'(x_0)$, where u_0 and u_1 are arbitrary generalized functions in $\mathcal{D}'(\mathbb{R}^n)$. The wave potentials

$$V_n^{(0)} = \left[u_1(\mathbf{x}) \times \delta(x_0)\right] * \mathcal{E}_n,$$
$$V_n^{(1)} = \left[u_0(\mathbf{x}) \times \delta'(x_0)\right] * \mathcal{E}_n$$

are called *surface wave potentials* (of the type of a simple and double layer with densities u_1 and u_0 respectively). The wave potential $V_n^{(1)}$ is the derivative, with respect to x_0, of $V_n^{(0)}$ with the same density:

$$V_n^{(1)} = \frac{\partial}{\partial x_0}\left\{\left[u_0(\mathbf{x}) \times \delta(x_0)\right] * \mathcal{E}_n\right\}. \tag{3.1}$$

---

[3]This is the reason why the wave potential $V_{2\nu+1}$ is also called the *retarded* potential.

The surface wave potentials $V_n^{(0)}$ *and* $V_n^{(1)}$ *belong to the class* $C^\infty([0,\infty))$ *with respect to* x_0; for $x_0 > 0$ and $k = 0, 1, \ldots,$

$$\frac{\partial^k}{\partial x_0^k} V_{nx_0}^{(0)}(\mathbf{x}) = u_1 * \frac{\partial^k \mathcal{E}_{nx_0}}{\partial x_0^k}, \tag{3.2}$$

$$\frac{\partial^k}{\partial x_0^k} V_{nx_0}^{(1)}(\mathbf{x}) = u_0 * \frac{\partial^{k+1} \mathcal{E}_{nx_0}}{\partial x_0^{k+1}}. \tag{3.3}$$

Indeed, using the lemma of Sec. 16.2, we have, for all $\varphi \in \mathcal{D}(\mathbb{R}^n)$ and $\psi \in \mathcal{D}(x_0 > 0)$ (see also Sec. 3.4),

$$\begin{aligned}
\left(\frac{\partial^k V_{n\varphi}^{(0)}}{\partial x_0^k}, \psi\right) &= (-1)^k (V_{n\varphi}^{(0)}, \psi^{(k)}) \\
&= (-1)^k (V_n^{(0)}, \varphi\psi^{(k)}) \\
&= (-1)^k ([u_1 \times \delta] * \mathcal{E}_n, \varphi\psi^{(k)}) \\
&= \left(\frac{\partial^k \mathcal{E}_n(x)}{\partial x_0^k} \times u_1(\xi) \times \delta(\xi_0), \eta(x)\eta_1(\xi)\varphi(\mathbf{x}+\xi)\psi(x_0+\xi_0)\right) \\
&= \left(\frac{\partial^k \mathcal{E}_n(x)}{\partial x_0^k}, \eta(x)\psi(x_0)(u_1(\xi), \varphi(\mathbf{x}+\xi))\right) \\
&= \int \left(\frac{\partial^k \mathcal{E}_{nx_0}(x)}{\partial x_0^k}, \eta(x)(u_1(\xi), \varphi(\mathbf{x}+\xi))\right) \psi(x_0)\, dx_0 \\
&= \int \left(u_1 * \frac{\partial^k \mathcal{E}_{nx_0}}{\partial x_0^k}, \varphi\right) \psi(x_0)\, dx_0,
\end{aligned}$$

so that

$$\frac{d^k V_{n\varphi}^{(0)}(x_0)}{dx_0^k} = \left(\frac{\partial^k \mathcal{E}_{nx_0}}{\partial x_0^k} * u_1, \varphi\right). \tag{3.4}$$

From this, using the theorem on the continuity of a convolution (see Sec. 4.3) and the lemma of Sec. 16.2, we conclude that $V_{n\varphi}^{(0)} \in C^\infty([0,\infty))$ and therefore $V_n^{(0)} \in C^\infty([0,\infty))$ with respect to x_0. The formula (3.2) follows from (3.4) since, by (4.1) and (4.3) of Sec. 3.4,

$$\frac{d^k V_{n\varphi}^{(0)}(x_0)}{dx_0^k} = \frac{d^k}{dx_0^k}(V_{nx_0}^{(0)}, \varphi) = \left(\frac{\partial^k V_{nx_0}^{(0)}}{\partial x_0^k}, \varphi\right).$$

From what has been proved for the potential $V_n^{(0)}$ follow all the statements concerning the potential $V_n^{(1)}$; here, use must be made of formula (3.1).

The following *limiting relations occur as* $x_0 \to +0$ *for the potentials* $V_n^{(0)}$ *and* $V_n^{(1)}$:

$$\begin{aligned}
&V_{nx_0}^{(0)}(\mathbf{x}) \to 0, &&\frac{\partial V_{nx_0}^{(0)}(\mathbf{x})}{\partial x_0} \to u_1(\mathbf{x}) \quad \text{in} \quad \mathcal{D}'(\mathbb{R}^n), \\
&V_{nx_0}^{(1)}(\mathbf{x}) \to u_0(\mathbf{x}), &&\frac{\partial V_{nx_0}^{(1)}(\mathbf{x})}{\partial x_0} \to 0 \qquad \text{in} \quad \mathcal{D}'(\mathbb{R}^n).
\end{aligned} \tag{3.5}$$

These follows from (2.2), (3.2) and (3.3) and from the continuity of the convolution (see Sec. 4.3).

If $u_1 \in \mathcal{L}^1_{\text{loc}}$, then for $x_0 > 0$,

$$V_n^{(0)}(x) = \frac{1}{2^{2\nu}\pi^\nu\Gamma(\nu)}\Box^{\nu-1}\frac{1}{x_0}\int\limits_{|\mathbf{x}-\boldsymbol{\xi}|=x_0} u_1(\boldsymbol{\xi})\,dS_{\boldsymbol{\xi}}, \qquad n = 2\nu+1 \geq 3,$$

$$V_n^{(0)}(x) = \frac{1}{2^{2\nu-1}\pi^\nu\Gamma(\nu)}\Box^{\nu-1}\int\limits_{|\mathbf{x}-\boldsymbol{\xi}|<x_0} \frac{u_1(\boldsymbol{\xi})\,d\boldsymbol{\xi}}{\sqrt{x_0^2-|\mathbf{x}-\boldsymbol{\xi}|^2}}, \qquad n = 2\nu, \tag{3.6}$$

$$V_1^{(0)}(x) = \frac{1}{2}\int\limits_{x_1-x_0}^{x_1+x_0} u_1(\xi_1)\,d\xi_1, \qquad n = 1.$$

To prove the formulae (3.6) for $n = 2\nu+1 \geq 3$, let us make use of equalities (3.2) (for $k = 0$) and (2.4):

$$\begin{aligned} V_n^{(0)}(x) &= V_{nx_0}^{(0)}(\mathbf{x}) = u_1 * \mathcal{E}_{nx_0} \\ &= \frac{1}{2^{2\nu}\pi^\nu\Gamma(\nu)} u_1 * \Box^{\nu-1}\frac{1}{x_0}\delta_{S_{x_0}} \\ &= \frac{1}{2^{2\nu}\pi^\nu\Gamma(\nu)}\Box^{\nu-1}\frac{1}{x_0}(u_1 * \delta_{S_{x_0}}). \end{aligned} \tag{3.7}$$

We will show that

$$u_1 * \delta_{S_{x_0}} = \int\limits_{|\mathbf{x}-\boldsymbol{\xi}|=x_0} u_1(\boldsymbol{\xi})\,dS_{\boldsymbol{\xi}}. \tag{3.8}$$

(By the Fubini theorem, the integral in (3.8), for each $x_0 > 0$, exists for almost all $\mathbf{x} \in \mathbb{R}^n$.)

Indeed, using (3.3) of Sec. 4.3, we obtain for all φ in $\mathcal{D}(\mathbb{R}^n)$

$$\begin{aligned} (u_1 * \delta_{S_{x_0}}, \varphi) &= (u_1(\boldsymbol{\xi}) \times \delta_{S_{x_0}}(\mathbf{y}), \eta(\mathbf{y})\varphi(\boldsymbol{\xi}+\mathbf{y})) \\ &= \int\limits_{|\mathbf{y}|=x_0}\int u_1(\boldsymbol{\xi})\varphi(\boldsymbol{\xi}+\mathbf{y})\,d\boldsymbol{\xi}\,dS_{\mathbf{y}} \\ &= \int\limits_{|\mathbf{y}|=x_0}\int u_1(\mathbf{x}-\mathbf{y})\varphi(\mathbf{x})\,d\mathbf{x}\,dS_{\mathbf{y}} \\ &= \int\varphi(\mathbf{x})\int\limits_{|\mathbf{y}|=x_0} u(\mathbf{x}-\mathbf{y})\,dS_{\mathbf{y}}\,d\mathbf{x} \\ &= \int\varphi(\mathbf{x})\int\limits_{|\mathbf{x}-\boldsymbol{\xi}|=x_0} u(\boldsymbol{\xi})\,dS_{\boldsymbol{\xi}}\,d\mathbf{x}, \end{aligned}$$

whence follows (3.8).

From (3.8) and (3.7) follows (3.6) for $n = 2\nu+1 \geq 3$. In the other cases, the proof is analogous and simpler.

16.4. The Cauchy problem for the wave equation. In accordance with the general theory (see Sec. 16.1), *the solution of the generalized Cauchy problem for the wave equation*

$$\Box u = F(x), \qquad F \in \mathcal{D}(\overline{\mathbb{R}}^1_+ \times \mathbb{R}^n) \tag{4.1}$$

exists and is unique in $\mathcal{D}'(\overline{\mathbb{R}}^1_+ \times \mathbb{R}^n)$ *and is given in the form of a wave potential*

$$u = V_n = F * \mathcal{E}_n$$

with density F. In particular, if

$$F(x) = u_1(\mathbf{x}) \times \delta(x_0) + u_0(\mathbf{x}) \times \delta'(x_0),$$

then the appropriate solution $u \in C^\infty([0,\infty))$ with respect to x_0; for $x_0 > 0$ it is given as the sum of two surface wave potentials:

$$u(x) = u_{x_0}(\mathbf{x}) = u_1 * \mathcal{E}_{nx_0} + \frac{\partial}{\partial x_0}(u_0 * \mathcal{E}_{nx_0}) \tag{4.2}$$

and, by (3.5), it satisfies the initial conditions as $x_0 \to +0$,

$$u_{x_0}(\mathbf{x}) \to u_0(\mathbf{x}), \qquad \frac{\partial u_{\mathbf{x}_0}(\mathbf{x})}{\partial x_0} \to u_1(\mathbf{x}) \quad \text{in} \quad \mathcal{D}'(\mathbb{R}^n). \tag{4.3}$$

A question arises when the solution of the generalized Cauchy problem is classical.

THEOREM. *If* $f \in C^{p_n}(\overline{\mathbb{R}}^1_+ \times \mathbb{R}^n)$, $u_0 \in C^{p_n+1}(\mathbb{R}^n)$ *and* $u_1 \in C^{p_n}(\mathbb{R}^n)$, *where* $p_n = 2[\frac{n}{2}]$, $n \geq 2$ *and* $p_1 = 1$, *then the solution of the classical Cauchy problem* (1.8)–(1.9) *exists and is representable in the form of a sum of three wave potentials (the Kirchhoff–Poisson–d'Alembert formula):*

$$u(x) = \frac{1}{2^{2\nu}\pi^\nu\Gamma(\nu)}\square^{\nu-1}\Bigg[\int\limits_{|\mathbf{x}-\xi|<x_0} \frac{f(x_0 - |\mathbf{x}-\xi|, \xi)}{|\mathbf{x}-\xi|}\,d\xi$$
$$+ \frac{1}{x_0}\int\limits_{|\mathbf{x}-\xi|=x_0} u_1(\xi)\,dS_\xi + \frac{\partial}{\partial x_0}\frac{1}{x_0}\int\limits_{|\mathbf{x}-\xi|=x_0} u_0(\xi)\,dS_\xi\Bigg], \quad n = 2\nu+1 \geq 3;$$

$$u(x) = \frac{1}{2^{2\nu-1}\pi^\nu\Gamma(\nu)}\square^{\nu-1}\Bigg[\int_0^{x_0}\int\limits_{|\mathbf{x}-\xi|<x_0-\xi_0} \frac{f(\xi)\,d\xi\,d\xi_0}{\sqrt{|\mathbf{x}-\xi|^2}}$$
$$+ \int\limits_{|\mathbf{x}-\xi|<x_0} \frac{u_1(\xi)\,d\xi}{\sqrt{x_0^2 - |\mathbf{x}-\xi|^2}} + \frac{\partial}{\partial x_0}\int\limits_{|\mathbf{x}-\xi|<x_0} \frac{u_0(\xi)\,d\xi}{\sqrt{x_0^2 - |\mathbf{x}-\xi|^2}}\Bigg], \quad n = 2\nu;$$

$$u(x) = \frac{1}{2}\int_0^{x_0}\int_{x_1-x_0+\xi_0}^{x_1+x_0-\xi_0} f(\xi)\,d\xi_1\,d\xi_0$$
$$+ \frac{1}{2}\int_{x_1-x_0}^{x_1+x_0} u_1(\xi_1)\,d\xi_1 + \frac{u_0(x_1+x_0) + u_0(x_1-x_0)}{2}, \quad n = 1. \tag{4.4}$$

PROOF. To prove the theorem, it remains to establish, by (4.3), that all wave potentials belong to the class $C^2(\overline{\mathbb{R}}^1_+ \times \mathbb{R}^n)$. Let us do that for $n = 2\nu + 1 \geq 3$. We have

$$\int\limits_{|\mathbf{x}-\boldsymbol{\xi}|<x_0} \frac{f(x_0 - |\mathbf{x}-\boldsymbol{\xi}|, \boldsymbol{\xi})}{|\mathbf{x}-\boldsymbol{\xi}|}\, d\boldsymbol{\xi}$$

$$= x_0^{n-1} \int\limits_{|\mathbf{y}|<1} f[x_0(1-|\mathbf{y}|), \mathbf{x} + x_0\mathbf{y}]\, \frac{d\mathbf{y}}{|\mathbf{y}|} \in C^{2\nu}(\overline{\mathbb{R}}_+^1 \times \mathbb{R}^n),$$

the substitution $\boldsymbol{\xi} = \mathbf{x} + x_0\mathbf{y}$, $d\boldsymbol{\xi} = x_0^n\, d\mathbf{y}$;

$$\frac{1}{x_0} \int\limits_{|\mathbf{x}-\boldsymbol{\xi}|=x_0} u_1(\boldsymbol{\xi})\, dS_{\xi} = x_0^{n-2} \int\limits_{|s|=1} u_1(\mathbf{x} + x_0 s)\, ds \in C^{2\nu}(\mathbb{R}^{n+1}),$$

the substitution $\boldsymbol{\xi} = \mathbf{x} + x_0 s$, $dS_{\xi} = x_0^{n-1}\, ds$;

$$\frac{\partial}{\partial x_0} \frac{1}{x_0} \int\limits_{|\mathbf{x}-\boldsymbol{\xi}|=x_0} u_0(\boldsymbol{\xi})\, dS_{\xi} = \frac{\partial}{\partial x_0} x_0^{n-2} \int\limits_{|s|=1} u_0(\mathbf{x} + x_0 s)\, ds \in C^{2\nu}(\mathbb{R}^{n+1}),$$

whence follow the required properties of smoothness of the wave potentials for $n = 2\nu + 1 \geq 3$. The remaining cases are considered in similar fashion. The theorem is proved. □

16.5. A statement of the generalized Cauchy problem for the heat equation. The Cauchy problem for the heat equation is studied by a method similar to that presented in Secs. 16.1 to 16.4 for the wave equation.

Let us consider the Cauchy problem

$$\frac{\partial u}{\partial t} = \Delta u + f(x,t), \qquad u\big|_{t=0} = u_0(x) \tag{5.1}$$

where $f \in C(\overline{\mathbb{R}}_+^1 \times \mathbb{R}^n)$, $u_0 \in C(\mathbb{R}^n)$. Assuming that the classical solution $u(x,t)$ of the Cauchy problem (5.1) exists, then extending it and the function f via zero onto $t < 0$, as in Sec. 16.1, we obtain that the extended functions $\tilde{u}$ and $\tilde{f}$ satisfy, in $\mathbb{R}^{n+1}$, the equation

$$\frac{\partial \tilde{u}}{\partial t} = \Delta \tilde{u} + \tilde{f}(x,t) + u_0(x) \times \delta(t). \tag{5.2}$$

This remark makes it possible to generalized the statement of the Cauchy problem for the heat equation in the following direction. Suppose $F \in \mathcal{D}'(\overline{\mathbb{R}}_+^1 \times \mathbb{R}^n)$. The *generalized Cauchy problem* for the heat equation with source F is the name we will give to the problem of finding, in $\mathbb{R}^{n+1}$, a generalized solution u, in $\mathcal{D}'(\overline{\mathbb{R}}_+^1 \times \mathbb{R}^n)$, of the equation

$$\frac{\partial u}{\partial t} = \Delta u + F(x,t). \tag{5.3}$$

16.6. Heat potential. First of all, let us make a study of the properties of the fundamental solution $\mathcal{E}(x,t)$ of the heat operator. In Sec. 15.4.5, it was shown that

$$\mathcal{E}(x,t) = \frac{\theta(t)}{(4\pi t)^{n/2}} e^{-\frac{|x|^2}{4t}}.$$

This function is nonnegative, vanishes for $t < 0$, is infinitely differentiable for $(x,t) \neq 0$ and is locally integrable in $\mathbb{R}^{n+1}$. What is more,

$$\int \mathcal{E}(x,t)\,dx = 1, \qquad t > 0, \tag{6.1}$$

by virtue of

$$\int \mathcal{E}(x,t)\,dx = \frac{1}{(4\pi t)^{n/2}} \int e^{-\frac{|x|^2}{4t}}\,dx = \frac{1}{\pi^{n/2}} \int e^{-|u|^2}\,du = 1.$$

The graph of the function $\mathcal{E}(x,t)$ for various t ($t_1 < t_2 < t_3$) is depicted in Fig. 41.

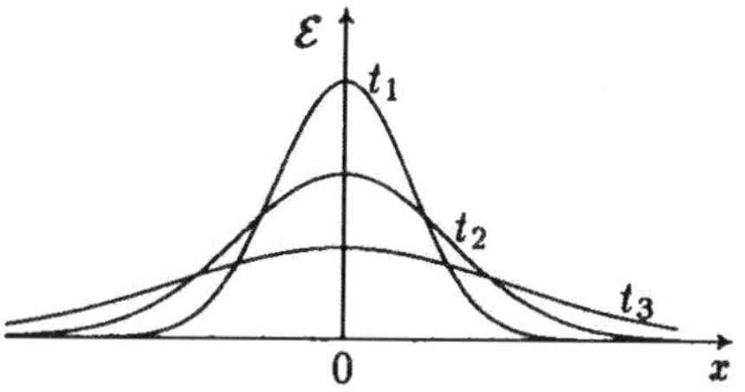

Figure 41

Let $F \in \mathcal{D}'(\overline{\mathbb{R}}^1_+ \times \mathbb{R}^n)$. The convolution $V = F * \mathcal{E}$ is termed the *heat potential* with density F. If the heat potential V exists in $\mathcal{D}'$, then it vanishes for $t < 0$, by virtue of (see Sec. 4.2.7)

$$\operatorname{supp} V \subset \overline{\operatorname{supp} F + \operatorname{supp} \mathcal{E}} = [(x,t) : t \geq 0],$$

so that $V \in \mathcal{D}'(\overline{\mathbb{R}}^1_+ \times \mathbb{R}^n)$, and it satisfies the heat equation (5.3).

Let us isolate the classes of densities F for which the heat potential definitely exists.

We denote by $\mathcal{M}$ the class of functions $\{f(x,t)\}$, measurable in $\mathbb{R}^{n+1}$, that vanish for $t < 0$ and that satisfy the following estimate in each strip $0 \leq t \leq T$, $x \in \mathbb{R}^n$:

$$|f(x,t)| \leq C_{T,\varepsilon}(f) e^{\varepsilon |x|^2} \tag{6.2}$$

for an arbitrary $\varepsilon > 0$. Similarly, by $\mathcal{M}_0$ we denote the class of functions $\{f(x)\}$ that are measurable in $\mathbb{R}^n$ and that satisfy the following estimate for arbitrary $\varepsilon > 0$:

$$|f(x)| \leq C_\varepsilon e^{\varepsilon |x|^2}, \qquad x \in \mathbb{R}^n. \tag{6.3}$$

In (6.2) it may be assumed that the quantity $C_{T,\varepsilon}$ does not decrease with respect to T.

If $f \in \mathcal{M}$, then the heat potential V exists in $\mathcal{M}$, is expressed by the integral

$$V(x,t) = \int_0^t \int \frac{f(\xi,\tau)}{\left[4\pi(t-\tau)\right]^{n/2}} e^{-\frac{|x-\xi|^2}{4(t-\tau)}}\,d\xi\,d\tau \tag{6.4}$$

satisfies the estimate: for arbitrary $\varepsilon > 0$,

$$|V(x,t)| \leq \frac{t C_{t,\varepsilon}(f)}{(1 - 8t\varepsilon)^{n/2}} e^{2\varepsilon |x|^2}, \qquad 0 < t < \frac{1}{8\varepsilon}, \tag{6.5}$$

and satisfies the initial condition: for arbitrary $R > 0$,

$$V(x,t) \overset{|x| \leq R}{\rightrightarrows} 0, \qquad t \to +0. \tag{6.6}$$

Indeed, since the functions $\mathcal{E}$ and f are locally integrable in $\mathbb{R}^{n+1}$, it follows that their convolution $V = f * \mathcal{E}$ exists, is expressed by the formula (6.4), and it is a locally integrable function in $\mathbb{R}^{n+1}$ if the function

$$h(x,t) = \int_0^t \int |f(\xi,\tau)|\,\mathcal{E}(x-\xi,t-\tau)\,d\xi\,d\tau$$

is locally integrable in $\mathbb{R}^{n+1}$ (see Sec. 4.1). We will prove that the function h satisfies the estimate (6.5). This estimate follows from the estimate (6.2), by virtue of the Fubini theorem,

$$h(x,t) \le C_{t,\varepsilon} \int_0^t \int e^{-\frac{|x-\xi|^2}{4(t-\tau)}+\varepsilon|\xi|^2} \frac{d\xi\,d\tau}{\left[4\pi(t-\tau)\right]^{n/2}}$$

$$\le C_{t,\varepsilon} \int_0^t \int e^{-\frac{|y|^2}{4s}+\varepsilon|y-x|^2} \frac{dy\,ds}{(4\pi s)^{n/2}}$$

$$\le C_{t,\varepsilon} e^{2\varepsilon|x|^2} \int_0^t \int e^{-|y|^2\left(\frac{1}{4s}-2\varepsilon\right)} \frac{dy\,ds}{(4\pi s)^{n/2}}.$$

Here we made use of the inequality $|y-x|^2 \le 2|y|^2 + 2|x|^2$. Making the following substitution in the inner integral,

$$u = y\sqrt{\frac{1}{4s} - 2\varepsilon}, \qquad du = \left(\frac{1}{4s} - 2\varepsilon\right)^{n/2} dy,$$

we continue our estimates:

$$h(x,t) \le C_{t,\varepsilon} e^{2\varepsilon|x|^2} \int_0^t \int \frac{e^{-u^2}\,du}{\pi^{n/2}} \frac{ds}{(1-8\varepsilon s)^{n/2}}$$

$$\le \frac{tC_{t,\varepsilon}}{(1-8\varepsilon t)^{n/2}} e^{2\varepsilon|x|^2}, \qquad 0 < t < \frac{1}{8\varepsilon},$$

which is what was required. Since $|V| \le h$, the potential V also satisfies the estimate (6.5). Furthermore, $V = h = 0$ for $t < 0$, so that $V \in \mathcal{M}$. From the estimate (6.5) it follows that V satisfies the initial conditions (6.6). □

Suppose the density $F(x,t) = u_0(x) \times \delta(t)$, where $u_0 \in \mathcal{D}'(\mathbb{R}^n)$. The heat potential

$$V^{(0)} = \left[u_0(x) \times \delta(t)\right] * \mathcal{E}$$

is called a *surface heat potential* (of the type of simple layer with density u_0).

If $u_0 \in \mathcal{M}_0$, then the surface heat potential $V^{(0)}$ exists in $\mathcal{M} \cap C^\infty(\overline{\mathbb{R}}^1_+ \times \mathbb{R}^n)$, is expressed by the integral

$$V^{(0)}(x,t) = \frac{\theta(t)}{(4\pi t)^{n/2}} \int u_0(\xi) e^{-\frac{|x-\xi|^2}{4t}}\,d\xi \tag{6.7}$$

and satisfies the estimate: for arbitrary $\varepsilon > 0$,

$$\left|V^{(0)}(x,t)\right| \le \frac{C_\varepsilon(u_0)}{(1-8\varepsilon t)^{n/2}} e^{2\varepsilon|x|^2}, \qquad 0 < t < \frac{1}{8\varepsilon}. \tag{6.8}$$

If, besides, $u_0 \in C$, then $V^{(0)} \in C(\overline{\mathbb{R}}^1_+ \times \mathbb{R}^n)$ and satisfies the initial condition

$$V^{(0)}\Big|_{t=+0} = u_0(x). \tag{6.9}$$

COROLLARY. *From (6.9) it follows that*

$$\mathcal{E}(x,t) \to \delta(x), \qquad t \to +0. \tag{6.10}$$

The representation (6.7) and the estimate (6.8) are proved in the same way as for the potential V. Here, use must be made of estimate (6.3). From the representation (6.7) it follows that $V^{(0)} \in \mathcal{M}$ and, besides, $V^{(0)} \in C^\infty$ for $t > 0$ and $x \in \mathbb{R}^n$. The latter again by virtue of (6.3).

It remains to prove that $V^{(0)}$ is a continuous function for $t \geq 0$, $x \in \mathbb{R}^n$ and satisfies the initial condition (6.9) if $u_0 \in \mathcal{M}_0 \cap C$. Suppose $(x,t) \to (x_0, 0)$, $t > 0$ and $\eta > 0$ is an arbitrary number. By continuity of the function $u_0(x)$, there is a number $\delta > 0$ such that

$$\big|u_0(\xi) - u_0(x_0)\big| < \eta \quad \text{for} \quad |\xi - x_0| < 2\delta.$$

Therefore if $|x - x_0| < \delta$ (then also $|x - y - x_0| < 2\delta$ for $|y| < \delta$), we will have, for $\varepsilon = 1$, by (6.1) and (6.3),

$$\begin{aligned}
\big|V^{(0)}(x,t) - u_0(x_0)\big| &\leq \int |u_0(\xi) - u_0(x_0)|\,\mathcal{E}(x-\xi,t)\,d\xi \\
&= \int\limits_{|y|<\delta} |u_0(x-y) - u_0(x_0)|\,\mathcal{E}(y,t)\,dy \\
&\quad + \int\limits_{|y|>\delta} |u_0(x-y) - u_0(x_0)|\,\mathcal{E}(y,t)\,dy \\
&\leq \eta \int \mathcal{E}(y,t)\,dy + \big|u_0(x_0)\big| \int\limits_{|y|>\delta} \mathcal{E}(y,t)\,dy \\
&\quad + C_1 \int\limits_{|y|>\delta} \mathcal{E}(y,t) e^{|x-y|^2}\,dy \\
&\leq \eta + \frac{\big|u_0(x_0)\big|}{(4\pi t)^{n/2}} \int\limits_{|y|>\delta} e^{-\frac{|y|^2}{4t}}\,dy + C_1 \frac{e^{2|x|^2}}{(4\pi t)^{n/2}} \int\limits_{|y|>\delta} e^{-|y|^2(\frac{1}{4t}-2)}\,dy \\
&\leq \eta + \frac{\big|u_0(x_0)\big|}{\pi^{n/2}} \int\limits_{|u|>\frac{\delta}{2\sqrt{t}}} e^{-|u|^2}\,du \\
&\quad + \frac{C_1 e^{|x|^2}}{\big[\pi(1-8t)\big]^{n/2}} \int\limits_{|u|>\frac{\delta}{2\sqrt{t}}\sqrt{1-8t}} e^{-|u|^2}\,du.
\end{aligned} \tag{6.11}$$

The second and the third summands in (6.11) may also be made $< \eta$ for all sufficiently small $t > 0$, $t < \delta_1$. Thus,

$$\Big|V^{(0)}(x,t) - u_0(x_0)\Big| < 3\eta, \qquad |x - x_0| < \delta, \qquad 0 < t < \delta_1,$$

which is what we require. □

16.7. Solution of the Cauchy problem for the heat equation.

THEOREM. *If $F(x,t) = f(x,t) + u_0(x) \times \delta(t)$, where $f \in \mathcal{M}$ and $u_0 \in \mathcal{M}_0$, then the solution of the generalized Cauchy problem* (5.3) *exists and is unique in the class $\mathcal{M}$ and is given as a sum of two heat potentials (Poisson's formula):*

$$\begin{aligned} u(x,t) &= V(x,t) + V^{(0)}(x,t) \\ &= \int_0^t \int \frac{f(\xi,\tau)}{[4\pi(t-\tau)]^{n/2}} e^{-\frac{|x-\xi|^2}{4(t-\tau)}} \, d\xi \, d\tau \\ &\quad + \frac{\theta(t)}{(4\pi t)^{n/2}} \int u_0(\xi) e^{-\frac{|x-\xi|^2}{4t}} \, d\xi. \end{aligned} \tag{7.1}$$

If, moreover, $f \in C^2(\overline{\mathbb{R}}^1_+ \times \mathbb{R}^n)$, $\partial^\alpha f \in \mathcal{M}$, $|\alpha| \le 2$, $u_0 \in \mathcal{M}_0 \cap C$, then the formula (7.1) *yields the classical solution to the Cauchy problem* (5.1).

PROOF. By what has been proved and in accordance with the general theory developed in Sec. 4.9.3, the solution of the equation

$$\frac{\partial u}{\partial t} = \Delta u + f(x,t) + u_0(x) \times \delta(t), \qquad (x,t) \in \mathbb{R}^{n+1}$$

exists and is unique in the class $\mathcal{M}$ and is given in the form of a sum of two heat potentials:

$$u = (f + u_0 \times \delta) * \mathcal{E} = f * \mathcal{E} + (u_0 \times \delta) * \mathcal{E} = V + V^{(0)}$$

whence and also from (6.4) and (6.7) follows formula (7.1).

Making a change of variables of integration in (6.4),

$$\xi = x - 2\sqrt{s}y, \qquad \tau = t - s,$$

we express the potential V in the form

$$V(x,t) = \frac{1}{\pi^{n/2}} \int_0^t \int f(x - 2\sqrt{s}y, t-s) e^{-|y|^2} \, dy \, ds. \tag{7.2}$$

Let $f \in C^2(\overline{\mathbb{R}}^1_+ \times \mathbb{R}^n)$ and $\partial^\alpha f \in \mathcal{M}$, $|\alpha| \le 2$. Using the theorems on the continuity and differentiability of integrals dependent on a parameter, we conclude, from formula (7.2) and from the equality

$$\begin{aligned} \frac{\partial V(x,t)}{\partial t} &= \frac{1}{\pi^{n/2}} \int_0^t \int \frac{\partial f(x - 2\sqrt{s}y, t-s)}{\partial t} e^{-|y|^2} \, dy \, ds \\ &\quad + \frac{1}{\pi^{n/2}} \int f(x - 2\sqrt{t}y, +0) e^{-|y|^2} \, dy, \end{aligned}$$

that all functions $\partial^\alpha V$, $|\alpha| \le 2$, with the exception of $\frac{\partial^2 V}{\partial t^2}$, are continuous for $t \ge 0$, $x \in \mathbb{R}^n$, and the function $\frac{\partial^2 V}{\partial t^2}$ is continuous for $t > 0$, $x \in \mathbb{R}^n$. Consequently, $V \in C^1(\overline{\mathbb{R}}^1_+ \times \mathbb{R}^n) \cap C^2(\overline{\mathbb{R}}^1_+ \times \mathbb{R}^n)$.

Finally, if $u_0 \in \mathcal{M}_0 \cap C$, then, by what has been proved, the potential $V^{(0)} \in C(\overline{\mathbb{R}}^1_+ \times \mathbb{R}^n) \cap C^\infty(\mathbb{R}^1_+ \times \mathbb{R}^n)$.

Thus, the generalized solution $u(x,t)$ defined by (7.1) belongs to the class $C(\overline{\mathbb{R}}^1_+ \times \mathbb{R}^n) \cap C^2(\mathbb{R}^1_+ \times \mathbb{R}^n)$ and therefore is the classical solution of the heat equation (5.1) for $t > 0$. Moreover, by (6.6) and (6.9), that solution satisfies the initial condition of (5.1) as well. Now this means that formula (7.1) will yield the solution to the classical Cauchy problem. The proof of the theorem is complete. □

REMARK. The uniqueness of the solution of the Cauchy problem for the heat equation may be established in a broader class, namely in the class of functions that satisfy in each strip $0 \le t \le T$, $x \in \mathbb{R}^n$, the estimate

$$|u(x,t)| \le C_T e^{a_T |x|^2},$$

(see, for example, Tikhonov [**101**]).

17. Holomorphic Functions with Nonnegative Imaginary Part in T^C

17.1. Preliminary remarks. We denote by $H_+(G)$ the class of functions that are holomorphic and have nonnegative imaginary part in the region G.

A function $u(x,y)$ of $2n$ variables (x,y) is said to be *plurisubharmonic* in the region $G \subset \mathbb{C}^n$ if it is semicontinuous above in G and its trace on every component of every open set $[\lambda\colon z^0 + \lambda a \subset G]$, $z^0 \in G$, $a \in \mathbb{C}^n$, $a \ne 0$, is a subharmonic function with respect to λ. The function $u(x,y)$ is said to be *pluriharmonic* in the region G if it is a real (or imaginary) part of some function that is holomorphic in G.

Concerning plurisubharmonic and convex functions, see, for example, Vladimirov [**105**, Chapter II].

The following statements are equivalent:

(1) A function $u(x,y)$ is pluriharmonic in G.

(2) A real generalized function $u(x,y)$ in $\mathcal{D}'(G)$ satisfies in G the system of equations

$$\frac{\partial^2 u}{\partial z_j\, \partial \bar{z}_k} = 0, \qquad 1 \le j,k \le n, \qquad z_j = x_j + iy_j.$$

(3) The functions $u(x,y)$ and $-u(x,y)$ are plurisubharmonic in G.

Here,

$$\frac{\partial}{\partial z_j} = \frac{1}{2}\left(\frac{\partial}{\partial x_j} - i\frac{\partial}{\partial y_j}\right), \qquad \frac{\partial}{\partial \bar{z}_j} = \frac{1}{2}\left(\frac{\partial}{\partial x_j} + i\frac{\partial}{\partial y_j}\right).$$

From this it follows that every pluriharmonic function in G is harmonic with respect to every pair of variables (x_j, y_j), $j = 1, \ldots, n$, separately and, hence, is a harmonic function in G,

$$\Delta u = \sum_{1\le j\le n} \left(\frac{\partial^2 u}{\partial x_j^2} + \frac{\partial^2}{\partial y_j^2}\right) = 4 \sum_{1\le j\le n} \frac{\partial^2 u}{\partial z_j\, \partial \bar{z}_j} = 0.$$

Therefore $u \in C^\infty(G)$ (see Sec. 15.6).

We denote by $\mathcal{P}_+(G)$ the class of nonnegative pluriharmonic functions in the region G.

Let the function $f(z)$ belong to the class $H_+(T^C)$ so that $\Im f \in \mathcal{P}_+(T^C)$, where the cone C is a domain. Without loss of generality, we may assume that the cone C is convex. Indeed, by the Bochner theorem the function $f(z)$ is holomorphic (and single-valued) in the hull of holomorphicity $T^{\operatorname{ch} C}$ of the domain T^C and assumes the same values in $T^{\operatorname{ch} C}$ as in T^C (see, for example, Vladimirov, [**105**, Sec. 17 and Sec. 20].

Furthermore, the cone C may be assumed to be different from the entire space $\mathbb{R}^n$. Otherwise, $f(z)$ is an entire function and the condition $\Im f(z) \geq 0$ in $\mathbb{C}^n$ leads via the Liouville theorem for harmonic functions to the equation $\Im f(z) = \text{const}$ in $\mathbb{C}^n$ and, hence, $f(z) = \text{const}$ in $\mathbb{C}^n$. Finally, we may assume that $\Im f(z) > 0$ in T^C. Indeed, if $\Im f(z^0) = 0$ in some point $z^0 \in T^C$, then, by the maximum principle for harmonic functions, $\Im f(z) \equiv 0$ in T^C, and then $f(z) = \text{const}$ in T^C.

The function $f(z)$ of the class $H_+(T^C)$ satisfies the following estimate (see Sec. 13.3): for any cone $C' \Subset C$ there is a number $M(C')$ such that

$$|f(z)| \leq M(C')\frac{1+|z|^2}{|y|}, \qquad z \in T^{C'}. \tag{1.1}$$

Consequently, $f \in H(C)$ (see Sec. 12.1).

Now let C be a (convex) acute cone (see Sec. 4.4) and let $f \in H_+(T^C)$. By virtue of the estimate (1.1), $f(z)$ possesses a spectral function $g(\xi)$ taken from $S'(C^*)$ (see Sec. 12.2), $f(z) = L[g]$. From this, using the definition of the Laplace transform (see Sec. 9.1), we have, for all $z \in T^C$,

$$\Im f(x+iy) = \frac{f(z) - \bar{f}(z)}{2i} = F\left[\frac{g(\xi)e^{-(y,\xi)} - g^*(\xi)e^{(y,\xi)}}{2i}\right](x) \tag{1.2}$$

where $g(\xi) \to g^*(\xi) = \overline{g(-\xi)}$. From (1.2) we derive the equation

$$\frac{1}{2i}\left[g(\xi)e^{-(y,\xi)} - g^*(\xi)e^{(y,\xi)}\right] = F_x^{-1}[\Im f(x+iy)](\xi), \qquad y \in C. \tag{1.3}$$

Let $f_+(x)$ be a boundary value of $f(z)$ in S', that is,

$$\int f(x+iy)\varphi(x)\,dx \to (f_+, \varphi), \qquad y \to 0, \quad y \in C, \quad \varphi \in S. \tag{1.4}$$

Then $g = F^{-1}[f_+]$ and $\Im f_+$ is a tempered nonnegative measure (see Sec. 5.3). We denote it by $\mu = \Im f_+$.

Passing to the limit in (1.3) as $y \to 0$, $y \in C$ in S' (see Sec. 12.2), and using (1.4), we obtain

$$\frac{g - g^*}{2i} = F^{-1}[\Im f_+] = F^{-1}[\mu], \tag{1.5}$$

so that

$$-ig(\xi) + ig^*(\xi) \gg 0 \qquad \text{(see (1.4) of Sec. 8.1)}$$

is a positive definite generalized function by virtue of the Bochner–Schwartz theorem (see Sec. 8.2).

Let us now prove the following uniqueness theorem for functions of the class $H_+(T^C)$ [and the class $\mathcal{P}_+(T^C)$].

THEOREM. *If $f \in H_+(T^C)$ and $\mu = \Im f_+ = 0$, then $f(z) = (a, z) + b$, where $a \in C^*$ and $\Im b = 0$.*

COROLLARY. *If $u \in \mathcal{P}_+(T^C)$ and its boundary value $\mu = 0$, then $u(x, y) = (a, y)$, where $a \in C^*$.*

PROOF. Since $\mu = 0$, it follows that, by (1.5), the spectral function g [in $S'(C^*)$] of the function f satisfies the condition $g = g^*$ and, hence, since $-C^* \cap C^* = \{0\}$

(cone C^* is acute!), the $\operatorname{supp} g = \{0\}$. By the theorem of Sec. 2.6,

$$g(\xi) = \sum_{|\alpha| \le N} c_\alpha \partial^\alpha \delta(\xi),$$

so that $f(z)$ is a polynomial. But $f \in H_+(T^C)$ and the estimate (1.1) shows that the degree of that polynomial cannot exceed one, so that $f(z) = (a, z) + b$, $z \in T^C$. But

$$\Im f(z) = (\Re a, y) + (\Im a, x) + \Im b \ge 0, \qquad z \in T^C$$

and therefore $\Re a \in C^*$ and $\Im a = 0$. Furthermore, from $\Im f_+(x) = 0$ it follows that $\Im b = 0$. The proof of the theorem is complete. □

REMARK. This theorem is an elementary variant of Bogolyubov's "edge-of-the-wedge" theorem (see, for example, Vladimirov [**105**, Sec. 27]).

EXAMPLES OF FUNCTIONS OF THE CLASS $H_+(G)$.

(1) If $f \in H_+(G)$, then $-\frac{1}{f} \in H_+(G)$ (see Sec. 13.3).
(2) If C is an acute cone, $\mu \not\equiv 0$ a nonnegative measure on the unit sphere, $\operatorname{supp} \mu \subset \operatorname{pr} C^*$, then

$$\left[\int_{\operatorname{pr} C^*} \frac{\mu(d\sigma)}{(z, \sigma)} \right]^{-1} \in H_+(T^{\operatorname{ch} C}).$$

(3) $\sqrt{z^2} \in H_+(T^{V^+})$ (see Example 2 of Sec. 10.2).

17.2. Properties of functions of the class $\mathcal{P}_+(T^C)$. Every function $u(x, y)$ of the class $\mathcal{P}_+(T^C)$ is an imaginary part of some function $f(z)$ of the class $H_+(T^C)$. Therefore it satisfies the estimate (1.1), and its boundary value in S' is a nonnegative tempered measure $\mu = \Im f_+ = u(x, +0)$, so that, by (1.4)

$$\int u(x, y)\varphi(x)\, dx \to \int \varphi(x)\mu(dx), \qquad y \to 0, \quad y \in C, \quad \varphi \in S. \tag{2.1}$$

However, for functions of the class $\mathcal{P}_+(T^C)$ more precise estimates of growth and boundary behaviour may be indicated in terms of the appropriate Poisson integral; namely, the following theorem holds.

THEOREM. *If $u \in \mathcal{P}_+(T^C)$, where C is an acute (convex) cone, then we have the estimate*

$$\int \mathcal{P}_C(x - x', y) u(x', y')\, dx' \le u(x, y + y'), \qquad (x, y) \in T^C, \quad y' \in C, \tag{2.2}$$

where $\mathcal{P}_C$ is the Poisson kernel of the tubular domain T^C. In particular, for a boundary value of the function $u(x, y)$, of the measure $\mu = u(x, +0)$, the estimate (2.2) takes the form

$$\int \mathcal{P}_C(x - x', y)\mu(dx') \le u(x, y) \qquad (x, y) \in T^C. \tag{2.2'}$$

The function $u(x,y)$ takes a boundary value μ in the following sense: for any $\varphi \in C \cap \mathcal{L}^\infty$,

$$\int u(x,y')\mathcal{P}_C(x,y)\varphi(x)\,dx \to \int \varphi(x)\mathcal{P}_C(x,y)\mu(dx), \qquad y' \to 0, \quad y' \in C', \quad \forall C' \Subset C, \quad y \in C. \tag{2.3}$$

COROLLARIES. *The following statements hold true under the hypotheses of the theorem:*

(1) *For arbitrary $\varepsilon > 0$ and for the compact $K \Subset T^C$, there is a number $R > 0$ such that*

$$\int_{|x'|>R} \mathcal{P}_C(x-x',y)\mu(dx') < \varepsilon, \qquad (x,y) \in K. \tag{2.4}$$

(2) *If $f \in C \cap \mathcal{L}^\infty$, then the integral*

$$\int f(x-x')\mathcal{P}_C(x-x',y)\mu(dx') \tag{2.5}$$

is a continuous function in T^C.

(3) *For the Poisson integral*

$$\int \mathcal{P}_C(x-x',y)\mu(dx') = \mu * \mathcal{P}_C$$

the Fourier transform formula

$$F[\mu * \mathcal{P}_C] = F[\mu]F[\mathcal{P}_C] \tag{2.6}$$

holds true.

(4) *The following limiting relations hold:*

$$\int \mathcal{P}_C(x-x',y)\mu(dx') \to \mu, \qquad y \to 0, \quad y \in C \quad \text{in} \quad \mathcal{S}', \tag{2.7}$$

$$\int \mathcal{P}_C(x-x',y')u(x',y)\,dx' \to u(x,y), \quad y' \to 0, \quad y' \in C, \quad (x,y) \in T^C. \tag{2.8}$$

(5) *There is a function $v_C(y)$ with the following properties:*
   (a) *$v_C(y)$ is nonnegative and continuous in C;*
   (b) *$v_C(y) \to 0$, $y \to 0$, $y \in C$;*
   (c) *the following representation holds*

$$u(x,y) = \int \mathcal{P}_C(x-x',y)\mu(dx') + v_C(y), \qquad (x,y) \in T^C. \tag{2.9}$$

(6) *If C is a regular cone, then*

$$\int u(x',y')\mathcal{S}_C(z-x';z^0-x')\,dx' \to \int \mathcal{S}_C(z-x';z^0-x')\mu(dx'), \qquad y' \to 0, \quad y' \in C', \quad C' \Subset C, \quad z \in T^C, \quad z^0 \in T^C, \tag{2.10}$$

where $\mathcal{S}_C$ is the Schwartz kernel of the tubular domain T^C (see Sec. 12.5).

REMARK 1. Since $u \in \mathcal{P}_+(T^C)$ implies that $u \in \mathcal{P}_+(T^{C_1})$, $C_1 \subset C$, it follows that all the above-enumerated statements hold true also for an arbitrary (open) convex cone $C_1 \subset C$.

REMARK 2. The limiting relation (2.7) also holds on functions of the form

$$\varphi(x) = \psi(x)\mathcal{P}_{C_1}(x, y'), \qquad \psi \in C \cap \mathcal{L}^\infty, \qquad y' \in C_1, \qquad C_1 \subset C. \tag{2.11}$$

REMARK 3. The estimates (2.2) and (2.2′), for $n = 1$, $C = (0, \infty)$ (upper half-plane), follow from the Herglotz–Nevanlinna representation (see Sec. 18.2 below). In the general case, they have been proved by Vladimirov (in **[111]** for $C = \mathbb{R}^n_+$; in **[114**, II] for $C = V^+$, $n = 4$; in **[116]** for the general case).

REMARK 4. The representation (2.9) was obtained in two special cases by Vladimirov in **[111]** ($C = \mathbb{R}^n_+$) and in **[114**, II] ($C = V^+$, $n = 4$).

To prove the theorem, fix $\varepsilon > 0$ and set

$$f_\varepsilon(z) = \frac{f(z)}{1 - i\varepsilon f(z)}, \tag{2.12}$$

where $f \in H_+(T^C)$ is such that $\Im f = u$. Put $\Re f = v$, so that $f = v + iu$. The function $f_\varepsilon(z)$ has the following properties:

(a) It belongs to the class $H_+(T^C)$ since

$$\Re(1 - i\varepsilon f) = 1 + \varepsilon u \geq 1, \qquad \Im f_\varepsilon = \frac{u + \varepsilon(v^2 + u^2)}{(1 + \varepsilon u)^2 + \varepsilon^2 v^2} \geq 0;$$

(b) It is bounded in T^C,

$$|f_\varepsilon(z)| \leq \min\left[\frac{1}{\varepsilon}, |f(z)|\right],$$

since

$$|f_\varepsilon|^2 = \frac{v^2 + u^2}{1 + 2\varepsilon u + \varepsilon^2(v^2 + u^2)};$$

(c) $f_\varepsilon(z) \overset{z \in K}{\Longrightarrow} f(z)$, $\quad \varepsilon \to 0$, $\quad K \Subset T^C$.

Let y' be an arbitrary fixed point in C. The function $f_\varepsilon(z + iy')$ is bounded and continuous in z on $\overline{T^C}$. Therefore, for all $z^0 \in T^C$,

$$f_\varepsilon(x + iy')\mathcal{K}_C(x - \overline{z^0}) \in \mathcal{L}^2 = \mathcal{H}_0,$$

so that the condition (5.5) of Sec. 12.5 is fulfilled. By the theorem of Sec. 12.5, the function $f_\varepsilon(z + iy')$ can be represented by the Poisson integral so that

$$\Im f_\varepsilon(z + iy') = \int \Im f_\varepsilon(x' + iy')\mathcal{P}_C(x - x', y)\, dx', \quad z \in T^C, \quad y' \in C. \tag{2.13}$$

Passing to the limit in (2.13) as $\varepsilon \to 0$, taking into account property (c), and using the Fatou lemma, we obtain inequality (2.2).

Now let us prove the estimate (2.2′). Let the sequence $\{\eta_k\}$ of functions taken from $\mathcal{D}(\mathbb{R}^n)$ converge to 1 in $\mathbb{R}^n$ (see Sec. 4.1), also $0 \leq \eta_k(x) \leq 1$, $\eta_k(x) \leq \eta_{k+1}(x)$, $k = 1, 2, \ldots$. Then from the inequality (2.2) it follows the inequality

$$\int \mathcal{P}_C(x - x', y)\eta_k(x')u(x', y')\, dx' \leq u(x, y + y'), \qquad k = 1, 2, \ldots,$$

whence and also from the limiting relation (2.1), as $y' \to 0$, $y' \in C$, we derive the inequality

$$\int \mathcal{P}_C(x - x', y)\eta_k(x')\mu(dx') \leq u(x, y), \qquad k = 1, 2, \ldots.$$

Passing to the limit here as $k \to \infty$ and using the theorem of B. Levi, we obtain the inequality (2.2′).

Now let us prove the limiting relation (2.3) for $y' \to 0$, $y' \in C$, on the functions $\varphi \in C$, $\varphi(\infty) = 0$, that is $\varphi \in \overline{C}_0$ (see Sec. 0.5). Since $\mathcal{D}$ is dense in $\overline{C}_0$ in norm in C (see Sec. 1.2), it follows that for $\forall \varepsilon > 0$, $\exists \psi \in \mathcal{D}$ such that $|\varphi(x) - \psi(x)| < \varepsilon$, $x \in \mathbb{R}^n$. From this, and also from the inequalities (2.2) and (2.2′), we derive the inequality

$$\begin{aligned}\left|\int u(x,y')\mathcal{P}_C(x,y)\varphi(x)\,dx - \int \varphi(x)\mathcal{P}_C(x,y)\mu(dx)\right| \\ \le \left|\int u(x,y')\mathcal{P}_C(x,y)\psi(x)\,dx - \int \psi(x)\mathcal{P}_C(x,y)\mu(dx)\right| \\ + \varepsilon\big[u(0,y+y') + u(0,y)\big],\end{aligned}$$

from which and from (2.1) we conclude that (2.3) holds on the functions $\varphi \in \overline{C}_0$ (if $y' \to 0$, $y' \in C$).

We now prove Corollary (1). It suffices to prove it for any sufficiently small ball K. Let $U(z_0; 2r_0) \Subset T^C$. By (2.2′) and (1.1) of Sec. 11.1 we have the inequality

$$\int \left|\mathcal{K}_C(x - x' + iy)\right|^2 \mu(dx') \le (2\pi)^n \mathcal{K}_C(2iy)u(x,y). \tag{2.14}$$

Since the function $|\mathcal{K}_C(z)|^2$ is plurisubharmonic in T^C, it follows, by the theorem on the spherical mean, that for all $x' \in \mathbb{R}^n$ and $z \in U(z_0; r_0)$ the following inequality holds (see, for example, Vladimirov [105, Sec. 10]):

$$\begin{aligned}\left|\mathcal{K}_C(z - x')\right|^2 &\le C_0 \int\limits_{U(z;r_0)} \left|\mathcal{K}_C(x'' - x' + iy'')\right|^2 dx''\,dy'' \\ &\le C_0 \int\limits_{U(x_0;2r_0)} \left|\mathcal{K}_C(x'' - x' + iy'')\right|^2 dx''\,dy''\end{aligned} \tag{2.15}$$

where $1/C_0 = \operatorname{mes} U(0; r_0)$. From the inequality (2.14) there follows, by the Fubini theorem, the existence of the integral

$$\begin{aligned}&\int \int\limits_{U(x_0;2r_0)} \left|\mathcal{K}_C(x'' - x' + iy'')\right|^2 dx''\,dy''\mu(dx') \\ &= \int\limits_{U(x_0;2r_0)} \int \left|\mathcal{K}_C(x'' - x' + iy'')\right|^2 \mu(dx')\,dx''\,dy'' \\ &\le (2\pi)^n \int\limits_{U(x_0;2r_0)} \mathcal{K}_C(2iy'')u(x'',y'')\,dx''\,dy'' < \infty.\end{aligned}$$

From this, by B. Levi's theorem,

$$\lim_{R\to\infty} \int\limits_{|x'|>R} \int\limits_{U(x_0;2r_0)} \left|\mathcal{K}_C(x'' - x' + iy'')\right|^2 dx''\,dy''\mu(dx') = 0. \tag{2.16}$$

Integrating the inequality (2.15) over the domain $|x'| > R$ in measure μ, we derive, from (2.16),

$$\int\limits_{|x'|>R} |\mathcal{K}_C(z - x')|^2 \mu(dx') \xrightarrow{z\in U(z_0;r_0)} 0, \qquad R \to \infty$$

and from this follows (2.4).

Let us prove Corollary (2). By Corollary (1), the integral (2.5) can be represented as a sum of two integrals in the neighbourhood of each point of the domain T^C: of a continuous function (for $|x'| \leq R$) and of an arbitrarily small function (for $|x'| > R$).

We now prove Corollary (3). Formula (2.6) follows from (2.2′) and from the Fubini theorem by virtue of the following operations:

$$\begin{aligned}
(\mu * \mathcal{P}_C, \varphi) &= \iint \mathcal{P}_C(x - x', y)\mu(dx')\varphi(x)\, dx \\
&= \iint \mathcal{P}_C(x - x', y)\varphi(x)\, dx\, \mu(dx') \\
&= (\mu, \mathcal{P}_C * \varphi) \\
&= (F|\mu|, F^{-1}[\mathcal{P}_C * \varphi]) \\
&= (F|\mu|, F[\mathcal{P}_C]F^{-1}[\varphi]) \\
&= (F|\mu|F[\mathcal{P}_C], F^{-1}[\varphi]) \\
&= (F^{-1}[F|\mu|F[\mathcal{P}_C]], \varphi), \qquad \varphi \in \mathcal{S}.
\end{aligned}$$

We now prove the limiting relation (2.7) on the functions $\varphi(x)$ of the form (2.11), in which it is also assumed that $\psi(\infty) = 0$. Without loss of generality, we can assume that $\varphi \geq 0$. We set

$$\chi(x) = \int \varphi(x + x')\mu(dx') = \int \psi(x + x')\mathcal{P}_{C_1}(x + x', y')\mu(dx'). \tag{2.17}$$

Taking into account Remark 1, we conclude that the function $\chi \geq 0$, which is continuous in $\mathbb{R}^n$ [Corollary (2)], satisfies, by virtue of (2.2′), the estimates

$$\begin{aligned}
\chi(x) &\leq \|\psi\|_{\mathcal{L}^\infty} \int \mathcal{P}_{C_1}(x + x', y')\mu(dx') \\
&\leq \|\psi\|_{\mathcal{L}^\infty} u(-x, y'), \quad y' \in C_1,
\end{aligned} \tag{2.18}$$

$$\begin{aligned}
\int \chi(x)\mathcal{P}_C(x, y)\, dx &\leq \|\psi\|_{\mathcal{L}^\infty} \int u(-x, y')\mathcal{P}_C(x, y)\, dx \\
&\leq \|\psi\|_{\mathcal{L}^\infty} u(0, y + y'), \quad y \in C.
\end{aligned} \tag{2.19}$$

The estimate (2.19) makes it possible to apply the Fubini theorem in the following chain of equalities:

$$\begin{aligned}
\int \varphi(x) \int \mathcal{P}_C(x - x', y)\mu(dx')\, dx &= \int \mathcal{P}_C(\xi, y) \int \varphi(x' + \xi)\mu(dx')\, d\xi \\
&= \int \mathcal{P}_C(\xi, y)\chi(\xi)\, d\xi.
\end{aligned}$$

From this, passing to the limit as $y \to 0$, $y \in C$, and using property 11.3.1 of Sec. 11.3, of the Poisson kernel $\mathcal{P}_C$, we obtain the inequality

$$\underline{\lim} \int \varphi(x) \int \mathcal{P}_C(x - x', y)\mu(dx')\, dx \geq \lim \int_{|\xi|<1} \mathcal{P}_C(\xi, y)\chi(\xi)\, d\xi$$

$$= \chi(0) = \int \varphi(x')\mu(dx'). \qquad (2.20)$$

On the other hand, by the estimate (2.2) for the cone C and by the limiting relation (2.3) (it was proven under condition $\varphi(\infty) = 0$), as $y \to 0$, $y \in C$, we have

$$\overline{\lim} \int \varphi(x) \int \mathcal{P}_C(x - x', y)\mu(dx')\, dx \leq \lim \int \varphi(x)u(x, y)\, dx$$

$$= \lim \int \psi(x)\mathcal{P}_{C_1}(x, y')u(x, y)\, dx$$

$$= \int \psi(x)\mathcal{P}_{C_1}(x, y')\mu(dx)$$

$$= \int \varphi(x)\mu(dx). \qquad (2.21)$$

The inequality (2.21), together with the opposite inequality (2.20), yields the limiting relation (2.7) on the functions φ of the kind under consideration.

The case $\varphi \in S$ is considered analogously and more simply.

The same method is used to prove the limiting relation (2.8) [compare (2.20) and (2.21)]: if $y' \to 0$, $y' \in C$, then

$$\overline{\lim} \int \mathcal{P}_C(x - x', y')u(x', y)\, dx' \leq \lim u(x, y + y') = u(x, y)$$

$$= \lim \int_{|x-x'|<1} \mathcal{P}_C(x - x', y')u(x', y)\, dx'$$

$$\leq \underline{\lim} \int \mathcal{P}_C(x - x', y')u(x', y)\, dx'$$

Here we again made use of (2.2) and Sec. 11.3.1.

We now prove Corollary (5). The function

$$v(x, y) = u(x, y) - \int \mathcal{P}_C(x - x', y)\mu(dx'), \qquad (x, y) \in T^C \qquad (2.22)$$

is nonnegative [see (2.2′)], continuous in T^C (see Corollary 2), and satisfies the estimate [see (2.2)]: for any cone $C' \subset C$

$$\int v(x, y)\mathcal{P}_{C'}(x, y')\, dx \leq \int u(x, y)\mathcal{P}_{C'}(x, y')\, dx \leq u(0, y + y'),$$
$$y \in C', \qquad y' \in C'. \qquad (2.23)$$

Applying the Fourier transform F_x^{-1} to (2.22) and using the formulae (1.3), (1.5), and (2.6), we obtain, for all $y \in C$,

$$2iF_x^{-1}[v](\xi) = g(\xi)e^{-(y,\xi)} - g^*(\xi)e^{(y,\xi)} - [g(\xi) - g^*(\xi)]F_x[\mathcal{P}_C](\xi), \qquad (2.24)$$

where $g(\xi)$ is the spectral function of the function $f(x)$,

$$\Im f(z) = u(x, y), \qquad g \in \mathcal{S}'(C^*) \quad \text{(see Sec. 17.1)}.$$

Taking into consideration the equality (1.10) of Sec. 11.1,

$$F_x[\mathcal{P}_C](\xi) = e^{-|(y,\xi)|}, \qquad \xi \in -C^* \cup C^*,$$

and noting that $\operatorname{supp} g \subset C^*$, $\operatorname{supp} g^* \subset -C^*$, we derive, from (2.24), $\operatorname{supp} F_x^{-1}[v] = \{0\}$.

From this, via the theorem of Sec. 2.6, it follows that

$$F_x^{-1}[v](\xi) = \sum_{|\alpha| \le N(y)} C_\alpha(y) \partial^\alpha \delta(\xi),$$

so that $v(x, y)$ is a polynomial in x. If in (2.23) we regard C' as an n-hedral cone, we conclude that $v(x, y) = v(y)$ does not depend on x [see Sec. 11.3.5]. Thus the properties (a) and (c) are proved. It remains to prove property (b). Suppose $\omega \in \mathcal{D}$, $\int \omega(x)\, dx = 1$. Taking into account the limiting relation (2.1) and (2.7), we obtain, from (2.9),

$$\begin{aligned} v_C(y) &= \int v(y)\omega(x)\, dx \\ &= \int u(x, y)\omega(x)\, dx - \int \omega(x) \int \mathcal{P}_C(x - x', y)\mu(dx')\, dx \\ &\to \int \omega(x)\mu(dx) - \int \omega(x)\mu(dx) = 0, \qquad y \to 0, \quad y \in C, \end{aligned}$$

which is what we set out to prove.

Let us now extend the limiting relation (2.3) to the functions $\varphi \in C \cap \mathcal{L}^\infty$. Suppose $C' \Subset C$. Since every cone $C' \Subset C$ may be covered with a finite number of n-hedral cones that are compact in C, it suffices to establish (2.3) for the n-hedral cone C'. Using the representation(2.9), where cone C changed by the cone C',

$$u(x, y') = \int \mathcal{P}_{C'}(x - x', y')\mu(dx') + v_{C'}(y'), \qquad (x, y') \in T^{C'},$$

where $v_{C'}(y') \to 0$, $y' \to 0$, $y' \in C'$, we have the equation

$$\begin{aligned} \int u(x, y')\mathcal{P}_C(x, y)\varphi(x)\, dx &= \int \mathcal{P}_{C'}(x, y')\psi(x, y)\, dx \\ &+ v_{C'}(y') \int \mathcal{P}_C(x, y)\varphi(x)\, dx, \qquad y \in C', \quad y' \in C', \end{aligned} \tag{2.25}$$

where

$$\psi(x, y) = \int \varphi(x + x')\mathcal{P}_C(x + x', y)\mu(dx').$$

Interchanging the order of integration in the integral on the right of (2.25) is possible by the Fubini theorem and the estimates (2.2) and (2.2′), which ensure the existence of the iterated integral

$$\begin{aligned} \int \mathcal{P}_{C'}(x, y')|\psi(x, y)|\, dx &\le \int \mathcal{P}_{C'}(x, y') \int |\varphi(x + x')|\mathcal{P}_C(x + x', y)\mu(dx')\, dx \\ &\le \|\varphi\|_{\mathcal{L}^\infty} \int \mathcal{P}_{C'}(x, y')u(-x, y)\, dx \\ &\le \|\varphi\|_{\mathcal{L}^\infty} u(0, y + y'), \qquad y \in C, \quad y' \in C'. \end{aligned}$$

Furthermore, the last estimate, together with the continuity of the function $\psi(x,y)$ in T^C [see Corollary (2)] permits applying the result of Sec. 11.3.4 to the integral on the right of (2.25). As a result, when $y' \to 0$, $y' \in C'$, we obtain (2.3):

$$\int u(x,y')\mathcal{P}_C(x,y)\varphi(x)\,dx \to \psi(0,y) = \int \varphi(x')\mathcal{P}_C(x',y)\mu(dx'), \quad y \in C',$$

for arbitrary cone $C' \Subset C$, containing y, and (2.3) follows.

The truth of the limiting relation (2.7) on functions of the form (2.11) follows from what has been proved, provided that $y \to 0$, $y \in C'$, $\forall C' \Subset C$ (Remark 2).

Let us now prove Corollary (6). By virtue of the estimates (5.4) of Sec. 12.5 and (2.2), the integral (2.10) exists.

Setting

$$\varphi(x') = \frac{\mathcal{S}_C(z-x';z^0-x')\left|\mathcal{K}_C(z-\overline{z^0})\right|}{\mathcal{K}_C(2iy)\mathcal{P}_C(x-x',y) + \left[\mathcal{K}_C(2iy^0) + \left|\mathcal{K}_C(z-\overline{z^0})\right|\right]\mathcal{P}_C(x^0-x',y^0)},$$

we have $|\varphi(x')| \le 1$ [see (5.4) of Sec. 12.5], $\varphi \in C^\infty$ and

$$\int \mathcal{S}_C(z-x';z^0-x')u(x',y')\,dx' = \int \left\{ \frac{\mathcal{K}_C(2iy)}{\left|\mathcal{K}_C(z-\overline{z^0})\right|}\mathcal{P}_C(x-x',y) \right.$$
$$\left. + \left[\frac{\mathcal{K}_C(2iy^0)}{\left|\mathcal{K}_C(z-\overline{z^0})\right|} + 1\right]\mathcal{P}_C(x^0-x',y^0)\right\}\varphi(x')u(x',y')\,dx.$$

From this and from (2.3) follows the limiting relation (2.10).

This completes the proof of the theorem and its corollaries. □

17.3. Estimates of the growth of functions of the class $H_+(T^C)$. Here we will establish that, together with the estimate (1.1), any function of the class $H_+(T^C)$ is estimated in terms of its imaginary part, the estimates of the growth and boundary behaviour of which are given in Sec. 17.2.

THEOREM. *If $f \in H_+(T^C)$, then*

$$\left|\mathcal{K}_C(z-\overline{z^0})[f(z)-\bar{f}(z^0)]\right|^2 \le 4\mathcal{K}_C(2iy^0)\mathcal{K}_C(2iy)\Im f(z^0)\Im f(z), \quad z \in T^C, \quad z^0 \in T^C. \tag{3.1}$$

COROLLARY 1. *If $f \in H_+(T^C)$, then*

$$\int \left|\mathcal{K}_C(z-\overline{z^0})\right|^4 |f(z)-\bar{f}(z^0)|^2\,dx$$
$$\le 4(2\pi)^n \mathcal{K}_C(2iy^0)\mathcal{K}_C(2iy)\mathcal{K}_C(2iy^0+2iy)\Im f(z^0)\Im f(z^0+2iy), \quad y \in C, \quad z^0 \in T^C. \tag{3.2}$$

COROLLARY 2. *If $f \in H_+(T^C)$ and the cone C is regular, then, for any $y^0 \in C$ there exists $M = M(y^0)$ such that*

$$|f(z)| \le M|f(iy^0)|\left[\frac{1+|z|^2}{\Delta_C(y)}\right]^n, \quad z \in T^C. \tag{3.3}$$

To prove this, we construct, via (2.12), a function $f_\varepsilon(z)$, $\varepsilon > 0$ and apply to the function $f_\varepsilon(z+iy')$, $y' \in C$, the representation (5.9) of Sec. 12.5:

$$\mathcal{K}_C(z-\overline{z^0})[f_\varepsilon(z+iy') - \bar{f}_\varepsilon(z^0+iy')] = \frac{2i}{(2\pi)^n}\int \Im f_\varepsilon(x'+iy')\mathcal{K}_C(z-x')\mathcal{K}_C(x'-\overline{z^0})\,dx', \qquad z^0 \in T^C, \quad z \in T^C, \quad y' \in C.$$

From this, using the Cauchy–Bunyakovsky inequality, we derive the inequality

$$\begin{aligned}\left|\mathcal{K}_C(z-\overline{z^0})[f_\varepsilon(z+iy') - \bar{f}_\varepsilon(z^0+iy')]\right|^2 &\le \frac{4}{(2\pi)^{2n}}\int \Im f_\varepsilon(x'+iy')\,|\mathcal{K}_C(z-x')|^2\,dx' \\ &\quad\times \int \Im f_\varepsilon(x''+iy')\left|\mathcal{K}_C(x''-\overline{z^0})\right|^2 dx' \\ &= 4\mathcal{K}_C(2iy)\mathcal{K}_C(2iy^0)\int \Im f_\varepsilon(x'+iy')\mathcal{P}_C(x-x',y)\,dx' \\ &\quad\times \int \Im f_\varepsilon(x''+iy')\mathcal{P}_C(x^0-x'',y^0)\,dx''.\end{aligned}$$

Applying the estimate (2.2) twice, we get

$$\left|\mathcal{K}_C(z-\overline{z^0})[f_\varepsilon(z+iy') - \bar{f}_\varepsilon(z^0+iy')]\right|^2 \le 4\mathcal{K}_C(2iy^0)\mathcal{K}_C(2iy)\Im f_\varepsilon(z+iy')\Im f_\varepsilon(z^0+iy'), \qquad z^0 \in T^C, \quad z \in T^C, \quad y' \in C.$$

Allowing $y' \to 0$, $y' \in C$, and then $\varepsilon \to 0$, we obtain the required estimate (3.1). □

To prove the inequality (3.2) we multiply the inequality (3.1) by $\left|\mathcal{K}_C(z-\overline{z^0})\right|^2$, integrate with respect to x, and take advantage of the inequality (2.2):

$$\begin{aligned}\int \left|\mathcal{K}_C(z-\overline{z^0})\right|^4 |f(z)-\bar{f}(z^0)|^2\,dx &\le 4(2\pi)^n\mathcal{K}_C(2iy^0)\mathcal{K}_C(2iy)\mathcal{K}_C(2iy^0+2iy)\Im f(z^0) \\ &\quad\times \int \mathcal{P}_C(x-x^0,y+y^0)\Im f(x+iy)\,dx \\ &\le 4(2\pi)^n\mathcal{K}_C(2iy^0)\mathcal{K}_C(2iy)\mathcal{K}_C(2iy^0+2iy)\Im f(z^0)\Im f(z^0+2iy).\end{aligned}$$

□

In order to prove inequality (3.3) we set $z^0 = iy^0$, $y^0 \in C$, in (3.1) and make use of estimate (2.4) of Sec. 10.2 for $\alpha = 0$. As a result, we obtain the inequality

$$|\mathcal{K}_C(z+iy^0)|^2\,|f(z)-\bar{f}(iy^0)|^2 \le C_1\Delta_C^{-n}(y)|f(iy^0)|\,|f(z)|, \qquad z \in T^C, \quad (3.4)$$

for some C_1. Let us estimate the function $|\mathcal{K}_C(z+iy^0)|$ from below. By virtue of (2.2) of Sec. 10.2, we have

$$\begin{gathered} |\mathcal{K}_C(z+iy^0)| = |z+iy^0|^{-n}\Gamma(n)\left|\int_{\operatorname{pr} C^*} \frac{d\sigma}{(p+iq,\sigma)^n}\right|, \\ p+iq = \frac{x+i(y+y^0)}{|x+i(y+y^0)|}. \end{gathered} \tag{3.5}$$

By the hypothesis (the cone C is regular!), the function

$$\left|\int_{\operatorname{pr} C^*} \frac{d\sigma}{(p+iq,\sigma)^n}\right|$$

is positive and continuous on a compact (the cone C is acute!)

$$\left[(p,q)\in\mathbb{R}^{2n}: p^2+q^2=1,\ q=\frac{y+y^0}{|z+iy^0|},\ y\in C\right];$$

hence, it is bounded from below by a positive number C_2 independent of z (y^0 is fixed); therefore, by virtue of (3.5)

$$|\mathcal{K}_C(z+iy^0)| \geq C_2\Gamma(n)|z+iy^0|^{-n}, \qquad z\in T^C.$$

This and inequality (3.4) imply the inequality

$$\begin{aligned} |f(z)-\bar{f}(iy^0)|^2 &\leq C_3|z+iy^0|^{2n}\Delta_C^{-n}(y)|f(iy^0)|\,|f(z)| \\ &\leq C_4A^n|f(iy^0)|\,|f(z)|, \qquad z\in T^C, \end{aligned}$$

where the notation

$$A = (1+|z|^2)\Delta_C^{-1}(y) \geq 1$$

is used. In turn, the last inequality implies the inequality

$$|f(z)|^2 \leq 2|f(iy^0)|^2 + 2|f(z)-f(iy^0)|^2 \leq (2|f(iy^0)| + 2C_4A^n|f(z)|)\,|f(iy^0)|,$$

which implies inequality (3.3):

$$|f(z)| \leq \left(C_4A^n + \sqrt{C_4^2A^{2n}+2}\right)|f(iy^0)| \leq CA^n|f(iy^0)|.$$

□

PROBLEM. Does estimate (3.3) hold, if n is replaced by 1?

17.4. Smoothness of the spectral function. The estimates (3.1) and (3.2) imply a definite smoothness of the spectral functions of functions of the class $H_+(T^C)$, namely:

THEOREM. *If $f\in H_+(T^C)$, then its spectral function $g(\xi)$ has the property*

$$\theta_{C^*}^2 * g \in \mathcal{L}_s^2(C^*), \qquad s < -\frac{3}{2}n-1. \tag{4.1}$$

COROLLARY. *If $f\in H_+(T^C)$, where C is a regular cone, then its spectral function g is uniquely representable as*

$$\theta_{C^*}^{-2} * g_1, \qquad g_1\in\mathcal{L}_s^2(C^*), \qquad s < -\frac{3}{2}n-1. \tag{4.2}$$

REMARK. The operators $\theta_C^{\alpha} *$ are introduced in Sec. 13.5.

EXAMPLE 1 (see (2.16) of Sec. 10.2).

$$\theta_{\overline{\mathbb{R}}_+^n}^{-2} * = \frac{\partial^{2n}}{\partial\xi_1^2 \ldots \partial\xi_n^2}.$$

EXAMPLE 2 (see Sec. 13.5).

$$\theta_{\overline{V}^+}^{-2} * = 4^{-n}\pi^{-n+1}\Gamma^{-2}\left(\frac{n+1}{2}\right)\Box^{n+1}.$$

To prove the theorem, in the inequality (3.2) substitute $f(z+iy^0)$ for $f(z)$ and then set $x^0 = 0$, $y^0 = y$.

After a few simple manipulations we obtain

$$\begin{aligned}\int |\mathcal{K}_C(x+2iy)|^4 \left|f(x+2iy)\right|^2 dx \\ \le 2\left|f(2iy)\right|^2 \int |\mathcal{K}_C(x+2iy)|^4\, dx \\ + 8\pi^n \mathcal{K}_C^3(2iy)\Im f(2iy)\Im f(4iy), \qquad y \in C. \quad (4.3)\end{aligned}$$

But by (1.7) of Sec. 11.1 we have, for $p = 2$, the estimate

$$\begin{aligned}\int |\mathcal{K}_C(x+2iy)|^4\, dx &= 2\pi\mathcal{K}_C^2(4iy)\int \mathcal{P}_C^2(x,2y)\, dx \\ &\le 2\pi\mathcal{K}_C^2(4iy)\frac{\mathcal{K}_C^2(2iy)}{(2\pi)^n\mathcal{K}_C(4iy)} \\ &= \pi^n\mathcal{K}_C^3(2iy).\end{aligned}$$

Therefore, inequality (4.3) takes the form

$$\begin{aligned}\left\|\mathcal{K}_C^2(x+2iy)f(x+2iy)\right\|^2 \\ \le 8\pi^n\mathcal{K}_C^3(2iy)\left|f(2iy)\right|\left(\left|f(2iy)\right| + \left|f(4iy)\right|\right), \qquad y \in C. \quad (4.4)\end{aligned}$$

Now suppose an arbitrary cone $C' \Subset C$. Taking into account the estimate (1.1),

$$\left|f(iy)\right| \le M(C')\frac{1+|y|^2}{|y|}, \qquad y \in C',$$

and the estimate (2.4) of Sec. 10.2,

$$0 < \mathcal{K}_C(iy) \le M_0\Delta^{-n}(y), \qquad y \in C,$$

from (4.4) we derive the following estimate:

$$\begin{aligned}\left\|\mathcal{K}_C^2(x+2iy)f(x+2iy)\right\|^2 \\ \le M_1(C')\Delta^{-3n}(y)\frac{1+4|y|^2}{2|y|}\left(\frac{1+4|y|^2}{2|y|} + \frac{1+16|y|^2}{4|y|}\right), \qquad y \in C'. \quad (4.5)\end{aligned}$$

But $|y| \ge \Delta(y) \ge \sigma|y|$, $y \in C'$, for some $\sigma > 0$ (see Lemma 1 of Sec. 4.4). Therefore the estimate (4.5) may be rewritten as

$$\left\|\mathcal{K}_C^2(x+2iy)f(x+2iy)\right\|^2 \le M_2(C')\left[|y|^2 + \Delta^{-3n-2}(y)\right], \qquad y \in C'. \quad (4.6)$$

The estimate (4.6) holds true if distance $\Delta(y)$ is replaced by the lesser distance $\Delta'(y)$ (from y to $\partial C'$). Applying the lemma of Sec. 10.5 (for $a = 0$, $s = 0$, $\gamma = \frac{3}{2}n + 1$

and $C = C'$), we conclude that the function $\mathcal{K}_C^2(z)f(z)$ is the Laplace transform of the function

$$g_1(\xi) = \theta_{C^*}^2 * g \equiv \theta_{C^*} * \theta_{C^*} * g$$

[see Sec. 9.2.7] taken from $\mathcal{L}_{s'}^2(C^*)$ for all $s' < -\frac{3}{2}n - 1$ and $C' \Subset C$, where g is the spectral function of the function f: $f(z) = L[g]$. Hence $g_1 \in \mathcal{L}_s^2(C^*)$ for all $s < -\frac{3}{2}n - 1$. The theorem is proved. □

To prove the corollary, set

$$g_1 = \theta_{C^*}^2 * g \in \mathcal{L}_s^2(C^*), \qquad s < -\frac{3}{2}n - 1.$$

Then in the convolution algebra $\mathcal{S}'(C^*)$ we have, in the case of a regular cone C [see Sec. 4.9.4 and Sec. 13.1],

$$\theta_{C^*}^{-2} * g_1 = \theta_{C^*}^{-2} * (\theta_{C^*}^2 * g) = (\theta_{C^*}^{-2} * \theta_{C^*}^2) * g = \delta * g = g.$$

The function g_1 with the indicated properties is unique. □

The results of Secs. 17.2–17.4 have been obtained by Vladimirov [**116**].

17.5. Indicator of growth of functions of the class $\mathcal{P}_+(T^C)$. In Sec. 17.2 we studied the growth of functions of the class $\mathcal{P}_+(T^C)$ as $y \to 0$, $y \in C$, and as $|x| \to \infty$. Here we will investigate the growth of such functions as $|y| \to \infty$, $y \in C$. First we will prove the following lemmas.

LEMMA 1. *If $u \in \mathcal{P}_+(T^C)$, where C is a convex cone, then for every bounded region $D \subset \mathbb{R}^n$ and for every point $y \in C$ there is a number $t_0 \geq 0$ such that for all $(x^0, y^0) \in T^D$ the function $u(x^0, y^0 + ty)(t - t_0)^{-1}$ does not increase with respect to t on (t_0, ∞).*

PROOF. Fix $z^0 = x^0 + iy^0 \in T^D$ and $y \in C$. Since the cone C is open and convex, there is a number $t_0 = t_0(y^0, y)$ such that $y^0 + ty \in C$ for all $t > t_0$. Therefore the function $u(x^0 + \sigma y, y^0 + (\tau + t_0)y)$ belongs to the class $\mathcal{P}_+(T^1)$ [with respect to the variables (σ, τ)], and so it can be represented by the formula (see Sec. 18.1)

$$u(x^0 + \sigma y, y^0 + (\tau + t_0)y) = \frac{\tau}{\pi} \int_{-\infty}^{\infty} \frac{\mu(x^0, y^0, y; d\sigma')}{(\sigma - \sigma')^2 + \tau^2} + a(x^0, y^0, y)\tau, \tag{5.1}$$

where $a \geq 0$ and the measure $\mu \geq 0$ satisfies the condition of growth [see (2.2′)]

$$\int_{-\infty}^{\infty} \frac{\mu(x^0, y^0, y; d\sigma')}{1 + \sigma'^2} < \infty.$$

Putting $\sigma = 0$ in (5.1), dividing by τ, and setting $\tau = t - t_0 > 0$, we get

$$\frac{u(x^0, y^0 + ty)}{t - t_0} = \frac{1}{\pi} \int_{-\infty}^{\infty} \frac{\mu(x^0, y^0, y; d\sigma')}{\sigma'^2 + (t - t_0)^2} + a(x^0, y^0, y),$$

whence, by the B. Levi theorem, we conclude that Lemma 1 holds true. □

LEMMA 2. *Suppose the function $f(x)$ is convex on the set A. Then for all $x^0 \in A$ and $z \in \mathbb{R}^n$ the function*

$$\frac{1}{t}[f(x^0+tx)-f(x^0)]$$

does not decrease with respect to t on the interval $[0, t_0]$ provided that all the points x^0+tx, $0 \le t \le t_0$, are contained in A.

PROOF. By the definition of a convex function (see Sec. 0.2), the function $f(x^0+tx)$ is convex with respect to t on $[0, t_0]$ and, hence, for arbitrary $0 \le t < t' \le t_0$,

$$\begin{aligned} f(x^0+tx) &= f\left[\frac{t}{t'}(t'x+x^0)+\left(1-\frac{t}{t'}\right)x^0\right] \\ &\le \frac{t}{t'}f(x^0+t'x)+\left(1-\frac{t}{t'}\right)f(x^0), \end{aligned}$$

that is

$$\frac{1}{t}[f(x^0+tx)-f(x^0)] \le \frac{1}{t'}[f(x^0+t'x)-f(x^0)]$$

which completes the proof of Lemma 2. □

LEMMA 3. *If the function $u(x,y)$ is plurisubharmonic in the tubular domain $T^D = \mathbb{R}^n + iD$ and is bounded from above on every subdomain $T^{D'}$, $D' \Subset D$, then the function*

$$M(y) = \sup_x u(x,y) \tag{5.2}$$

is convex and, hence, continuous in D.

PROOF. Suppose the points y' and y'' in D are such that $\tau y' + (1-\tau)y'' \Subset D$ for all $0 \le \tau \le 1$. Then the function

$$v(\sigma,\tau) = u\big(x+\sigma(y'-y''), y''+\tau(y'-y'')\big) \tag{5.3}$$

is subharmonic in the neighbourhood of the strip $0 \le \tau \le 1$, $\sigma \in \mathbb{R}^1$, is bounded from above and, by virtue of (5.3),

$$\begin{aligned} v(\sigma,0) &= u\big(x+\sigma(y'-y''), y''\big) \le M(y''), \\ v(\sigma,1) &= u\big(x+\sigma(y'-y''), y'\big) \le M(y'). \end{aligned}$$

But then the function

$$\chi(\sigma,\tau) = v(\sigma,\tau) - \tau M(y') - (1-\tau)M(y'') \tag{5.4}$$

is subharmonic in the neighbourhood of the strip $0 \le \tau \le 1$, $\sigma \in \mathbb{R}^1$, is bounded above and is nonpositive on the boundary of the strip. By the Phragmén–Lindelöf theorem for subharmonic functions, $\chi(\sigma,\tau) \le 0$, $0 \le \tau \le 1$, $\sigma \in \mathbb{R}^1$, so that by (5.4) and (5.3) (for $\sigma = 0$),

$$u\big(x, \tau y' + (1-\tau)y''\big) \le \tau M(y') + (1-\tau)M(y'').$$

From this, by (5.2), we derive the inequality

$$M\big(\tau y' + (1-\tau)y''\big) \le \tau M(y') + (1-\tau)M(y''),$$

which completes the proof of Lemma 3. □

Let $u \in \mathcal{P}_+(T^C)$, where C is a convex cone. We introduce the growth indicator $h(u;y)$ of the function u via the formula

$$h(u;y) = \lim_{t\to+\infty} \frac{u(0,ty)}{t}, \qquad y \in C. \tag{5.5}$$

By Lemma 1, the limit in (5.5) exists and is nonnegative.

We introduce the function

$$\lambda(u;y) = \lim_{t\to+\infty} \frac{m(ty)}{t}, \tag{5.6}$$

where the quantity $m(y)$ is given by

$$m(y) = \inf_x u(x,y), \qquad y \in C.$$

By Lemma 3, the function $m(y)$ is nonnegative and concave (see Sec. 0.2) in C and, hence, such is the function $\frac{m(ty)}{t}$ for all $t > 0$. By Lemma 2, for all $\varepsilon > 0$ and $y \in C$, the function

$$\frac{1}{t}\big[m(\varepsilon y + ty) - m(\varepsilon y)\big]$$

does not increase with respect to t. Therefore the limit in (5.6) exists and defines a (nonnegative) concave function $\lambda(u;y)$ in C that satisfies the estimate

$$\lambda(u;y) \le \frac{1}{t}\big[m(\varepsilon y + ty) - m(\varepsilon y)\big] \le \frac{m(\varepsilon y + ty)}{t}, \qquad y \in C, \quad t > 0.$$

Setting $t = 1$ here and allowing $\varepsilon \to 0$, we obtain the estimate

$$\lambda(u;y) \le m(y) \le u(x,y), \qquad (x,y) \in T^C. \tag{5.7}$$

Finally, note that the functions h and λ are homogeneous of degree of homogeneity 1; for example,

$$\lambda(u;ry) = \lim_{t\to\infty} \frac{m(try)}{t} = r \lim_{t\to\infty} \frac{m(try)}{tr}$$
$$= r \lim_{t'\to\infty} \frac{m(t'y)}{t'} = r\lambda(u;y), \qquad r > 0.$$

From this, and also from (5.7) and (5.5) follows the inequality

$$\lambda(u;y) \le h(u;y) \qquad y \in C. \tag{5.8}$$

We will now prove the following theorem.

THEOREM. *If $u \in \mathcal{P}_+(T^C)$, where C is a convex cone, then the growth indicator $h(u;y)$ is nonnegative, concave, homogeneous of degree of homogeneity 1 in C, and*

$$\lambda(u;y) = h(u;y) = \lim_{t\to\infty} \frac{u(x^0, y^0 + ty)}{t}, \qquad (x^0,y^0) \in \mathbb{C}^n, \quad y \in C. \tag{5.9}$$

For $(x'', y'') \in T^C$ the function $\frac{1}{t}u(x^0, y^0 + ty)$ does not increase with respect to $t \in (0,\infty)$ and the following inequality holds true:

$$h(u;y) \le u(x^0, y^0 + y), \qquad (x^0,y^0) \in T^C, \quad y \in C. \tag{5.10}$$

PROOF. We will prove that for every $y \in C$ the function

$$\lim_{t\to\infty} \frac{u(x^0, y^0 + ty)}{t} = \lim_{t\to\infty} \frac{u(x^0, y^0 + ty)}{t - t_0} \tag{5.11}$$

does not depend on (x^0, y^0). For this it suffices to prove, by virtue of the Liouville theorem, that for every $y \in C$ the nonnegative function (5.11) is pluriharmonic with respect to (x^0, y^0) in $\mathbb{C}^n$. That is, it is pluriharmonic in every tubular domain $T^D = \mathbb{R}^n + iD$, where $D \Subset \mathbb{R}^n$. By Lemma 1, the function (5.11) in the domain T^D is the limit of a nonincreasing sequence of functions $u(x^0, y^0 + ty)(t - t_0)^{-1}$, $t \to \infty$, $t > t_0$, of the class $\mathcal{P}_+(T^D)$ and therefore is itself pluriharmonic in T^D (see Sec. 17.1). Thus, by (5.5), the second of the equalities (5.9) holds, and, by Lemma 1, the function $\frac{1}{t}u(x^0, y^0 + ty)$ does not increase with respect to t for $t > 0$ if $y^0 \in C$. Therefore

$$h(u; y) \leq \frac{1}{t}u(x^0, y^0 + ty), \qquad (x^0, y^0) \in T^C, \quad y \in C, \quad t > 0.$$

Putting $t = 1$ here, we obtain the estimate (5.10). From this estimate we derive

$$h(u; y) \leq m(y), \qquad y \in C,$$

so that, by (5.6),

$$h(u; y) \leq \lambda(u; y), \qquad y \in C.$$

This inequality together with the inverse inequality (5.8) is what yields the first of the equalities (5.9), from which fact it follows that the indicator $h(u; y)$ is a convex function in C. This completes the proof of all assertions of the theorem. □

REMARK. A more general theory of growth of plurisubharmonic functions in tubular domains over convex cones is developed in Vladimirov [**117**].

17.6. An integral representation of functions of the class $H_+(T^C)$. We established here that a function of the class $H_+(T^C)$, where C is an acute regular cone, is representable in the form of a sum of the Schwartz integral and a linear term if and only if the corresponding Poisson integral is a pluriharmonic function in T^C. We first prove a lemma that generalizes the Lebesgue theorem on the limiting passage under the sign of the Lebesgue integral (see Vladimirov [**114**, IV]).

LEMMA. *Suppose the sequences $u_k(x)$ and $v_k(x)$, $k = 1, 2, \ldots$, of functions in $\mathcal{L}^1$ have the following properties:*

(1) $|u_k(x)| \leq v_k(x)$, $k = 1, 2, \ldots$, *almost everywhere in* $\mathbb{R}^n$;
(2) $u_k(x) \to u(x)$, $v_k(x) \to v(x) \in \mathcal{L}^1$, $k \to \infty$, *almost everywhere in* $\mathbb{R}^n$;
(3) $\int v_k(x)\,dx \to \int v(x)\,dx, \qquad k \to \infty.$

Then $u \in \mathcal{L}^1$ and

$$\int u_k(x)\,dx \to \int u(x)\,dx, \qquad k \to \infty. \tag{6.1}$$

PROOF. From (1) and (2) it follows that $u \in \mathcal{L}^1$ and $v_k(x) \pm u_k(x) \geq 0$, $k = 1, 2, \ldots$, almost everywhere in $\mathbb{R}^n$. Applying the Fatou lemma to the sequences of functions $v_k \pm u_k$, $k \to \infty$, and making use of (3), we derive the following chain of inequalities:

$$\int [v(x) \pm u(x)]\,dx \le \underline{\lim}_{k\to\infty} \int [v_k(x) \pm u_k(x)]\,dx$$
$$= \underline{\lim}_{k\to\infty} \int v_k(x)\,dx + \underline{\lim}_{k\to\infty} \int \pm u_k(x)]\,dx$$
$$= \int v(x)\,dx + \underline{\lim}_{k\to\infty} \int \pm u_k(x)]\,dx,$$

whence we derive

$$\overline{\lim}_{k\to\infty} \int u_k(x)\,dx \le \int u(x)\,dx \le \underline{\lim}_{k\to\infty} \int u_k(x)\,dx,$$

which is equivalent to the limiting relation (6.1). □

THEOREM (Vladimirov [**118**]). *Let $f \in H_+(T^C)$, where C is an acute (convex) cone. Then the following statements are equivalent:*

(1) *The Poisson integral*

$$\int \mathcal{P}_C(x - x', y)\mu(dx'), \qquad \mu = \Im f_+, \tag{6.2}$$

is a pluriharmonic function in T^C.

(2) *The function $\Im f(z)$ is representable in the form*

$$\Im f(z) = \int \mathcal{P}_C(x - x', y)\mu(dx') + (a, y), \qquad z \in T^C, \tag{6.3}$$

for a certain $a \in C^*$.

(3) *For all $y' \in C$ the following representation holds:*

$$\Im f(z + iy') = \int \mathcal{P}_C(x - x', y)\Im f(x' + iy')\,dx' + (a, y), \qquad z \in T^C. \tag{6.4}$$

(4) *If C is a regular cone, then for an arbitrary $z^0 \in T^C$, the function $f(z)$ can be represented as*

$$f(z) = i \int \mathcal{S}_C(z - x'; z^0 - x')\mu(dx') + (a, z) + b(z^0), \qquad z \in T^C, \tag{6.5}$$

where $b(z^0)$ is a real number.

Here, $b(z^0) = \Re f(z^0) - (a, x^0)$ and (a, y) is the best linear minorant of the growth indicator $h(\Im f; y)$ in the cone C.

REMARK. Under the hypothesis of the theorem, the best linear minorant of the nonnegative convex function $h(\Im f; y)$ of degree of homogeneity 1 exists in the cone C (see Sec. 17.5). For example, $h(\Im\sqrt{z^2}; y) = \sqrt{y^2}$ and $(a, y) = 0$ in V^+.

PROOF. Let $f \in H_+(T^C)$. (1) → (2). The function

$$v(x, y) = \Im f(x) - \int \mathcal{P}_C(x - x', y)\mu(dx')$$

belongs to the class $\mathcal{P}_+(T^C)$ and its boundary value, as $y \to 0$, $y \in C$, is equal to 0 (see Sec. 17.2). By a corollary to the theorem of Sec. 17.1, $v(x, y) = (a, y)$ for some $a \in C^*$. The representation (6.3) is proved.

We now prove that (a, y) is the best linear minorant of the function $h(\Im f; y)$ in the cone C. From (6.3) and (5.5) it follows that (a, y) is a linear minorant of h in C. Suppose (a', y) is another linear minorant of h in C, that is

$$(a', y) \le h(\Im f; y), \qquad y \in C. \tag{6.6}$$

The function

$$f_1(z) = f(z) - (a', z), \qquad \Im f_1(z) = \Im f(z) - (a', y)$$

belongs to the class $H_+(T^C)$ since

$$\Im f(z) \ge h(\Im f; y) \ge (a', y), \qquad z \in T^C,$$

by the theorem of Sec. 17.5 and by virtue of (6.6). Furthermore, since $\Im f_{1+} = \Im f_+ = \mu$, it follows that condition (1) is fulfilled for $f_1(z)$. Applying the representation (6.3) to $\Im f_1(z)$, we obtain

$$\begin{aligned}\Im f_1(z) &= \Im f(z) - (a', y)\\ &= \int \mathcal{P}_C(x - x', y)\mu(dx') + (a'', y), \qquad z \in T^C,\end{aligned}$$

for some $a'' \in C^*$. Comparing that with (6.3), we derive

$$(a, y) = (a', y) + (a'', y) \ge (a', y), \qquad y \in C,$$

which is what we set out to prove.

(2) → (3). The function

$$\varphi(z) = f(z) - (a, z), \qquad \Im\varphi(z) = \Im f(z) - (a, y) \tag{6.7}$$

belongs to the class $H_+(T^C)$. Therefore the function

$$\begin{aligned}v(x, y, y') &= \Im\varphi(x + iy + iy') - \int \mathcal{P}_C(x - x', y)\Im\varphi(x' + iy')\, dx'\\ &= \Im f(x + iy + iy') - \int \mathcal{P}_C(x - x', y)\Im f(x' + iy')\, dx' - (a, y), \qquad (6.8)\\ &\qquad (x, y) \in T^C, \qquad y' \in C,\end{aligned}$$

is a nonnegative function [see (2.2)] that is pluriharmonic with respect to (x, y') in T^C and, by (2.3) (for $\varphi = 1$) and (6.3),

$$\begin{aligned}v(x, y, y') \to \Im f(x + iy) - \int \mathcal{P}_C(x - x', y)\mu(dx') - (a, y) = 0,\\ y' \to 0, \qquad y' \in C' \Subset C.\end{aligned}$$

By the corollary to the theorem of Sec. 17.1, $v(x, y, y') = (A_y, y')$, $(x, y') \in T^C$, $A_y \in C^*$ for every $y \in C$. Therefore (6.8) takes the form

$$\begin{aligned}\Im f(x + iy + iy') - \int \mathcal{P}_C(x - x', y)\Im f(x' + iy')\, dx' - (a, y) - (A_y, y') = 0, \qquad (6.9)\\ (x, y) \in T^C, \qquad y \in C.\end{aligned}$$

For y' substitute ty', $t > 0$, divide by t and allow t to go to ∞. As a result, using the theorem of Sec. 17.5 and the B. Levi theorem, we obtain

$$\begin{aligned}
0 &= \lim_{t\to\infty} \frac{1}{t} \Im f(x+iy+ity') - \lim_{t\to\infty} \int \mathcal{P}_C(x-x',y) \frac{\Im f(x'+ity')}{t}\, dx' \\
&\quad - \lim_{t\to\infty} \left[\frac{(a,y)}{t} + (A_y, y')\right] \\
&= h(\Im f; y') - \int \mathcal{P}_C(x-x',y) \lim_{t\to\infty} \frac{\Im f(x'+ity')}{t}\, dx' - (A_y, y') \\
&= h(\Im f; y') - h(\Im f; y') \int \mathcal{P}_C(x-x',y)\, dx' - (A_y, y') \\
&= (A_y, y'),
\end{aligned}$$

from which, and also from (6.9), follows the representation (6.4).

(3) $\to$ (4). Using (6.7) we introduce the function $\varphi(z)$ and then, by (2.12), we construct the function $\varphi_\varepsilon(z)$, $\varepsilon > 0$, with the properties (a)–(c). Let $y' \in C$. To the functions $\Im\varphi_\varepsilon(z+iy')$ and $\varphi_\varepsilon(z+iy')$ we apply the representations of Poisson and Schwartz, respectively (see Secs. 11.2 and 12.5):

$$\Im\varphi_\varepsilon(z+iy') = \int \mathcal{P}_C(x-x',y)\Im\varphi_\varepsilon(x'+iy')\, dx', \qquad z \in T^C, \tag{6.10}$$

$$\varphi_\varepsilon(z+iy') = i\int \mathcal{S}_C(z-x'; z^0-x')\Im\varphi_\varepsilon(x'+iy')\, dx' + \Re\varphi_\varepsilon(z^0+iy'), \tag{6.11}$$
$$z \in T^C, \qquad z^0 \in T^C.$$

The representation (6.4) shows that passage to the limit under the integral sign is possible in (6.10) as $\varepsilon \to 0$, since

$$\begin{aligned}
\lim_{\varepsilon\to 0} \Im\varphi_\varepsilon(z+iy') &= \Im\varphi(z+iy') \\
&= \Im f(z+iy') - (a, y+y') \\
&= \int \mathcal{P}_C(x-x',y)\, [\Im f(x'+iy') - (a,y')]\, dx \\
&= \int \mathcal{P}_C(x-x',y) \lim_{\varepsilon\to 0} \varphi_\varepsilon(x'+iy')\, dx.
\end{aligned}$$

But then, by virtue of the inequality (5.4) of Sec. 12.5, the lemma on the possibility of passing to the limit as $\varepsilon \to 0$ under the integral sign in (6.11) is applicable. As a result, using (5.3) of Sec. 12.5, we obtain the equalities

$$\begin{aligned}
\varphi(z+iy') &= f(z+iy') - (a, z+iy') \\
&= i\int \mathcal{S}_C(z-x'; z^0-x')\Im\varphi(x'+iy')\, dx' + \Re\varphi(z^0+iy') \\
&= i\int \mathcal{S}_C(z-x'; z^0-x')\, [\Im f(x'+iy') - (a,y')]\, dx' \\
&\quad + \Re f(z^0+iy') - (a, x^0) \\
&= i\int \mathcal{S}_C(z-x'; z^0-x')\Im f(x'+iy')\, dx' - i(a,y') \\
&\quad + \Re f(z^0+iy') - (a, x^0).
\end{aligned}$$

That is

$$f(z+iy') = i\int \mathcal{S}_C(z-x';z^0-x')\Im f(x'+iy')\,dx' + (a,z)$$
$$+\Re f(z^0+iy') - (a,x^0), \qquad z\in T^C, \qquad z^0\in T^C, \qquad y'\in C.$$

Passing to the limit here as $y'\to 0$, $y'\in C'\Subset C$, and making use of the limiting relation (2.10), we obtain the representation (6.5).

(4) $\to$ (1). Putting $z^0=z$ in the representation (6.5) and making use of (5.2) of Sec. 12.5, and also separating the imaginary part, we obtain the representation (6.3), from which follows the pluriharmonicity in T^C of the Poisson integral (6.2).

The theorem is proved. □

18. Holomorphic Functions with Nonnegative Imaginary Part in T^n

In the case of the cone $C=\mathbb{R}^n_+$, the results of Sec. 17 admit of being strengthened. Here we will obtain integral representations for all functions of the classes $H_+(T^n)$ and $\mathcal{P}_+(T^n)$. We will first prove some lemmas.

18.1. Lemmas. Set

$$\mathcal{E}_n(\xi) = \theta_n(\xi)\xi_1\cdots\xi_n.$$

Here, $\theta_n(\xi)$ is the characteristic function of the cone $\overline{\mathbb{R}^n_+}$, see Sec. 0.2. Then $\mathcal{E}_n(\xi)$ is the fundamental solution of the operator $\partial_1^2\ldots\partial_n^2$.

Suppose $f\in C^{2n}$. Then the following equation holds:

$$e^{-(y,\xi^+)}\partial_1^2\ldots\partial_n^2 f(\xi) = T_1\ldots T_n\left[e^{-(y,\xi^+)}f(\xi)\right], \tag{1.1}$$

where $\xi\to\xi^+ = (|\xi_1|,\ldots,|\xi_n|)$ and

$$T_j = \partial_j^2 + y_j^2 + 2y_j\partial_j\operatorname{sgn}\xi_j - 2y_j\delta(\xi_j), \qquad j=1,\ldots,n.$$

The right-hand side of (1.1) is meaningful in $\mathcal{D}'$ and for $f\in C$, and we will take it for a definition of the generalized function in the left-hand member of (1.1).

Note that if $f\in C$ and $\operatorname{supp} f\subset\overline{\mathbb{R}^n_+}$, then

$$e^{-(y,\xi)}\partial_1^2\ldots\partial_n^2 f(\xi) = e^{-(y,\xi^+)}\partial_1^2\ldots\partial_n^2 f(\xi). \tag{1.2}$$

Indeed, for any $\varphi\in\mathcal{D}$, we have, by (1.1),

$$\begin{aligned}
\left(e^{-(y,\xi^+)}\partial_1^2\ldots\partial_n^2 f,\varphi\right) &= \left(T_1\ldots T_n\left[e^{-(y,\xi^+)}f\right],\varphi\right)\\
&= \int e^{-(y,\xi^+)}f(\xi)\prod_{1\le j\le n}\left(\partial_j^2+y_j^2-2y_j\operatorname{sgn}\xi_j\partial_j\right)\varphi(\xi)\,d\xi\\
&= \int e^{-(y,\xi)}f(\xi)\prod_{1\le j\le n}\left(\partial_j^2+y_j^2-2y_j\partial_j\right)\varphi(\xi)\,d\xi\\
&= \int f(\xi)\partial_1^2\ldots\partial_n^2\left[e^{-(y,\xi)}\varphi(\xi)\right]d\xi\\
&= \left(e^{-(y,\xi)}\partial_1^2\ldots\partial_n^2 f,\varphi\right),
\end{aligned}$$

which is equivalent to (1.2). □

LEMMA 1. *Suppose $f(\xi)$ is a continuous positive definite function in $\mathbb{R}^n$. Then for all $y \in \mathbb{R}^n_+$, the generalized function*

$$e^{-(y,\xi^+)}(1-\partial_1^2)\dots(1-\partial_n^2)f(\xi)$$

is positive definite and the following equation holds:

$$F\left[e^{-(y,\xi^+)}(1-\partial_1^2)\dots(1-\partial_n^2)f\right] = \int \mathcal{P}_n(x-x',y)(1+x_1'^2)\dots(1+x_n'^2)\sigma(dx'), \quad (1.3)$$

where the measure $\sigma = F[f]$, and $\mathcal{P}_n(x,y)$ is the Poisson kernel of the domain T^n (see (1.2) of Sec. 11.1).

PROOF. Note that by the Bochner theorem (see Sec. 8.2), $\sigma = F[f]$ is a measure that is nonnegative and with compact support on $\mathbb{R}^n$, and equalities of the following type hold:

$$f(\xi_1,\dots,\xi_k,0,\dots,0) = \frac{1}{(2\pi)^n}\int e^{-ix_1\xi_1-\dots-ix_k\xi_k}\sigma(dx). \quad (1.4)$$

Let us prove (1.3) for $n = 1$. From (1.1), for all $y > 0$, we have

$$\begin{aligned} F\left[e^{-y|\xi|}(1-\partial^2)f\right] &= F\left[(1-T)(e^{-y|\xi|}f)\right] \\ &= F\left[(1-\partial^2-y^2-2y\partial\operatorname{sgn}\xi+2y\delta(\xi))(e^{-y|\xi|}f(\xi))\right] \\ &= (1+x^2-y^2)F\left[e^{-y|\xi|}f(\xi)\right]+2ixyF\left[\operatorname{sgn}\xi e^{-y|\xi|}f(\xi)\right]+2yf(0). \quad (1.5) \end{aligned}$$

Taking into account the equalities

$$F[e^{-y|\xi|}](x) = 2\int_0^\infty e^{-y\xi}\cos x\xi\,d\xi = \frac{2y}{x^2+y^2},$$

$$F[\operatorname{sgn}\xi e^{-y|\xi|}](x) = 2i\int_0^\infty e^{-y\xi}\sin x\xi\,d\xi = \frac{2xi}{x^2+y^2},$$

we obtain

$$\begin{aligned} F\left[f(\xi)e^{-y|\xi|}\right] &= \frac{1}{2\pi}F[f]*F[e^{-y|\xi|}] \\ &= \frac{y}{\pi}\int\frac{\sigma(dx')}{(x-x')^2+y^2}, \\ F\left[f(\xi)\operatorname{sgn}\xi e^{-y|\xi|}\right] &= \frac{1}{2\pi}F[f]*F[\operatorname{sgn}\xi e^{-y|\xi|}] \\ &= \frac{1}{\pi}\int\frac{(x-x')\sigma(dx')}{(x-x')^2+y^2}. \end{aligned}$$

Substituting the resulting expressions into (1.5) and taking into account (1.4) for $k = 0$, we obtain (1.3) for $n = 1$.

The case of $n > 1$ is considered in similar fashion if one notices that every operator T_j operates only on its own variable ξ_j, if one applies the Fourier transform technique with respect to some of the variables (see Sec. 6.2 and Sec. 6.3), and if one takes advantage of equations of the type (1.4). The proof of Lemma 1 is complete. □

LEMMA 2. *Let the function $v(\xi)$ be continuous and bounded in $\mathbb{R}^n$ and let* $\operatorname{supp} v \subset \overline{\mathbb{R}^n_+}$. *Then the solution of the equation*

$$\partial_1^2 \dots \partial_n^2 u(\xi) = (1-\partial_1^2)\dots(1-\partial_n^2)v(\xi) \tag{1.6}$$

exists and is unique in the class of continuous functions in $\mathbb{R}^n$ with support in $\overline{\mathbb{R}^n_+}$, which functions satisfy the estimate

$$|u(\xi)| \le C(1+\xi_1^2)\dots(1+\xi_n^2). \tag{1.7}$$

PROOF. The solution of equation (1.6) is unique even in the algebra $\mathcal{D}'(\overline{\mathbb{R}^n_+})$ and is representable in the form (see Sec. 4.9.4)

$$u = \mathcal{E}_n * (1-\partial_1^2)\dots(1-\partial_n^2)v. \tag{1.8}$$

Let us represent the right-hand side of (1.8) as

$$\begin{aligned} u(\xi) &= (1-\partial_1^2)\dots(1-\partial_n^2)\mathcal{E}_n * v \\ &= \{[\theta(\xi_1)\xi_1 - \delta(\xi_1)] \times \dots \times [\theta(\xi_n)\xi_n - \delta(\xi_n)]\} * v \\ &= (-1)^n v(\xi) + \sum_{\substack{1\le k\le n \\ j_1<\dots<j_k}} (-1)^{n-k} \int_0^{\xi_{j_1}} \dots \int_0^{\xi_{j_k}} v(\dots,\xi'_{j_1},\dots,\xi'_{j_k},\dots) \\ &\quad \times (\xi_{j_1}-\xi'_{j_1})\dots(\xi_{j_k}-\xi'_{j_k})\, d\xi'_{j_1}\dots d\xi'_{j_k}. \end{aligned}$$

It remains to note that each summand in the last sum is a continuous function that satisfies the estimate (1.7). This completes the proof of Lemma 2. □

LEMMA 3. *Suppose $u(\xi)$ is a continuous tempered function in $\mathbb{R}^n$. Then the solution of the equation*

$$(1-\partial_1^2)\dots(1-\partial_n^2)v(\xi) = \partial_1^2\dots\partial_n^2 u(\xi) \tag{1.9}$$

exists and is unique in the class of continuous tempered functions.

PROOF. The solution of equation (1.9) is unique even in the class $\mathcal{S}'$ since the Fourier transform of the generalized function $(1-\partial_1^2)\dots(1-\partial_n^2)\delta(\xi)$, equal to $(1+x_1^2)\dots(1+x_n^2)$, does not vanish anywhere in $\mathbb{R}^n$. We will prove its existence:

$$\mathcal{E}(\xi) = \frac{1}{2}e^{-|\xi_1|-\dots-|\xi_n|}$$

is the fundamental solution of the operator $(1-\partial_1^2)\dots(1-\partial_n^2)$. Since $u(\xi)$ is tempered, the convolution $\mathcal{E} * u$ exists (see Sec. 4.1). Therefore, the solution v of equation (1.9) can be expressed in the form of a convolution:

$$\begin{aligned} v &= \mathcal{E} * \partial_1^2\dots\partial_n^2 u \\ &= \partial_1^2\dots\partial_n^2\mathcal{E} * u \\ &= \left\{\left[-\delta(\xi_1) + \frac{1}{2}e^{-|\xi_1|}\right] \times \dots \times \left[-\delta(\xi_n) + \frac{1}{2}e^{-|\xi_n|}\right]\right\} * u \\ &= (-1)^n u(\xi) + \sum_{\substack{1\le k\le n \\ j_1<\dots<j_k}} \frac{(-1)^{n-k}}{2^k} \int_{\mathbb{R}^k} u(\dots,\xi'_{j_1},\dots,\xi'_{j_k},\dots) \\ &\quad \times e^{-|\xi_{j_1}-\xi'_{j_1}|-\dots-|\xi_{j_k}-\xi'_{j_k}|}\, d\xi'_{j_1}\dots d\xi'_{j_k}. \end{aligned}$$

It remains to note that each term in the last sum is a continuous tempered function. Lemma 3 is proved. □

LEMMA 4. *If the function* $v(\xi)$ *is continuous, bounded in* $\overline{\mathbb{R}}^2_+ \times \mathbb{R}^{n-2}$, $n \geq 2$, *and satisfies the equation*

$$(1-\partial_1^2)\dots(1-\partial_n^2)v(\xi) = 0, \qquad \xi \in \mathbb{R}^2_+ \times \mathbb{R}^{n-2}, \tag{1.10}$$

then it can be expressed as

$$v(\xi) = e^{-\xi_1}v(0,\xi_2,\tilde{\xi}) + e^{-\xi_2}v(\xi_1,0,\tilde{\xi}) - e^{-\xi_1-\xi_2}(0,0,\tilde{\xi}) \tag{1.11}$$

where $\tilde{\xi} = (\xi_3,\dots,\xi_n)$.

PROOF. We continue the function $v(\xi)$ by zero onto the whole space $\mathbb{R}^n$ and we construct for the function the mean function

$$v_\varepsilon(\xi) = \int v(\xi')\omega_\varepsilon(\xi-\xi')\,d\xi' = v * \omega_\varepsilon$$

with the following properties (see Sec. 1.2 and Sec. 4.6):

$$\partial^\alpha v_\varepsilon \in C^\infty \cap \mathcal{L}^\infty, \quad \forall\alpha; \qquad v_\varepsilon(\xi) \to v(\xi), \quad \varepsilon \to 0, \quad \xi \in \mathbb{R}^2_+ \times \mathbb{R}^{n-2},$$

$$(1-\partial_1^2)\dots(1-\partial_n^2)v_\varepsilon(\xi) = 0, \qquad \xi_1 > 2\varepsilon, \quad \xi_2 > 2\varepsilon, \quad \tilde{\xi} \in \mathbb{R}^{n-2}.$$

Put

$$\chi_\varepsilon(\xi) = (1-\partial_1^2)\dots(1-\partial_n^2)v_\varepsilon(\xi). \tag{1.12}$$

The $\partial^\alpha \chi_\varepsilon \in C^\infty \cap \mathcal{L}^\infty$ for $\forall\alpha$, and χ_ε satisfies the equation

$$(1-\partial_1^2)(1-\partial_2^2)\chi_\varepsilon(\xi) = 0, \qquad \xi_1 > 2\varepsilon, \quad \xi_2 > 2\varepsilon, \quad \tilde{\xi} \in \mathbb{R}^{n-2}. \tag{1.13}$$

Fix $\delta > 0$ and let $2\varepsilon < \delta$. From the equation (1.13) and from the boundedness of the function $(1-\partial_2^2)\chi_\varepsilon(\xi)$ we derive the relation

$$(1-\partial_2^2)\chi_\varepsilon(\xi_1,\xi_2,\tilde{\xi}) = (1-\partial_2^2)\chi_\varepsilon(\delta,\xi_2,\tilde{\xi})e^{-(\xi_1-\delta)},$$

that is,

$$(1-\partial_2^2)\left[\chi_\varepsilon(\xi_1,\xi_2,\tilde{\xi}) - \chi_\varepsilon(\delta,\xi_2,\tilde{\xi})e^{-(\xi_1-\delta)}\right] = 0, \qquad \xi_1 \geq \delta, \qquad \xi_2 \geq \delta. \tag{1.14}$$

Similarly, from the equation (1.14) we derive the relation

$$\chi_\varepsilon(\xi_1,\xi_2,\tilde{\xi}) - \chi_\varepsilon(\delta,\xi_2,\tilde{\xi})e^{-(\xi_1-\delta)} = \left[\chi_\varepsilon(\xi_1,\delta,\tilde{\xi}) - \chi_\varepsilon(\delta,\delta,\tilde{\xi})e^{-(\xi_1-\delta)}\right]e^{-(\xi_2-\delta)}.$$

That is, by (1.12),

$$(1-\partial_3^2)\dots(1-\partial_n^2)\left[v_\varepsilon(\xi_1,\xi_2,\tilde{\xi}) - v_\varepsilon(\delta,\xi_2,\tilde{\xi})e^{-(\xi_1-\delta)} - v_\varepsilon(\xi_1,\delta,\tilde{\xi})e^{-(\xi_2-\delta)} + v_\varepsilon(\delta,\delta,\tilde{\xi})e^{-(\xi_1+\xi_2-2\delta)}\right] = 0,$$
$$\xi_1 \geq \delta, \qquad \xi_2 \geq \delta, \qquad \tilde{\xi} \in \mathbb{R}^{n-2}.$$

From this, by uniqueness of the solution of the last equation (via Lemma 3), follows the equality

$$v_\varepsilon(\xi) = v_\varepsilon(\delta,\xi_2,\tilde{\xi})e^{-(\xi_1-\delta)} + v_\varepsilon(\xi_1,\delta,\tilde{\xi})e^{-(\xi_2-\delta)} - v_\varepsilon(\delta,\delta,\tilde{\xi})e^{-(\xi_1+\xi_2-2\delta)},$$
$$\xi_1 \geq \delta, \qquad \xi_2 \geq \delta, \qquad \tilde{\xi} \in \mathbb{R}^{n-2}.$$

Passing to the limit here as $\varepsilon \to 0$ and, furthermore, as $\delta \to 0$, we obtain the representation (1.10). This completes the proof of Lemma 4. □

LEMMA 5. *The equation*

$$\partial_1^2 \dots \partial_n^2 u(\xi) + \sum_{1\le|\alpha|\le N} a_\alpha \partial^\alpha \delta(\xi) = 0, \tag{1.15}$$

provided that $u \in C$, $\operatorname{supp} u \subset \overline{\mathbb{R}^n_+}$, *is possible only when* $u(\xi) = 0$ *and* $a_\alpha = 0$, $1 \le |\alpha| \le N$.

PROOF. In the algebra $\mathcal{D}'(\overline{\mathbb{R}^n_+})$, the equation (1.15) is equivalent (see Sec. 4.9.4) to

$$\begin{aligned} u(\xi) &= -\mathcal{E}_n * \sum_{1\le|\alpha|\le N} a_\alpha \partial^\alpha \delta \\ &= -\sum_{1\le|\alpha|\le N} a_\alpha \partial^\alpha \mathcal{E}_n(\xi) \\ &= -\sum_{j=1}^{n} \sum_{\substack{\alpha_1=\dots=\alpha_j=1\\ \alpha_{j+1}=\dots=\alpha_n=0}} a_\alpha \theta(\xi_{\alpha_1})\dots\theta(\xi_{\alpha_j})\xi_{\alpha_{j+1}}\theta(\xi_{\alpha_{j+1}})\dots\xi_{\alpha_n}\theta(\xi_{\alpha_n}) \\ &\quad -\sum_{j=1}^{n}\sum_{\alpha_j\ge 0} a_\alpha \partial^\alpha \mathcal{E}_n(\xi). \end{aligned} \tag{1.16}$$

Each term in the second sum of (1.16) contains at least one δ-function or their derivatives with respect to any one of the variables ξ_j, $1 \le j \le n$, and the combinations of those δ-functions and their derivatives in all terms are distinct. The other summands in (1.16) are locally integrable functions, whence we conclude that $a_\alpha = 0$ if there is a f such that $\alpha_j \ge 2$, and (1.16) takes the form

$$u(\xi) = -\sum_{j=1}^{n} \sum_{\substack{\alpha_1=\dots=\alpha_j=1\\ \alpha_{j+1}=\dots=\alpha_n=0}} a_\alpha \theta(\xi_{\alpha_1})\dots\theta(\xi_{\alpha_j})\xi_{\alpha_{j+1}}\theta(\xi_{\alpha_{j+1}})\dots\xi_{\alpha_n}\theta(\xi_{\alpha_n}).$$

From this, taking into account the properties of the function u, it is easy to derive, by induction on n, that all $a_\alpha = 0$ and $u(\xi) = 0$. Lemma 5 is proved. □

LEMMA 6. *The general solution of the equation*

$$\partial_1^2 \dots \partial_n^2 u(\xi) = 0 \tag{1.17}$$

in the class of continuous functions with support in $-\overline{\mathbb{R}^n_+} \cup \overline{\mathbb{R}^n_+}$ *is expressed by the formula*

$$u(x) = C\big[\mathcal{E}_n(x) - \mathcal{E}_n(-x)\big], \tag{1.18}$$

where C *is an arbitrary constant.*

PROOF. Function (1.18) satisfies (1.17) since $\partial_1^2 \dots \partial_n^2 \mathcal{E}_n(\pm\xi) = \delta(\xi)$. Let $u(\xi)$ be an arbitrary solution to (1.17) taken from the class under consideration. Then the function $u^+(\xi) = \theta_n(\xi)u(\xi)$ satisfies (1.17) in $\mathbb{R}^n \setminus \{0\}$ and hence (see Sec. 2.6)

$$\begin{aligned} \partial_1^2 \dots \partial_n^2 u^+(\xi) &= \sum_{0\le|\alpha|\le N} c_\alpha \partial^\alpha \delta(\xi) \\ &= c_0 \partial_1^2 \dots \partial_n^2 \mathcal{E}_n(\xi) + \sum_{1\le|\alpha|\le N} c_\alpha \partial^\alpha \delta(\xi) \end{aligned} \tag{1.19}$$

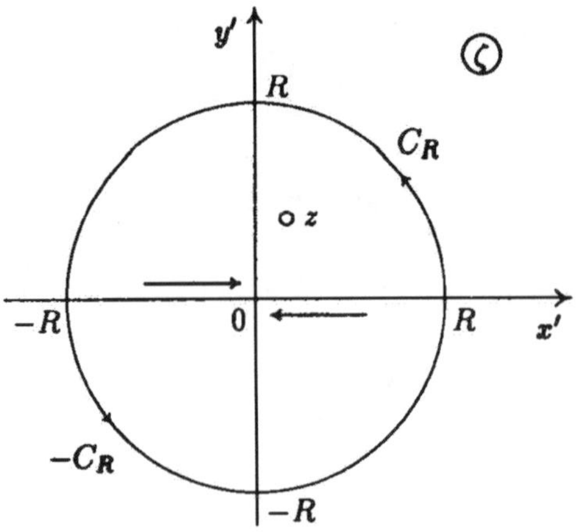

Figure 42

for certain N and c_α. By Lemma 5, the equation (1.19) is possible only for $c_\alpha = 0$, $|\alpha| \geq 1$ and $u^+(\xi) = c_0\mathcal{E}_n(\xi)$. Similarly, we derive that $u^-(\xi) = \theta_n(-\xi)u(\xi) = c_0'\mathcal{E}_n(-\xi)$ so that $u(\xi) = u^+(\xi) + u^-(\xi) = c_0\mathcal{E}_n(\xi) + c_0'\mathcal{E}_n(-\xi)$. But by virtue of (1.17)

$$\begin{aligned}\partial_1^2 \ldots \partial_n^2 u(\xi) &= c_0\partial_1^2 \ldots \partial_n^2\mathcal{E}_n(\xi) + c_0'\partial_1^2 \ldots \partial_n^2\mathcal{E}_n(-\xi)\\ &= (c_0 + c_0')\delta(\xi) = 0,\end{aligned}$$

so that $c_0' = -c_0$ and the representation (1.18) is proved. The proof of Lemma 6 is complete. □

18.2. Functions of the classes $H_+(T^1)$ and $\mathcal{P}_+(T^1)$. We first consider the case $n = 1$. Suppose the function $f \in H_+(T^1)$, that is, $f(z)$ is holomorphic and $\Im f(z) = u(x, y) \geq 0$ in the upper half-plane T^1 so that $\Im f \in \mathcal{P}_+(T^1)$.

Recall that $f(z)$ satisfies the estimate (see Sec. 13.3)

$$|f(z)| \leq M\frac{1 + |z|^2}{y}, \qquad y > 0, \tag{2.1}$$

and the measure $\mu = \Im f_+ = u(x, +0)$ satisfies the condition (see Sec. 17.2)

$$\int \frac{\mu(dx)}{1 + x^2} < \infty. \tag{2.2}$$

Let $\varepsilon > 0$ and $R > 1$ and denote by C_R and $-C_R$ semicircles of radius R centred at 0, as depicted in Fig. 42. By the residue theorem we have

$$\frac{f(z + i\varepsilon)}{1 + z^2} = \frac{1}{2\pi i}\left(\int_{-R}^{R} + \int_{C_R}\right)\frac{f(\zeta + i\varepsilon)\,d\zeta}{(1 + \zeta^2)(\zeta - z)} + \frac{f(i + i\varepsilon)}{2i(z - i)}, \qquad y > 0, \quad |z| < R. \tag{2.3}$$

Analogously, for the function $\frac{\bar{f}(z+i\varepsilon)}{1+z^2}$, which is meromorphic in the lower half-plane $y<0$ with the sole simple pole $-i$, we have

$$0 = \frac{1}{2\pi i}\left(\int_{-R}^{R} + \int_{-C_R}\right)\frac{\bar{f}(\zeta+i\varepsilon)\,d\zeta}{(1+\zeta^2)(\zeta-z)} - \frac{\bar{f}(i+i\varepsilon)}{2i(z+i)}, \qquad y>0, \quad |z|<R. \tag{2.4}$$

Sending R to ∞ in (2.3) and (2.4), and using the estimate (2.1), according to which

$$\begin{aligned}\left|\int_{C_R}\frac{f(\zeta+i\varepsilon)\,d\zeta}{(1+\zeta^2)(\zeta-z)}\right| &= \left|\int_0^\pi \frac{f(Re^{i\varphi}+i\varepsilon)iRe^{i\varphi}\,d\varphi}{(1+R^2e^{2i\varphi})(Re^{i\varphi}-z)}\right| \\ &\le M\int_0^\pi \frac{\left(1+\left|Re^{i\varphi}+i\varepsilon\right|^2\right)R\,d\varphi}{(R\sin\varphi+\varepsilon)\left|1+R^2e^{2i\varphi}\right|\left|Re^{i\varphi}-z\right|} \\ &\le \frac{MR[1+(R+\varepsilon)^2]}{(R^2-1)(R-|z|)}\int_0^\pi \frac{d\varphi}{R\sin\varphi+\varepsilon} \to 0, \qquad R\to\infty\end{aligned}$$

(and similarly for the contour $-C_R$), we obtain

$$f(z+i\varepsilon) = \frac{1+z^2}{2\pi i}\int_{-\infty}^{\infty}\frac{f(x'+i\varepsilon)\,dx'}{(1+x'^2)(x'-z)} + \frac{f(i+i\varepsilon)}{2i}(z+i), \qquad y>0,$$

$$0 = -\frac{1+z^2}{2\pi i}\int_{-\infty}^{\infty}\frac{\bar{f}(x'+i\varepsilon)\,dx'}{(1+x'^2)(x'-z)} - \frac{\bar{f}(i+i\varepsilon)}{2i}(z-i), \qquad y>0.$$

Adding together the resulting equalities, we derive an integral representation for the function $f(z+i\varepsilon)$:

$$f(z+i\varepsilon) = \frac{1+z^2}{\pi}\int_{-\infty}^{\infty}\frac{u(x',\varepsilon)\,dx'}{(1+x'^2)(x'-z)} + zu(0,1+\varepsilon) + \Re f(i+i\varepsilon), \qquad y>0. \tag{2.5}$$

Separating the imaginary part in (2.5), we obtain an integral representation for the function $u(x,y+\varepsilon)$:

$$u(x,y+\varepsilon) = \frac{y}{\pi}\int_{-\infty}^{\infty}u(x',\varepsilon)\left[\frac{1}{(x-x')^2+y^2} - \frac{1}{1+x'^2}\right]dx' + yu(0,1+\varepsilon), \qquad y>0. \tag{2.6}$$

Passing to the limit in (2.5) and (2.6) as $\varepsilon\to 0$, and making use of the limiting relation (2.3) of Sec. 17.2, we obtain the necessity of the conditions in the Herglotz–Nevanlinna theorem (see Nevanlinna [**78**]).

THEOREM I. *For the function $f(z)$ to belong to the class $H_+(T^1)$, it is necessary and sufficient that it be representable in the form*

$$f(z) = \frac{1}{\pi}\int_{-\infty}^{\infty} \frac{(1+x'z)\mu(dx')}{1+x'^2)(x'-z)} + az + b$$

$$= i\int_{-\infty}^{\infty} \mathcal{S}_1(z-x'; i-x')\mu(dx') + az + b, \qquad y>0, \tag{2.7}$$

where the measure μ is nonnegative and satisfies the condition (2.2), $a \geq 0$, and b is a real number. The representation (2.7) is unique, and $\mu = \Im f_+$, $b = \Re f(i)$,

$$a = \Im f(i) - \frac{1}{\pi}\int_{-\infty}^{\infty} \frac{\mu(dx)}{1+x^2} = \lim_{y\to\infty} \frac{\Im f(iy)}{y}, \tag{2.8}$$

$$\Im f(z) = \frac{y}{\pi}\int_{-\infty}^{\infty} \frac{\mu(dx')}{(x-x')^2+y^2} + ay, \qquad y>0. \tag{2.9}$$

The sufficiency of the conditions of Theorem I is straightforward.

COROLLARY. *For the function $u(x, y)$ to belong to the class $\mathcal{P}_+(T^1)$, it is necessary and sufficient that it be representable in the form*

$$u(x,y) = \frac{y}{\pi}\int_{-\infty}^{\infty} \frac{\mu(dx')}{(x-x')^2+y^2} + ay, \qquad y>0, \tag{2.9}$$

where $a \geq 0$ and the measure μ is nonnegative and satisfies the condition (2.2); here,

$$\mu = u(x, +0) \quad \textit{and} \quad a = \lim_{y\to\infty} \frac{u(0,y)}{y}.$$

REMARK. From the representation (2.9) it follows that the Poisson integral is a harmonic function in T^1. By the theorem of Sec. 17.6, the representation with the Schwartz kernel holds with respect to any point $z^0 \in T^1$ [formula (2.7) for $z^0 = 1$].

In terms of the spectral function $g(\xi)$ of the function $f(z)$ (see Sec. 17.1), the class $H_+(T^1)$ is characterized by the following theorem (König and Zemanian [62]).

THEOREM II. *For a function $f(z)$ to belong to the class $H_+(T^1)$, it is necessary and sufficient that its spectral function $g(\xi)$ have the following properties:*

(a) $-ig(\xi) + ig^*(\xi) \gg 0$,
(b) $g(\xi) = iu''(\xi) + ia\delta'(\xi)$,
where $a \geq 0$ and $u(\xi)$ is a continuous function with support in $[0, \infty)$, which function satisfies the growth condition

$$|u(\xi)| \leq C(1+\xi^2). \tag{2.10}$$

Here, the expansion (b) is unique, the number a is defined by (2.8), and $\Im f(z)$ is defined by (2.9).

COROLLARY. *For the measure μ to be a boundary value of the function $u(x,y)$ of the class $\mathcal{P}_+(T^1)$, $\mu = u(x,+0)$, it is necessary and sufficient that $\mu = F[v'']$, where $v'' \gg 0$, v is a continuous $*$-Hermitian function satisfying the growth condition (2.10), and $v(0) = 0$. In this case the function v with the indicated properties is unique to within the summand $ic\xi$, where c is an arbitrary real number.*

This follows from Theorem II for $v = u + u^*$ (necessity) and for $u = \theta v$ (sufficiency) if we take advantage of (1.5) of Sec. 17.1, $\mu = \frac{1}{2}F[-ig + ig^*]$.

PROOF OF THEOREM II. NECESSITY. Let $f \in H_+(T^1)$. Condition (a) was proved in Sec. 17.1. To prove condition (b), rewrite representation (2.7) as [compare with (2.5)]

$$f(z) = \frac{1+z^2}{\pi}\int_{-\infty}^{\infty}\frac{\mu(dx')}{(1+x'^2)(x'-z)} + \Im f(i)z + b$$

$$= i(1+z^2)(\sigma * \mathcal{K}_1(x'+iy)) + \Im f(i)z + b, \qquad \sigma = \frac{\mu}{\pi(1+x'^2)}. \tag{2.11}$$

Since $\mathcal{K}_1(x+iy) \in \mathcal{H}_s$ (for all s and $y > 0$) [see (2.5) of Sec. 11.2] and $\int \sigma(dx') < \infty$ [see (2.2)], the Fourier transform formula of the convolution $\sigma * \mathcal{K}_1$ holds:

$$F^{-1}[\sigma * \mathcal{K}_1] = F[\sigma](-\xi)F^{-1}[\mathcal{K}_1](\xi) = e^{-y\xi}\theta(\xi)v(\xi), \tag{2.12}$$

where $v(\xi) = F[\sigma](-\xi)$ a continuous positive definite (and, hence, bounded) function (see Sec. 8). Now, using (2.11) and (2.12), we compute the spectral function $g(\xi)$ (see Sec. 9):

$$g(\xi) = i(1-\partial^2)[\theta(\xi)v(\xi)] + i\Im f(i)\delta'(\xi) + b\delta(\xi). \tag{2.13}$$

By Lemma 2 of Sec. 18.1 there exists a continuous function $u_1(\xi)$ with support in $[0,\infty)$ that satisfies the estimate (2.10) and is such that

$$(1-\partial^2)\{\theta(\xi)[v(\xi) - v(0)]\} = \partial^2 u_1(\xi).$$

Therefore, (2.13) takes the form

$$g(\xi) = i\partial^2\left[u_1(\xi) + \frac{v(0)}{2}\xi^2\theta(\xi) - ib\xi\theta(\xi)\right] - i[\Im f(i) - v(0)]\delta'(\xi). \tag{2.14}$$

Setting

$$u(\xi) = u_1(\xi) + \frac{v(0)}{2}\xi^2\theta(\xi) - ib\xi\theta(\xi)$$

and noting that, by (2.8),

$$\Im f(i) - v(0) = \Im f(i) - \frac{1}{\pi}\int_{-\infty}^{\infty}\frac{\mu(dx)}{1+x^2} = a,$$

we obtain, from (2.14), the representation (b). The uniqueness of expansion (b) follows from the uniqueness of the spectral function g and from Lemma 5 of Sec. 18.1.

SUFFICIENCY. Suppose the generalized function $g(\xi)$ satisfies conditions (a) and (b). Then $g \in \mathcal{S}'(\mathbb{R}^1_+)$ and its Laplace transform $f(z) = L[g]$ is a function of the class $H(\mathbb{R}^1_+)$ (see Sec. 12.2). It remains to prove that $\Im f(z) \geq 0$, $y > 0$.

Using the formulae (1.2) and (1.5) of Sec. 17.1, we have

$$\begin{aligned}\Im f(z) &= \frac{1}{2i}F\left[g(\xi)e^{-y\xi} - g^*(\xi)e^{y\xi}\right] \\ &= \frac{1}{2}F\left[u''(\xi)e^{-y\xi} + u^{*\prime\prime}(\xi)e^{y\xi}\right] \\ &\quad + \frac{a}{2}F\left[\delta'(\xi)e^{-y\xi} + \delta'(-\xi)e^{y\xi}\right], \qquad y > 0, \qquad (2.15)\end{aligned}$$

$$\Im f_+ = \frac{1}{2}F[-ig + ig^*] = \frac{1}{2}F[u'' + u^{*\prime\prime}]. \qquad (2.16)$$

The equation (2.16) shows, by virtue of the Bochner–Schwartz theorem (see Sec. 8.2), that $\mu = \Im f_+$ is a nonnegative measure and $(u+u^*)'' \gg 0$. By Lemma 3 of Sec. 18.1 there exists a continuous tempered function v such that

$$(1-\partial^2)v(\xi) = \frac{1}{2}\left[u(\xi) + u^*(\xi)\right]''. \qquad (2.17)$$

From (2.17) it follows that $v(\xi)$ is a continuous positive definite function, and, by (2.16),

$$(1+x^2)F[v] = \Im f_+ = \mu. \qquad (2.18)$$

Using the formulae (1.2), (1.3) and (2.17), we now obtain the following chain of equalities:

$$\begin{aligned}\frac{1}{2}F\left[u''(\xi)e^{-y\xi} + u^{*\prime\prime}(\xi)e^{y\xi}\right] &= \frac{1}{2}F\left[e^{-y|\xi|}(u+u^*)''\right] \\ &= F\left[e^{-y|\xi|}(1-\partial^2)v\right] \\ &= \frac{y}{\pi}\int \frac{\mu(dx')}{(x-x')^2+y^2}. \qquad (2.19)\end{aligned}$$

Finally, taking into account

$$a(\xi)\delta'(\xi) = -a'(0)\delta(\xi) + a(0)\delta'(\xi), \qquad \delta'(-\xi) = -\delta'(\xi),$$

we obtain

$$\delta'(\xi)e^{-y\xi} + \delta'(-\xi)e^{y\xi} = \delta'(\xi)(e^{-y\xi} - e^{y\xi}) = 2y\delta(\xi). \qquad (2.20)$$

Substituting the expressions (2.19) and (2.20) into (2.15), we obtain the representation (2.9) for $\Im f(z)$, from which it follows that $\Im f(z) \geq 0$, $y > 0$. Theorem II is proved. □

18.3. Functions of the class $\mathcal{P}_+(T^n)$. The case $n > 1$ may be considered in similar fashion to Sec. 18.2 with use made of the residue theorem (see Vladimirov [111]). However, here we will apply a different method, one which makes use of Lemma 4 of Sec. 18.1 on the general form of a bounded continuous solution of the differential equation (1.10).

Suppose the function $u(x,y)$ belongs to the class $\mathcal{P}_+(T^n)$ and the measure $\mu = u(x,+0) \geq 0$ is its boundary value (see Sec. 17.2).

By the theorem of Sec. 17.2, the measure μ has the following properties

$$\frac{1}{\pi^n}\int \frac{\mu(dx)}{(1+x_1^2)\dots(1+x_n^2)} \leq u(0,1), \qquad (3.1)$$

where $\mathbf{1} = (1,\dots,1)$; for any $\varphi \in C \cap \mathcal{L}^\infty$,

$$\int \frac{u(x,y)\varphi(x)\,dx}{(1+x_1^2)\dots(1+x_n^2)} \to \int \frac{\varphi(x)\mu(dx)}{(1+x_1^2)\dots(1+x_n^2)}, \qquad (3.2)$$
$$y \to 0, \qquad y \in C' \Subset \mathbb{R}^n_+.$$

For the measure μ we construct $2^n - 2$ measures $\mu_{j_1\dots j_k}$, $1 \le k \le n-1$, $1 \le j_1 < \dots < j_k \le n$, of the variables $x_{j_1\dots j_k} = (x_{j_1},\dots,x_{j_k}) \in \mathbb{R}^k$ via the following formula: for arbitrary $\varphi \in C \cap \mathcal{L}^\infty$,

$$\int \frac{\varphi(x_{j_1\dots j_k})\mu_{j_1\dots j_k}(dx_{j_1\dots j_k})}{(1+x_{j_1}^2)\dots(1+x_{j_k}^2)} = \int \frac{\varphi(x_{j_1\dots j_k})\mu(dx)}{(1+x_1^2)\dots(1+x_n^2)}. \qquad (3.3)$$

From this definition (for $\varphi = 1$) and from (3.1) it follows, by the Fubini theorem, that the measures $\mu_{j_1\dots j_k}$ are nonnegative and satisfy the condition

$$\frac{1}{\pi^n}\int \frac{\mu_{j_1\dots j_k}(dx_{j_1\dots j_k})}{(1+x_{j_1}^2)\dots(1+x_{j_k}^2)} \le u(0,\mathbf{1}). \qquad (3.4)$$

Set

$$\sigma_{j_1\dots j_k} = \frac{\mu_{j_1\dots j_k}}{(1+x_{j_1}^2)\dots(1+x_{j_k}^2)}, \qquad (3.5)$$

$$\chi_{j_1\dots j_k}(\xi_{j_1\dots j_k}) = (2\pi)^{n-k}F^{-1}[\sigma_{j_1\dots j_k}], \qquad (3.6)$$

$$\mu_{1\dots n} = \mu, \qquad \sigma_{1\dots n} = \sigma, \qquad \chi_{1\dots n} = \chi. \qquad (3.7)$$

By virtue of (3.4) to (3.7), the functions $\chi_{j_1\dots j_k}$ are continuous positive definite in $\mathbb{R}^k$ and, hence, are bounded in $\mathbb{R}^k$ (see Sec. 8). Furthermore, the following equalities hold true:

$$\chi_{j_1\dots j_k}(\xi_{j_1\dots j_k}) = \chi(\xi)\Big|_{\xi_{j_{k+1}}=\dots=\xi_{j_n}=0}. \qquad (3.8)$$

Indeed, using (3.3) to (3.7), we obtain (3.8):

$$\begin{aligned}\chi(\xi)\Big|_{\xi_{j_{k+1}}=\dots=\xi_{j_n}=0} &= \frac{1}{(2\pi)^n}\int \frac{\exp(-i\xi_{j_1}x_{j_1} - \dots - i\xi_{j_k}x_{j_k})\mu(dx)}{(1+x_1^2)\dots(1+x_n^2)}\\ &= \frac{1}{(2\pi)^n}\int \frac{\exp(-i\xi_{j_1}x_{j_1} - \dots - i\xi_{j_k}x_{j_k})\mu_{j_1\dots j_k}(dx_{j_1\dots j_k})}{(1+x_{j_1}^2)\dots(1+x_{j_k}^2)}\\ &= \frac{1}{(2\pi)^n}\int \exp(-i\xi_{j_1}x_{j_1} - \dots - i\xi_{j_k}x_{j_k})\sigma_{j_1\dots j_k}(dx_{j_1\dots j_k})\\ &= \chi_{j_1\dots j_k}(\xi_{j_1\dots j_k}).\end{aligned}$$

□

We now prove that

$$F_k^{-1}[\mu_{j_1\dots j_k}](\xi_{j_1\dots j_k}) = 0, \qquad \xi_{j_1\dots j_k} \notin -\overline{\mathbb{R}}^k_+ \cup \overline{\mathbb{R}}^k_+, \qquad (3.9)$$

where F_k is the Fourier transform operation with respect to k variables $\xi_{j_1\dots j_k} = (\xi_{j_1},\dots,\xi_{j_k})$.

For the measure $\mu = \mu_{1\dots n}$, (3.9) follows from (1.5) of Sec. 17.1, where $g \in \mathcal{S}'(\overline{\mathbb{R}}^n_+)$ (g is the spectral function of the function f taken from $H_+(T^n)$, for which function $\Im f = u$).

Now suppose

$$\varphi(x_{j_1\dots j_k}) = (1+x_{j_1}^2)\dots(1+x_{j_k}^2)F_k^{-1}[\alpha], \qquad (3.10)$$

where $\alpha(\xi_{j_1\dots j_k})$ is an arbitrary function in $\mathcal{D}(\mathbb{R}^k)$ with support outside $-\overline{\mathbb{R}}^k_+ \cup \overline{\mathbb{R}}^k_+$. Substituting the expression (3.10) into (3.3) and rewriting it in terms of Fourier transforms, we obtain

$$\begin{aligned}(\mu_{j_1\dots j_k}, F_k^{-1}[\alpha]) &= (F_k^{-1}[\mu_{j_1\dots j_k}], \alpha) \\ &= \left(F^{-1}[\mu], F\left[F_k^{-1}[\alpha]\frac{1}{(1+x_{j_{k+1}}^2)\dots(1+x_{j_n}^2)}\right]\right) \\ &= \left(F^{-1}[\mu], \alpha(\xi_{j_1\dots j_k})F_{n-k}\left[\frac{1}{(1+x_{j_{k+1}}^2)\dots(1+x_{j_n}^2)}\right]\right).\end{aligned}$$

Using the formula

$$F\left[\frac{1}{1+x^2}\right] = \pi e^{-|\xi|},$$

we can rewrite the last equalities in the form

$$(F_k^{-1}[\mu_{j_1\dots j_k}], \alpha) = \pi^{n-k}\left(F^{-1}[\mu], \alpha(\xi_{j_1\dots j_k})\exp(-|\xi_{j_{k+1}}| - \dots - |\xi_{j_n}|\right).$$

The right-hand side of this equation vanishes because the support of the function

$$\alpha(\xi_{j_1\dots j_k})\exp(-|\xi_{j_{k+1}}| - \dots - |\xi_{j_n}|)$$

lies outside $(-\overline{\mathbb{R}}^k_+ \cup \overline{\mathbb{R}}^k_+) \times \mathbb{R}^{n-k}$, and $F^{-1}[\mu]$, by what has been proved, vanishes outside $-\overline{\mathbb{R}}^n_+ \cup \overline{\mathbb{R}}^n_+$. That precisely is what proves (3.9)

From the equations (3.9), (3.5) to (3.7) there follow differential equations for the function $\chi_{j_1\dots j_k}$:

$$\begin{gathered}(1-\partial_{j_1}^2)\dots(1-\partial_{j_k}^2)\chi_{j_1\dots j_k}(\xi_{j_1\dots j_k}) = 0, \\ \xi_{j_1\dots j_k} \notin -\overline{\mathbb{R}}^k_+ \cup \overline{\mathbb{R}}^k_+.\end{gathered} \tag{3.11}$$

THEOREM I. *If $u \in \mathcal{P}_+(T^n)$, $n \geq 2$, then the function*

$$\chi(\xi) = F^{-1}\left[\frac{\mu}{(1+x_1^2)\dots(1+x_n^2)}\right], \qquad \mu = u(x, +0), \tag{3.12}$$

may be uniquely represented in the form

$$\begin{aligned}\chi(\xi) = &\sum_{2\leq k\leq n}\ \sum_{1\leq j_1<\dots<j_k\leq n} \exp(-|\xi_{j_{k+1}}| - \dots - |\xi_{j_n}|)\Phi_{j_1\dots j_k}(\xi_{j_1\dots j_k}) \\ &+ \exp(-|\xi_2| - \dots - |\xi_n|)\chi(\xi_1, 0, \dots, 0) + \dots \\ &+ \exp(-|\xi_1| - \dots - |\xi_{n-1}|)\chi(0, \dots, 0, \xi_n) \\ &- (n-1)\exp(-|\xi_1| - \dots - |\xi_n|)\chi(0), \qquad \xi \in \mathbb{R}^n,\end{aligned} \tag{3.13}$$

where $\Phi_{j_1\dots j_k}$ are continuous bounded functions in $\mathbb{R}^k$ with support in $-\overline{\mathbb{R}}^k_+ \cup \overline{\mathbb{R}}^k_+$.

PROOF. The uniqueness of the representation (3.13) in each octant follows from the properties of continuity and of the support of the functions $\Phi_{j_1\dots j_k}$. We carry out the existence proof of the representation (3.13) by induction on n. The function $\chi(\xi) = \chi_{1\dots n}(\xi)$ and all the functions $\chi_{j_1\dots j_k}(\xi_{j_1\dots j_k})$ are, by (3.8), continuous and bounded in $\mathbb{R}^k$ and satisfy equation (3.11) outside $-\overline{\mathbb{R}}^k_+ \cup \overline{\mathbb{R}}^k_+$. Therefore when

$n = 2$ Theorem I holds, by virtue of Lemma 4, if in the representation (3.13) we put

$$\Phi_{12}(\xi) = \chi(\xi) + e^{-|\xi_1|}\chi(0,\xi_2) - e^{-|\xi_2|}\chi(\xi_1,0) + e^{-|\xi_1|-|\xi_2|}\chi(0).$$

Suppose the representation (3.13) holds for all dimensions $k < n$, so that the functions $\chi_{j_1\dots j_k}$ in $\mathbb{R}^k$ are representable in the form of the corresponding formulae (3.13). We now prove the representation (3.13) in the domain

$$G_{+-} = \left[\xi : \xi_1 > 0,\ \xi_2 < 0,\ \tilde{\xi} \in \mathbb{R}^{n-2}\right].$$

By Lemma 4 of Sec. 18.1, the function $\chi(\xi)$ can be represented as

$$\chi(\xi) = e^{-|\xi_1|}\chi(0,\xi_2,\tilde{\xi}) + e^{-|\xi_2|}\chi(\xi_1,0,\tilde{\xi}) - e^{-|\xi_1|-|\xi_2|}\chi(0,0,\tilde{\xi}), \quad \xi \in G_{+-}. \tag{3.14}$$

In accordance with the induction hypothesis, for the functions that follow

$$\chi(0,\xi_2,\tilde{\xi}) = \chi_{2\dots n}(\xi_2,\tilde{\xi}),$$
$$\chi(\xi_1,0,\tilde{\xi}) = \chi_{13\dots n}(\xi_1,\tilde{\xi}),$$
$$\chi(0,0,\tilde{\xi}) = \chi_{3\dots n}(\tilde{\xi}),$$

the corresponding representations (3.13) hold true. Substituting them into (3.14), we obtain (3.13) in the domain G_{+-}. The representation (3.13) occurs also in other domains of the type G_{+-} that do not contain $-\overline{\mathbb{R}}^n_+ \cup \overline{\mathbb{R}}^n_+$. From the uniqueness of the representation (3.13) in the indicated domains of the type G_{+-} it follows that the appropriate representations (3.13) coincide in the intersections of those domains. Hence, the representation (3.13) holds true everywhere outside $-\overline{\mathbb{R}}^n_+ \cup \overline{\mathbb{R}}^n_+$. By introducing the function

$$\begin{aligned}\Phi_{1\dots n}(\xi) = \chi(\xi) - \sum_{2\le k\le n-1}\ \sum_{1\le j_1<\dots<j_k\le n} &\exp(-|\xi_{j_{k+1}}| - \dots - |\xi_{j_n}|)\Phi_{j_1\dots j_k}(\xi_{j_1\dots j_k})\\ &- \exp(-|\xi_2| - \dots - |\xi_n|)\chi(\xi_1,0,\dots,0) - \dots\\ &- \exp(-|\xi_1| - \dots - |\xi_{n-1}|)\chi(0,\dots,0,\xi_n)\\ &+ (n-1)\exp(-|\xi_1| - \dots - |\xi_n|)\chi(0),\end{aligned}$$

which is continuous and bounded in $\mathbb{R}^n$ with support in $-\overline{\mathbb{R}}^n_+ \cup \overline{\mathbb{R}}^n_+$, we are convinced of the truth of (3.13) throughout the space $\mathbb{R}^n$. This completes the proof of Theorem I. □

THEOREM II. *For the measure μ to be a boundary value of the function $u(x,y)$ of the class $\mathcal{P}_+(T^n)$, $\mu = u(x,+0)$, it is necessary and sufficient that*

$$\mu = F[\partial_1^2 \dots \partial_n^2 v], \tag{3.15}$$

where $\partial_1^2 \dots \partial_n^2 v$ is a positive definite generalized function, v is a continuous $*$-Hermitian function that satisfies the growth conditions (1.7), and $\operatorname{supp} v \subset -\overline{\mathbb{R}}^n_+ \cup \overline{\mathbb{R}}^n_+$.

Here, the function v having the indicated properties is unique up to the additive term $iC\left[\mathcal{E}_n(\xi) - \mathcal{E}_n(-\xi)\right]$, where C is an arbitrary real number.

PROOF. For $n = 1$, Theorem II has already been proved in Sec. 18.2. Sufficiency for $n \ge 2$ follows from the theorem of Sec. 18.4

NECESSITY FOR $n \geq 2$. Suppose $u \in \mathcal{P}_+(T^n)$ and $\mu = u(x, +0)$. From (3.5) to (3.7) we obtain

$$F^{-1}[\mu] = F^{-1}\big[(1+x_1^2)\dots(1+x_n^2)\sigma\big] = (1-\partial_1^2)\dots(1-\partial_n^2)\chi(\xi). \tag{3.16}$$

Noting that

$$(1-\partial^2)e^{-|\xi|} = 2\delta(\xi)$$

and using (3.13), we continue the equalities (3.16):

$$\begin{aligned} F^{-1}[\mu] = &\sum_{2\le k\le n} 2^{n-k} \sum_{1\le j_1<\dots<j_k\le n} \delta(\xi_{j_{k+1}}) \times\dots\times \delta(\xi_{j_n}) \\ &\times (1-\partial_{j_1}^2)\dots(1-\partial_{j_k}^2)\Phi_{j_1\dots j_k}(\xi_{j_1},\dots,\xi_{j_k}) \\ &+ 2^{n-1}\delta(\xi_2)\times\dots\times\delta(\xi_n) \\ &\times (1-\partial_1^2)\big[\chi(\xi_1,0,\dots,0) - \chi(0)e^{-|\xi_1|}\big] \\ &+\dots+ 2^{n-1}\delta(\xi_1)\times\dots\times\delta(\xi_{n-1}) \\ &\times (1-\partial_n^2)\big[\chi(0,\dots,0,\xi_n) - \chi(0)e^{-|\xi_n|}\big] \\ &+ 2^n\chi(0)\delta(\xi), \end{aligned} \tag{3.17}$$

where $\Phi_{j_1\dots j_k}$ are continuous bounded functions in $\mathbb{R}^k$ with support in $-\overline{\mathbb{R}}_+^k \cup \overline{\mathbb{R}}_+^k$. Every term under the summation sign on the right in (3.17) can, by Lemma 2 of Sec. 18.1, be represented as

$$\begin{aligned} &2^{n-k}\partial_{j_{k+1}}^2\dots\partial_{j_n}^2\big[\theta(\xi_{j_{k+1}})\xi_{j_{k+1}}\dots\theta(\xi_{j_n})\xi_{j_n}\big] \\ &\quad\times(1-\partial_{j_1}^2)\dots(1-\partial_{j_k}^2)\big[\theta(\xi_{j_1})\dots\theta(\xi_{j_k})\Phi_{j_1\dots j_k}(\xi_{j_1},\dots,\xi_{j_k})\big] \\ &\quad+(-2)^{n-k}\partial_{j_{k+1}}^2\dots\partial_{j_n}^2\big[\theta(-\xi_{j_{k+1}})\xi_{j_{k+1}}\dots\theta(-\xi_{j_n})\xi_{j_n}\big] \\ &\quad\times(1-\partial_{j_1}^2)\dots(1-\partial_{j_k}^2)\big[\theta(-\xi_{j_1})\dots\theta(-\xi_{j_k})\Phi(\xi_{j_1},\dots,\xi_{j_k})\big] \\ &\quad= \partial_1^2\dots\partial_n^2 v_{j_1\dots j_k}(\xi), \end{aligned} \tag{3.18}$$

where $v_{j_1\dots j_k}$ is a continuous function in $\mathbb{R}^n$ with support in $-\overline{\mathbb{R}}_+^n \cup \overline{\mathbb{R}}_+^n$, the function satisfying (1.7).

By the same reasoning, the other terms on the right of (3.17) can also be represented in the form of (3.18):

$$\begin{aligned} &2^{n-1}\partial_{j_2}^2\dots\partial_{j_n}^2\big[\theta(\xi_{j_2})\xi_{j_2}\dots\theta(\xi_{j_n})\xi_{j_n}\big] \\ &\quad\times(1-\partial_{j_1}^2)\Big\{\theta(\xi_{j_1})\big[\chi(0,\dots,\xi_{j_1},\dots,0) - \chi(0)e^{-|\xi_1|}\big]\Big\} \\ &\quad+(-2)^{n-1}\partial_{j_2}^2\dots\partial_{j_n}^2\big[\theta(-\xi_{j_2})\xi_{j_2}\dots\theta(-\xi_{j_n})\xi_{j_n}\big] \\ &\quad\times(1-\partial_{j_1}^2)\Big\{\theta(-\xi_{j_1})\big[\chi(0,\dots,\xi_{j_1},\dots,0) - \chi(0)e^{-|\xi_1|}\big]\Big\} \\ &\quad= \partial_1^2\dots\partial_n^2 v_{j_1}(\xi); \end{aligned} \tag{3.19}$$

$$\begin{aligned} 2^n\chi(0)\delta(\xi) &= 2^n\chi(0)\partial_1^2\dots\partial_n^2\big[\theta_n(\xi)\xi_1\dots\xi_n\big] \\ &= \partial_1^2\dots\partial_n^2 v_0(\xi). \end{aligned} \tag{3.20}$$

Putting

$$v(\xi) = \sum_{0\le k\le n}\ \sum_{1\le j_1<\dots<j_k\le n} v_{j_1\dots j_k}(\xi),$$

we obtain, by (3.17) to (3.20), the representation (3.15), where the function v is continuous with support in $-\overline{\mathbb{R}}^n_+ \cup \overline{\mathbb{R}}^n_+$ and satisfies (1.7). If v is not $*$-Hermitian, then it may be replaced by $\frac{1}{2}(v+v^*)$, since the measure μ is real.

The conclusion that the function v is unique follows from Lemma 6 of Sec. 18.1. The proof of Theorem II is complete. □

18.4. Functions of the class $H_+(T^n)$. We recall that the Poisson kernel $\mathcal{P}_n(x,y)$ and the Schwartz kernel $\mathcal{S}_n(z;z^0)$ for the domain T^n have been written out in Sec. 11.1 and Sec. 12.5 respectively.

THEOREM. *The following conditions are equivalent:*

(1) *The function $f(z)$ belongs to the class $H_+(T^n)$.*

(2) *Its spectral function $g(\xi)$ has the following properties:*

 (a) $-ig(\xi) + ig^*(\xi) \gg 0$,

 (b) $g(\xi) = i\partial_1^2 \ldots \partial_n^2 u(\xi) + i(a,\partial)\delta(\xi)$,

 where $a \in \overline{\mathbb{R}}^n_+$ and $u(\xi)$ is a continuous function in $\mathbb{R}^n$ with support in $\overline{\mathbb{R}}^n_+$, which function satisfies the growth condition (1.7). The expansion (b) is unique.

(3) *The following representation holds:*

$$\Im f(z) = \int \mathcal{P}_n(x-x',y)\mu(dx') + (a,y), \qquad z \in T^n. \tag{4.1}$$

(4) *For all $z^0 \in T^n$ the following representation holds:*

$$f(z) = i\int \mathcal{S}_n(z-x';z^0-x')\mu(dx') + (a,z) + b(z^0), \qquad z \in T^n. \tag{4.2}$$

Here, $\mu = \Im f_+$, $b(z^0) = \Re f(z^0) - (a,x^0)$,

$$a_j = \lim_{y_j\to\infty} \frac{\Im f(iy)}{y_j}, \qquad j=1,\ldots,n, \qquad y \in \overline{\mathbb{R}}^n_+, \tag{4.3}$$

(a,y) is the best linear minorant of the indicator $h(\Im f;y)$ in the cone $\overline{\mathbb{R}}^n_+$.

PROOF. For $n=1$, the theorem has already been proved in Sec. 18.2. Suppose $n \geq 2$.

(1) → (2). Suppose $f \in H_+(T^n)$. Then $f(z) = L[g]$, $g \in \mathcal{S}'(\overline{\mathbb{R}}^n_+)$, $f_+ = f[g]$, $\mu = \Im f_+$ and

$$F^{-1}[\mu] = \frac{g-g^*}{2i} \tag{4.4}$$

(see Sec. 17.2). From (4.4) follows the condition (a) (see Sec. 8).

To prove condition (b), let us make use of Theorem II of Sec. 18.3 (the necessity of its hypotheses has already been proved). By (3.15) equation (4.4) takes the form

$$\frac{1}{2i}[g(\xi) - g^*(\xi)] = \partial_1^2 \ldots \partial_n^2 v(\xi), \tag{4.5}$$

where $v = v^* \in C(\mathbb{R}^n)$, $\operatorname{supp} v \subset -\overline{\mathbb{R}}^n_+ \cup \overline{\mathbb{R}}^n_+$ and v satisfies the growth condition (1.7). The generalized function

$$g_0(\xi) = 2i\partial_1^2 \ldots \partial_n^2 [\theta_n(\xi)v(\xi)] \tag{4.6}$$

satisfies (4.5) in $\mathbb{R}^n$. The general solution of the homogeneous equation (4.5), $g-g^* = 0$, has, in the class $\mathcal{S}'(-\overline{\mathbb{R}}^n_+ \cup \overline{\mathbb{R}}^n_+)$, support 0 and, hence, can be represented in the form (see Sec. 2.6)

$$a_0\delta(\xi) + \sum_{1\le|\alpha|\le N} i^{|\alpha|} a_\alpha \partial^\alpha \delta(\xi),$$

where a_α are arbitrary real constants. From this and from (4.6) it follows that the spectral function g is representable as

$$g(\xi) = i\partial_1^2 \dots \partial_n^2 [2\theta_n(\xi)v(\xi) - ia_0\mathcal{E}_n(\xi)] + \sum_{1\le|\alpha|\le N} i^{|\alpha|} a_\alpha \partial^\alpha \delta(\xi). \tag{4.7}$$

Set

$$u(\xi) = 2\theta_n(\xi)v(\xi) - ia_0\mathcal{E}_n(\xi).$$

the function $u(\xi)$ satisfies conditions (b) of the theorem. Here, (4.7) takes the form

$$g(\xi) = i\partial_1^2 \dots \partial_n^2 u(\xi) + \sum_{1\le|\alpha|\le N} i^{|\alpha|} a_\alpha \partial^\alpha \delta(\xi). \tag{4.8}$$

Thus (see Sec. 9.2)

$$f(z) = L[g] = i(-1)^n z_1^2 \dots z_n^2 \int_{\mathbb{R}^n_+} u(\xi) e^{i(z,\xi)}\, d\xi + \sum_{1\le|\alpha|\le N} a_\alpha z^\alpha, \qquad z \in T^n. \tag{4.9}$$

Setting $z = iy$, $y \in \mathbb{R}^n_+$ in (4.9), we obtain

$$f(iy) = iy_1^2 \dots y_n^2 \int_{\mathbb{R}^n_+} u(\xi) e^{-(y,\xi)}\, d\xi + \sum_{1\le|\alpha|\le N} i^{|\alpha|} a_\alpha y^\alpha. \tag{4.10}$$

We now prove that for every $j = 1, \dots, n$,

$$\lim_{y_j\to\infty} y_1 \dots y_n \int_{\mathbb{R}^n_+} u(\xi) e^{-(y,\xi)}\, d\xi = 0. \tag{4.11}$$

Indeed, from the properties of the function $u(\xi)$ it follows (by the Lebesgue theorem) that passage to the limit under the integral sign is valid:

$$\lim_{y_j\to\infty} y_1 \dots y_n \int_0^\infty \dots \int_0^\infty |u(\xi)| e^{-(y,\xi)}\, d\xi$$

$$= \lim_{y_j\to\infty} \int_0^\infty \dots \int_0^\infty \left| u\left(\frac{x_1}{y_1}, \dots, \frac{x_n}{y_n}\right)\right| e^{-x_1-\dots-x_n}\, dx = 0.$$

Taking into account the estimate (3.1) of Sec. 13.3, we obtain, from (4.10), the inequality

$$\left| iy_1^2 \dots y_n^2 \int_{\mathbb{R}^n_+} u(\xi) e^{-(y,\xi)}\, d\xi + i(a,y) + \sum_{2\le|\alpha|\le N} i^{|\alpha|} a_\alpha y^\alpha \right| \le M(C') \frac{1+|y|^2}{|y|},$$

$$y \in C \Subset \mathbb{R}^n_+,$$

from which, and also from the limiting relations (4.11), we conclude that $a_\alpha = 0$, $|\alpha| \geq 2$, and for the numbers a_j the formula (4.3) holds, so that $a_j \geq 0$, $j = 1, \ldots, n$, that is, $a \in \overline{\mathbb{R}^n_+}$. This, by (4.8), proves the representation (b). Its uniqueness follows from Lemma 5 of Sec. 18.1.

(2) → (3). The proof is literally the same as for the one-dimensional case in the sufficiency proof in Theorem II of Sec. 18.2. There, use is made of Lemma 1 and Lemma 3 of Sec. 18.1.

(3) → (4) → (1). From the representation (4.1) it follows that the corresponding Poisson integral is a pluriharmonic function in T^n. All other assertions of the theorem follows from this and from the theorem of Sec. 17.6. This completes the proof of the theorem. □

COROLLARY. *If $f \in H_+(T^n)$, then the best linear minorant (a, y) of the indicator $h(\Im f; y)$ in the cone $\mathbb{R}^n_+$ is given by*

$$a_j = \lim_{y \to e_j,\ y \in \mathbb{R}^n_+} h(\Im f; y), \qquad j = 1, \ldots, n, \tag{4.12}$$

where e_j are unit vectors in $\mathbb{R}^n$, $e_1 = (1, 0, \ldots, 0)$, $e_2 = (0, 1, \ldots, 0)$, ...

Indeed, from the inequality

$$(a, y) \leq h(\Im f; y)$$

it follows that

$$a_j \leq \lim_{y \to e_j,\ y \in \mathbb{R}^n_+} h(\Im f; y). \tag{4.13}$$

The function

$$\frac{1}{t} \Im f(\mathrm{i} + ity), \qquad y \in \mathbb{R}^n_+ \cup \{e_1, \ldots, e_n\}, \qquad t > 0,$$

is continuous in y, does not increase with respect to $t > 0$, and tends (as $t \to \infty$) to the (semicontinuous above) function

$$\hbar(y) = \begin{cases} h(\Im f; y), & y \in \mathbb{R}^n_+, \\ a_j, & y = e_j, \qquad j = 1, \ldots, n. \end{cases} \tag{4.14}$$

For $y \in \mathbb{R}^n_+$ this assertion has been proved (see the theorem in Sec. 17.5). For $y = e_j$ it follows from the representation (4.1),

$$\frac{1}{t} \Im f(\mathrm{i} + ite_j) = \frac{1}{\pi^n} \left(1 + \frac{1}{t}\right) \int \frac{1}{x_j^2 + (1+t)^2} \prod_{\substack{1 \leq k \leq n \\ k \neq j}} \frac{1}{1 + x_k^2} \mu(dx)$$
$$+ \frac{1}{t} \sum_{1 \leq k \leq n} a_k + a_j, \tag{4.15}$$

because, by virtue of B. Levi's theorem, no summand in the right member of (4.15) increases with respect to $t > 0$. From the fact that the function $\hbar(y)$ is semicontinuous above on the set $\mathbb{R}^n_+ \cup \{e_1, \ldots, e_n\}$ and from (4.14) follows the inequality

$$\overline{\lim}_{y \to e_j,\ y \in \mathbb{R}^n_+} h(\Im f; y) \leq \overline{\lim}_{y \to e_j,\ y \in \mathbb{R}^n_+ \cup \{e_j\}} \hbar(y) = \hbar(e_j) = a_j,$$

which, together with the inequality (4.13), implies equation (4.12). □

REMARK 1. The representation (b) strengthens the results of Sec. 17.4 concerning the smoothness of the spectral function and worsens the estimate of its growth in the case of the cone $\mathbb{R}^n_+$:

$$g(\xi) = \partial_1^2 \ldots \partial_n^2 g_1(\xi),$$
$$g_1(\xi) = iu(\xi) + i\theta_n(\xi) \sum_{1\le j\le n} a_j \xi_1 \ldots \xi_{j-1}\xi_{j+1}\ldots \xi_n,$$
$$g_1 \in \mathcal{L}^2_s(\overline{\mathbb{R}^n_+}), \qquad s < -\frac{5}{2}n \qquad \left(\text{in Sec. 17.4} \quad s < -\frac{3}{2}n - 1\right).$$

REMARK 2. A description of the functions of the class $H_+(G)$ in the polycircle G has been given by Korányi and Pukánszky [63] and Vladimirov and Drozhzhinov [121]; in the "generalized unit circle" (in the set of 2×2 matrices w that satisfy the condition $ww^* < I$) by Vladimirov [114]; in bounded strictly star domains, in particular, in the classical symmetric domains, by Aizenberg and Dautov [1]; in the "future tube" $\tau^+ = T^{V^+}$ ($n = 3$, see Sec. 4.4) by Vladimirov [114]. In the last case it was established that the Poisson integral for $f \in H_+(\tau^+)$ is a pluriharmonic function (and, hence, the theorem of Sec. 17.6 holds) if and only if the indicator $h(\Im f; f)$ possesses the properties

$$h(\Im f; y) = h_0(y) + (a, y), \qquad h_0(y) \ge 0, \qquad a \in \overline{V^+}; \qquad y \in V^+;$$
$$\lim_{|y| \to 1-0} \int_{|s|=1} h_0(1, s|\mathbf{y}|)\, ds = 0.$$

REMARK 3. One can prove (see Drozhzhinov, Zavialov [26]) that the function $u(x)$ in the representation (2) (b) satisfies the condition $\Re u(x) \ge 0$.

19. Positive Real Matrix Functions in T^C

Suppose $A(x) = (A_{kj}(x))$ is a square matrix with elements A_{kj} taken from $\mathcal{D}'$. We use the following terminology: $A^*(x) = \bar{A}^T(-x)$ is the *$*$-Hermitian conjugate of A*; $A^+(x) = \bar{A}^T(x)$ is the *$+$-Hermitian conjugate of A*;

$$\Re A = \frac{1}{2}(A + A^+), \qquad \Im A = \frac{1}{2i}(A - A^+)$$

are the *real part* and *imaginary part* of A (compare Sec. 1.3).

If $A = A^*$ or $A = A^+$, then A will be called *$*$-Hermitian* (compare Sec. 8.1) or *$+$-Hermitian* respectively. For constant matrices, both concepts of Hermitian conjugacy coincide, and in that case we will simply call them *Hermitian matrices* or *Hermitian conjugate matrices*.

Clearly, if $A(x)$ is a tempered matrix (that is, $A_{kj} \in \mathcal{S}'$), then

$$F[A^+] = F[A]^*, \qquad F[A^*] = F[A]^+,$$

where the Fourier transform $F[A]$ of matrix A signifies a matrix with components $F[A_{kj}]$.

The matrix function $A(z)$, holomorphic in the tubular domain T^C is said to be *positive real in* T^C if it satisfies the conditions:

(a) $\Re A(z) \ge 0$, $z \in T^C$,

(b) $A(iy)$ is real for all $y \in C$ [and then $A(z) = \bar{A}(-\bar{z})$, $z \in T^C$, by virtue of the Schwartz symmetry principle].

It is clear that if $A(z)$ is positive real in T^C, then $A(z)$ is positive real in any $T^{C'}$, $C' \subset C$, as well.

We term the matrix $Z(\xi)$, $Z_{kj} \in \mathcal{D}'$, for which $A(z) = L[Z]$, the *spectral matrix function* of the matrix $A(z)$.

Our problem is to give a description of positive real matrix functions in T^C, where C is an acute convex cone. Let us first consider the scalar case, that is, positive real functions in T^C.

19.1. Positive real functions in T^C. A function $f(z)$ is positive real in T^C if and only if $H_+(T^C) \ni if$ and its spectral function g is real. The last assertion is due to the equalities

$$f(z) = L[g] = F\left[g(\xi)e^{-(y,\xi)}\right] = \bar{f}(-\bar{z}) = F\left[\bar{g}(\xi)e^{-(y,\xi)}\right].$$

Suppose $C' = \left[y\colon (e_1, y) > 0, \ldots, (e_n, y) > 0\right]$ is an n-hedral acute cone. Then (see Sec. 4.4)

$$C'^* = \left[\xi : \xi = \sum_{1\le j\le n} \lambda_j e_j,\ \lambda_j \ge 0\right].$$

We denote by A the (nonsingular) linear transformation

$$z \to \zeta = \left(\zeta_1 = (e_1, z), \ldots, \zeta_n = (e_n, z)\right) = Az. \tag{1.1}$$

The transformation $\zeta = Az$ maps biholomorphically the domain $T^{C'}$ onto the domain T^n, and the transformation $\xi' = A^{-1T}\xi$ maps the cone C'^* onto the cone $\overline{\mathbb{R}}^n_+$. In the process, the derivatives $\partial = (\partial_1, \ldots, \partial_n)$ pass into the derivatives $\partial' = (\partial'_1, \ldots, \partial'_n)$, $\partial'_j = \frac{\partial}{\partial \xi'_j}$, via the formulae

$$\partial'_j = \sum_{1\le k\le n} \frac{\partial \xi_k}{\partial \xi'_j}\partial_k = (e_j, \partial) = (A\partial)_j, \qquad j = 1, \ldots, n. \tag{1.2}$$

That is, $\partial' = A\partial$.

Furthermore (see Sec. 1.9),

$$\delta(\xi) = \delta(A^T\xi') = \frac{\delta(\xi')}{|\det A|}. \tag{1.3}$$

LEMMA. *If the vectors $e_1, \ldots, e_n$ define an n-hedral acute cone C', then*

$$(e_1, \partial)^2 \ldots (e_n, \partial)^2 u(\xi) + (a, \partial)\delta(\xi) = 0, \tag{1.4}$$

where $u \in C(\mathbb{R}^n)$, $\operatorname{supp} u \subset C'^*$, is possible only for $u(\xi) = 0$ and $a = 0$.

PROOF. In the variables $\xi' = A^{-1T}\xi$, the equation (1.4) becomes, by virtue of (1.2) and (1.3),

$$\partial_1'^2 \ldots \partial_n'^2 \tilde{u}(\xi') + (\tilde{a}, \partial')\delta(\xi') = 0, \tag{1.5}$$

where

$$\tilde{u}(\xi') = |\det A| u(A^T\xi'), \qquad \tilde{a} = A^{-1T}a, \tag{1.6}$$

and $\tilde{u} \in C(\mathbb{R}^n)$, $\operatorname{supp}\tilde{u} \subset \overline{\mathbb{R}}^n_+$. By Lemma 5 of Sec. 18.1, $\tilde{u}(\xi') = 0$ and $\tilde{a} = 0$, whence, by (1.6), we obtain $u(\xi) = 0$ and $a = 0$, which proves the lemma. □

THEOREM. *For a function $f(z)$ to be positive real in T^C, where C is an acute (convex) cone in $\mathbb{R}^n$, it is necessary and sufficient that its spectral function $g(\xi)$ have the following properties:*

(a) $g(\xi) + g^*(\xi) \gg 0$,

(b) *for any n-hedral cone $C' = [y\colon (e_1, y) > 0, \ldots, (e_n, y) > 0]$ contained in the cone C, that it be (uniquely) representable in the form*

$$g(\xi) = (e_1, \partial)^2 \ldots (e_n, \partial)^2 u_{C'}(\xi) + (a_{C'}, \partial)\delta(\xi), \tag{1.7}$$

where $a_{C'} \in C'^*$ and $u_{C'}(\xi)$ is a real continuous tempered function in $\mathbb{R}^n$ with support in the cone C'^*.

PROOF. NECESSITY. Let $f(z)$ be positive real in T^C so that $-if \in H_+(T^C)$ and $f(z) = L[g]$, and the spectral function $g(\xi)$ is real in $S'(C^*)$. From this fact and from (1.5) of Sec. 17.1 there follows the condition (a). To prove the representation (1.7) for the n-hedral cone C', let us perform a biholomorphic mapping $\zeta = Az$ [see (1.1)] of the domain $T^{C'}$ onto T^n; in the process, the function $f(z)$ passes into the positive real function $f(A^{-1}\zeta)$ in T^n. By the theorem of Sec. 17.4 we conclude that there exist a vector $a_1 \in \overline{\mathbb{R}^n_+}$ and a continuous tempered function $u_1(\xi')$ with support in $\overline{\mathbb{R}^n_+}$ such that the spectral function $g_1(\xi')$ of the function $f(A^{-1}\zeta)$ is representable as

$$g_1(\xi') = \partial_1'^2 \ldots \partial_n'^2 u_1(\xi') + (a_1, \partial')\delta(\xi'). \tag{1.8}$$

Let us now pass to the old variables $z = A^{-1}\zeta$ and $\xi = A^T\xi'$. The spectral functions $g(\xi)$ and $g_1(\xi')$ are connected by the relation (see Sec. 9.2.5)

$$|\det A| g(\xi) = g_1(\xi') = g_1(A^{-1T}\xi). \tag{1.9}$$

Using the formulae (1.2) and (1.3), we derive from (1.8) and (1.9) the representation (1.7) for $g(\xi)$ in which

$$u_{C'}(\xi) = \frac{1}{|\det A|} u_1(A^{-1T}\xi), \qquad a_{C'} = A^T a_1. \tag{1.10}$$

Taking into account that the transformation A^T carries the cone $\overline{\mathbb{R}^n_+}$ onto the cone C'^*, we conclude from (1.10) that $a_{C'} \in C'^*$ and $u_{C'} \in C(\mathbb{R}^n)$ is tempered, $\operatorname{supp} u_{C'} \subset C'^*$.

The uniqueness of the expansion (1.7) and the real nature of the function $u_{C'}$ follow from the reality of the spectral function g and the vector a by virtue of the lemma of Sec. 18.1.

SUFFICIENCY. Suppose the generalized function $g(\xi)$ has properties (a) and (b). Then from the representation (1.7) it follows that g is real and $g \in S'(C'^*)$ for all n-hedral cones $C' \subset C$, so that $g \in S'(C^*)$. Therefore the function $f(z) = L[g]$ is holomorphic in T^C and $f(iy)$ is real in C. It remains to prove that $\Re f(z) \geq 0$, $z \in T^C$. Let us take an arbitrary n-hedral cone $C' \subset C$ and pass to the new variables $\zeta = Az$ and $\xi' = A^{-1T}\xi$. Then, as in the proof of necessity, we conclude that the spectral functions $g(\xi)$ and $g_1(\xi')$ of the functions $f(z)$ and $f(A^{-1}\zeta)$ are connected by the relation (1.9) and therefore $g_1(\xi')$ can be represented by the relation (1.9), where $u_1(\xi')$ and a_1 are expressed in terms of $u_{C'}(\xi)$ and $a_{C'}$ via the formulae (1.10), so that $a_1 \in \overline{\mathbb{R}^n_+}$ and $u_1 \in C(\mathbb{R}^n)$ are tempered, $\operatorname{supp} u_1 \subset \overline{\mathbb{R}^n_+}$. Besides, by (1.9)

$$g_1(\xi') + g_1^*(\xi') = |\det A| [g(\xi) + g^*(\xi)] \gg 0.$$

From this, by the theorem of Sec. 17.4, we conclude that $if(A^{-1}\zeta) \in H_+(T^n)$, that is, $\Re f(z) \geq 0$ in T^C, whence, by the arbitrariness of $C' \subset C$, it follows that $\Re f(z) \geq 0$ in T^C, which is what we set out to prove. The theorem is proved. □

19.2. Positive real matrix functions in T^C.

THEOREM. *For an $N \times N$ matrix function $A(z)$ to be positive real in T^C, where C is an acute (convex) cone in $\mathbb{R}^n$, it is necessary and sufficient that its spectral matrix function $Z(\xi)$ have the following properties:*

(a)

$$\big(Z(\xi)a + Z^*(\xi)a, a\big) \gg 0, \qquad a \in \mathbb{C}^N, \tag{2.1}$$

(b) *for any n-hedral cone $C' = \big[y\colon (e_1, y) > 0, \dots, (e_n, y) > 0\big]$ contained in the cone C, it is (uniquely) representable in the form*

$$Z(\xi) = (e_1, \partial)^2 \dots (e_n, \partial)^2 Z_{C'}(\xi) + \sum_{1 \leq j \leq n} Z_{C'}^{(j)} \partial_j \delta(\xi), \tag{2.2}$$

where the matrix function $Z_{C'}(\xi)$ is a continuous tempered function in $\mathbb{R}^n$ with support in $\overline{\mathbb{R}}_+^n$, and the matrices $Z_{C'}^{(j)}$, $j = 1, \dots, n$, are real symmetric and such that

$$\sum_{1 \leq j \leq n} y_j Z_{C'}^{(j)} \geq 0, \qquad y \in \overline{C}'. \tag{2.3}$$

Here the following equation holds:

$$\Re \int \langle Z * \varphi, \varphi \rangle \, d\xi \geq 0, \qquad \varphi = (\varphi_1, \dots, \varphi_N) \in \mathcal{S}^{\times N}. \tag{2.4}$$

PROOF. SUFFICIENCY. From (b) of the theorem it follows that the spectral function $Z(\xi)$ is real and its elements $Z_{kj} \in \mathcal{S}'(C^*)$, so that the matrix function $A(z) = L[Z]$ is holomorphic in the domain T^C, where $C = \operatorname{int} C^{**}$ (see Sec. 12.2), and satisfies the condition of reality $A(z) = \bar{A}(-\bar{z})$.

Let us now verify that the generalized function $g_a(\xi) = \big\langle Z(\xi)a, a\big\rangle$ satisfies, for all $a \in \mathbb{C}^N$, the conditions (a) and (b) of the theorem of Sec. 18.1.

Condition (a) is fulfilled by virtue of (2.1):

$$g_a(\xi) + g_a^*(\xi) = \langle Z(\xi)a + Z^*(\xi)a, a\rangle \gg 0.$$

Conditions (b) are fulfilled by virtue of (2.2) and (2.3):

$$g_a(\xi) = (e_1, \partial)^2 \dots (e_n, \partial)^2 \langle Z_{C'}(\xi)a, a\rangle + \sum_{1 \leq j \leq n} \langle Z_{C'}^{(j)} a, a\rangle \partial_j \delta(\xi),$$

where $\langle Z_{C'}(\xi)a, a\rangle$ is a continuous tempered function in $\mathbb{R}^n$ with support in C'^*, and

$$\sum_{1 \leq j \leq n} y_j \langle Z_{C'}^{(j)} a, a\rangle \geq 0, \qquad y \in \overline{C}'.$$

That is, the vector

$$\Big(\langle Z_{C'}^{(1)} a, a\rangle, \dots, \langle Z_{C'}^{(n)} a, a\rangle\Big) \in C'^*.$$

Noting that $g_a(\xi)$ is the spectral function of the function $\langle A(z)a, a\rangle$, we derive from the theorem of Sec. 18.1 that $\Re\langle A(z)a, a\rangle \geq 0$, $z \in T^C$, $a \in \mathbb{C}^N$. That is, $\Re A(z) \geq 0$, $z \in T^C$. Thus, the matrix function $A(z)$ is positive real in T^C.

NECESSITY. Let $A(z)$ be a positive real matrix function in T^C. Then for every vector $a \in \mathbb{C}^N$ the function $\langle A(z), a\rangle$ is positive real in T^C. By the theorem of Sec. 18.1, its spectral function $g_a(\xi)$ taken from $S'(C^*)$ has the following properties:

(a′) $g_a(\xi) + g_a^*(\xi) \gg 0$,

(b′) for any n-hedral cone $C' \subset C$ it can be represented as

$$g_a(\xi) = (e_1, \partial)^2 \dots (e_n, \partial)^2 U_{C'}(\xi; a) + \sum_{1\le j\le n} A_{C'}^{(j)}(a)\partial_j \delta(\xi), \tag{2.5}$$

where the function $U_{C'}(\xi; a)$ is a continuous tempered function in $\mathbb{R}^n$ with support in the cone C'^*, and the vector

$$\left(A_{C'}^{(1)}(a), \dots, A_{C'}^{(n)}(a)\right) \in C'^*. \tag{2.6}$$

Furthermore, since $g_a(\xi)$ is the spectral function of the quadratic form $\langle A(z)a, a\rangle$, $a \in \mathbb{C}^N$, it follows that $g_a(\xi)$ is a quadratic form with respect to the vector a, so that there exists an $N \times N$ matrix $Z(\xi)$ [the spectral matrix function of the matrix $A(z)$] such that

$$\langle Z(\xi)a, a\rangle = g_a(\xi), \qquad Z_{kj} \in S'(C^*). \tag{2.7}$$

From this and from the condition (a′) it follows that the matrix $Z(\xi)$ satisfies the condition (a). Furthermore, from the equality $A(z) = \bar{A}(-\bar{z})$ follows the real nature of the matrix $Z(\xi)$.

Now, using the equation

$$\begin{aligned}\langle Z(\xi)a, b\rangle &= \frac{1}{4}\langle Z(\xi)(a+b), a+b\rangle \\ &\quad - \frac{1}{4}\langle Z(\xi)(a-b), a-b\rangle \\ &\quad + \frac{i}{4}\langle Z(\xi)(a+ib), a+ib\rangle \\ &\quad - \frac{i}{4}\langle Z(\xi)(a-ib), a-ib\rangle, \qquad a, b \in \mathbb{C}^N,\end{aligned}$$

we derive, from (2.5) and (2.6),

$$\begin{aligned}\langle Z(\xi)a, b\rangle &= \frac{1}{4}(e_1, \partial)^2 \dots (e_n, \partial)^2 \big[U_{C'}(\xi; a+b) \\ &\quad + U_{C'}(\xi; a-b) + iU_{C'}(\xi; a+ib) - iU_{C'}(\xi; a-ib)\big] \\ &\quad + \frac{1}{4}\sum_{1\le j\le n} \big[A_{C'}^{(j)}(a+b) + A_{C'}^{(j)}(a-b) + iA_{C'}^{(j)}(a+ib) \\ &\quad - iA_{C'}^{(j)}(a-ib)\big]\partial_j\delta(\xi).\end{aligned}$$

This implies the existence of the $N \times N$ matrix function $Z_{C'}(\xi)$, which is a continuous tempered function in $\mathbb{R}^n$ with support in the cone C'^*, and the existence of $N \times N$ matrices $Z_{C'}^{(j)}$, $j = 1, \dots, n$, such that the representation (2.2) holds. From the lemma of Sec. 18.1 it follows the uniqueness of the representation (2.2), the reality of the matrices $Z_{C'}(\xi)$ and $Z_{C'}^{(j)}$, $j = 1, \dots, n$ [by virtue of the reality of the matrix $Z(\xi)$], and the equalities [by virtue of (2.5) and (2.7)]

$$U_{C'}(\xi; a) = \langle Z_{C'}(\xi)a, a\rangle, \qquad A_{C'}^{(j)} = \langle Z_{C'}^{(j)}a, a\rangle,$$
$$j = 1, \dots, n, \qquad a \in \mathbb{C}^N.$$

From this and from (2.6) it follows that the matrices $Z^{(j)}_{C'}$ are symmetric and satisfy the condition (2.3):

$$\left\langle \sum_{1\le j\le n} y_j Z^{(j)}_{C'} a, a \right\rangle = \sum_{1\le j\le n} y_j (Z^{(j)}_{C'} a, a) = \sum_{1\le j\le n} y_j A^{(j)}_{C'}(a) \ge 0.$$

Thus, the spectral function $Z(\xi)$ satisfies conditions (b) as well.

It remains to prove the inequality (2.4). Let $\varphi \in \mathcal{S}^{\times N}$; set $\psi = F[\varphi] \in \mathcal{S}^{\times N}$. Taking into account the equalities

$$A_+(x) = F[Z], \qquad \Re A_+(x) = \lim_{y\to 0,\ y\in C} \Re A(x+iy) \quad \text{in } \mathcal{S}'$$

and using the properties of the Fourier transform (see Sec. 6.3 and Sec. 6.5), we have the following chain of equalities:

$$\begin{aligned}
\Re \int \langle Z * \varphi, \varphi\rangle \, d\xi &= \Re \sum_{1\le k,j\le N} \int (Z_{kj} * \varphi_j, \bar\varphi_k)\, d\xi \\
&= \Re \sum_{1\le k,j\le N} (Z_{kj}(-\xi), \varphi_j * \varphi_k^*) \\
&= \Re \sum_{1\le k,j\le N} (F[Z_{kj}], F^{-1}[(\varphi_j * \varphi_k^*)(-\xi)]) \\
&= \frac{1}{(2\pi)^n} \Re \sum_{1\le k,j\le N} (A_{+kj}, \psi_j \bar\psi_k) \\
&= \frac{1}{2(2\pi)^n} \sum_{1\le k,j\le N} (A_{+kj} + \bar A_{+jk}, \psi_j \bar\psi_k) \\
&= \frac{1}{2(2\pi)^n} \lim_{y\to 0,\ y\in C} \sum_{1\le k,j\le N} \int [A_{kj}(x+iy) \\
&\quad + \bar A_{jk}(x+iy)]\, \psi_j(x) \bar\psi_k(x)\, dx \\
&= \frac{1}{2(2\pi)^n} \lim_{y\to 0,\ y\in C} \int \langle \Re A(x+iy)\psi(x), \psi(x)\rangle \, dx
\end{aligned}$$

which is greater than zero. The proof of the theorem is complete. □

REMARK 1. For $n = 1$ the theorem has been proved by König and Zemanian [**62**]; for $n \ge 2$ by Vladimirov [**113**].

REMARK 2. In $\mathbb{R}^2$, any convex open cone C is dihedral, that is,

$$C = [y\colon (e_1, y) > 0,\ (e_2, y) > 0],$$

and for that reason we can take the cone C itself for the cone C' in the representation (2.2).

20. Linear Passive Systems

20.1. Introduction. We consider a physical system obeying the following scheme. Suppose the original in-perturbation $u(x) = (u_1(x), \dots, u_N(x))$ is acting on the system, as a result of which there arises an out-perturbation (response of the system) $f(x) = (f_1(x), \ldots, f_N(x))$. Here, by $x = (x_1, \ldots, x_n)$ are to be understood the temporal, spatial and other variables. Suppose the following conditions have been fulfilled:

(a) *Linearity*: if to the original perturbations u_1 and u_2 there correspond perturbations f_1 and f_2 then their linear combination $\alpha u_1 + \beta u_2$ is associated with the perturbation $\alpha f_1 + \beta f_2$.

(b) *Reality*: if the original perturbation u is real, then the response perturbation f is real.

(c) *Continuity*: if all components of the original perturbations $u(x)$ tend to 0 in $\mathcal{E}'$, then so do all components of the response perturbation $f(x)$ tend to 0 in $\mathcal{D}'$.

(d) *Translational invariance*: if a response perturbation $f(x)$ is associated with the original perturbation $u(x)$, then, for any translation $h \in \mathbb{R}^n$, to the original perturbation $u(x+h)$ there corresponds a response perturbation $f(x+h)$.

The conditions (a)–(d) are equivalent to the existence of a unique $N \times N$ matrix $Z(x) = (Z_{kj}(x))$, $Z_{kj} \in \mathcal{D}'(\mathbb{R}^n)$, which connects the original $u(x)$ perturbation and the response perturbation $f(x)$ via the formula (see Sec. 4.8)

$$Z * u = f. \tag{1.1}$$

Let us impose on the system (1.1) yet another requirement, the so-called condition of *passivity relative to the cone* Γ. Suppose Γ is a closed, convex, solid cone in $\mathbb{R}^n$ (with vertex at 0).

(e) *Passivity relative to the cone* Γ: for any vector function $\varphi(x)$ in $\mathcal{D}^{\times N}$ the following inequality holds:

$$\Re \int_{-\Gamma} (Z * \varphi, \varphi)\, dx \geq 0. \tag{1.2}$$

Note that the function $(Z * \varphi, \varphi) \in \mathcal{D}$ (see Sec. 4.6), so that the integral in (1.2) always exists. Furthermore, because of the reality of the matrix $Z(x)$ the condition of passivity (1.2) is equivalent to the condition

$$\int_{-\Gamma} (Z * \varphi, \varphi)\, dx \geq 0, \qquad \varphi \in \mathcal{D}_r^{\times N} \tag{1.2'}$$

where $\mathcal{D}_r^{\times N}$ consists of real N vectors with components in $\mathcal{D}$.

The inequality (1.2′) is of the energy type: it reflects the ability of a physical system to absorb and redistribute energy, but not generate it. Here, causality relative to the cone Γ is taken into account (see below, Sec. 20.2).

The convolution operator $Z*$ is termed a *passive operator relative to the cone* Γ, and the corresponding matrix function $\tilde{Z}(\zeta)$ — the Laplace transform of the matrix $Z(x)$ — is called the *impedance* of the physical system.

To illustrate the proposed scheme, let us consider an one-dimensional passive system ($n = N = 1$): an elementary electric circuit consisting of a resistance R, a self-inductance L, a capacitance C, and a source of electromotive force $e(t)$ that is switched on at time $t = 0$ (Fig. 43). Then, by the Kirchhoff law, the current $i(t)$ in the circuit satisfies the integro-differential equation

$$L\frac{di}{dt} + Ri + \frac{1}{C}\int_0^t i(\tau)\,d\tau = e(t),$$

that is,

$$Z * i = e$$

where

$$Z(t) = L\delta'(t) + R\delta(t) + \frac{1}{C}\theta(t)$$

is the generalized "resistance" of the circuit. We now verify that the operator Z_* satisfies the condition of passivity (1.2′) relative to the cone $\Gamma = [0, \infty)$:

$$\int_{-\infty}^{0} (Z * \varphi)\varphi\,dt = \int_{-\infty}^{0} \left[L\varphi'(t) + R\varphi(t) + \frac{1}{C}\int_0^t \varphi(\tau)\,d\tau\right]\varphi(t)\,dt$$

$$= \frac{L}{2}\varphi^2(0) + R\int_{-\infty}^{0} \varphi^2(t)\,dt + \frac{1}{2C}\left[\int_{-\infty}^{0} \varphi(t)\,dt\right]^2 \geq 0, \qquad \varphi \in \mathcal{D}_r.$$

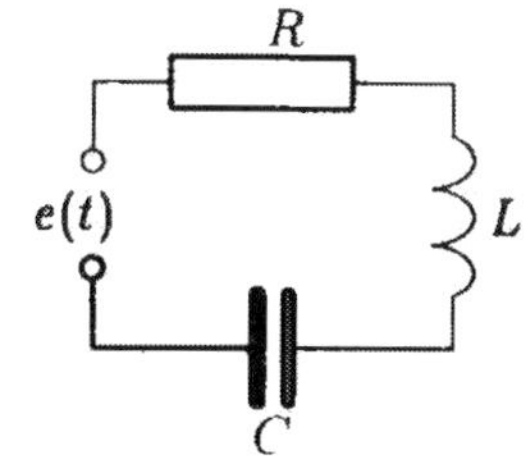

Figure 43

One-dimensional ($n = 1$) linear passive systems describe the relationship between currents and voltage in complex electric circuits. They also describe linear thermodynamic systems, the scattering of electromagnetic waves and elementary particles (see König and Meixner [**61**], Youla, Castriota and Carlin [**128**], Wu [**127**], Zemanian [**130**], Beltrami and Wohlers [**4**], Güttinger [**42**]). One-dimensional passive operators have been studied by many authors and the results of their investigations have been summarized in two monographs (1965–1966): Zemanian [**130**] and Beltrami and Wohlers [**4**]. This theory has been extended by Hackenbroch [**43**] and Zemanian [**131**] from the matrix case to the case of operators in Hilbert space.

Multidimensional ($n \geq 2$) linear passive systems are frequently encountered in mathematical physics: they describe physical systems with account taken of their space-time dynamics (some instances of such systems are given below in Sec. 20.7). The theory of multidimensional linear passive systems has been elaborated by Vladimirov [**113, 115**] on the basis of the theory of positive real matrix functions (see Sec. 19).

The passive systems of several variables with values in Hilbert spaces have been considered by Galeev [**34, 35**]. Nontranslation-invariant passive systems have been investigated by Drozhzhinov [**22**].

20.2. Corollaries to the condition of passivity.

20.2.1. The condition of passivity (1.2) is fulfilled in the strong form:

$$\Re \int_{-\Gamma + x_0} \langle Z * \varphi, \varphi \rangle\,dx \geq 0, \qquad \varphi \in \mathcal{D}^{\times N}, \quad x_0 \in \mathbb{R}^n. \tag{2.1}$$

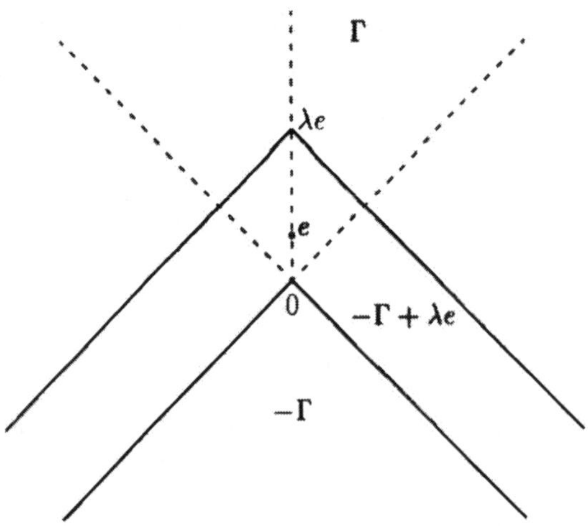

Figure 44

Indeed, if $\varphi \in \mathcal{D}^{\times N}$, for every $x_0 \in \mathbb{R}^n$ the vector function $\varphi_{x_0}(x) = \varphi(x + x_0) \in \mathcal{D}^{\times N}$ and therefore, by virtue of the property of translational invariance of a convolution (see Sec. 4.2.3), from the inequality (1.2) there follows the inequality (2.1):

$$0 \leq \Re \int\limits_{-\Gamma} \langle Z * \varphi_{x_0}, \varphi_{x_0} \rangle \, dx = \Re \int\limits_{-\Gamma} \langle (Z * \varphi)(x + x_0), \varphi(x + x_0) \rangle \, dx$$
$$= \Re \int\limits_{-\Gamma + x_0} \langle Z * \varphi, \varphi \rangle \, dx', \qquad x' = x + x_0.$$

□

20.2.2. *Dissipation:*

$$\Re \int \langle Z * \varphi, \varphi \rangle \, dx \geq 0, \qquad \varphi \in \mathcal{D}^{\times N}. \tag{2.2}$$

Indeed, putting $x_0 = \lambda e$, $e \in \operatorname{int} \Gamma$, in (2.1) and passing to the limit as $\lambda \to +\infty$ (so that $-\Gamma + \lambda e \to \mathbb{R}^n$, Fig. 44), we obtain from (2.1) the inequality (2.2). □

20.2.3. *Causality with respect to the cone* Γ*:*

$$\operatorname{supp} Z(x) \subset \Gamma. \tag{2.3}$$

Indeed, let φ and $\psi \in \mathcal{D}_r^{\times N}$ and let λ be a real number. Substituting $\varphi + \lambda\psi$ for φ in the inequality (1.2′), we get the inequality

$$\int\limits_{-\Gamma} \langle Z * \varphi, \varphi \rangle \, dx + \lambda \int\limits_{-\Gamma} [\langle Z * \psi, \varphi \rangle + \langle Z * \varphi, \psi \rangle] \, dx + \lambda^2 \int\limits_{-\Gamma} \langle Z * \psi, \psi \rangle \, dx \geq 0,$$

which holds true for all real λ. We therefore have the inequality

$$\left[\int\limits_{-\Gamma} \langle Z * \varphi, \psi \rangle \, dx + \int\limits_{-\Gamma} \langle Z * \psi, \varphi \rangle \, dx \right]^2 \leq 4 \int\limits_{-\Gamma} \langle Z * \varphi, \varphi \rangle \, dx \int\limits_{-\Gamma} \langle Z * \psi, \psi \rangle \, dx. \tag{2.4}$$

Suppose $\operatorname{supp}\varphi \subset \mathbb{R}^n \setminus (-\Gamma)$. Then it follows from the inequality (2.4) that

$$\int_{-\Gamma} \langle Z * \varphi, \psi \rangle \, dx = 0$$

for all $\psi \in \mathcal{D}_r^{\times N}$. By the Du Bois Reymond lemma (see Sec. 1.6), we conclude from this that $Z * \varphi = 0$, $x \in -\Gamma$, and therefore (see Sec. 4.6)

$$(Z_{kj} * \varphi_0)(x) = \big(Z_{kj}(x'), \varphi_0(x - x')\big) = 0, \qquad x \in -\Gamma$$

for all $\varphi_0 \in \mathcal{D}_r(\mathbb{R}^n \setminus (-\Gamma))$. Putting $x = 0$ in the last equality, we obtain $\big(Z_{kj}(-x'), \varphi_0(x')\big) = 0$ so that $Z_{kj}(-x) = 0$, $x \in \mathbb{R}^n \setminus (-\Gamma)$, and therefore $\operatorname{supp} Z_{kj} \subset \Gamma$ (see Sec. 1.5). The inclusion (2.3) is proved. □

20.2.4. *Positive definiteness:*

$$\langle Za + Z^* a, a \rangle \gg 0, \qquad a \in \mathbb{C}^N \tag{2.5}$$

or, in an equivalent form

$$\Re\left(\langle Za, a \rangle, \varphi * \varphi^*\right) \geq 0, \qquad a \in \mathbb{C}^N, \qquad \varphi \in \mathcal{D}. \tag{2.5'}$$

It follows from the inequality (2.2) for $\varphi = a\varphi_0(-x)$, where $a \in \mathbb{C}^N$ and $\varphi_0 \in \mathcal{D}$, that

$$\begin{aligned}
0 &\leq \Re \int \langle Z * a\varphi_0(-x), a\varphi_0(-x) \rangle \, dx \\
&= \Re \int \left[\langle Za, a \rangle * \varphi_0(-x)\right] \varphi_0^*(x) \, dx \\
&= \Re\left(\langle Za, a \rangle, \varphi_0 * \varphi_0^*\right) \\
&= \frac{1}{2}\left(\langle Z(x)a, a \rangle, \varphi_0 * \varphi_0^*\right) + \frac{1}{2}\left(\overline{\langle Z(x)a, a \rangle}, \overline{\varphi_0 * \varphi_0^*}\right) \\
&= \frac{1}{2}\left(\langle Z(x)a + Z^T(-x)a, a \rangle, \varphi_0 * \varphi_0^*\right),
\end{aligned}$$

which proves the inequalities (2.5) and (2.5′) (see Sec. 8.1). Here we made use of the property of the convolution (6.4) of Sec. 4.6. □

In what follows we assume that the cone Γ is acute.

20.2.5. *Restriction to growth:* $Z \in (\mathcal{S}')^{\times N^2}$.

Indeed, by the Bochner–Schwartz theorem (see Sec. 8.2), the generalized function $\langle Z(x)a + Z^*(x)a, a \rangle$ belongs to $\mathcal{S}'$ for all $a \in \mathbb{C}^N$, and from this it follows that the generalized function $\langle Z(x)a + Z^*(x)a, b \rangle \in \mathcal{S}'$ for all a and b in $\mathbb{C}^N$, so that

$$f_{kj}(x) = Z_{kj}(x) + Z_{jk}(-x) \in \mathcal{S}', \qquad 1 \leq k, j \leq N.$$

From the causality condition (2.3), $\operatorname{supp} Z_{kj} \subset \Gamma$, it follows that $\operatorname{supp} f_{kj} \subset -\Gamma \cup \Gamma$. Suppose $\eta \in C^\infty$, $\eta(t) = 1$, $t > 1$, $\eta(t) = 0$, $t < 0$ and $e \in \operatorname{int}\Gamma^*$. Then the function $\eta\big((e, x)\big) \in \theta_M$ and for that reason $\eta\big((e, x)\big) f_{kj} \in \mathcal{S}'$ (see Sec. 5.3). Furthermore, the support of the generalized function

$$s_{kj}(x) = Z_{kj} - \eta\big((e, x)\big) f_{kj}(x)$$

is compact, by virtue of Lemma 1 of Sec. 4.4 (see Fig. 22; the cone Γ is assumed to be acute!), so that $s_{kj} \in \mathcal{S}'$ (see Sec. 5.3). Conclusion: $Z_{kj} \in \mathcal{S}'$. □

20.2.6. The condition of passivity (1.2) holds in the *strong form*:

$$\Re \int_{-\Gamma} \langle Z * \varphi, \varphi \rangle \, dx \geq 0, \qquad \varphi \in S^{\times N}. \tag{2.6}$$

Indeed, fix φ taken from $S^{\times N}$ and let $\varphi_\nu \in \mathcal{D}^{\times N}$, $\varphi_\nu \to \varphi$, $\nu \to \infty$ in $S^{\times N}$ (see Sec. 5.1). Then by (1.2)

$$\Re \int_{-\Gamma} \langle Z * \varphi_\nu, \varphi_\nu \rangle \, dx \geq 0, \qquad \nu = 1, 2, \ldots . \tag{2.7}$$

By 20.2.5, $Z_{kj} \in S'$, $1 \leq k, j \leq N$, and therefore Z_{kj} is of finite order (see Sec. 5.2). Denoting by m the largest of the orders and using the estimate (6.4) of Sec. 5.6, we obtain, for $k = 1, \ldots, N$,

(α) $|(Z * \varphi_\nu)_k(x)| \leq C\left(1 + |x|^2\right)^{m/2} \max_{1 \leq j \leq N} \|\varphi_{\nu j}\|_m, \qquad \nu = 1, 2, \ldots .$

(β) $(Z * \varphi_\nu)_k(x) \overset{|x| \leq R}{\Longrightarrow} (Z * \varphi)_k(x), \qquad \nu \to \infty$

for arbitrary $R > 0$, from which we conclude that passage to the limit is possible as $\nu \to \infty$ under the integral sign in the inequality (2.7); we thus obtain the inequality (2.6). □

20.2.7. *The existence of impedance:*

$$\tilde{Z}(\zeta) = L[Z] = F\left[Z(x)e^{-(q,x)}\right](p), \qquad \zeta = p + iq$$

is a holomorphic matrix function in the domain $T^C = \mathbb{R}^n + iC$, where $C = \operatorname{int} \Gamma^*$.

This follows from the properties 20.2.3 and 20.2.5: $Z_{kj} \in S'(\Gamma)$ (see Sec. 9.1). □

20.2.8. *The condition of reality of impedance:*

$$\tilde{Z}(\zeta) = \bar{\tilde{Z}}(-\bar{\zeta}), \qquad \zeta \in T^C. \tag{2.8}$$

This follows from the reality of the matrix $Z(x)$. □

20.2.9. *The property of positivity of impedance:*

$$\Re \tilde{Z}(\zeta) \geq 0, \qquad \zeta \in T^C. \tag{2.9}$$

Indeed, let the function $\eta_\epsilon(x)$ be such that $\eta_\epsilon \in C^\infty$; $\eta_\epsilon(x) = 1$, $x \in \Gamma^{\epsilon/2}$; $\eta_\epsilon(x) = 0$, $x \notin \Gamma^\epsilon$; $|\partial^\alpha \eta_\epsilon(x)| \leq C_{\alpha\epsilon}$. Then, for all $\zeta \in T^C$ (see Sec. 9.1),

$$\varphi(\zeta; x) = \eta_\epsilon(-x)e^{-i(\zeta, x)} \in S,$$

and for all $a \in \mathbb{C}^N$ the vector function $a\varphi \in S^{\times N}$; therefore, using the formula (6.4) of Sec. 4.6, we have

$$\begin{aligned} \langle Z * a\varphi, a\varphi \rangle &= \left[\langle Za, a \rangle * \varphi\right](x)\bar{\varphi}(\zeta; x) \\ &= \left(\langle Z(x')a, a \rangle, \varphi(\zeta; x - x')\right)\bar{\varphi}(\zeta; x) \\ &= e^{2(q,x)}\eta_\epsilon(-x)\left(\langle Z(x')a, a \rangle, \eta_\epsilon(x' - x)e^{i(\zeta, x')}\right). \end{aligned} \tag{2.10}$$

But $\eta_\epsilon(x' - x) = 1$ for $x' \in \Gamma^{\epsilon/2}$ and $x \in -\Gamma$ because by Lemma 2 of Sec. 4.4 $x' - x \in \Gamma + U_{\epsilon/2} + \Gamma = \Gamma + U_{\epsilon/2} = \Gamma^{\epsilon/2}$. From this, taking into account that $\operatorname{supp} Z(x') \subset \Gamma$ and using (10.2) of Sec. 1.10, we continue the chain of equalities (2.10) for $x \in -\Gamma$:

$$\langle Z * (a\varphi), a\varphi \rangle = e^{2(q,x)}\left(\langle Z(x')a, a \rangle, \eta_\epsilon(x')e^{i(\zeta, x')}\right). \tag{2.11}$$

Now let us take advantage of (1.4) of Sec. 9.1 for the Laplace transform and let us integrate (2.11) with respect to the cone $-\Gamma$. As a result, using property 20.2.6, we obtain the inequality

$$0 \leq \Re \int_{-\Gamma} \langle Z * (a\varphi), a\varphi \rangle \, dx = \Re \langle \tilde{Z}(\zeta)a, a \rangle \int_{-\Gamma} e^{2(q,x)} \, dx. \tag{2.12}$$

Noting that for all $q \in C$ the last integral exists and is positive (see Sec. 10.2), we conclude from (2.12) that

$$\Re \langle \tilde{Z}(\zeta)a, a \rangle = \frac{1}{2} \langle \tilde{Z}(\zeta)a + \tilde{Z}^{+}(\zeta)a, a \rangle \geq 0,$$

which is equivalent to (2.9). □

From the properties 20.2.7, 20.2.8 and 20.2.9 it follows that the impedance $\tilde{Z}(\zeta)$ belongs to the class of positive real matrix functions in T^C, the description of which was given in Sec. 19.2 (the cone Γ is assumed to be acute).

REMARK. It is readily seen, if we slightly modify the foregoing reasoning, that the corollaries 20.2.3, 20.2.4, 20.2.5, 20.2.7, 20.2.8, 20.2.9 remain true even when the weak condition of passivity is carried out relative to the cone Γ:

$$\Re \int_{-\Gamma} [\langle Z(x)a, a \rangle * \varphi] \bar{\varphi} \, dx \geq 0, \qquad \varphi \in \mathcal{D}, \qquad a \in \mathbb{C}^N. \tag{2.13}$$

20.3. The necessary and sufficient conditions for passivity.

THEOREM I. *For a matrix $Z(x)$ to define a passive operator relative to an acute cone Γ, it is necessary and sufficient that its impedance $\tilde{Z}(\zeta)$ be a positive real matrix function in the domain T^C, where $C = \operatorname{int} \Gamma^*$.*

COROLLARY. *If a system is passive relative to an acute cone Γ, then it is also passive relative to any acute cone containing Γ.*

REMARK. For $n = 1$, Theorem I was proved by Zemanian [**132**], and for $n \geq 2$ it was proved by Vladimirov [**113**].

First let us prove the following lemma.

LEMMA. *Suppose an $N \times N$ matrix $Z(x)$ has the following properties:*

(a) *it defines a passive operator relative to a certain cone Γ_0 containing the cone C^*; the boundary of Γ_0 is assumed to be piecewise smooth;*

(b) *for any n-hedral cone $C' = [q\colon (e_1, q) > 0, \ldots, (e_n, q) > 0]$ contained in a convex acute cone C, it is given in the form*

$$Z(x) = (e_1, \partial)^2 \ldots (e_n, \partial)^2 Z_{C'}(x) + \sum_{1 \leq j \leq n} Z_{C'}^{(j)} \partial_j \delta(x), \tag{3.1}$$

where the matrix function $Z_{C'}(x)$ is continuous, real, and tempered in $\mathbb{R}^n$ with support in the cone C'^*; the matrices $Z_{C'}^{(j)}$, $j = 1, \ldots, n$, are real, symmetric and such that

$$\sum_{1 \leq j \leq n} q_j Z_{C'}^{(j)} \geq 0, \qquad q \in \bar{C}'. \tag{3.2}$$

Then the matrix $Z(x)$ defines a passive operator relative to the cones $\Gamma_e = [x\colon (e, x) \geq 0, \ x \in \Gamma_0]$, where e is any unit vector taken from the cone $\bar{C}$.

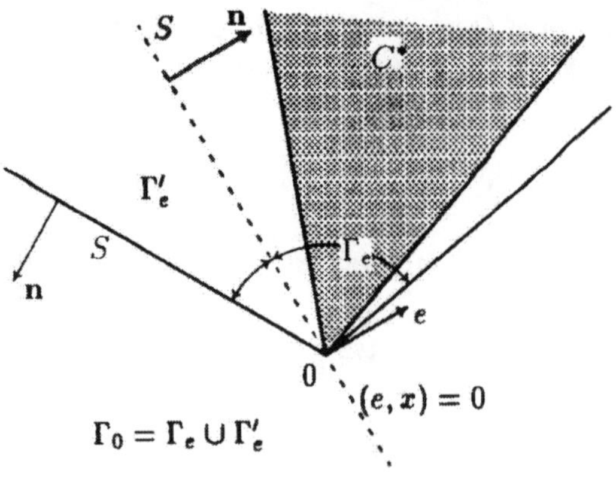

Figure 45

PROOF. From conditions (b) it follows that the matrix $Z(x)$ is real and tempered with support in the cone C^*.

The lemma is non-trivial if the cone

$$\Gamma'_e = \Gamma_0 \setminus \Gamma_e = [x\colon (e,x) < 0,\ x \in \Gamma_0]$$

is a solid cone. Clearly, $\Gamma_0 = \Gamma_e \cup \Gamma'_e$ (Fig. 45).

Let $\eta \in C^\infty$; $0 \le \eta(t) \le 1$; $\eta(t) = 1, t < \frac{1}{2}$; $\eta(t) = 0, t > 1$. We set

$$\eta_\varepsilon(x) = \eta\left[\frac{(e,x)}{\varepsilon}\right].$$

Then for all $\varphi \in \mathcal{D}_r^{\times N}$ we have

$$\begin{aligned}\int_{-\Gamma_e} \langle Z * \varphi, \varphi\rangle\, dx = &\int_{-\Gamma_0} \langle Z * (\varphi\eta_\varepsilon), \varphi\eta_\varepsilon\rangle\, dx \\ &+ \int_{-\Gamma_e} [\langle Z * \varphi, \varphi\rangle - \langle Z * (\varphi\eta_\varepsilon), \varphi\eta_\varepsilon\rangle]\, dx \\ &- \int_{-\Gamma'_e} \langle Z * (\varphi\eta_\varepsilon), \varphi\eta_\varepsilon\rangle\, dx. \end{aligned} \tag{3.3}$$

The first summand in the right-hand member of (3.3) is non-negative for all $\varepsilon > 0$ by virtue of the condition (a). Furthermore, since $\operatorname{supp} Z(x') \subset C^*$ and $\eta_\varepsilon(x - x') = 1$, $x \in -\Gamma_e$, $x' \in (C^*)^{\varepsilon/2}$ [because $(e, x - x') \le -(e, x') = -(e, x_1) - (e, x_2) \le |(e, x_2)| \le \varepsilon/2$, where $x' = x_1 + x_2$, $x_1 \in C^*$, $|x_2| < \varepsilon/2$], it follows that

$$\begin{aligned}(Z * (\varphi\eta_\varepsilon))_k(x) &= \sum_{1\le j\le N} (Z_{kj}(x'), \varphi_j(x - x')\eta_\varepsilon(x - x')) \\ &= \sum_{1\le j\le N} (Z_{kj}(x'), \varphi_j(x - x')) \\ &= (Z * \varphi)_k(x), \qquad x \in -\Gamma_e,\end{aligned}$$

and, hence, the second summand on the right of (3.3) is zero. Thus, for all $\varepsilon > 0$ we have the inequality

$$\int_{-\Gamma_\varepsilon} \langle Z * \varphi, \varphi \rangle \, dx \geq - \int_{-\Gamma'_\varepsilon} \langle Z * (\varphi\eta_\varepsilon), \varphi\eta_\varepsilon \rangle \, dx, \qquad \varphi \in \mathcal{D}_r^{\times N}. \tag{3.4}$$

We choose linearly independent vectors $e'_1 = e, e'_2, \ldots, e'_n$ [from the cone $\bar{C}$] and let $\{e_k\}$ be a system of vectors that is biorthogonal to the system $\{e'_j\}$, $(e_k, e'_j) = \delta_{kj}$. We set $C' = [q\colon (e_1, q) > 0, \ldots, (e_n, q) > 0]$. Then $C' \subset C$, $e \in \bar{C}'$ and $C'^* = [x\colon (e, x) \geq 0, \ldots, (e'_n, x) \geq 0]$. For the cone C', the representation (3.1) holds true.

Taking into account that representation, we transform the right-hand side of the inequality (3.4) to the following form:

$$\begin{aligned} - \int_{-\Gamma'_\varepsilon} \langle Z * (\varphi\eta_\varepsilon), \varphi\eta_\varepsilon \rangle \, dx &= - \int_{-\Gamma'_\varepsilon} (e_1, \partial)^2 \ldots (e_n, \partial)^2 \langle Z_{C'} * (\varphi\eta_\varepsilon), \varphi\eta_\varepsilon \rangle \, dx \\ &\quad - \sum_{1 \leq j \leq n} \int_{-\Gamma'_\varepsilon} \left\langle Z_{C'}^{(j)} \frac{\partial}{\partial x_j} (\varphi\eta_\varepsilon), \varphi\eta_\varepsilon \right\rangle dx \\ &= I_1(\varepsilon) + I_2(\varepsilon). \end{aligned} \tag{3.5}$$

For the quantity $I_1(\varepsilon)$ we have the estimate

$$\begin{aligned} |I_1(e)| \leq c_1 \sum_{1 \leq k,j \leq N} \int_{\substack{|x| < R \\ 0 < (e,x) < \varepsilon}} \Bigg| (e_1, \partial)^2 \ldots (e_n, \partial)^2 \\ \times \int_{C'^*} Z_{C',kj}(x') \varphi_j(x - x') \eta \left[\frac{(e, x - x')}{\varepsilon} \right] dx' \Bigg| \, dx, \end{aligned} \tag{3.6}$$

where $c_1 = \max_{1 \leq k \leq N} |\varphi_k(x)|$ and the number $R > 0$ is such that $\operatorname{supp} \varphi \subset U_R$. In the inner and outer integrals in (3.6) we make a change of the variables of integration via the following formulae, respectively:

$$x \to Bx = y = [y_1 = (e, x), \ldots, y_n = (e'_n, x)], \qquad x' \to Bx' = y'.$$

Then the cone C'^* goes into the cone $\overline{\mathbb{R}^n_+} = [y'\colon y'_1 \geq 0, \ldots, y'_n \geq 0]$, the ball U_R goes into a bounded domain contained in some ball U_{R_1}, the strip $0 < (x, e) < \varepsilon$ goes into the strip $0 < y_1 < \varepsilon$, and the derivative (e_k, ∂) into the derivative ∂_k (see Sec. 19.1).

Setting

$$\begin{gathered} Z_{C',kj}(B^{-1}y') = v_{kj}(y'), \qquad \varphi_1(B^{-1}y) = \psi_1(y), \\ \partial_2^2 \ldots \partial_n^2 \psi_j(y) = u_j(y), \end{gathered}$$

we obtain from (3.6) the estimate

$$|I_1(\varepsilon)| \leq \frac{c_1}{(\det B)^2} \sum_{1\leq k,j\leq N} \int_{\substack{|y|<R_1\\ 0<y_1<\varepsilon}} \left| \partial_1^2 \int_{\substack{\mathbb{R}^n_+\\ |y-y'|<R_1}} v_{kj}(y')u_j(y-y')\eta\left(\frac{y_1-y_1'}{\varepsilon}\right) dy' \right| dy$$

$$= \frac{c_1}{(\det B)^2} \sum_{1\leq k,j\leq N} \int_{\substack{|y|<R_1\\ 0<y_1<\varepsilon}} \int_{\substack{|y'|<2R_1\\ y'\in\mathbb{R}^n_+}} |v_{kj}(y')|$$

$$\times \left| \eta\left(\frac{y_1-y_1'}{\varepsilon}\right) \frac{\partial^2 u_j(y-y')}{\partial y_1^2} + \frac{2}{\varepsilon}\eta'\left(\frac{y_1-y_1'}{\varepsilon}\right) \frac{\partial u_j(y-y')}{\partial y_1} \right.$$

$$\left. + \frac{1}{\varepsilon^2}\eta''\left(\frac{y_1-y_1'}{\varepsilon}\right) u_j(y-y') \right| dy'\, dy.$$

We set

$$\chi(y_1') = \sum_{1\leq k,j\leq N} \int_{\substack{y_2'>0,\dots,y_n'>0\\ y_2'^2+\dots+y_n'^2<4R_1^2}} [v_{kj}(y_1', y_2', \dots, y_n)]\, dy_2' \dots dy_n'.$$

Since the functions v_{kj} are continuous in $\mathbb{R}^n$ with supports in $\overline{\mathbb{R}}^n_+$, it follows that the function $\chi(y_1')$ is continuous in $\mathbb{R}^1$ and is zero for $y_1' < 0$. Using this notation, we continue our estimate:

$$|I_1(\varepsilon)| \leq c_2 \int_0^\varepsilon \left[\int_0^{2R_1} \chi(y_1')\, dy_1' \right] dy_1 + \left(\frac{c_3}{\varepsilon} + \frac{c_4}{\varepsilon^2}\right) \int_0^\varepsilon \left[\int_0^{y_1+\varepsilon/2} \chi(y_1')\, dy_1' \right] dy_1$$

$$\leq c_5\varepsilon + \frac{c_4}{\varepsilon} \int_0^{3\varepsilon/2} \chi(y_1')\, dy_1',$$

whence it follows that

$$\lim_{\varepsilon\to+0} I_1(\varepsilon) = 0. \tag{3.7}$$

We now consider the quantity $I_2(\varepsilon)$. Taking into account that the matrices $Z_{C'}^{(j)}$, $j = 1, \dots, n$ are real and symmetric, we have

$$I_2(\varepsilon) = -\sum_{1\leq j\leq n} \int_{-\Gamma_\varepsilon'} \left\langle Z_{C'}^{(j)} \frac{\partial}{\partial x_j}(\varphi\eta_\varepsilon), \varphi\eta_\varepsilon \right\rangle dx$$

$$= \sum_{1\leq j\leq n} \sum_{1\leq k,s\leq N} Z_{C',ks}^{(j)} \int_{-\Gamma_\varepsilon'} \frac{\partial}{\partial x_j}(\varphi_s\eta_\varepsilon)\varphi_k\eta_\varepsilon\, dx$$

$$= -\frac{1}{2} \sum_{1\leq j\leq n} \sum_{1\leq k,s\leq N} Z_{C',ks}^{(j)} \int_{-\Gamma_\varepsilon'} \frac{\partial}{\partial x_j}(\varphi_s\varphi_k\eta_\varepsilon^2)\, dx$$

$$= -\frac{1}{2} \int_{-\Gamma_\varepsilon'} \sum_{1\leq j\leq n} \frac{\partial}{\partial x_j} \left[\langle Z_{C'}^{(j)}\varphi, \varphi\rangle \eta_\varepsilon^2 \right] dx. \tag{3.8}$$

The cone Γ'_e has a piecewise smooth boundary, which we denote by S; let $\mathbf{n}$ be an outer normal to S (see Fig. 45). Applying the Gauss–Ostrogradsky formula to the integral in (3.8), we obtain

$$I_2(\varepsilon) = -\frac{1}{2}\int_S \sum_{1\le j\le n} \langle Z_{C'}^{(j)}\varphi, \varphi\rangle \eta_\varepsilon^2 \cos(\widehat{\mathbf{n}x_j})\, dS.$$

Passing to the limit here as $\varepsilon \to +0$ and noting that

$$\eta_\varepsilon(x) = \eta\left[\frac{(e,x)}{\varepsilon}\right] \to \theta[-(e,x)], \qquad 0 \le \eta_\varepsilon(x) \le 1,$$

we obtain

$$\lim_{\varepsilon\to+0} I_2(\varepsilon) = -\frac{1}{2}\int_{S\cap[(e,x)\le 0]} \sum_{1\le j\le n} \langle Z_{C'}^{(j)}\varphi, \varphi\rangle \cos(\widehat{\mathbf{n}x_j})\, dS. \tag{3.9}$$

But $(e,x) < 0$ for interior points of the cone Γ'_e (see Fig. 45), and so $(e,x) \le 0$ on $\partial\Gamma'_e$ and then $(e,x) \ge 0$ on S. For this reason, actually only that part of the boundary S is left in (3.9), where $(e,x) = 0$ [and there, $e = -\mathbf{n} = (-\cos(\widehat{\mathbf{n}x_1}), \ldots, -\cos(\widehat{\mathbf{n}x_n}))$] (see Fig. 45), so that (3.9) becomes

$$\lim_{\varepsilon\to+0} I_2(\varepsilon) = \frac{1}{2}\int_{S\cap[(e,x)=0]} \left\langle \sum_{1\le j\le n} e_j Z_{C'}^{(j)}\varphi, \varphi\right\rangle dS.$$

Now the last quantity is non-negative by virtue of the condition (3.2). From this and also from (3.7), (3.5) and (3.4) follows the condition for passivity relative to the cone Γ_e. The proof of the lemma is complete. □

Proof of Theorem I. Necessity was proved in Sec. 20.2. We will prove sufficiency. Suppose the matrix function $\tilde{Z}(\zeta)$ is positive real in T^C. Then by the theorem of Sec. 19.2 it is the Laplace transform of the matrix $Z(x)$ that satisfies the conditions of the lemma for $\Gamma_0 = \mathbb{R}^n$. Therefore the matrix $Z(x)$ defines a passive operator relative to the half-plane $\Gamma_1 = [x\colon (e_1,x) \ge 0]$, where e_1 is any unit vector in $\bar{C}$. Again applying the lemma to the cone Γ_1 and to any vector $e_2 \in \bar{C}$, $|e_2| = 1$, we obtain the passivity of $Z(x)$ relative to the cone $\Gamma_2 = [x\colon (e_1,x) \ge 0,\ (e_2,x) \ge 0]$ and so forth. By means of an m-fold repetition of that process we obtain that the matrix $Z(x)$ defines a passive operator relative to the cone $\Gamma_m = [x\colon (e_1,x) \ge 0, \ldots, (e_m,x) \ge 0]$,

$$\int_{-\Gamma_m} \langle Z * \varphi, \varphi\rangle\, dx \ge 0, \qquad \varphi \in \mathcal{D}_r^{\times N}. \tag{3.10}$$

But the convex cone $C^* = [x\colon (x,q) \ge 0,\ q \in \bar{C}]$ may be approximated from above by arbitrarily close m-hedral cones Γ_m as $m \to \infty$. Therefore, passing to the limit as $\Gamma_m \to C^*$ under the condition of passivity (3.10), we obtain the condition for passivity for the cone $C^* = (\operatorname{int}\Gamma^*)^* = \Gamma$, which is what we set out to prove. □

Combining Theorem I, the theorem of Sec. 19.2, and the remark of Sec. 20.2, we obtain

Theorem II. *The following conditions are equivalent:*

(a) *The matrix $Z(x)$ defines a passive operator relative to an acute cone Γ.*

(b) *The matrix* $Z(x)$ *satisfies the weak condition of passivity* (2.13) *relative to the cone* Γ.
(c) *The matrix* $Z(x)$ *satisfies the condition* (2.5) *and the conditions* (b) *of the lemma.*
(d) *The matrix* $Z(x)$ *satisfies the condition of dissipation* (2.2) *and the conditions* (b) *of the lemma.*

20.4. Multidimensional dispersion relations. The results obtained in Sec. 20.3 permit deriving (multidimensional) dispersion relations (see Sec. 10.6) that connect the real and imaginary parts of the matrix $\tilde{Z}(p)$ — the boundary value of the impedance $\tilde{Z}(\zeta)$. For the sake of simplicity of exposition, we confine ourselves to the case of the cone $C = \mathbb{R}^n_+$.

Let us first prove the following lemma.

LEMMA. *The general solution of the matrix equation*

$$\partial_1^2 \dots \partial_n^2 Z(x) = 0 \tag{4.1}$$

in the class of real continuous $*$-Hermitian matrix functions in $\mathbb{R}^n$ *with support in* $-\overline{\mathbb{R}^n_+} \cup \overline{\mathbb{R}^n_+}$ *is given by the formula*

$$Z(x) = Z_0\big[\mathcal{E}_n(x) - \mathcal{E}_n(-x)\big], \tag{4.2}$$

where Z_0 *is an arbitrary constant real skew-symmetric matrix.*

PROOF. By Lemma 6 of Sec. 18.1 we have

$$Z_{kj}(x) = Z_{0,kj}\big[\mathcal{E}_n(x) - \mathcal{E}_n(-x)\big], \qquad 1 \le k, j \le N,$$

where $Z_{0,kj}$ are arbitrary real numbers. From this and from the conditions $Z_{jk}(x) = Z_{kj}(-x)$ it follows that $Z_{0,kj} = -Z_{0,jk}$, that is, $Z_0 = -Z_0^T$. The representation (4.2) is proved. The lemma is proved. □

We denote by $N(-\overline{\mathbb{R}^n_+} \cup \overline{\mathbb{R}^n_+})$ the class of $+$-Hermitian matrices that are the Fourier transforms of real continuous tempered matrix functions in $\mathbb{R}^n$ with support in $-\overline{\mathbb{R}^n_+} \cup \overline{\mathbb{R}^n_+}$.

For a matrix of the class $N(-\overline{\mathbb{R}^n_+} \cup \overline{\mathbb{R}^n_+})$, all matrix elements belong to the space of generalized functions $\mathcal{D}'_{\mathcal{L}^2}$ (see Sec. 10.1).

From the lemma just proved it follows that *the generalized solution of the matrix equation*

$$p_1^2 \cdots p_n^2 M(p) = 0$$

in the class $N(-\overline{\mathbb{R}^n_+} \cup \overline{\mathbb{R}^n_+})$ *is given by the formula*

$$M(p) = iZ^{(0)}\partial_1 \cdots \partial_n \Im\big[i^n \mathcal{K}_n(p)\big], \tag{4.3}$$

where $Z^{(0)}$ *is an arbitrary constant real skew-symmetric matrix.*

Indeed, passing to the Fourier transforms in (4.2) and using the definition of the kernel $\mathcal{K}_n(p)$ (see Sec. 10.2), we have

$$\begin{aligned} M(p) &= Z_0\{F[\mathcal{E}_n] - \bar{F}[\mathcal{E}_n]\} \\ &= 2iZ_0 \Im F[\mathcal{E}_n](p) \\ &= 2iZ_0 \Im F[\theta_n(x)x_1 \cdots x_n] \\ &= iZ^{(0)}\partial_1 \cdots \partial_n \Im\big[i^n \mathcal{K}_n(p)\big], \end{aligned} \tag{4.4}$$

where $Z^{(0)} = 2(-1)^n Z_0$. □

THEOREM. *In order that the matrix $Z(x)$ should define a passive operator relative to the cone $\overline{\mathbb{R}}^n_+$, it is necessary and sufficient that its Fourier transform $\tilde{Z}(p)$ satisfy the dispersion relation*

$$\Im\tilde{Z}(p) = \frac{2}{(2\pi)^n} p_1^2 \cdots p_n^2 (M * \Im\mathcal{K}_n) + iZ^{(0)} - \sum_{1 \le j \le n} Z^{(j)} p_j, \tag{4.5}$$

where the matrix $M(p)$ is a solution in the class $N(-\overline{\mathbb{R}}^n_+ \cup \overline{\mathbb{R}}^n_+)$ of the equation

$$p_1^2 \cdots p_n^2 M(p) = \Re\tilde{Z}(p). \tag{4.6}$$

Here the matrix $\Re\tilde{Z}(p)$ is such that for all $a \in \mathbb{C}^N$ the generalized function $\langle \Re\tilde{Z}(p)a, a\rangle$ is a non-negative tempered measure in $\mathbb{R}^n$; the matrix $Z^{(0)}$ is real, constant, skew-symmetric, and the matrices $Z^{(j)}$, $j = 1, \ldots, n$, are real, constant and positive.

In the dispersion relation (4.5), *matrices $[M(p), Z^{(0)}, Z^{(1)}, \ldots, Z^{(n)}]$ are unique up to additive terms of the form*

$$\left[iA\partial_1 \cdots \partial_n \Im\left[i^n \mathcal{K}_n(p)\right], A, 0, \ldots, 0\right], \tag{4.7}$$

where A is an arbitrary constant real skew-symmetric matrix.

REMARK 1. For $n = 1$ the theorem was proved by Beltrami and Wohlers [**4**]; for $n \ge 2$, it was proved by Vladimirov [**113**].

REMARK 2. The actual growth of the measure $\langle \Re\tilde{Z}(p)a, a\rangle$ is such that the measure

$$\frac{\langle \Re Z(p)a, a\rangle}{(1 + p_1^2) \cdots (1 + p_n^2)}$$

is finite on $\mathbb{R}^n$ (see the theorem of Sec. 18.4).

PROOF OF THE THEOREM. NECESSITY. Suppose the matrix $Z(x)$ defines a passive operator relative to the cone $\overline{\mathbb{R}}^n_+$. By Theorem I of Sec. 20.3, the matrix $Z(x)$ has the following properties:

(a)

$$\langle Z(x)a + Z^*(x)a, a\rangle \gg 0, \qquad a \in \mathbb{C}^N; \tag{4.8}$$

(b)

$$Z(x) = \partial_1^2 \cdots \partial_n^2 Z_0(x) + \sum_{1 \le j \le n} Z^{(j)} \partial_j \delta(x), \tag{4.9}$$

where the matrix-function $Z_0(x)$ is continuous, real, and tempered in $\mathbb{R}^n$ with support in the cone $\overline{\mathbb{R}}^n_+$; the matrices $Z^{(j)}$, $j = 1, \ldots, n$, are real constant and positive. Passing to the Fourier transform in (4.8) and (4.9), we conclude that for all $a \in \mathbb{C}^n$ the generalized function

$$\langle \Re\tilde{Z}(p)a, a\rangle = \frac{1}{2} F\left[\langle Z(x)a + Z^*(x)a, a\rangle\right] \tag{4.10}$$

is a non-negative tempered measure in $\mathbb{R}^n$ (by the Bochner–Schwartz theorem; see Sec. 8.2) and

$$\tilde{Z}(p) = (-1)^n p_1^2 \cdots p_n^2 F[Z_0](p) - i \sum_{1\le j\le n} Z^{(j)} p_j, \tag{4.11}$$

$$\Re\tilde{Z}(p) = \frac{(-1)^n}{2} p_1^2 \cdots p_n^2 F\big[Z_0(x) + Z_0^*(x)\big](p). \tag{4.12}$$

We set

$$M(p) = \frac{(-1)^n}{2} F\big[Z_0(x) + Z_0^*(x)\big](p). \tag{4.13}$$

The matrix $M(p)$ belongs to the class $N(-\overline{\mathbb{R}^n_+} \cup \overline{\mathbb{R}^n_+})$ and by (4.12) it satisfies the equation (4.6). Furthermore, taking into account the equalities (see Sec. 10.1 and Sec. 10.2)

$$\begin{aligned} F[Z_0](p) &= F\left[(Z_0(x) + Z_0^*(x))\theta_n(x)\right] \\ &= \frac{1}{(2\pi)^n} F\left[Z_0(x) + Z_0^*(x)\right] * F[\theta_n] \\ &= 2(-1)^n M * \mathcal{K}_n, \end{aligned}$$

we rewrite relations (4.11) as

$$\tilde{Z}(p) = \frac{2}{(2\pi)^n} p_1^2 \cdots p_n^2 (M * \mathcal{K}_n) - i \sum_{1\le j\le n} Z^{(j)} p_j. \tag{4.14}$$

Separating the real and imaginary parts in (4.14), we obtain the dispersion relation (4.5) (for $Z^{(0)} = 0$) and the relation

$$\Re\tilde{Z}(p) = \frac{2}{(2\pi)^n} p_1^2 \cdots p_n^2 (M * \Re\mathcal{K}_n), \tag{4.15}$$

which is equivalent to the relation (4.6) by virtue of (4.6) of Sec. 10.4:

$$M = \frac{2}{(2\pi)^n} M * \Re\mathcal{K}_n. \tag{4.16}$$

SUFFICIENCY. Suppose the matrix $Z(x)$ is such that its Fourier transform $\tilde{Z}(p)$ satisfies the dispersion relation (4.5), where the matrix $M(p)$ is a solution in the class $N(-\overline{\mathbb{R}^n_+} \cup \overline{\mathbb{R}^n_+})$ of the equation (4.6), and that matrix is such that for any $a \in \mathbb{C}^n$ the generalized function $(\Re\tilde{Z}(p)a, a)$ is a non-negative tempered measure; the matrix $Z^{(0)}$ is real and skew-symmetric, and the matrices $Z^{(j)}$, $j = 1, \ldots, n$, are real, constant and positive.

By (4.16), the equation (4.6) is equivalent to equation (4.15), which, together with the dispersion relation (4.5), yields

$$\tilde{Z}(p) = \frac{2}{(2\pi)^n} p_1^2 \cdots p_n^2 (M * \mathcal{K}_n) - Z^{(0)} - i \sum_{1\le j\le n} Z^{(j)} p_j, \tag{4.17}$$

whence, using the inverse Fourier transform, we obtain

$$Z(x) = \partial_1^2 \cdots \partial_n^2 Z_1(x) - Z^{(0)}\delta(x) + \sum_{1\le j\le n} Z^{(j)} \partial_j \delta(x), \tag{4.18}$$

where $Z_1(x) = 2(-1)^n F^{-1}[M](x)\theta(x)$ is a real continuous tempered function in $\mathbb{R}^n$ with support in the cone $\overline{\mathbb{R}^n_+}$. Noticing that

$$\partial_1^2 \cdots \partial_n^2 Z_1(x) - Z^{(0)}\delta(x) = \partial_1^2 \cdots \partial_n^2 [Z_1(x) - Z^{(0)}\mathcal{E}_n(x)],$$

we obtain that the matrix $Z(x)$ satisfies the condition (4.9). The condition (4.8) is also fulfilled, by virtue of (4.10) and the Bochner–Schwartz theorem (see Sec. 8.2). By Theorem II of Sec. 20.3, the matrix $Z(x)$ defines a passive operator relative to the cone $\overline{\mathbb{R}^n_+}$.

We now prove the uniqueness of the dispersion relation (4.5) up to additive terms of the form (4.7). Suppose the representation (4.5) occurs with other matrices $[M_1, Z_1^{(0)}, Z_1^{(1)}, \ldots, Z_1^{(n)}]$. Then, by what has been proved,

$$M(p) - M_1(p) = iA\partial_1 \cdots \partial_n \Im[i^n \mathcal{K}_n(p)],$$

where A is some constant real skew-symmetric matrix. From this, by subtracting the distinct representations (4.5) for $\Im\tilde{Z}(p)$, we obtain

$$\frac{2i}{(2\pi)^n} Ap_1^2 \cdots p_n^2 [\partial_1 \cdots \partial_n \Im(i^n \mathcal{K}_n) * \Im\mathcal{K}_n] \\ + i[Z^{(0)} - Z_1^{(0)}] - \sum_{1 \le j \le n} [Z^{(j)} - Z_1^{(j)}]p_j = 0. \quad (4.19)$$

Passing to the inverse Fourier transform in (4.19) and using the formulae (4.4) and (2.8) of Sec. 10.2, we obtain the equality

$$\begin{aligned} -\frac{i}{2}A\partial_1^2 \cdots \partial_n^2 &\{[\mathcal{E}_n(x) - \mathcal{E}_n(-x)][\theta_n(x) - \theta_n(-x)]\} \\ &+ i[Z^{(0)} - Z_1^{(0)}]\delta(x) + \sum_{1 \le j \le n} [Z^{(j)} - Z_1^{(j)}]\partial_j\delta(x) \\ &= -\frac{i}{2}A\partial_1^2 \cdots \partial_n^2[\mathcal{E}_n(x) + \mathcal{E}_n(-x)] \\ &+ i[Z^{(0)} - Z_1^{(0)}]\delta(x) + \sum_{1 \le j \le n} [Z^{(j)} - Z_1^{(j)}]\partial_j\delta(x) \\ &= i[Z^{(0)} - Z_1^{(0)} - A]\delta(x) + \sum_{1 \le j \le n} [Z^{(j)} - Z_1^{(j)}]\partial_j\delta(x) = 0, \end{aligned}$$

which is only possible for

$$Z^{(0)} = Z_1^{(0)} + A, \qquad Z^{(j)} = Z_1^{(j)}, \qquad j = 1, \ldots, n.$$

This completes the proof of the theorem. □

20.5. The fundamental solution and the Cauchy problem. The fundamental solution of the passive operator $Z*$ relative to the cone Γ is any matrix $A(x)$, $A_{kj} \in \mathcal{D}'$, that satisfies the convolution matrix equation

$$Z * A = I\delta(x). \quad (5.1)$$

The operator $A*$ is also said to be the *inverse* of $Z*$ (compare Sec. 4.9.4), and the matrix function $\tilde{A}(\zeta)$ — the Laplace transform of the matrix $A(x)$ — is called the *admittance* of the physical system.

The passive operator $Z*$ relative to the cone Γ is said to be ***non-singular*** (respectively, ***completely non-singular***) if Γ is an acute solid cone and $\det \tilde{Z}(\zeta) \neq 0$, $\zeta \in T^C$, where $C = \operatorname{int} \Gamma^*$ (and, respectively, if for any $a \in \mathbb{C}^N$, $a \neq 0$, there exists a point $\zeta_0 \in T^C$ such that

$$\Re \left\langle \tilde{Z}(\zeta_0)a, a \right\rangle > 0.) \tag{5.2}$$

The equivalent definition of a non-singular passive operator $Z*$ is: $Z*$ is passive if there exists a point $\zeta_0 \in T^C$ such that $\det \tilde{Z}(\zeta_0) \neq 0$ (Drozhzhinov [22]).

If the operator $Z*$ that is passive relative to the cone Γ is completely non-singular, then

$$\Re \tilde{Z}(\zeta) > 0, \qquad \zeta \in T^C. \tag{5.3}$$

Indeed, by Theorem I of Sec. 20.3, the function $\langle \tilde{Z}(\zeta)a, a\rangle$ is holomorphic and $\Re\langle \tilde{Z}(\zeta)a, a\rangle \geq 0$ in T^C. But then, by (5.2), the inequality $\Re\langle \tilde{Z}(\zeta)a, a\rangle > 0$ holds if $a \neq 0$ (see the reasoning in Sec. 17.1), which is equivalent to (5.3). □

From this it follows that any completely non-singular passive operator is also a non-singular passive operator relative to the same cone.

Furthermore, for an operator $Z*$ that is passive relative to an (acute solid) cone Γ to be completely non-singular, it is necessary and sufficient that the equality

$$\langle Z(x)a, a\rangle = ig\delta(x) \tag{5.4}$$

be impossible for any $a \in \mathbb{C}^N$, $a \neq 0$, and for any real g.

Indeed, if the operator $Z*$ that is passive relative to the cone Γ is completely non-singular, then (5.4), which is equivalent to the equality

$$\langle \tilde{Z}(\zeta)a, a\rangle = ig, \qquad \zeta \in T^C,$$

is impossible by (5.2) for any $a \neq 0$ and for any real g.

Conversely, suppose the operator $Z*$ that is passive relative to the acute solid cone Γ is not completely non-singular. Then, for some $a \neq 0$, we would have $\Re\langle \tilde{Z}(\zeta)a, a\rangle \leq 0$, $\zeta \in T^C$. On the other hand, by Theorem I of Sec. 20.3, the function $\langle Z(\zeta)a, a\rangle$ is holomorphic and $\Re\langle \tilde{Z}(\zeta)a, a\rangle \geq 0$ in T^C and therefore $\Re\langle \tilde{Z}(\zeta)a, a\rangle = 0$ in T^C. Hence, $\langle \tilde{Z}(\zeta)a, a\rangle = ig$, where g is a real number so that (5.4) holds for certain $a \neq 0$ and for certain real g. □

THEOREM I. *Every non-singular passive operator relative to a cone Γ has a unique fundamental solution that determines a non-singular passive operator relative to that same cone Γ.*

PROOF. Let $Z*$ be a non-singular passive operator relative to a cone Γ so that $\tilde{Z}(\zeta)$ is a positive real matrix in T^C (by Theorem I of Sec. 20.3) and $\det \tilde{Z}(\zeta) \neq 0$, $\zeta \in T^C$. We will prove the existence and uniqueness of the solution of equation (5.1) in the class of matrices $A(x)$ that define non-singular passive operators relative to Γ. Applying the Laplace transform to equation (5.1), we obtain an equivalent matrix equation

$$\tilde{Z}(\zeta)\tilde{A}(\zeta) = I, \qquad \zeta \in T^C. \tag{5.5}$$

Equation (5.5) is uniquely solvable for all $\zeta \in T^C$ and its solution — the matrix function $\tilde{A}(\zeta) = \tilde{Z}^{-1}(\zeta)$ — is holomorphic and $\det \tilde{A}(\zeta) \neq 0$ in T^C. Furthermore, from the equality $\tilde{Z}(\zeta) = \bar{\tilde{Z}}(-\bar{\zeta})$, $\zeta \in T^C$, and from (5.5) it follows that

$\tilde{Z}(\zeta)\tilde{\bar{A}}(-\bar{\zeta}) = I$, that is,

$$\tilde{Z}^{-1}(\zeta) = \tilde{A}(\zeta) = \tilde{\bar{A}}(-\bar{\zeta}), \qquad \zeta \in T^C.$$

Finally, from the condition $\Re\tilde{Z}(\zeta) \geq 0$, $\zeta \in T^C$, and from (5.5) we derive

$$\Re\tilde{A}(\zeta) = \tilde{A}^+(\zeta)\left[\Re\tilde{Z}(\zeta)\right]\tilde{A}(\zeta) \geq 0, \qquad \zeta \in T^C. \tag{5.6}$$

Consequently, the matrix $\tilde{A}(\zeta)$ is positive real in T^C. By Theorem I of Sec. 20.3 the matrix $A(x)$ defines a non-singular passive operator relative to the cone Γ. The matrix $A(x)$ is unique. The proof of Theorem I is complete. □

COROLLARY. *If the passive operator $Z*$ is completely non-singular, then its inverse operator $A*$ is completely non-singular.*

Indeed, since $\Re\tilde{Z}(\zeta) > 0$ and $\det\tilde{A}(\zeta) \neq 0$, it follows, by (5.6), that $\Re\tilde{A}(\zeta) > 0$, $\zeta \in T^C$. □

Let Γ be a closed convex acute cone, $C = \operatorname{int}\Gamma^*$, let S be a C-like surface, and let S_+ be a region lying above S (see Sec. 4.4).

By analogy with Sec. 16.1 we introduce the following definition. By the generalized Cauchy problem for an operator $Z*$ that is passive relative to the cone Γ with source $f \in \mathcal{D}'(\bar{S}_+)^{\times N}$ we call the problem of finding, in $\mathbb{R}^n$, a solution $u(x)$ taken from $\mathcal{D}'(\bar{S}_+)^{\times N}$ of the system (1.1).

As in Sec. 16.1, the following theorem is readily proved.

THEOREM II. *If a passive operator $Z*$ is non-singular relative to a (solid) cone Γ, then the solution of its generalized Cauchy problem exists for any f in $\mathcal{D}'(\bar{S}_+)^{\times N}$, is unique, and is given by the formula*

$$u = A * f. \tag{5.7}$$

COROLLARY. *If S is a strictly C-like surface and $f \in \mathcal{S}'(\bar{S}_+)^{\times N}$, then the solution of the generalized Cauchy problem for the operator $Z*$ exists and is unique in the class $\mathcal{S}'(\bar{S}_+)^{\times N}$* [and is given by the formula (5.7)].

This follows from Theorem II and from the results of Sec. 5.6.2. □

Thus, passive systems behave in similar fashion to hyperbolic systems (see Sec. 16.1, Hörmander [**46**, Chapter 5], Friedrichs [**33**], Dezin [**15**]).

20.6. What differential and difference operators are passive operators? A system of N linear differential equations of order at most m (with constant coefficients) is determined by the matrix (compare Sec. 15.1)

$$Z(x) = \sum_{0 \leq |\alpha| \leq m} Z_\alpha \partial^\alpha \delta(x), \tag{6.1}$$

where Z_α are (constant) $N \times N$ matrices.

THEOREM I. *For a system of N linear differential equations with constant coefficients to be passive relative to an acute cone Γ, it is necessary and sufficient that*

$$Z(x) = \sum_{1 \leq j \leq n} Z_j \partial_j \delta(x) + Z_0 \delta(x), \tag{6.2}$$

where $Z_1, \ldots, Z_n$ are real symmetric $N \times N$ matrices such that $\sum_{1 \leq j \leq n} q_j Z_j \geq 0$ for all $q \in C = \operatorname{int}\Gamma^*$; the matrix Z_0 is real and $\Re Z_0 \geq 0$.

PROOF. NECESSITY. Suppose the differential operator $Z*$ defined by formula (6.1) is passive relative to the cone Γ. Then by Theorem I of Sec. 20.3 the matrix function

$$\tilde{Z}(\zeta) = \sum_{0\le|\alpha|\le m} (-i\zeta)^\alpha Z_\alpha \tag{6.3}$$

is positive real in T^C. Therefore, for every $a \in \mathbb{C}^N$ the function $\langle\tilde{Z}(\zeta)a, a\rangle$ is holomorphic and $\Re\langle\tilde{Z}(\zeta)a, a\rangle \ge 0$ in T^C. Therefore, that function satisfies the estimate (1.1) of Sec. 17.1 and, hence, all matrix elements $\tilde{Z}_{kj}(\zeta)$ satisfy that estimate:

$$\left|\sum_{0\le|\alpha|\le m} (-i\zeta)^\alpha Z_{\alpha,kj}\right| \le M(C')\frac{1+|\zeta|^2}{|q|}, \qquad \zeta \in T^{C'},$$

which is only possible for $Z_{\alpha,kj} = 0$, $1 \le k, j \le N$, $|\alpha| \ge 2$. Therefore, the matrix (6.3) takes the form

$$\tilde{Z}(\zeta) = -i \sum_{1\le j\le n} Z_j\zeta_j + Z_0 \tag{6.4}$$

and the representation (6.2) is proved. Writing out the conditions for positive reality of the matrix $\tilde{Z}(\zeta)$,

$$\sum_{1\le j\le n} q_j Z_j + Z_0 = \sum_{1\le j\le n} q_j\bar{Z}_j + \bar{Z}_0, \qquad q \in C,$$

$$-i\sum_{1\le j\le n} \zeta_j Z_j + Z_0 + i\sum_{1\le j\le n} \bar{\zeta}_j Z_j^+ + Z_0^+ \ge 0, \qquad \zeta \in T^C,$$

we conclude that the matrices Z_j, $j = 0, 1, \ldots, n$, are real, and the matrices Z_j, $j = 1, \ldots, n$, are symmetric, $\sum_{1\le j\le n} q_j Z_j \ge 0$ and $\Re Z_0 \ge 0$.

SUFFICIENCY. Let the matrix $Z(x)$ satisfy the conditions of Theorem I. Then its Laplace transform $\tilde{Z}(\zeta)$ is of the form (6.4) and

$$\Re\tilde{Z}(\zeta) = \sum_{1\le j\le n} q_j Z_j + \Re Z_0 \ge 0, \qquad \zeta \in T^C. \tag{6.5}$$

Therefore the matrix function $\tilde{Z}(\zeta)$ is positive real in T^C and by Theorem I of Sec. 20.3 the operator $Z*$ is passive relative to the cone Γ. The proof of the theorem is complete. □

Suppose we are given real symmetric $N \times N$ matrices $Z_1, \ldots, Z_n$ having the property that for a certain vector $l \in \mathbb{R}^n$, $\sum_{1\le j\le n} l_j Z_j > 0$. We set

$$\Gamma_c = \left[x : x_1 = \langle Z_1 a, a\rangle, \ldots, x_n = \langle Z_n a, a\rangle, a \in \mathbb{C}^N\right],$$
$$\Gamma_r = \left[x : x_1 = \langle Z_1 a, a\rangle, \ldots, x_n = \langle Z_n a, a\rangle, a \in \mathbb{R}^N\right].$$

Under the mapping

$$a \to x = (\langle Z_1 a, a\rangle, \ldots, \langle Z_n a, a\rangle) \tag{6.6}$$

the pre-image of 0 is 0 by virtue of the inequality

$$(l, x) = \left\langle \sum_{1\le j\le n} l_j Z_j a, a \right\rangle \ge \varkappa|a|^2, \qquad \varkappa > 0. \tag{6.7}$$

Clearly, Γ_c and Γ_r are cones with vertex at 0, and $\Gamma_r \subset \Gamma_c$.

LEMMA. *The cones* Γ_c *and* Γ_r *are closed and acute:* $\Gamma_c = \Gamma_r + \Gamma_r$; $l \in \operatorname{int} \Gamma_c^*$.

PROOF. The mapping (6.6) is continuous from $\mathbb{C}^N$ (from $\mathbb{R}^N$) into $\mathbb{R}^n$ and, by virtue of the inequality (6.7), is of a compact nature, that is, the pre-image of any compact set is a compact set. Therefore the cones Γ_c and Γ_r are closed. Furthermore, from the equalities

$$\langle Z_j a, a\rangle = \langle Z_j b, b\rangle + \langle Z_j c, c\rangle, \qquad a = b + ic, \qquad j = 1, \ldots, n,$$

we conclude that $\Gamma_c = \Gamma_r + \Gamma_r$. Finally, by the inequality (6.7) the plane $(l, x) = 0$ has only one point in common with the cone Γ_c — its vertex. Therefore Γ_c and Γ_r are acute cones and $l \in \operatorname{int} \Gamma_c^*$ (see Lemma 1 of Sec. 4.4). The lemma is proved. □

Notice that the cones Γ_c and Γ_r may not be solid, as the following example shows: $Z_1 > 0$, $Z_2 = 0$ and Γ_c and Γ_r lie in the plane $x_2 = 0$.

THEOREM II. *In order that the matrix* (6.2) *define a passive completely non-singular operator, it is necessary and sufficient that the matrices* $Z_1, \ldots, Z_n$ *be real symmetric, that the matrix* Z_0 *be real and* $\Re Z_0 \geq 0$, *and that there exists a vector* $l \in \mathbb{R}^n$ *such that*

$$\sum_{1 \leq j \leq n} l_j Z_j + \Re Z_0 > 0. \tag{6.8}$$

Here, the passivity and the complete non-singularity of the operator $Z*$ *occur in the case of any acute cone* Γ *that contains the cone* Γ_c, *and* $l \in \operatorname{int} \Gamma^*$.

PROOF. NECESSITY. Suppose the matrix (6.2) defines a passive and completely non-singular operator with respect to a certain (acute) cone Γ. Then the conditions of Theorem I are fulfilled and, by (5.3) and (6.5),

$$\Re \tilde{Z}(\zeta) = \sum_{1 \leq j \leq n} q_j Z_j + \Re Z_0 > 0, \qquad \zeta \in T^C, \qquad C = \operatorname{int} \Gamma^*,$$

so that the condition (6.8) holds for all $q \in C$.

SUFFICIENCY. Let the matrices $Z_0, \ldots, Z_n$ in (6.2) satisfy the conditions of Theorem II. Suppose Γ is an acute cone containing the cone Γ_c and such that $l \in C = \operatorname{int} \Gamma^*$. From this if follows that $(q, x) \geq 0$ for all $q \in C$, $x \in \Gamma_c \subset \Gamma$, that is,

$$(q, x) = \sum_{1 \leq j \leq n} q_j \langle Z_j a, a\rangle \geq 0, \qquad q \in C, \qquad a \in \mathbb{C}^N.$$

This means that

$$\sum_{1 \leq j \leq n} q_j Z_j \geq 0, \qquad q \in C.$$

By Theorem I, the matrix $Z(x)$ defines a passive operator relative to the cone Γ. Furthermore, it is given that $l \in C$ and so, by (6.5) and (6.8),

$$\Re Z(il) = \sum_{1 \leq j \leq n} l_j Z_j + \Re Z_0 > 0,$$

so that the operator $Z*$ is completely non-singular relative to the cone Γ (see Sec. 20.5). The proof of Theorem II is complete. □

REMARK. Theorem II states that the matrices $\sum_{1\le j\le n} Z_j\partial_j\delta(x)$, which define passive completely non-singular differential operators

$$\sum_{1\le j\le n} Z_j\frac{\partial}{\partial x_j},$$

coincide with the principal parts of Friedrichs-symmetric (see Friedrichs **[33]**) differential operators with constant coefficients.

A system of N linear difference equations with the number of steps at most m is given by the matrix

$$Z(x) = Z_0\delta(x) + \sum_{1\le\nu\le m} Z_\nu\delta(x-h_\nu). \tag{6.9}$$

THEOREM III. *For a system of N linear difference equations (when $h_\nu \ne 0$, $\nu = 1,\dots,m$, and $h_\nu \ne h_k$, $\nu \ne k$) to be completely non-singular passive relative to an acute cone Γ, it is necessary and sufficient that $h_\nu \in \Gamma$, $\nu = 1,\dots,m$ and the $N\times N$ matrices $Z_0, Z_1,\dots,Z_m$ be real and for all $\zeta \in T^C$, $C = \operatorname{int}\Gamma^*$, the matrix*

$$\Re Z_0 + \sum_{1\le\nu\le m} e^{-(q,h_\nu)}\left[\cos(p.h_\nu)\Re Z_\nu - \sin(p,h_\nu)\Im Z_\nu\right] > 0. \tag{6.10}$$

PROOF. NECESSITY. Suppose the difference operator $Z*$ that is defined by the matrix (6.9) is completely non-singular passive relative to the acute cone Γ. From the fact that the matrix $Z(x)$ is real and from the condition $h_\nu \ne 0$, $\nu = 1,\dots,m$, and $h_\nu \ne h_k$ for $\nu \ne k$ it follows that the matrices $Z_0, Z_1,\dots,Z_m$ are real. From the condition $\operatorname{supp} Z \subset \Gamma$ (see Sec. 20.2.2) it follows that $h_\nu \in \Gamma$, $\nu = 1,\dots,m$. By Theorem I of Sec. 20.3 and (5.3) of Sec. 20.5

$$\Re\tilde{Z}(\zeta) = \Re Z_0 + \Re\sum_{1\le\nu\le m} e^{i(\zeta,h_\nu)}Z_\nu > 0, \qquad \zeta\in T^C, \tag{6.11}$$

and the condition (6.10) is fulfilled.

SUFFICIENCY. Suppose that $h_\nu \in \Gamma$, $\nu = 1,\dots,m$ and the matrices $Z_0,\dots,Z_m$ in (6.9) satisfy the conditions of Theorem III. Then, by (6.11), the matrix $\tilde{Z}(\zeta)$ is positive real in T^C and by Theorem I of Sec. 20.3 the matrix $Z(x)$ defines a passive (completely non-singular) operator relative to the cone $C^* = \Gamma$. Theorem III is proved. □

REMARK. We make a special note of the necessity conditions of passivity of the matrix (6.9): the smallest convex cone containing the points $\{0, h_1,\dots,h_m\}$ must be acute and $Z_0 > 0$.

20.7. Examples. Let us denote by $V^+(a) = [(x,t)\colon at > |x|]$ the future light cone in $\mathbb{R}^4$, which corresponds to the speed of propagation a: $V^+ = V^+(1)$ (compare Sec. 4.4).

20.7.1. *Maxwell's equations.* The principal part of the approximate differential operator is of the form[4]

$$\frac{\partial\mathbf{D}}{\partial x_0} - \operatorname{rot}\mathbf{H}, \qquad \frac{\partial\mathbf{B}}{\partial x_0} + \operatorname{rot}\mathbf{E}, \tag{7.1}$$

[4]Specification of div D and div B in the system of Maxwell's equations is not essential for our purposes; actually, these are consistency conditions.

where $x_0 = ct$, c is the speed of light in vacuum, $x = (x_0, \mathbf{x})$ and

$$\mathbf{D} = \varepsilon * \mathbf{E}, \qquad \mathbf{B} = \mu * \mathbf{H}, \tag{7.2}$$

where ε and μ are 3×3 matrices called tensors of dielectric and magnetic permeability respectively.

If ε and μ are constant matrices that are multiples of the unit matrix, $\varepsilon = \varepsilon I\delta(x)$, $\mu = \mu I\delta(x)$, then the system (7.1)–(7.2) becomes

$$\varepsilon\frac{\partial \mathbf{E}}{\partial x_0} - \operatorname{rot}\mathbf{H}, \qquad \mu\frac{\partial \mathbf{H}}{\partial x_0} + \operatorname{rot}\mathbf{E}. \tag{7.3}$$

The system (7.3) is *passive with respect to the cone* $\overline{V}^{+}(1/\sqrt{\varepsilon\mu})$ by virtue of the inequality

$$\int\limits_{-V^{+}(1/\sqrt{\varepsilon\mu})} \left[\varepsilon\left(\frac{\partial \mathbf{E}}{\partial x_0}, \mathbf{E}\right) - (\mathbf{E}, \operatorname{rot}\mathbf{H}) + \mu\left(\frac{\partial \mathbf{H}}{\partial x_0}, \mathbf{H}\right) + (\mathbf{H}, \operatorname{rot}\mathbf{E})\right] dx \geq 0, \tag{7.4}$$

which holds for all $\mathbf{E} \in \mathcal{D}_r(\mathbb{R}^4)^{\times 3}$ and $\mathbf{H} \in \mathcal{D}_r(\mathbb{R}^4)^{\times 3}$. Here, $N = 6$, $n = 4$.

To prove the inequality (7.4) we make use of the identity

$$(\mathbf{H}, \operatorname{rot}\mathbf{E}) - (\mathbf{E}, \operatorname{rot}\mathbf{H}) = \operatorname{div}(\mathbf{E} \times \mathbf{H}),$$

by virtue of which the left-hand member of (7.4) is equal to

$$\begin{aligned}
&\int\limits_{\mathbb{R}^3} \int\limits_{-\infty}^{-\sqrt{\varepsilon\mu}|x|} \frac{\partial}{\partial x_0}\left(\varepsilon|\mathbf{E}|^2 + \mu|\mathbf{H}|^2\right) dx_0\, d\mathbf{x} + \int\limits_{-\infty}^{0} \int\limits_{|x|<-x_0/\sqrt{\varepsilon\mu}} \operatorname{div}(\mathbf{E} \times \mathbf{H})\, d\mathbf{x} \\
&\quad = \frac{1}{2}\int\limits_{\mathbb{R}^3} \left[\varepsilon|\mathbf{E}|^2 + \mu|\mathbf{H}|^2 + 2\sqrt{\varepsilon\mu}((\mathbf{E} \times \mathbf{H}), \mathbf{n})\right]\Big|_{x_0=-\sqrt{\varepsilon\mu}|x|} d\mathbf{x} \\
&\quad \geq \frac{1}{2}\int\limits_{\mathbb{R}^3} \left(\varepsilon|\mathbf{E}|^2 + \mu|\mathbf{H}|^2 - 2\sqrt{\varepsilon\mu}|\mathbf{E}|\,|\mathbf{H}|\right)\Big|_{x_0=-\sqrt{\varepsilon\mu}|x|} d\mathbf{x} \\
&\quad = \frac{1}{2}\int\limits_{\mathbb{R}^3} \left(\sqrt{\varepsilon}|\mathbf{E}| - \sqrt{\mu}|\mathbf{H}|\right)^2\Big|_{x_0=-\sqrt{\varepsilon\mu}|x|} d\mathbf{x} \geq 0.
\end{aligned}$$

Let us verify that when $\varepsilon = \mu = 1$ the cone $\Gamma_c = \Gamma_r = \overline{V}^{+}$. Indeed, if $a \in \mathbb{R}^6$, then the mapping (6.6) takes the form

$$\begin{aligned}
x_0 &= a_1^2 + \cdots + a_6^2, \\
x_1 &= -2a_2a_6 + 2a_3a_5, \\
x_2 &= 2a_1a_6 - 2a_3a_4, \\
x_3 &= -2a_1a_5 + 2a_2a_4,
\end{aligned}$$

so that $x_0 \geq 0$ and

$$\begin{aligned}
x^2 &= x_0^2 - |\mathbf{x}|^2 \\
&= (a_1^2 + a_2^2 + a_3^2 - a_4^2 - a_5^2 - a_6^2)^2 + 4(a_1a_4 + a_2a_5 + a_3a_6)^2 \geq 0.
\end{aligned}$$

But if the tensors ε and μ are non-trivial, then, depending on the properties of the medium, it is natural to postulate passivity with respect to an acute cone of certain of the operators

$$\varepsilon *, \quad \mu *, \quad \frac{\partial \varepsilon}{\partial x_0} *, \quad \frac{\partial \mu}{\partial x_0} *, \qquad (N = 3, \; n = 4).$$

Relative to appropriate impedance and admittance matrices, all propositions of the theory developed in Secs. 20.3 to 20.5 hold true; in particular, the four-dimensional dispersion relations (see Sec. 20.4, compare Silin and Rukhadze [95]).

20.7.2. *Dirac's equation.* The appropriate operator is

$$i \sum_{0 \le \mu \le 3} \gamma^\mu \partial_\mu - m, \tag{7.5}$$

where γ^μ is a 4×4 Dirac matrix; in the Majorans basis (see, for example, Bogolyubov, Logunov, Oksak, and Todorov [7, Chapter 2]), they are of the form

$$\gamma^0 = \begin{pmatrix} 0 & i\sigma_1 \\ -i\sigma_1 & 0 \end{pmatrix}, \quad \gamma^1 = \begin{pmatrix} iI & 0 \\ 0 & -iI \end{pmatrix}, \quad \gamma^2 = \begin{pmatrix} 0 & \sigma_2 \\ -\sigma_2 & 0 \end{pmatrix}, \quad \gamma^3 = \begin{pmatrix} 0 & iI \\ iI & 0 \end{pmatrix},$$

where σ_k are 2×2 Pauli matrices:

$$\sigma_1 = \begin{pmatrix} 0 & 1 \\ 1 & 0 \end{pmatrix}, \qquad \sigma_2 = \begin{pmatrix} 0 & -i \\ i & 0 \end{pmatrix}, \qquad \sigma_3 = \begin{pmatrix} 1 & 0 \\ 0 & -1 \end{pmatrix}.$$

We now prove that the operator (7.5), after multiplication by the matrix $-i\gamma^0$,

$$\sum_{0 \le \mu \le 3} \gamma^0 \gamma^\mu \partial_\mu + im\gamma^0, \tag{7.6}$$

is passive and completely non-singular relative to the cone $\overline{V^+}$.

Indeed, the matrices $\gamma^0\gamma^\mu$ are real and symmetric, and the matrix $im\gamma^0$ is real and skew-symmetric. Furthermore, the cone Γ_r coincides with the boundary of the cone V^+ since the mapping (6.6) is, for $a \in \mathbb{R}^4$, of the form

$$\begin{aligned} x_0 &= \langle \gamma^0\gamma^0 a, a \rangle = a_1^2 + \cdots + a_4^2, \\ x_1 &= \langle \gamma^0\gamma^1 a, a \rangle = 2a_1a_4 + 2a_2a_3, \\ x_2 &= a_1^2 - a_2^2 + a_3^2 - a_4^2, \\ x_3 &= -2a_1a_2 + 2a_3a_4, \end{aligned}$$

so that $x_0 \ge 0$ and

$$\begin{aligned} x_0^2 - |\mathbf{x}|^2 = {} & \left(a_1^2 + \cdots + a_4^2\right)^2 - 4\left(a_1a_4 + a_2a_3\right)^2 \\ & - \left(a_1^2 - a_2^2 + a_3^2 - a_4^2\right)^2 - 4\left(a_1a_2 - a_3a_4\right)^2 = 0. \end{aligned}$$

By Theorem II of Sec. 20.6 the operator (7.6) is passive and completely non-singular relative to the cone $\Gamma_c = \overline{V^+}$, the convex hull of the cone Γ_r (compare with the lemma of Sec. 20.6).

20.7.3. *The equations of a rotating fluid and acoustics.* These equations have the following form:[5]

$$\alpha \frac{\partial p}{\partial t} + \rho \operatorname{div} \mathbf{v}, \qquad \rho \frac{\partial \mathbf{v}}{\partial t} + \operatorname{grad} p + \mathbf{v} \times \mathbf{w}. \tag{7.7}$$

[5]See Maslennikova [75] and Drozhzhinov and Galeev [23].

Here $N = n = 4$. For all $\alpha > 0$, the system (7.7) is passive and completely non-singular relative to the cone $\overline{V}^{+}(1/\sqrt{\alpha})$. The mapping (6.6) is, for $a \in \mathbb{R}^4$, of the form

$$t = \alpha a_1^2 + a_2^2 + a_3^2 + a_4^2, \qquad x_1 = 2a_1a_2, \qquad x_2 = 2a_1a_3, \qquad x_3 = 2a_1a_4,$$

so that $t \geq 0$ and

$$\begin{aligned} t^2 - \alpha|x|^2 &= \left(\alpha a_1^2 + a_2^2 + a_3^2 + a_4^2\right)^2 - 4\alpha a_1^2(a_2^2 + a_3^2 + a_4^2) \\ &= \left(\alpha a_1^2 - a_2^2 - a_3^2 - a_4^2\right)^2 \geq 0. \end{aligned}$$

Therefore, $\Gamma_c = \Gamma_r = \overline{V}^{+}(1/\sqrt{\alpha})$.

20.7.4. *Equations of magnetic hydrodynamics.* These equations have the following form:[6]

$$\begin{gathered} \frac{\partial p}{\partial t} + a^2 \rho \operatorname{div} \mathbf{v}, \qquad \frac{\partial \mathbf{H}}{\partial t} - \operatorname{rot}(\mathbf{v} \times \mathbf{B}), \\ \rho \frac{\partial \mathbf{v}}{\partial t} + \operatorname{grad} p - \frac{1}{4\pi}(\operatorname{rot} \mathbf{H}) \times \mathbf{B}, \end{gathered} \tag{7.8}$$

where $\mathbf{B}$ is a specified vector; here, $N = 7$, $n = 4$. The system (7.8) is passive and completely non-singular relative to a certain cone.

20.7.5. *Equations of the theory of elasticity:* [7]

$$\frac{\partial^2 w_i}{\partial t^2} = \sum_{1 \leq j,m,n \leq 3} c_{mn}^{ij} \frac{\partial^2 w_m}{\partial x_j\, \partial x_n}, \qquad i = 1,2,3, \tag{7.9}$$

where $c_{mn}^{ij} = c_{mn}^{ji} = c_{nm}^{ij} = c_{ji}^{nm}$. If we introduce the velocity vector ($v_i = \frac{\partial w_i}{\partial t}$, $i = 1,2,3$) and the stress tensor

$$\left\{ \Sigma_{ij} = \sum_{1 \leq n,m \leq 3} c_{mn}^{ij} \frac{\partial w_m}{\partial x_n}, \ 1 \leq i,j \leq 3 \right\},$$

then the operator (7.9) becomes passive and completely non-singular relative to a certain cone. Here, $N = 9$, $n = 4$.

20.7.6. *Transfer equation:*

$$\frac{\partial}{\partial t} + (\Omega, \operatorname{grad}) \tag{7.10}$$

where Ω is a constant vector in $\mathbb{R}^3$; here, $N = 1$, $n = 4$. The operator (7.10) is passive and completely non-singular relative to any (acute) cone containing the vector $(1, \Omega)$ in $\mathbb{R}^4$. If we apply the spherical harmonics method to the operator (7.10), then to any $\mathcal{P}_N$-approximation what is obtained is a passive completely non-singular system relative to a certain cone (dependent on N). The appropriate operators are written out in Godunov and Sultangazin [**40**].

---

[6] See Leonard [**66**] and Drozhzhinov [**21**].

[7] See Wilcox [**126**].

20.8. Quasiasymptotics of the solutions of systems of equations in convolutions. Consider the system of equations in convolutions

$$\mathcal{K} * u = f \tag{8.1}$$

in the algebra $\mathcal{S}'(\Gamma)$ (see Sec. 5.6). Here $\mathcal{K}(\xi) = (\mathcal{K}_{ij}(\xi))$ is a fixed $N \times N$-matrix and $f(\xi) = (f_i(\xi))$ is a fixed N-vector whose coordinates belong to the convolution algebra $\mathcal{S}'(\Gamma)$; $u(\xi) = (u_i(\xi))$ is an unknown N-vector which is also assumed to belong to $\mathcal{S}'(\Gamma)$.

The set of $N \times N$-matrices whose entries belong to the algebra $\mathcal{S}'(\Gamma)$ forms the convolution algebra $\mathcal{S}'_{N\times N}(\Gamma)$ which is with unity, associative and commutative.

A matrix $\mathcal{K}(\xi)$ is called *non-singular* (*non-singular in the algebra* $\mathcal{S}'(\Gamma)$) if the discriminant of its Laplace transform is non-zero in T^C (is a divisor of unity in the algebra $H(C)$, respectively). Here $C = \operatorname{int}\Gamma^*$, $T^C = \mathbb{R}^n + iC$.

For the sake of brevity, we denote the Laplace transform of a generalized function (matrix, vector, etc.) $f \in \mathcal{S}'(\Gamma)$ by $\tilde{f}(z)$. If $\mathcal{K}(\xi)$ is a matrix non-singular in the algebra $\mathcal{S}'_{N\times N}(\Gamma)$, then its inverse matrix $\mathcal{K}^{-1}(\xi)$is defined as the inverse Laplace transform of the matrix $\tilde{\mathcal{K}}^{-1}(z)$, $\mathcal{K}^{-1} = L^{-1}[\tilde{\mathcal{K}}^{-1}]$. In this case

$$\mathcal{K}^{-1} \in \mathcal{S}'_{N\times N}(\Gamma), \qquad \mathcal{K} * \mathcal{K}^{-1} = \mathcal{K}^{-1} * \mathcal{K} = I\delta, \qquad \tilde{\mathcal{K}}^{-1} \in H(C).$$

First, we prove the following lemma.

LEMMA. *If generalized functions f and g taken from $\mathcal{S}'(\Gamma)$ have quasiasymptotics f_0 and g_0 of orders α and β at ∞, respectively, then their convolution $f * g$ has the quasiasymptotic $f_0 * g_0$ of the order $\alpha + \beta + n$ at ∞.*

PROOF follows from the general Tauberian theorem (see Sec. 14.2) and from the formula for the Laplace transform of the convolution (see Sec. 9.2),

$$(\widetilde{f * g})(z) = \tilde{f}(z)\tilde{g}(z), \qquad z \in T^C.$$

One can see from this formula that the function $(\widetilde{f * g})(z)$ has the asymptotic $\tilde{f}_0(z)\tilde{g}_0(z)$ of the order $\alpha+\beta+2n$ at 0 and satisfies an estimate of the type (2.4) of Sec. 14.2. Applying the inverse Laplace transform and using once again the general Tauberian theorem, we obtain the statement of the lemma. □

Now we prove a more complicated theorem.

THEOREM. *Let a matrix $\mathcal{K}$ non-singular in the algebra $\mathcal{S}'_{N\times N}(\Gamma)$ and a vector f taken from $\mathcal{S}'_N(\Gamma)$ have the quasiasymptotics $\mathcal{K}_0$ and f_0 of orders α and β at ∞, respectively. Suppose that the conditions*

$$\det \tilde{\mathcal{K}}_0(z) \neq 0, \qquad z \in T^C, \tag{8.2}$$

$$\rho^{N(\alpha+n)} \left|\det \tilde{\mathcal{K}}(\rho x + i\rho\lambda e)\right| \geq m\lambda^q, \qquad 0 < \rho \leq 1, \quad 0 < \lambda \leq 1, \quad |x| \leq \lambda^\gamma, \tag{8.3}$$

hold for some $m > 0$, $q \geq 0$ and $\gamma \in [0,1)$. Then the (unique) solution $u \in \mathcal{S}'_N(\Gamma)$ of equation (8.1) has the quasiasymptotic u_0 of the order $\beta - \alpha - n$ at ∞ and the following equalities are valid:

$$\mathcal{K}_0 * u_0 = f_0, \qquad u_0 = \mathcal{K}_0^{-1} * f_0. \tag{8.4}$$

PROOF. Applying the formulae for the Laplace transform of the convolution to equation (8.1), we obtain

$$\tilde{\mathcal{K}}(z)\tilde{u}(z) = \tilde{f}(z), \qquad z \in T^C, \tag{8.5}$$

and using the condition $\mathcal{K}^{-1}(z) \in H_{N\times N}(C)$ we rewrite equation (8.5) in the equivalent form

$$\tilde{u}(z) = \tilde{\mathcal{K}}^{-1}(z)\tilde{f}(z), \qquad z \in T^C. \tag{8.6}$$

Since $\det \tilde{\mathcal{K}}_0(z)$ is a homogeneous function in T^C, condition (8.2) implies that this function is a divisor of unity in the algebra $H(C)$; hence, $\det^{-1} \tilde{\mathcal{K}}_0(z) \in H(C)$ (see Sec. 13.1), which implies that $\tilde{\mathcal{K}}_0^{-1}(z) \in H_{N\times N}(C)$.

Further, the matrix function $\tilde{\mathcal{K}}(z)$ has the asymptotic $\tilde{\mathcal{K}}_0$ of order $\alpha + n$ at 0 (see Sec. 14.2); therefore, $\det \tilde{\mathcal{K}}(z)$ has the asymptotic of order $N(\alpha+n)$ in 0. From here it follows that matrix $\tilde{\mathcal{K}}^{-1}(z)$ has the asymptotics $\tilde{\mathcal{K}}_0^{-1}(z)$ of order $-\alpha - n$ at 0. However, in this case equality (8.6) implies that the vector function $\tilde{u}(z)$ has the asymptotic $\tilde{\mathcal{K}}_0^{-1}(z)\tilde{f}_0(z)$ of order $\beta - \alpha$ at 0. By the general Tauberian theorem of Sec. 14.2 and by virtue of (8.3), the function $\tilde{u}(z)$ satisfies an estimate of type (2.4) of Sec. 14.2. Therefore, by the same theorem, $u(\xi)$ has the quasiasymptotic $\mathcal{K}_0^{-1} * f_0$ of order $\beta - \alpha - n$ at ∞ and equalities (8.4) hold. The theorem is proven. □

COROLLARY 1. *If the matrix $\mathcal{K}$ satisfies the conditions of the theorem, then the fundamental solution $\mathcal{E}$ of the operator $\mathcal{K}*$, $\mathcal{K} * \mathcal{E} = I\delta$, has a quasiasymptotics $\mathcal{E}_0$ of order $-\alpha - 2n$ at ∞, and*

$$\mathcal{K}_0 * \mathcal{E}_0 = I\delta.$$

This follows from the theorem for $f = \delta$ and $\beta = -n$. □

COROLLARY 2. *Let the matrix $Z(x)$ defining a nondegenerate passive operator $Z*$ with respect to an acute cone Γ have a quasiasymptotics $Z_0(x)$ of order α at ∞ and $\det Z_0(\xi_0) \neq 0$ at a point $\xi_0 \in T^C$. Then the admittance $A(x)$ of the operator $Z*$ has the quasiasymptotics $A_0(x) = Z_0^{-1}(x)$ of order $-\alpha - 2n$ at ∞; so $Z_0 * \mathcal{E}_0 = I\delta$.*

This follows from the theorem and the theorem of Sec. 14.5. □

In the book by Vladimirov, Drozhzhinov and Zavialov [**122**, Chap. IV] some other Tauberian theorems concerning the solutions of specific differential equations, equations in convolutions and passive systems are given.

21. Abstract Scattering Operator

Let us apply the results concerning linear passive systems to the study of a finite-dimensional scattering matrix. As above, Γ is a closed convex acute solid cone in $\mathbb{R}^n$, $C = \operatorname{int}\Gamma^*$. For $n = 1$, see Beltrami and Wohlers [**4**], Lax and Phillips [**65**], Güttinger [**42**], and Rañada [**83**].

21.1. The definition and properties of an abstract scattering matrix. We use the term *abstract scattering matrix* relative to a cone Γ for the real $N \times N$ matrix $S(x) = (S_{kj}(x))$, $S_{kj} \in \mathcal{D}'(\mathbb{R}^n)$ that satisfies the conditions:

of causality relative to the cone Γ,

$$\operatorname{supp} S(x) \subset \Gamma; \tag{1.1}$$

of boundedness,

$$\int \langle S * \varphi, S * \varphi\rangle\, dx \le \int \langle \varphi, \varphi\rangle\, dx, \qquad \varphi \in \mathcal{D}^{\times N}. \tag{1.2}$$

The corresponding operator $S*$ is called the *scattering operator* (relative to the cone Γ).

Properties of an abstract scattering matrix.

21.1.1. *The operator $S*$ admits of extension onto $(\mathcal{L}^2)^{\times N}$ with the inequality* (1.2) *preserved:*

$$\sum_{1\le k\le N}\left\|\sum_{1\le j\le N} S_{kj}*\varphi_j\right\|^2 \le \sum_{1\le k\le N}\|\varphi_k\|^2, \qquad \varphi_k\in\mathcal{L}^2. \tag{1.2'}$$

This follows from (1.2) and from the density $\mathcal{D}$ in $\mathcal{L}^2$ (see Sec. 1.2).

21.1.2. *Restriction to growth:*

$$S\in(\mathcal{D}'_{\mathcal{L}^2})^{\times N^2}. \tag{1.3}$$

Indeed, from (1.2′) it follows that if $\varphi\in\mathcal{L}^2$, then also $S_{kj}*\varphi\in\mathcal{L}^2$. Let $\mathcal{E}_{n,m}(x)$ be a fundamental solution of the operator Δ^m, which solution belongs to $\mathcal{L}^2_{\text{loc}}$. [By virtue of Sec. 15.4.7, such m exist for every n, and $\mathcal{E}_{n,m}\in C^\infty(\mathbb{R}^n\setminus\{0\})$.] Suppose $\alpha\in\mathcal{D}$, $\alpha(x)=1$ in the neighbourhood of the point 0. Then

$$\Delta^m(\alpha\mathcal{E}_{n,m})=\delta(x)+\eta(x),$$

where $\eta\in\mathcal{D}$ and $\alpha\mathcal{E}_{n,m}\in\mathcal{L}^2$. Therefore

$$\begin{aligned}S_{kj}=S_{kj}*\delta&=S_{kj}*\Delta^m(\alpha\mathcal{E}_{n,m})-S_{kj}*\eta\\&=\Delta^m(S_{kj}*(\alpha\mathcal{E}_{n,m}))-S_{kj}*\eta.\end{aligned}$$

But $S_{kj}*(\alpha\mathcal{E}_{n,m})\in\mathcal{L}^2$, $S_{kj}*\eta\in\mathcal{L}^2$, and therefore $S_{kj}\in\mathcal{H}_s$ for some $s<0$ (see Sec. 10.1). The inclusion (1.3) is proved. □

21.1.3. *The following inequality holds true:*

$$\int_{-\Gamma}[\langle\varphi,\varphi\rangle-\langle S*\varphi,S*\varphi\rangle]\,dx\ge 0,\qquad \varphi\in(\mathcal{L}^2)^{\times N}. \tag{1.4}$$

Indeed, let $\varphi\in(\mathcal{L}^2)^{\times N}$; then $\psi=\theta_{-\Gamma}\varphi\in(\mathcal{L}^2)^{\times N}$, $\operatorname{supp}\psi\subset-\Gamma$ and, hence, $S*\varphi=S*\psi$ almost everywhere in $-\Gamma$ because, by Sec. 4.2.7 and (1.1),

$$\begin{aligned}\operatorname{supp}(S*\varphi-S*\psi)&=\operatorname{supp}S*[(1-\theta_{-\Gamma})\varphi]\\&\subset\overline{\operatorname{supp}S+\operatorname{supp}(1-\theta_{-\Gamma})}\subset\Gamma+\overline{\mathbb{R}^n\setminus(-\Gamma)}=\overline{\mathbb{R}^n\setminus(-\Gamma)}.\end{aligned}$$

From this and from (1.2′) follows the inequality (1.4):

$$\begin{aligned}\int_{-\Gamma}[\langle\varphi,\varphi\rangle-\langle S*\varphi,S*\varphi\rangle]\,dx&=\int\langle\psi,\psi\rangle-\int_{-\Gamma}\langle S*\psi,S*\psi\rangle\,dx\\&\ge\int[\langle\psi,\psi\rangle-\langle S*\psi,S*\psi\rangle]\,dx\ge 0.\end{aligned}$$

□

21.1.4. *The inequality* (1.4) *holds true in strong form*

$$\int_{-\Gamma+x_0}[\langle\varphi,\varphi\rangle-\langle S*\varphi,S*\varphi\rangle]\,dx\ge 0,\qquad \varphi\in(\mathcal{L}^2)^{\times N},\qquad x_0\in\mathbb{R}^n. \tag{1.5}$$

This follows from the inequality (1.4) via reasoning similar to that given in Sec. 20.2.1. □

From the inequality (1.5), as in Sec. 20.2.2, follows the inequality (1.2). □

REMARK. It is also interesting to find out whether the causality condition (1.1) follows from the inequality (1.4), as in the case for passive operators (see Sec. 20.2.3).

21.1.5. *Positive definiteness:*

$$|a|^2\delta - \langle (S^* * S)a, a\rangle \gg 0, \qquad a \in \mathbb{C}^N. \tag{1.6}$$

Indeed, setting $\varphi = a\varphi_0$, $a \in \mathbb{C}^N$, and $\varphi_0 \in \mathcal{D}$ in (1.2), we obtain (1.6) (see Sec. 8.1):

$$\begin{aligned} 0 &\le \int \left[|a|^2|\varphi_0|^2 - \langle S * a\varphi_0, S * a\varphi_0\rangle\right] dx \\ &= \left(\delta, |a|^2\varphi_0 * \varphi_0^* - \sum_{1\le j\le N} (S * a\varphi_0)_j * (S * a\varphi_0)_j^*\right) \\ &= \left(\delta, |a|^2\varphi_0 * \varphi_0^* - \sum_{1\le j\le N} (S * a\varphi_0)_j * (S(-x) * \bar{a}\varphi_0^*)_j\right) \\ &= \left(\delta, |a|^2\varphi_0 * \varphi_0^* - \sum_{1\le j,k,l\le N} S_{jk} * S_{jl}^* * \varphi_0 * \varphi_0^* a_k \bar{a}_l\right) \\ &= \left(|a|^2\delta - \langle (S^* * S)a, a\rangle, \varphi_0 * \varphi_0^*\right). \end{aligned}$$

Here we made use of the existence of convolutions for generalized functions taken from $\mathcal{D}'_{\mathcal{L}^2}$ [property 21.1.2] and of the properties of convolutions (see Sec. 10.1, Sec. 4.2, and Sec. 4.6).

21.1.6. *The matrix $S(x)$ has the Laplace transform $\tilde{S}(\zeta)$ that is holomorphic in T^C, $\tilde{S}_{kj} \in H(C)$, and that satisfies the reality condition:*

$$\tilde{S}(\zeta) = \bar{\tilde{S}}(-\bar{\zeta}), \qquad \zeta \in T^C. \tag{1.7}$$

This follows from

21.1.7. *The boundary value $\tilde{S}(p) = F[S]$ of the matrix $\tilde{S}(\zeta)$ satisfies the inequality (for almost all $p \in \mathbb{R}^n$)*

$$I - \tilde{S}^+(p)\tilde{S}(p) \ge 0. \tag{1.8}$$

In particular, for almost all $p \in \mathbb{R}^n$,

$$\sum_{1\le k\le N} \left|\tilde{S}_{kj}(p)\right|^2 \le 1, \qquad j = 1, \dots, N. \tag{1.9}$$

This follows from (1.6) by the Bochner–Schwartz theorem (see Sec. 8.2):

$$\langle \tilde{S}^+(p)\tilde{S}(p)a, a\rangle = \left|\tilde{S}(p)a\right|^2 \le |a|^2, \qquad a \in \mathbb{C}^N. \tag{1.8'}$$

□

21.1.8. *The matrix function $\tilde{S}(\zeta)$ satisfies the inequality*

$$I - \tilde{S}^*(\zeta)\tilde{S}(\zeta) \geq 0, \qquad \zeta \in T^C. \tag{1.10}$$

In particular,

$$\sum_{1\leq k\leq N} \left|\tilde{S}_{kj}(\zeta)\right|^2 \leq 1, \qquad \zeta \in T^C, \qquad j = 1, \ldots, N. \tag{1.11}$$

Indeed, from (1.8′) for all a and b in $\mathbb{C}^N$ we have

$$\langle \tilde{S}(p)a, b\rangle \leq |a|\,|b| \quad \text{for almost all} \quad p \in \mathbb{R}^n. \tag{1.12}$$

Furthermore, the function $\langle \tilde{S}(\zeta)a, b\rangle$ belongs to the class $H(C)$ [see 21.1.6], and its boundary value $\langle \tilde{S}(p)a, b\rangle$ satisfies the estimate (1.12). By the Phragmén–Lindelöf theorem (see Sec. 12.6) the function $\langle \tilde{S}(\zeta)a, b\rangle$ satisfies the estimate

$$\left|\langle \tilde{S}(\zeta)a, b\rangle\right| \leq |a|\,|b|, \qquad \zeta \in T^C,$$

from which, for $b = \tilde{S}(\zeta)a$, follows the estimate

$$\left|\tilde{S}(\zeta)a\right|^2 \leq |a|^2, \qquad \zeta \in T^C, \qquad a \in \mathbb{C}^N, \tag{1.10′}$$

which is equivalent to the estimate (1.10). □

Every matrix that is holomorphic in T^C and satisfies the conditions of reality (1.7) and boundedness (1.10) is said to be *bounded-real* in T^C.

We have thus proved that the matrix function $\tilde{S}(\zeta)$ is bounded-real in T^C.

21.2. A description of abstract scattering matrices.

THEOREM. *In order that the matrix $S(x)$ should define a scattering operator relative to the cone Γ, it is necessary and sufficient that its Laplace transform $\tilde{S}(\zeta)$ be a bounded-real matrix function in the domain of T^C, where $C = \operatorname{int}\Gamma^*$.*

PROOF. Necessity was proved in Sec. 21.1. We now prove sufficiency. Suppose $\tilde{S}(\zeta)$ is a bounded-real matrix function in T^C. Then it is a Laplace transform,

$$\tilde{S}(\zeta) = L[S] = F[Se^{-(q,x)}],$$

of a real matrix $S(x)$ with elements taken from $\mathcal{S}'$ that satisfies the causality condition (1.1) relative to the cone $C^* = \Gamma$; here $\tilde{S}(p) = F[S]$, where $\tilde{S}(p)$ is the boundary value of $\tilde{S}(\zeta)$ as $q \to 0$, $q \in C$, in $\mathcal{S}'$ (see Sec. 12.2). Furthermore, since the elements of the matrix $\tilde{S}(\zeta)$ are uniformly bounded in T^C and converge weakly as $q \to 0$, $q \in C$, on the set $\mathcal{S}$ that is dense in $\mathcal{L}^1$ (see Sec. 1.2), it follows that the elements of the boundary matrix $\tilde{S}(p)$ may be identified with functions taken from $\mathcal{L}^\infty$.

Suppose $\varphi \in \mathcal{D}^{\times N}$ and then $F[\varphi] = \tilde{\varphi} \in \mathcal{S}^{\times N}$. From the condition (1.10′) we have, for every $q \in C$, the inequality

$$\int \langle \tilde{S}(\zeta)\tilde{\varphi}(p), \tilde{S}(\zeta)\tilde{\varphi}(p)\rangle\, dp \leq \int \langle \tilde{\varphi}, \tilde{\varphi}\rangle\, dp.$$

From this, using the Parseval–Steklov equation (see Sec. 6.6.3) and, also the Fourier transform theorem of a convolution (see Sec. 6.5), we obtain

$$\int \langle [Se^{-(q,x)}] * \varphi, [Se^{-(q,x)}] * \varphi\rangle\, dx \leq \int \langle \varphi, \varphi\rangle\, dx, \qquad q \in C. \tag{2.1}$$

We now apply formula (2.9) of Sec. 9.2,

$$[Se^{-(q,x)}] * \varphi = e^{-(q,x)}\left(S * [\varphi e^{(q,x)}]\right).$$

As a result, inequality (2.1) becomes

$$\int e^{-2(q,x)}\langle S * [\varphi e^{(q,x)}], S * [\varphi e^{(q,x)}]\rangle\, dx \le \int \langle \varphi, \varphi\rangle\, dx, \qquad q \in C. \tag{2.2}$$

We now prove the possibility of passing to the limit as $q \to 0$, $q \in C$, under the sign of the integral on the left-hand side of (2.2). For this purpose, note that since all elements of the matrix $\tilde{S}(p)$ belong to $\mathcal{L}^\infty$ and, hence, all elements of the matrix $S = F^{-1}[\tilde{S}(p)]$ belong to $\mathcal{D}'_{\mathcal{L}^2}$ (see Sec. 10.1), and the representation (1.2) of Sec. 10.1 holds for the latter, it suffices to consider the case where all elements of the matrix S belong to $\mathcal{L}^2$. In that case, the following properties hold true:

(a) $|S_{kj} * [\varphi_j e^{(q,x)}]| \le |S_{kj}| * [|\varphi_j| e^{|x|}] \in \mathcal{L}^2$, $|q| \le 1$ (see Sec. 4.1.2);

(b) $S * [\varphi e^{(q,x)}] = \int S(x')\varphi(x - x')e^{(q,x-x')}\, dx' \to \int S(x')\varphi(x - x')\, dx' = S * \varphi$, $q \to 0$;

(c) $\operatorname{supp} S * [\varphi e^{(q,x)}] \subset \overline{\operatorname{supp} S + \operatorname{supp}\varphi} \subset \Gamma + \overline{U}_n$, if $\operatorname{supp}\varphi \subset \overline{U}_n$ (see Sec. 4.2.7);

(d) $e^{-2(q,x)} \le e^{2R}$, $x \in \Gamma + \overline{U}_n$, $|q| \le 1$, $q \in C$.

The properties from (a) to (d) are what ensure, by the Lebesgue theorem, the possibility of passing to the limit as $q \to 0$, $q \in C$, under the integral sign in the left-hand member of (2.2). We thus obtain the boundedness condition (1.2). The theorem is proved. □

REMARK 1. This theorem was proved for $n = 1$ by Beltrami and Wohlers [**4**] and for $n = 2$ by Vladimirov [**113**].

REMARK 2. The proof of sufficiency in the theorem is simplified if use is made of the well-known Fatou theorem, according to which $\tilde{S}(p + iq) \to \tilde{S}(p)$, $q \to 0$, $q \in C$, for almost all $p \in \mathbb{R}^n$. It will be noticed that we did not make use of the Fatou theorem.

21.3. The relationship between passive operators and scattering operators. In order to establish the relationship between the scattering matrix $S(x)$ and passive operators, it is convenient to introduce a new matrix $T(x)$ via the formula

$$S(x) = I\delta(x) + 2iT(x). \tag{3.1}$$

It is called the *scattering amplitude*.

The scattering operator $S*$ relative to the cone Γ is said to be *nonsingular* if

$$\det[I - \tilde{S}(\zeta)] \ne 0, \qquad \zeta \in T^C. \tag{3.2}$$

In (3.1) let us pass to the Laplace transform

$$\tilde{S}(\zeta) = I + 2i\tilde{T}(\zeta), \qquad \zeta \in T^C. \tag{3.3}$$

By the theorem of Sec. 21.2 the matrix $\tilde{S}(\zeta)$ is bounded real in T^C, that is, it is holomorphic in T^C and satisfies the conditions (1.7) and (1.10). Therefore, it follows from (3.2) and (3.3) that the matrix $-i\tilde{T}(\zeta)$ is holomorphic in T^C, satisfies the reality condition (1.7), $\det \tilde{T}(\zeta) \neq 0$, $\zeta \in T^C$, and

$$\Re[-i\tilde{T}(\zeta)] = -\frac{i}{2}\left[\tilde{T}(\zeta) - \tilde{T}^+(\zeta)\right] \geq \tilde{T}^+(\zeta)\tilde{T}(\zeta) > 0, \qquad \zeta \in T^C. \tag{3.4}$$

Thus the matrix $-i\tilde{T}(\zeta)$ is positive real in T^C and, by Theorem I of Sec. 20.3, the matrix $-iT(x)$ defines the passive operator $-iT*$ relative to the cone Γ; by virtue of (3.4), that operator is completely nonsingular (see Sec. 20.5). By Theorem I of Sec. 20.5, there exists an inverse operator $(-iT)^{-1}*$ that is a completely nonsingular passive operator relative to the cone Γ.

Let us now introduce new vector quantities: j for "current" and v for "voltage" — via the formulae

$$j = u - S * u, \qquad v = u + S * u. \tag{3.5}$$

By (3.1), $S * u = u + 2iT * u$, whence

$$j = -2iT * u = -iT * (v + j)$$

and therefore j and v are connected by the relation

$$v = Z * j \tag{3.6}$$

where

$$\begin{aligned} Z* &= (-iT)^{-1} * -I\delta * \\ &= 2(I\delta - S)^{-1} * -I\delta * \\ &= (I\delta - S)^{-1} * (I\delta + S) * . \end{aligned} \tag{3.7}$$

We now prove that the matrix $Z(x)$ defines a passive operator relative to the cone Γ.

Indeed, the matrix $Z(x)$ is real and, by (3.7) and (3.4), for all $\zeta \in T^C$,

$$\begin{aligned} \Re\tilde{Z}(\zeta) &= \frac{1}{2}\left[\tilde{Z}(\zeta) + \tilde{Z}^+(\zeta)\right] \\ &= \frac{1}{2i}\left[\tilde{T}^{+-1}(\zeta) - \tilde{T}^{-1}(\zeta)\right] - I \\ &= \tilde{T}^{+-1}(\zeta)\left\{-\frac{i}{2}[\tilde{T}(\zeta) - \tilde{T}^+(\zeta)] - \tilde{T}^+(\zeta)\tilde{T}(\zeta)\right\}\tilde{T}^{-1}(\zeta) \geq 0. \end{aligned}$$

Thus, the matrix function $\tilde{Z}(\zeta)$ is positive real in T^C and, hence, the matrix $Z(x)$ defines a passive operator relative to the cone Γ. □

REMARK. If, in addition to the condition (3.2), we require fulfillment of the condition

$$\det[I + \tilde{S}(\zeta)] \neq 0, \qquad \zeta \in T^C,$$

then the operator $Z*$ is nonsingular by (3.3) and (3.7),

$$\det \tilde{Z}(\zeta) = \left(\frac{i}{2}\right)^n \frac{\det[I + \tilde{S}(\zeta)]}{\det \tilde{T}(\zeta)} \neq 0, \qquad \zeta \in T^C.$$

Conversely, suppose we have a passive operator $Z*$ relative to the cone Γ. Then the operator $Z * + I\delta*$ is passive and completely nonsingular relative to the cone Γ. Therefore there exists the inverse operator

$$Q* = (Z + I\delta)^{-1} * \tag{3.8}$$

which is passive, completely nonsingular operator relative to Γ (see Sec. 20.5). We now prove the inequality

$$\Re\tilde{Q}(\zeta) \geq \tilde{Q}^{+}(\zeta)\tilde{Q}(\zeta), \qquad \zeta \in T^{C}. \tag{3.9}$$

Indeed, by (3.8), the matrix $Q(x)$ is real and (see Sec. 20.3)

$$\tilde{Z}(\zeta) = \tilde{Q}^{-1}(\zeta) - I, \qquad \Re\tilde{Z}(\zeta) \geq 0, \qquad \zeta \in T^{C},$$

whence follows the inequality (3.9):

$$\begin{aligned}\Re\tilde{Q}(\zeta) &= \frac{1}{2}[\tilde{Q}(\zeta) + \tilde{Q}^{+}(\zeta)] \\ &= \tilde{Q}^{+}(\zeta)\tilde{Q}(\zeta) + \frac{1}{2}\{\tilde{Q}^{+}(\zeta)[\tilde{Q}^{-1}(\zeta) - I]\tilde{Q}(\zeta) \\ &\quad + \tilde{Q}^{+}(\zeta)[\tilde{Q}^{-1}(\zeta) - I]\tilde{Q}(\zeta)\} \\ &= \tilde{Q}^{+}(\zeta)\tilde{Q}(\zeta) + \Re\tilde{Q}^{+}(\zeta)\tilde{Z}(\zeta)\tilde{Q}(\zeta) \\ &\geq \tilde{Q}^{+}(\zeta)\tilde{Q}(\zeta).\end{aligned}$$

□

Let us now prove that the operator

$$S* = I\delta * -2Q* = (Z + I\delta)^{-1} * (Z - I\delta) * \tag{3.10}$$

is a scattering operator relative to the cone Γ.

Indeed, $S(x)$ is a real matrix and

$$\tilde{S}(\zeta) = I - 2\tilde{Q}(\zeta), \qquad \zeta \in T^{C}$$

and so, by virtue of (3.9), we have

$$\begin{aligned}\tilde{S}^{+}(\zeta)\tilde{S}(\zeta) &= [I - 2\tilde{Q}^{+}(\zeta)][I - 2\tilde{Q}(\zeta)] \\ &= I - 4\Re\tilde{Q}(\zeta) + 4\tilde{Q}^{+}(\zeta)\tilde{Q}(\zeta) \leq I.\end{aligned}$$

Thus the matrix function $\tilde{S}(\zeta)$ is bounded real in T^{C}. By the theorem of Sec. 21.2, the matrix $S(x)$ defines a scattering operator relative to the cone Γ.

We have thus proved the following theorem.

THEOREM. *Every nonsingular abstract scattering operator $S*$ defines, by the formula*

$$Z* = (I\delta - S)^{-1} * (I\delta + S)*,$$

a passive operator; conversely, every passive operator $Z*$ defines, via the formula

$$S* = (Z + I\delta)^{-1} * (Z - I\delta)*,$$

an abstract scattering operator.

BIBLIOGRAPHY

[1] Aizenberg, L.A., and Dautov, Sh.A., "Holomorphic functions of several complex variables with nonnegetive real part. Traces of holomorphic and pluriharmonic functions on the Shilov boundary", *Mat. Sb.* 99 (141) (1976), 343–355, 479 (in Russian).

[2] Antosik, P., Mikusinski, J., and Sikorski, R., *Theory of Distributions: The Sequential Approach*, Elsevier, Amsterdam, 1973.

[3] Arsac, J., *Fourier Transforms and the Theory of Distributions*, Prentice-Hall, Englewood Cliffs, N.J., 1966.

[4] Beltrami, E.J., and Wohlers, M.R., *Distributions and the Boundary Values of Analytic Functions*, Academic Press, New York, 1966.

[5] Bochner, S., *Lectures on Fourier Integrals*, Princeton University Press, Princeton, N.J., 1959.

[6] ———, "Group invariance of Cauchy's formula in several variables", *Ann. of Math.* 45 (1944), 687–707.

[7] Bogolyubov, N.N., Logunov, A.A., Oksak, A.I., and Todorov, I.T., *General Principles of Quantum Field Theory*, Nauka, Moscow, 1987 (in Russian).

[8] Bogolyubov, N.N., Medvedev, B.V., and Polivanov, M.K., *Theory of Dispersion Relations*, Lawrence Radiation Laboratory, Berkley, Calif., 1961.

[9] Bogolyubov, N.N., and Shirkov, D.V., *Introduction to the Theory of Quantized Fields*, 4rd ed. Nauka, Moscow, 1984 (in Russian) [English transl. of 3rd Russian ed.: Wiley-Interscience, New York, 1980].

[10] Bogolyubov, N.N., and Vladimirov, V.S., "Representation of n-point functions", *Trudy Mat. Inst. Steklov* 112 (1971), 5–21 [English transl.: *Proc. Steklov Inst. Math.* 112 (1971), 1–18 (1973)].

[11] Bourbaki, N., *Eléments de mathématique*, Livre V, *Espaces vectoriels topologiques*, Hermann, Paris, 1953, 1955.

[12] Bremermann, H., *Distributions, Complex Variables, and Fourier Transforms*, Addison-Wesley, Reading, Mass., 1965.

[13] Colombeau, J.F., *Elementary Introduction to New Generalized Functions*, North Holland, 1985.

[14] Danilov, L.I., "On regularity of an acute cone in $\mathbf{R}^n$", *Sibirian Math. J.* 26 (1985) 198–201 (in Russian).

[15] Dezin, A.A., "Boundary-value problems for certain first-order symmetric linear systems", *Mat. Sb.* 49 (91) (1959), 459–484 (in Russian).

[16] Dierolf, P., Voigt, J., "Convolution and S'-convolution of distributions", *Collect. Math.*, 29, fasc. 3, (1978), 3–14.

[17] Dieudonné, J., and Schwartz, L., "La dualité dans les espaces (F) et (LF)", *Ann. Inst. Fourier (Grenoble)*, 1 (1949), 61–101 (1950).

[18] Dirac, P.A.M., "Quantised singularities in the electromagnetic field", *Proc. Roy. Soc. London, Ser. A* 133 (1931), 60–72.

[19] ———, *The Principles of Quantum Mechanics*, 4th ed. Clarendon Press, Oxford, 1958.

[20] ———, "The physical interpretation of the quantum dynamics", *Proc. Roy. Soc. London Ser. A* 113 (1927), 621–641.

[21] Drozhzhinov, Yu.N., "Asymptotic behavior of the solution of the Cauchy problem for the linearized system of equations of magnetohydrodynamics", *Dokl. Akad. Nauk SSSR* 212 (1973), 831–833 [English transl.: *Soviet Physics Dokl.* 18 (1973), 638–639 (1974)].

[22] ———, "Linear passive systems of partial differential equation", *Math. Sbornik*, **116** (1981), 299–309.

[23] Drozhzhinov, Yu.N., and Galeev, R.Kh., "Asymptotic behavior of the solution of the Cauchy problem for the two-dimensional system of a rotating compressible fluid", *Differentsial'nye Uravnenija* 10 (1974), 53–58, 179 [English transl.: *Differential Equations* 10 (1974), 37–40 (1975)].

[24] Drozhzhinov, Yu.N., and Zavialov, B.I., "Tauberian theorems for generalized functions with support in cones", *Mat. Sb.*, **108** (1979), 79–90 [English transl.: *Math. USSR Sb.* **36** (1980)]

[25] ———, "Multi-dimensional Tauberian theorems for holomorphic functions of bounded argument and the quasi-asymptotics of passive systems", *Mat. Sb.*, **117** (1982), 44–59 [English transl.: *Math. USSR Sb.* **45**, no. 1, (1983), 45–61]

[26] ———, "Multi-dimensional Tauberian comparison theorems for holomorphic functions of bounded argument", *Izv. AN SSSR Ser. Mat.*, **55**, no. 6, (1991), 1139–1155 [English transl.: *Math. USSR Izv.* **39**, no. 3, (1992), 1097–1112]

[27] ———, "Multi-dimensional Abelian and Tauberian comparison theorems", *Mat. Sb.*, **180**. no. 9 (1989), 1234–1258 [English transl.: *Math. USSR Sb.* **68**, no. 1, (1991), 85–110]

[28] ———, "The Wiener-type Tauberian theorem for tempered generalized functions", *Mat. Sb.* **189**, no. 7 (1998), 90–130 (in Russian).

[29] Egorov, Yu.V., "To the theory of generalized functions", *Uspekhi Mat. Nauk* **45**, no. 5, (1990), 3–40 (in Russian).

[30] Ehrenpreis, L., *Fourier Analysis in Several Complex Variables*, Wiley-Interscience, New York, 1970.

[31] ———, "Solutions of some problems of division. I. Division by a polynomial of distribution", *Amer. J. Math.*, **76** (1954), 883–903.

[32] Euler, L., *Integralrechnung*, B. III, Verlag Carl Gerold, Wien, 1830.

[33] Friedrichs, K.O., "Symmetric hyperbolic linear differential equations", *Comm. Pure Appl. Math.*, **7** (1954), 345–392.

[34] Galeev, R.Kh., "Multi-dimensional linear passive systems in a Hilbert space" *Differentsial'nye Uravneniya*, **17**, no. 2, (1981), 278–285 [English transl.: *Differential Equations*, **17** (1981), 191–196].

[35] ———, "The Cauchy problem for passive systems in a Hilbert space" *Differentsial'nye Uravneniya*, **18**, no. 10, (1982), 1718–1724 [English transl.: *Differential Equations*, **18** (1982), 000–000].

[36] Gårding, L., "Linear hyperbolic partial differential equations with constant coefficients", *Acta Math.*, **85** (1951), 1–62.

[37] Garsoux, J., *Espaces vectoriels topologiques et distributions*, Dunod, Paris, 1963.

[38] Gel'fand, I.M., and Shilov, G.E., *Generalized Functions*, vols. 1–3, Academic Press, New York, 1964, 1968, 1967.

[39] Gel'fand, I.M., and Vilenkin, N.Ya., *Generalized Functions*, vol. 4: Applications to Harmonic Analysis, Academic Press, New York, 1964.

[40] Godunov, S.K., and Sultangazin, U.M., "The dissipativity of V.S. Vladimirov's boundary conditions for a symmetric system of the method of spherical harmonics", *Zh. Vychisl. Mat. i Mat. Fiz.* 11 (1971), 688–704 (in Russian).

[41] Green, M.B., Schwarz, J.H., Witten, E., *Superstring Theory*, v. 1, *Introduction*, Cambridge Univ. Press, 1987.

[42] Güttinger, W., "Generalized functions in elementary particle physics and passive system theory. Recent trends and problems", Sympos. Applications of Generalized Functions (Stony Brook, N.Y., 1966), *SIAM J. Appl. Math.* **15** (1967), 964–1000.

[43] Hackenbroch, W., "Integraldarstellung einer Klasse dissipativer linearer Operatoren", *Math. Z.* 109 (1969), 273–287.

[44] Hadamard, J., *Lectures on Cauchy's Problem in Linear Partial Differential Equations*, Yale University Press, New Haven, Conn. (1923) [Dover Publications, New York (1952)]. There is a later French edition: *Le problème de Cauchy et les equations aux dérivées partielles lineaires hyperboliques*, Hermann, Paris, 1932.

[45] Hirata, Y., Ogata, H., "On the exchange formula for distributions", *J. Sci. Hiroshima Univ., Ser. A*, **22** (1958), 147–152.

[46] Hörmander, L., *Linear Differential Operators*, Springer, Berlin, 1963.

[47] ———, "On the division of distributions by polynomials,,. *Ark. Mat.* 3 (1958), 558–568.

[48] ———, "Pseudo-differential operators", *Comm. Pure Appl. Math.* 18 (1965), 501–517.

[49] ———, "Pseudo-differential operators and non-elliptic boundary problems", *Ann. of Math.* 83 (1966), 129–209.

[50] ———, "Pseudo-differential operators and hyperbolic equations", *Singular Integrals* (ed. A.P. Calderon) (University of Chicago, April 1966), American Mathematical Society, Providence, R.I., 1967.

[51] ———, *The Analysis of Linear Partial Differential Operators* vols. I–IV, Berlin ect.: Springer, 1983–85 [M.: Mir, In Russian t. I–IV, 1986-1988].

[52] Itano, M., "On the theory of multiplicative products of distributions", *J. Sci. Hiroshima Univ., Ser. A-I*, 30 (1966), 151–181.

[53] Jost, R., *The General Theory of Quantized Fields*, American Mathematical Society, Providence, R.I., 1965.

[54] Kaminski, A., "Convolution, product and Fourier transform of distributions", *Studia Math.*, 74 (1982), 83–96.

[55] Kantorovich, L.V., and Akilov, G.P., *Functional Analysis in Normed Spaces*, Pergamon Press, Oxford, 1964.

[56] Kohn, J.J., and Nirenberg, L., "An algebra of pseudo-differential operators", *Comm. Pure Appl. Math.* 18 (1965), 269–305.

[57] ———, "Non-coercive boundary value problems", *Comm. Pure Appl. Math.* 18 (1965), 443–492.

[58] Kolmogorov, A.N., and Fomin, S.V., *Elements of the Theory of Functions and Functional Analysis*, 4th ed., Nauka, Moscow, 1976, (in Russian) [English transl. of 2nd Russian ed.: *Introductory Real Analysis*, Prentice-Hall, Englewoo-Cliffs, N.J., 1970; a corrected reprinting: Dover Publications, New York, 1975].

[59] Komatsu, H., "Ultradistributions and hyperfunctions", *Hyperfunctions and Pseudo-differential Equations* (Proc. Conf. on the Theory of Hyperfunctions and Analytic Functionals and Applications, R.I.M.S., Kyoto Univ., Kyoto, 1971: dedicated to the memory of André Martineau), pp. 164–179. Lecture Notes in Math., vol. 287, Springer, Berlin, 1973.

[60] ———, "Ultradistributions", I. "Structure theorems and a characterization", *I. Fac. Sci. Univ. Tokyo, Section IA*, 20 (1973), 25–105; II. "The kernel theorem and ultradistributions with support in a submanifold", *ibid.* 24 (1977), 607–628.

[61] König, von H., and Meixner, T., "Lineare Systeme und lineare Transformationen", *Math. Nachr.*, 19 (1958), 265–322.

[62] König, von H., and Zemanian, A.H., "Necessary and sufficient conditions for a matrix distribution to have a positive-real Laplace transform", *SIAM J. Appl. Math.* 13 (1965), 1036–1040.

[63] Korányi, A., and Pukánsky, J., "Holomorphic functions with positive real part on polycylinders", *Trans. Amer. Math. Soc.* 108 (1963), 449–456.

[64] Köthe, G., "Die Randverteilungen analytischer Funktionen", *Math. Z.* 57 (1952), 13–33.

[65] Lax, P.D., and Phillips, R.S., *Scattering Theory*, Academic Press, New York, 1967.

[66] Leonard, P., "Problèmes aux limites pour les opérateurs matriciels de dérivation hyperboliques des premier et deuxième ordres", *Mem. Soc. Roy. Sci. Liège Coll. in-8°* (5) 11 (1965), No. 3, 5–131.

[67] Lighthill, M.J., *Fourier Transform and Generalized Functions*, Cambridge Tracts on Mechanics and Applied Mathematics, 1958.

[68] Lions, J.-L., "Supports dans la transformation de Laplace", *J. Analyse Math.* 2 (1953), 369–380.

[69] Lojasiewicz, S., "Sur le problème de division", *Studia Math.* 18 (1959), 87–136.

[70] Luszczki, Z., and Zieleźny, Z., "Distributionen der Räume $\mathcal{D}'_{\mathcal{L}^p}$ als Randverteilungen analytischer Funktionen", *Colloq. Math.* 8 (1961), 125–131.

[71] Lützen, J., *The Prehistory of the Theory of Distributions*, Springer, 1982.

[72] Malgrange, B., *Ideals of Differentiable Functions*, Oxford University Press, London, 1966.

[73] ———, "Equations aux dérivées partielles à coefficients constants. I. Solution élémentaire", *Compt. rend. acad. sci. Paris* 237 (1953), 1620–1622.

[74] Martineau, A., "Distributions et valeurs au bord des fonctions holomorphes", *Theory of Distributions* (Proc. Internat. Summer Inst., Lisbon, 1964), Inst. Gulbenkiau Ci., Lisbon, 1964, 193–326.

[75] Maslennikova, V.N., "An explicit representation and asymptotic behavior for $t \to \infty$ of the solution of the Cauchy problem for a linearized system of a rotating compressible fluid", *Dokl. Akad. Nauk SSSR* 187 (1969), 989–992, [English transl.: *Soviet Math. Dokl.* 10 (1969), 978–982].

[76] Mikusinski, J., "Criteria of the existence and the associativity of the product of distributions", *Studia Math.*, 21 (1962), 253–259.

[77] ———, "Irregular operations on distributions", *Studia Math.*, 20 (1961), 163–169.

[78] Nevanlinna, R., *Eindeutige analytische Funktionen*, 2nd ed. Springer, Berlin, 1953.

[79] Palamodov, V.P., *Linear Differential Operators with Constant Coefficients*, Springer, Berlin, 1970.

[80] Petrovski, I.G., "Sur l'analyticité des solutions des systémes d'équations differetielles", *Mat. Sb.* 5 (47) (1939), 3–70.

[81] Plamenevskii, B.A., *Hyperfunctions and Pseudo-differential Operators*, Springer, Lecture Notes in Math., v. 287, 1973.

[82] Prudnikov, A.P., Brychkov, Yu.A., Marichev, O.I., *Integrals, Series, Special Functions*, Nauka, Moscow, 1983 (in Russian).

[83] Rañada, A.F., "Causality and the *S*-matrix", *J. Mathematical Phys.* 8 (1967), 2321–2326.

[84] Reed, M., and Simon, B., *Methods of Modern Mathematical Physics*, vols. I–IV, Academic Press, New York, 1972–1978. [M.: Mir, in Russian, t. I–IV, 1977-1982].

[85] Riesz, M., "L'intégrale de Riemann–Liouville et le probléme de Cauchy", *Acta Math.* 81 (1949), 1–223.

[86] Rothaus, O.S., "Domains of positivity", *Abh. Math. Sem. Univ. Hamburg* 24 (1960), 189–235.

[87] Sato, M., "Theory of hyperfunctions", *J. Fac. Sci. Univ. Tokyo Sect. I* 8 (1959–1960), 139–193, 387–437.

[88] Sato, M., Kawai, T., and Kashiwara, M., *Microfunctions and Pseudo-differential Equations*, Lectures notes in Math., 287. Hyperfunctions and Pseudo-differential Equations, Berlin etc.: Springer, 1973, 265–529.

[89] Schwartz, L., *Théorie des distributions*, 2 vols., Hermann, Paris, 1950, 1951.

[90] ———, *Méthodes mathématiques pour les sciences physiques*, Hermann, Paris, 1961.

[91] ———, "Transformation de Laplace des distributions", *Medd. Lunds Univ. Mat. Sem.* (Supplementband) (1952), 196–206.

[92] Seneta, E., *Regularly Varying Functions*, Lecture Notes in Math., 508, Springer, Berlin, 1976.

[93] Shiraishi, R., "On the definition of convolution for distributions", *J. Sci. Hiroshima Univ., Ser. A*, 23 (1959), 19–32.

[94] Shiraishi, R., Itano, M., "On the multiplicative products of distributions", *J. Sci. Hiroshima Univ., Ser. A-I*, 28 (1964), 223–235.

[95] Silin, V.P., and Rukhadze, A.A., *Electromagnetic Properties of Plasma and Plasma-like Media*, Gosatomizdat, Moscow, 1961 (in Russian).

[96] Sobolev, S.L., "Méthods nouvelle à resoudre le probème de Cauchy pour les équations linéaires hyperboliques normales", *Mat. Sb.* 1 (1935), 39–72.

[97] ———, *Applications of Functional Analysis in Mathematical Physics*, Leningrad University Press, Leningrad, 1950 [English transl.: *Amer. Math. Soc. Transl. Math. Mono.* 7 (1963)].

[98] Sochozki, J.W., *On Definite Integrals Used in Series Expansions*, M. Stalyusevich's Printing House, St. Petersburg, 1873 (in Russian).

[99] Stein, E.M., and Weiss, G., *Introduction to Fourier Analysis on Euclidean Spaces*, Princeton University Press, Princeton, N.J., 1971.

[100] Streater, R.F., and Wightman, A.S., *PCT, Spin and Statistics and All That*, W.A. Benjamin, New York, 1964.

[101] Tikhonov, A.N., "Théorèmes d'unicité pour l'équation de la chaleur", *Mat. Sb.* 42 (1935), 199–216.

[102] Tillmann, H.G., "Darstellung der Schwartzschen Distributionen durch analytische Funktionen", *Math. Z.* 77 (1961), 106–124.

[103] ———, "Distributionen als Randverteilungen analytischer Funktionen", II: *Math. Z.* 76 (1961), 5–21.

[104] Treves, J.F., *Lectures on Linear Partial Differential Equations with Constant Coefficients* (Notas de Matemática, No. 27), Instituto de Matemática Pura e Aplicada do Conselho Nacional de Pesquisas, Rio de Janeiro, 1961.

[105] Vladimirov, V.S., *Methods of the Theory of Functions of Many Complex Variables*, M.I.T. Press, Cambridge, Mass., 1966.

[106] ——, *Equations of Mathematical Physics*, 5th ed., Nauka, Moscow, 1988, (in Russian) [English transl. of 2nd Russian ed.: Marcel Dekker, New York, 1971 and 4th Russian ed.: Mir, Moscow, 1984].

[107] ——, "A generalization of the Cauchy–Bochner integral representation", Izv. Akad. Nauk SSSR Ser. Mat. 33 (1969), 90–108 [English transl.: *Math. USSR-Izv.* 3 (1969), 87–104].

[108] ——, "Generalized functions whose supports are bounded outside a convex acute cone", *Sibirsk. Mat. Zh.* 9 (1968), 1238–1247 [English transl.: *Sibirian Math. J.* 9 (1968), 930–938].

[109] ——, "Construction of envelopes of holomorphy for regions of a special type and their application', *Trudy Mat. Inst. Steklov* 60 (1961), 101–144 [English transl.: *Amer. Math. Soc. Transl.* (2) 48 (1965), 107–150].

[110] ——, "On Cauchy–Bochner representation", *Izv. Akad. Nauk SSSR Ser. Mat.* 36 (1972), 534–539 [English transl.: *Math. UssR-Izv.* 6 (1972), 529–535].

[111] ——, "Holomorphic functions with nonnegative imaginary part in a tubular domain over a cone", *Mat. Sb.* 79 (121) (1969), 128–152 [English transl.: *Math. USSR-Sb.* 8 (1969), 125–146].

[112] ——, "Construction of shells of holomorphy for a special kind of region", *Dokl. Akad. Nauk SSSR* 134 (1960), 251–254 [English transl.: *Soviet Math. Dokl.* 1 (1960), 1039–1042].

[113] ——, "Linear passive systems", *Transl. Mat. Fiz.* 1 (1969), 67–94 [English transl.: *Theoretical and Math. Phys.* 1, (1969), 51–72].

[114] ——, "Holomorphic functions with positive imaginary part in the future tube", I: *Mat. Sb.* 93 (135) (1974), 3–17 [English transl.: *Math. USSR-Sb.* 22 (1974), 1–16]; II: *Mat. Sb.* 94 (136) (1974), 499–515 [English transl.: *Math. USSR-Sb.* 23 (1974), 467–482]; IV: *Mat. Sb.* 104 (146) (1977), 341–370 [English transl.: *Math. USSR-Sb.* 33 (1977), 301–303].

[115] ——, "Multidimensional linear passive systems", *Continuum Mechanics and Related Problem of Analysis* (a collection of papers dedicated to the eightieth birthday of Academician I.A. Muskhelishvili), Nauka, Moscow, 1972, pp. 121–134, (in Russian).

[116] ——, "Growth estimates of boundary values of nonnegative pluriharmonic functions in a tubular region over an acute cone", *Complex Analysis and Its Applications* (a collection of papers dedicated to the seventieth birthday of Academician I.N. Vecua), Nauka, Moscow, 1978, pp. 137–148, (in Russian).

[117] ——, "On plurisubharmonic functions in tubular radial regions", *Izv. Akad. Nauk SSSR Ser. Mat.* 29 (1965), 1123–1146 [English transl.: *Amer. Math. Soc. Transl.* (2) 76 (1968), 179–205].

[118] ——, "Holomorphic functions with nonnegative imaginary part in tube domains over cones", *Dokl. Akad. Nauk SSSR* 239 no. 1, (1978), 26–29 [English transl.: *Soviet Math. Dokl.* 19, no. 2, (1978), 254–258].

[119] ——, "Many-dimensional generalization of the Hardy–Littlewood Tauberian theorem", *Izv. Akad. Nauk SSSR Ser. Mat.* 40 no. 6, (1976), 1084–1101 [English transl.: *Math. USSR Izv.*, 10, no. 5 (1978)].

[120] ——, *Generalized Functions in Mathematical Physics*, 2nd ed., Nauka, Moscow, 1979, [English transl.: Mir, Moscow, 1979].

[121] Vladimirov, V.S., and Drozhzhinov, Yu.N., "Holomorphic functions in a polydisc with nonnegative imaginary part", *Mat. Zametki* 15 (1974), 55–61 [English transl.: *Mat. Notes* 15 (1974), 31–34].

[122] Vladimirov, V.S., Drozhzhinov, Yu.N., and Zavialov, B.I., *Tauberian Theorems for Generalized Functions*, Kluwer Ac. Press, 1988 [Russian ed. M.: Nauka, 1986].

[123] Vladimirov, V.S., Volovich, I.V., and Zelenov, E.I., *p-Adic Analysis and Mathematical Physics*, Nauka, Moscow, 1994 [English transl.: World Scientific, Singapore, 1994].

[124] Weyl, H., "The method of orthogonal projection in potential theory", *Duke Math. J.* 7 (1940), 411–444.

[125] Wiener, N., *The Fourier Integral and Certain of its Applications*, Cambridge University Press, Cambridge, Mass., 1935.

[126] Wilcox, C.H., "Wave operators and asymptotic solutions of wave propagation problems of classical physics", *Arch. Rational Mech. Anal.* 22 (1966), 37–78.

[127] Wu, T.T., "Some properties of impendance as a causal operator", *J. Mathematical Phys.* 3 (1962), 262–271.

[128] Youla, D.C., Castriota, L.J., and Carlin, H.J., "Bounded real scattering matrices and the foundations of linear passive network theory", *IRE Trans. Circuit Theory* CT-6 (1959), 102–124.

[129] Zharinov, V.V., "Distributive structures and their applications in complex analysis", *Trudy Mat. Inst. Steklov* 162 (1983), [English transl.: *Proc. Steklov Inst. Math.* 162 (1983)].

[130] Zemanian, A.H., *Distribution Theory and Transform Analysis*, McGrow-Hill, New York, 1965.

[131] ———, "The Hilbert port", *SIAM J. Appl. Math.* 18 (1970), 98–138.

[132] ———, "An N-port realizability theory based on the theory of distributions", IEEE Trans. Circuit Theory CT-10 (1963), 265–274.

[133] ———, *Generalised Integral Transformations*, Wiley (Interscience) N.Y., 1968.

Index